Xth International
Seaweed Symposium
Proceedings

Xth International Seaweed Symposium

Proceedings

Göteborg, Sweden, August 11 – 15, 1980

Editor Tore Levring

Walter de Gruyter · Berlin · New York 1981

Editor
Professor Dr. Tore Levring
University of Göteborg, Sweden
Marine Botanical Institute
Carl Skottsbergs Gata 22
S-413 19-Göteborg, Sweden

CIP-Kurztitelaufnahme der Deutschen Bibliothek

International Seaweed Symposium <10. 1980. Göteborg>:
Proceedings / Xth [Tenth] International Seaweed Symposium : Göteborg, Sweden, August 11 – 15, 1980 / ed. Tore Levring. – Berlin; New York: de Gruyter, 1981.
ISBN 3-11-008389-2
NE: Levring, Tore [Hrsg.]

Printing: Hildebrand, Berlin, Binding: Verlagsbuchbinderei Dieter Mikolai, Berlin.

SPONSORS

Swedish Government, Ministry of Education
Marcus Wallenberg Foundation for International Cooperation in Science
Wilhelm and Martina Lundgren Foundation
The City of Göteborg
Norwegian Research Council for Technology and Natural Science
UNESCO, Paris
Alginate Industries Ltd, London, England
Atlantic Container Line, Göteborg
CECA S.A., Vélizy-Villacoublay, France
The Copenhagen Pectin Factory, Denmark
Kelco Division of Merck & Co Inc., San Diego, USA
A/S Litex, Glostrup, Denmark
Marcel Trading Corporation, Manila, Philippines
Marine Colloids Division, FMC Corporation, Rockland, Maine, USA
Protan & Fagertun A.S., Drammen, Norway
Scandinavian Airlines System, SAS

INTERNATIONAL SEAWEED ASSOCIATION

International Seaweed Symposia
Proceedings (1980) Vol. X

International Advisory Committee

W.A.P. Black	T. Levring
J. Christensen	J.L. McLachlan
R. Delépine	J.R. Moss
M.S. Doty	E.C. de Oliveira Filho
A. Jensen	Y. Tsuchiya

Scandinavian Organizing Committee for the Xth
International Seaweed Symposium

J. Christensen (Denmark), A. Jensen (Norway), T. Levring (Sweden) President

U. Gustavsson (S, General Secretary)

T. Christensen (DK)	S. Nygren (S)
E. Jaasund (S)	H. Torgersen (N)
G. Michanek (S)	T. von Wachenfeldt (S)

CONTENTS

2. Ecology

3. Biochemistry, Physiology

4. Cultures

GREETINGS

Ladies and Gentlemen
Dear Colleagues and Friends

It is a great honour for me to wish you all a hearty welcome to Scandinavia and to the Xth International Seaweed Symposium. These meetings started in Edinburgh 28 years ago. The idea was to bring representatives responsible for various aspects of basic seaweed research together with those who have practical applications of seaweeds as their main interest.

The first symposium was a great success and has been followed by eight others held in different countries. A growing interest and an increasing number of participants has been one of the results. In the first circular, which was distributed by the organizing committee, it was pointed out that one intended to give priority to papers, more or less, lying within the scope of the original aim of the seaweed symposia. We have invited seven distinguished phycologists who in their plenar lectures will present facts about different fields of marine botany, which may be of current interest today. We will also be able to listen to 99 papers and to view and discuss 15 posters. Three special sessions have also been arranged: Marine Algae and Fouling, Marine Algae in Pharmacy and a discussion - Methods and Terminology in Phycology. Altogether, 304 participants from 38 countries have gathered here for the conference.

I should also like to take this opportunity of thanking all those - persons, corporations etc.-who have made this meeting possible.

I will now ask Professor Georg Lundgren, Rector of Göteborg University, to open the Xth International Seaweed Symposium.

Tore Levring

Opening address to the Symposium

Ladies and Gentlemen

One hundred years ago, the Swedish naval officer, Adolf Erik Nordenskjöld, sailed his ship, the "Vega", north of the Scandinavian peninsula and north of Russia and Siberia until he finally safely reached the Pacific Ocean. This was a very daring expedition at that time and the aim was to find a sea route north of Europe and Asia from the Atlantic Ocean to the Pacific. It was one of the economic-geographical explorations of that century.

About a month ago, another Swedish expedition started out in order to celebrate the 100-year anniversary of the "Vega" expedition. For special reasons it cannot follow the same route as the "Vega", but its destination is the waters of the Arctic Ocean, north of Spitsbergen, Iceland and Greenland. This time the ship is the Swedish ice-breaker "Ymer" and it carries, besides its crew, an international body of scientists specialized in marine biology, marine chemistry, marine geology, oceanography and glaciology. The aim of the expedition this time is not, therefore, an economic-geographic one, but its purpose is to explore and investigate the waters themselves, their currents, their composition and their contents of living matter.

The "Ymer" expedition demonstrates the growing interest shown by our government, our research foundations and our society as a whole, in the marine sciences. They realise the importance of and the need for better knowledge of the conditions necessary for the use of the resources of the seas, and amidst this lies the main object of your symposium, - the basic research on seaweed and its cultivation and practical use in industry, in pharmacy and as a food product.

I think this situation can be found in most countries located close to the ocean. In a situation where the world population is increasing very fast, and where agriculture in many places is threatened by the lack of fertilizers and ground water, they feel it necessary to increase their efforts to support research on the resources of the seas.

This symposium, the Xth International Seaweed Symposium, is thus an important one and we are very glad that so many prominent scientists, from so many nearby and far-away countries, have found an opportunity to come to Göteborg in order to participate in it. For those at our young and active, although large university, who are involved in your field of research, it is a stimulating experience that your international advisory committee has chosen to hold its Xth Symposium in our city and to organise it in close cooperation with Professor Tore Levring and our Department of Marine Botany.

An essential element in the development of a scientific field is, in my opinion, to enable scientists from different departments and institutions, and of varying experiences, to meet in not too large a group and to discuss their experiments and observations prior to publication. It makes it possible for people from not-so-advanced laboratories to form contacts and to learn the ingenious methods and the results from the more prominent ones. But it also gives an opportunity for old friends to meet and to compare the progress they have made since they last met.

That your field of research is a very active one is apparent from the large number of papers to be discussed and the high scientific value they represent. I therefore hope that this will be a successful meeting and that you will all benefit from the lectures and discussions at your symposium.

However, hard work at the sessions also demands time to relax, to form new personal contacts and to strengthen old ones. I therefore hope that you will also in this way enjoy yourselves during the social functions and excursions.

On behalf of the University of Göteborg I would thus like to welcome you most warmly to Sweden, to Göteborg and to the Xth International Seaweed Symposium.

With this, I declare this symposium open.

Georg Lundgren

Part I

PLENARY LECTURES

BIOSYNTHESIS OF ALGINATE

Bjørn Larsen
Institute of Marine Biochemistry, University of Trondheim,
N-7034 Trondheim-NTH, Norway.

Marine algae show a number of special features in carbohydrate biosynthesis at the monosaccharide as well as at the polysaccharide level. Green algae synthesise sulphated polysaccharides of very complex structure, red algae produce galactans with an alternating type of structure, and brown algae produce sulphated, fucose-containing polysaccharides - and alginate (1). Among green and red algae there are considerable differences in the structures elaborated within genera, and in some cases even within families. In brown algae, on the other hand, alginate appears to be of general occurrence. It may therefore be useful for chemotaxonomic purposes since it does not occur anywhere else in the plant Kingdom, except in certain strains of two bacteria, *Azotobacter vinelandii* and *Pseudomonas sp.*

Alginate was first isolated in the 1880's from brown algae by E.E.C. Stanford (2). Structural studies suggested initially that it was a polyuronide consisting only of D-mannuronic acid residues linked β-1,4. This view changed when Fischer and Dörfel, in 1955, demonstrated the presence of a second kind of uronic acid residue, L-guluronic acid (3). This put a new dimension into the alginate molecule and, as we shall see later, it creates considerable problems in arriving at a mechanism of biosynthesis.

There have been many problems in establishing the chemical structure of the alginate molecule, but today we may summarize our knowledge by the simple formulae given in Fig. 1.

Xth International Seaweed Symposium
Printed in Germany

Fig. 1.

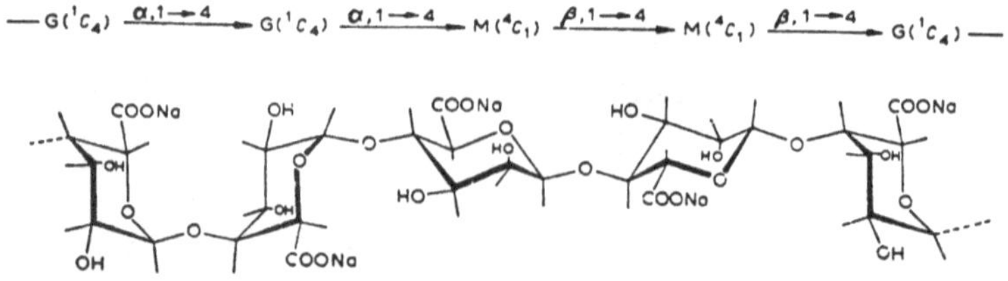

Alginate is a linear copolymer of the two uronic acids, D-mannuronic and L-guluronic acid, the former linked β-1.4, the latter α-1.4. To non-carbohydrate chemists the linkages may appear to be the same, which indeed they are, but according to the rules of nomenclature, β-D corresponds to α-L. The conformations of the sugar rings in the polymer chains are, however, different. The D-mannuronic acid is in what is usually termed the $^4C_1$ conformation. For the present purpose, the important feature is that both linkages "entering" and "leaving" the sugar are equatorial, *i.e.* parallel to the reference plane of the sugar ring. L-guluronic acid units, on the other hand, are in the $^1C_4$ conformation, and in this case both linkages are axial. Wherever along the chain two guluronic acid units occupy adjacent positions, a kind of "cavity" is formed, the dimensions of which fit very nicely with the size of ions having a strong affinity for the alginate molecule in solution (4). This suggest that the sequence of units along the chain is important in determining the ion-exchange selectivities of alginates.

This brings two more descriptive parameters into focus: the monomer composition and the monomer sequence. I presume you are all fimiliar with the fact that alginate isolated from various sources can vary considerably in monomeric composition (5). The usual range of guluronic-acid content is between

30 and 60 %, but more extreme values may be found in certain algae or certain tissues. *Mesogloia* alginate, for example, contains about 80 % guluronic acid, while the alginate in the intercellular spaces of *Ascophyllum* receptacles contain only 15 %.

The first approach to the study of the sequence of the two uronic acids along the chain was carried out in our laboratory many years ago (6, 7). Upon heterogeneous hydrolysis we observed that the partially degraded polymer could be separated into three fractions of widely different composition. This demonstrated that the two uronic acids were not distributed along the chain in a random manner, or in any type of repeating structure, but in three different types of blocks: the M-block, containing approximately 90 % mannuronic acid; the G-block with approximately 90 % guluronic acid; and a block of intermediate composition, usually called the MG-block. There is controversy as to the arrangement of M and G-units within the MG-block. With a purified fraction of MG-blocks from *Laminaria digitata*, which had a guluronic acid content close to 50 %, we inferred that the sequence was predominantly alternating (8). That this idea is an oversimplification is suggested by studies carried out by Professor Nisizawa's and Professor Turvey's groups (9, 10, 11). With selective guluronide lyases they demonstrated that trimers of the type MMG occur more frequently than GMG. When considering the whole, or at least a major fraction, of the interspacing MG-blocks, this is most probably true. One has to remember that the block concept is based on hydrolysis and a physico-chemical fractionation procedure, and it would be astonishing if these polysaccharide fractions were pure, chemical entities. Table 1 emphasizes this very clearly. By nmr spectroscopy we can determine the nearest-neighbour frequencies MM, MG, GM and GG (12). This is a counting procedure which tells us, on an average, the number of places along the chain where an M-unit is followed by an M, an M is followed by a G, and so on.

Table 1. Doublet frequencies of isolated block-fractions obtained by chemical fractionation of alginate from *L.digitata*.

Block type		MG+GM[c]	MM	GG	Mannuronic acid content, F_M
GG	a	0.10	0.07	0.83	0.17
	b			1.0	0.08
MM	a	0.19	0.71	0.10	0.83
	b		1.0		0.83
MG	a	0.67	0.21	0.13	0.56
	b	1.0			0.62

[a]Doublet frequency from ^{13}C-NMR.
[b]Block distribution obtained by chemical fractionation.
[c]Acid-soluble fraction in chemical method.

Three purified fractions were subjected to analysis by nmr, and the results (13) are given in Table 1. As isolated, neither the G- nor the M-block is a pure homopolymer but contains a considerable proportion of the other type of unit. From this we may conclude that the average block-length must be fairly small. The MG-block has a high occurence of MG and GM doublets, but contains appreciable amounts also of MM and even GG. This clearly invalidates any method for block determination involving assumptions about block structure.

I should like to draw two conclusions from the structural data presented so far:

1) They clearly support the concept of a block structure but, at the same time, they serve as a warning against looking upon these blocks as having the ideal structure of a homopolymer or an alternating polymer.

2) The structure of the alginate molecule can only be described by polymer statistics.

I have gone into such detail about the structure and sequense of the alginate molecule because I want to emphasise these factors in discussing biosynthesis. To explain how alginate is synthesised, it is not sufficient to explain the chemical reactions putting each type of unit into the chain. It is at least equally important to obtain information on how the particular distribution of units is created and how it is controlled. At this stage, I must warn you not to be too optimistic. These problems are very difficult to approach both experimentally and theoretically. Most of the work is plowing new soil, and much more is needed before we can approach a complete solution. However, as far as we know today, the process of alginate biosyntheses is unique, and very interesting. I am glad to have this opportunity to tell you about the advances so far, and to emphasize some points that we in Trondheim would like to be considered by everybody working on the structure and biosynthesis of binary copolymers.

Finally, I should like to remind you of the strong correlation between the physical properties of alginate and monomer composition and sequence. While viscosity is mainly related to molecular size and not *per se* to chemical composition, the ability to bind divalent cations is fundamental to the process of gel-formation with alginate (14). As mentioned above, the ion-binding site is the "cavity" formed between two neighbouring guluronic-acid residues in the chain. The physical properties will thus be directly related to the sequence of G's in the polymer, showing an increase in gel-strength with higher G-blocks contents (14, 15). It is generally agreed that alginate is located in the cell wall and in the matrix, and that it makes a considerable contribution to the mechanical strength of the tissue. Through the biosynthetic process the algae may thus be able to control the physical strength and flexibility of the tissue by controlling the distribution of G-units along the chain.

Biosynthesis

The classical work on biosynthesis was carried out by Lin and Hassid in 1966 (16, 17). In an extensive investigation of *Fucus gardneri* they isolated and characterized the guanosinedi-phosphate D-mannuronic acid (GDP-D-mannuronic acid, Fig. 2) which would, according to the usual mechanism of polysaccharide biosynthesis, be the expected precursor, or activated intermediate, for incorporation of mannuronic-acid units into a polysaccharide.

Fig. 2. Uronic acid nucleotides.

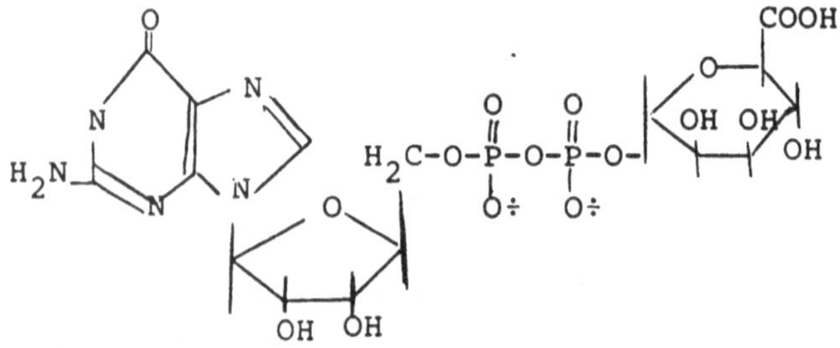

GDP-D-MANNURONIC ACID.

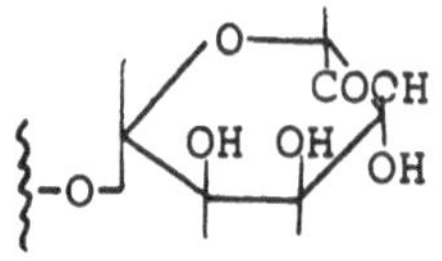

GDP-L-GULURONIC ACID

They also isolated a small amount of another acidic sugar nucleotide. Based mainly on lactonization and reduction of the sugar part of the nucleotide, they concluded that this compound was most probably GDP-L-guluronic acid (Fig. 2). In the subsequent investigation they demonstrated that extracts of the algae were capable of transforming ^{14}C-D-mannose into the isolated GDP-D-mannuronate through the steps shown in Fig. 3.

Fig. 3. Biosynthesis of GDP-D-mannuronic acid.

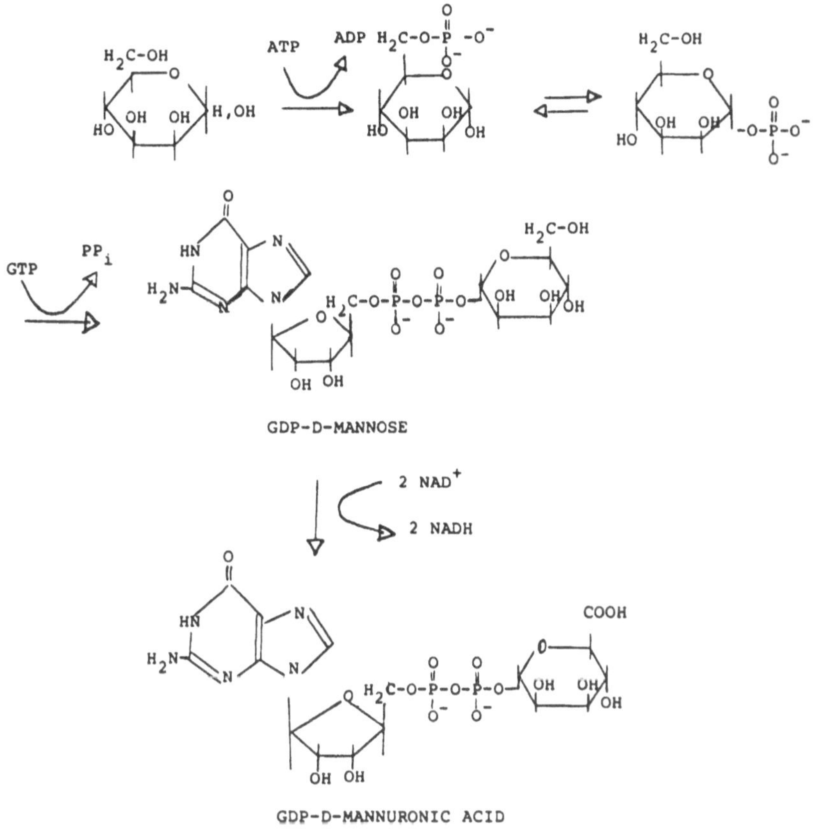

This implies that the alga contains all the necessary enzymes for converting D-mannose into the activated form of D-mannuronic acid, ready for the transferase reaction leading to chain elongation. (Fig. 4).

The presence of the enzymes catalysing these reactions has also been demonstrated in *Azotobacter* by Pindar and Bucke (18). Furthermore, Lin and Hassid demonstrated with a particulate preparation from the algae, that incubation with purified tritiated GDP-D-mannuronic acid led to the incorporation of tritiated uronic acid into the alginate fraction. There is thus every reason to believe that the active precursor really is GDP-D-mannuronic acid and that the transferase reaction uses this compound for chain elongation, thus creating a linear β-1.4 linked polymer. In conclusion, Lin and Hassid state: "Inasmuch as L-guluronate is the minor component of alginic acid, further studies on the conversion of GDP-D-mannuronate to GDP-L-guluronate will be necessary - - -". This statement implies that D-mannuronic acid is converted into L-guluronic acid at the sugar nucleotide level and that tranferases utilize both precursors to build the alginate molecule. In view of later developments, this statement has led to considerable discussion.

Fig. 4. The polymerization reaction.

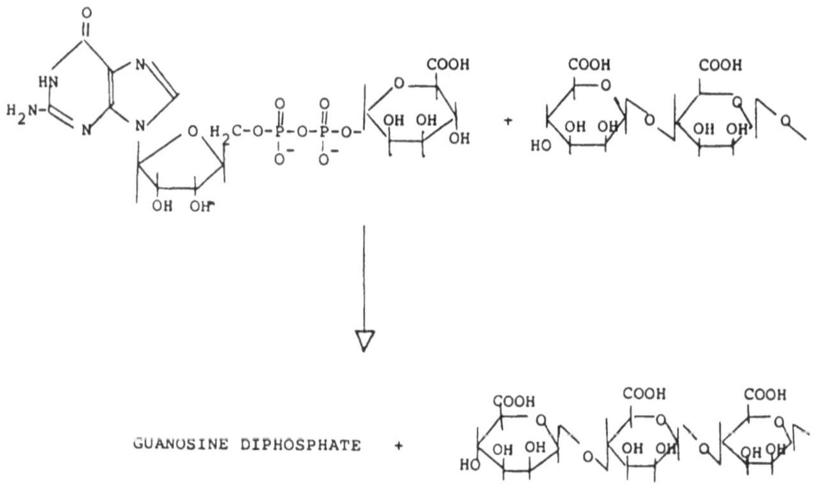

I regret that this problem is still unsolved. There is still some doubt as to the existence and/or participation of GDP-L-guluronate in the process, and this cannot be conclusively decided until somebody succeeds in isolating this compound, if present, and measuring its incorporation into alginic acid.

The publication by Gorin and Spencer (19) on the production of an extracellular polyuronide resembling alginate by the soil bacterium *Azotobacter vinelandii* aroused our curiosity. We investigated the production of alginate by this bacterium with special reference to the effect of conditions on monomeric composition (20). One very important factor was the calcium-ion concentration in the growth medium. The extracellular alginate produced at low calcium-ion concentrations is characterized by a high mannuronic-acid content, and an increase in calcium concentration leads to increased guluronic-acid content (Table 2).

Table 2. Yield of extracelular polysaccharide from *Azotobacter vinelandii* in media containing NH_4^+-ions and various amounts of calcium ions.

Ca^{2+} (mM)	Extracellular polysaccharide[a] (mg)	F_M	Degree of acetylation
0.03	171	0.82	0.35
0.1	228	0.82	0.20
0.34	360	0.26	0.24
1.0	275	0.62	0.23
3.4	312	0.60	0.20
0.1	210	0.80	0.19
0.34	273	0.47	0.09
3.4	302	0.52	0.03

[a]Yield from 150 ml of medium.

The variation in composition of the polyuronide with increasing time of incubation indicated that at least part of the extracellular polyuronide must increase in guluronic-acid content after polymerization has taken place. After partial purification by ammonuium sulphate precipitation, an enzyme preparation was isolated from the bacteria-free supernatant which was capable of converting in-chain D-mannuronic acid into $\underline{\underline{L}}$-guluronic acid (21). Time will not allow any discussion of the details of these studies of the enzyme from *Azotobacter*, but we will concentrate on three main conclusions. Firstly the activity of the enzyme depends upon the calcium-ion concentration in solution as measured at the start of the experiment (Fig. 5).

Fig. 5. Activity as a function of calcium concentration.

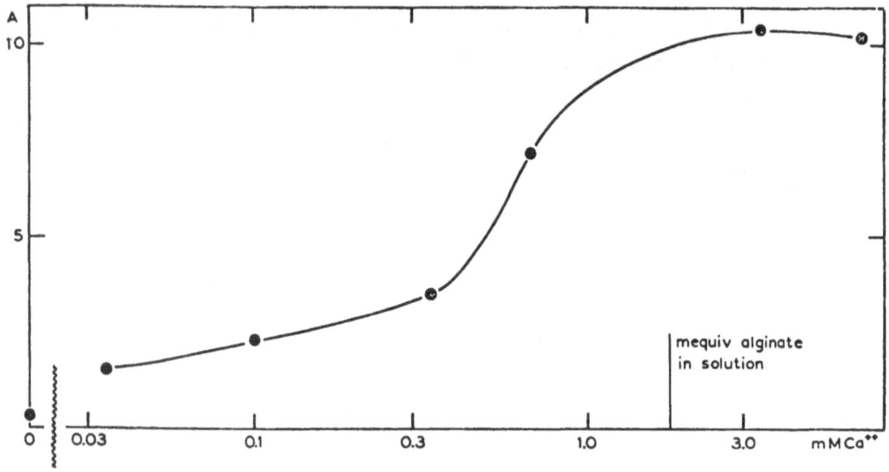

In other words, the number of M→G-conversions carried out by an enzyme molecule increases with increasing concentration of calcium ions, and a maximum rate is reached at approximately equimolar concentrations of alginate and Ca^{2+}. Secondly, the enzyme can use other types of alginate as substrate. Some examples are given in Table 3, with the usual high-M substrate included for comparison. In these cases, all epimerizations were carried out over an extended period to approach the end point, if it exists. The alginate from *Laminaria digitata*, which has a high content of M-blocks, is clearly attacked by the enzyme, as demonstrated by a decrease in M-content from 62 to 29 %. Similar results were obtained for the alginate isolated from old tissue of *Ascophyllum* , which has a very high content of MG-blocks. With alginate from *L. hyperborea*, no epimerization could be detected, but as the guluronic-acid content of this alginate is already very high a small degree of further conversion would be difficult to detect by the colorimetric method.

Table 3. End-point of the enzyme reaction.

Alginate sample	Ca^{2+} (mM)	Mannuronic acid (%)	
		Before treatment	After treatment
A. nodosum high-M alginate	0	92	89
	0.07	92	28
	0.34	92	32
	3.4	92	39
L. digitata	0	62	59
	0.07	62	29
	3.4	62	39
A. nodosum old tissue	0.07	61	43
	0.14	61	37
L. hyperborea	0	27.5	27.5
	0.07	27.5	27.0
	3.4	27.5	26.5

A further conclusion is indicated by these results: if they really represent end-points, there must be some M-units in the substrate molecules that cannot be attacked by the enzyme. This seems to contradict the observation that alginate from *L. digitata* and *A. nodosum*, old tissue, behave very similarly as substrates although they differ widely in distribution of units. A comparison was therefore made of the block-distribution in M-rich alginate and old tissue-alginate before and after treatment with epimerase. The results (Table 4) demonstrate that the distribution of units among blocks is very different in the epimerized alginates; when one starts with long M-sequences, the main product is G-blocks, where as, when one starts with a high MG-content, this is further increased and the increase in G-blocks is only moderate. It seems possible to explain these results only by assuming that most of the M-units occurring in the MG-block are protected against the epimerase action. This has been confirmed later with purified MG-blocks (22).

Table 4. Partial hydrolysis of enzyme-treated alginates.

	Mannuronic acid (%)	Soluble 0.3M HCl, 2h, 100°	Insoluble Soluble pH 2.85	Insoluble Insoluble pH 2.85
High-M alginate				
Before treatment	92	9.1	80.5	10.4
After treatment	49	31.0	17.2	51.8
A.nodosum, old tissue				
Before treatment	60	59.0	27.0	14.0
After treatment	45	65.0	12.0	23.0

Thirdly, we have demonstrated that the epimerase reaction involves an exchange of protons between carbohydrate and solvent (23). By incubating the enzyme preparation with high-mannuronic acid alginate in water containing large amounts of T_2O, incorporation of tritium was easily detected. After addition of "cold" alginate, complete hydrolysis with acid was performed and the uronic acids were separated on an anion-exchange column. The results are shown in Table 5.

Table 5. Recovery and specific activity of uronic acids after hydrolysis and separation.

	Tritium activity (c.p.m.)			Uronic acid (mg)		
	D-Mannuronic	L-Guluronic	Total	D-Mannuronic	L-Guluronic	Total
Before hydrolysis			236.000	11.7	19.1	30.8[a]
Eluate from column	6.100	70.000	76.100	5.3	6.9	12.2
Recovery (%)			32.3			39.5
Specific activity (c.p.m./mg)	8.000[b]	173.000[b]				

[a]After dilution with inactive alginate (28.0 mg).

[b]Corrected for the dilution with inactive alginate.

Recoveries measured by isotope counting or by colorimetry of uronic acids are very similar. The specific activity demonstrates that 92 % of the activity recovered was present in the guluronic acid. The significance of the 8 % found in the mannuronic acid is still not clear. We have domonstrated that this exchange takes place at the C-5 atom (24), in agreement with the results obtained by Jacobson *et.al.* for the glucuronic-acid C5-epimerase active during the biosynthesis of heparan (25).

To sum up at this stage: the bacterial epimerase is an enzyme modifying the polyuronide after polymerization. The mechanism of the enzymic reaction involves an H5-abstraction, and the activity depends on two factors: the concentration of calcium ions and the distribution of already existing guluronic acid units. A prerequisite for activity is that the polymer molecule contains at least 10-20 units (26).

The next question is whether this enzyme exists in marine algae. Dr. John Madgwick from Australia tackled this problem at our Institute. It was a very difficult task, because phenolic constituents destroyed most of the enzymic activity, and there was also an alginate lyase in the alga. He finally succeeded in isolating an active preparation from *Pelvetia canaliculata*, and the results are shown in Tables 6 and 7 (27). There appeared a significant increase in carbazole reactivity after only 90 minutes incubation, demonstrating an increase in guluronic-acid content of the substrate molecule, the high mannuronic-acid alginate. Inevitably, the colour reaction at the start of the experiment is higher with these enzyme preparations because of contamination with algal alginate. To verify the results, it was therefore necessary to establish that the same chemical reaction took place. As is seen from Table 7, tritium is incorporated when the reaction was carried out in T_2O, thus confirming that a polymannuronic acid C5-epimerase is present in marine algae. You will also hear about this from Professor Nisizawa during this meeting.

Table 6. Assay of epimerase activity in *P. canaliculata* using the carbazole reaction.

Incubation time (h)	Carbazole reactives (extinction at 530 nm) Without borate	With borate	Ratio With borate/ without borate
0	0.58	0.72	1.24
0.5	0.64	0.80	1.25
1.5	0.72	0.74	1.03
4.5	0.96	0.70	0.73
18.0	1.36	0.61	0.45

Table 7. Incorporation of tritium by uronic acids during enzymic epimerisation of polymannuronic acid.

Time (h)	Total	Activity, (counts per minute) Guluronic acid	Mannuronic acid
0	14840	5	33
5	49630	2539	672

Pindar and Bucke (18) studied the biosynthesis of alginic acid in *Azotobacter vinelandii* and arrived at the following conclusion: "Results indicate that the first polymeric substance formed in the synthesis is polymannuronic acid and that mannuronic acid units are epimerized to guluronic acid at the polymer level. Guluronic acid does not appear to be formed at the monomer level, either free or in combination with GDP." It is exceedingly difficult to disprove the participation of GDP-L-guluronic acid in the biosynthesis, and for the events taking place in algae we would modify the conclusion to say that the epimerase probably is the key enzyme for creating the specific composition of the alginate molecule as it is found in various tissues of marine, brown algae. The rest of this lecture will therefore be concerned with studies on the relation between enzyme properties and the structure of the polymer created by its action.

Table 8. Fractionation of mannuronic-rich alginate from *Ascophyllum* with calcium ions.

	F_M	F_G	F_{MM}	F_{MG}	F_{GM}	F_{GG}
Whole alginate	0.91	0.09	0.87	0.04	0.04	0.05
Soluble, Ca_L	0.978	0.022	0.965	0.013	0.013	0.009
Insoluble, Ca_F	0.45	0.55	0.39	0.06	0.06	0.49

I have already stressed the fact that the sequence of units along the chain is an important parameter for characterizing a linear copolymer, and, in the case of alginate, for determining the physical properties of the molecule. The next question related to biosynthesis is obviously: what kind of sequence is produced when the bacterial enzyme works on a substrate molecule? Before we can investigate a problem like this, we must have an analytical method capable of measuring a parameter related to the sequence of units. We have demonstrated that the resonance signals from protons and carbon atoms in a uronic acid are influenced by the neighbouring units in the chain, and that this technique may consequently be used to determine neighbour frequencies (12, 13). To introduce the terminology, I have drawn four molecules in Fig. 6, each 20 units long and having equal numbers of M's and G's (F_M=0.5). Starting from the left, we may count how many times an M is followed by an M, and M is followed by a G, a G by an M, and finally, a G by a G. Dividing each of these figures by the total number of units, we get the nearest-neighbour frequencies, or "diads" as we usually call them. Similarly, we may extend this system to counting all the 8 possible combinations of three units in sequence and we get the next-nearest neighbour frequencies, or triads.

Fig. 6. Description of sequence of units in a binary copolymer by diad and triad frequencies.

		Diads				G-Triads			
		MM	MG	GM	GG	GGG	GGM	MGG	MGM
1.	M-M-M-M-M-M-M-M-M-M-G-G-G-G-G-G-G-G-G	0.45	0.05	0.05	0.45	0.40	0.05	0.05	0.00
2.	M-G-M-G-M-G-M-G-M-G-M-G-M-G-M-G-M-G-M-G	0	0.50	0.50	0	0	0	0	0.50
3.	M-G-G-M-M-G-G-M-M-G-G-M-M-G-G-M-M-G-G-M	0.20	0.30	0.25	0.25	0	0.25	0.25	0
4.	G-M-M-M-G-G-M-G-G-G-M-G-M-M-G-M-G-G-M-M	0.20	0.30	0.30	0.20	0.05	0.15	0.15	0.15

In the Figure, only half of these triad frequencies are given, namely those having a G as the centre unit. Nmr allows us to determine these diad and triad-frequencies for a population of molecules, and the values obtained will obviously be the average frequencies for the molecules in solution. As you can see from the figures given, the four molecules give very different values for diad and triad frequencies, triads being more discriminatory than diads.

Using this as our analytical tool, we may now investigate what happens when the enzyme modifies a substrate molecule. Ideally, the substrate should be a polymannuronic acid. The closest we can get to this is the soluble alginate found inside the fruit bodies of *Ascophyllum nodosum*. Even this alginate contains approximately 10 % guluronic acid, and questions could arise as to the effect of these pre-existing units on the enzymic reaction. We therefore fractionated this alginate, after a very mild, homogenous acid hydrolysis, by precipitation with calcium ions. Table 8 demonstrates that this procedure gave a calcium soluble fraction still having a high $\overline{DP}_n$ (usual range 50-100) and with a very low guluronic-acid content. One may tentatively assume that the effect of pre-existing guluronic-acid units may be neglected when using the soluble fraction as substrate.

The activity of the enzyme, *i.e.* the number of units transformed per unit time, is strongly influenced by the calcium-ion concentration (cfr. Fig. 5). It seemed reasonable to assume that this was an effect on the turnover-number of the enzyme as might be expected if, for instance, calcium was an effector. In that case, the product formed at a certain degree of conversion would be the same, regardless of the calcium-ion level used. Much to our surprise, this turned out not to be true. In another lecture at this Symposium, my colleague, Miss Ofstad, will give more detailed information of her studies on the effect of calcium-ion concentration on the course of the enzymic reaction. Suffice it now to illustrate this effect by the figures given in Table 9.

Table 9. Composition of products of the epimerase reaction at low and high levels of calcium ions.

mM Ca^{2+}	F_M	F_G	F_{MM}	F_{MG}	F_{GM}	F_{GG}
0.85	0.62	0.38	0.51	0.11	0.11	0.27
3.40	0.69	0.31	0.43	0.23	0.23	0.08
Theory a=1	0.62	0.38	0.32	0.30	0.30	0.08

Despite the small difference in total conversion to guluronic acid between the two samples, there is a dramatic difference in distribution. As a general conclusion we may say that a low calcium-ion concentration favours epimerization at a position adjacent to a pre-existing G-unit, while an increased calcium-ion concentration favours epimerization at positions adjacent to M-units. To have a system of reference, we will consider the polymer as having a direction, starting at the non-reducing end. In Table 9, we have arbitrarly chosen the enzyme to work in this direction, and the first unit in a given diad is the unit whose signal we are observing.

We are thus forced to conclude that calcium ions are not an activator of the enzyme in the classical sense, since in addition to increasing the rate of reaction, they also influence the course of reaction with respect to the polymeric substrate. To try to understand this effect, it is necessary to devise a model capable of describing the type of product formed by an epimerase working on a polymer. We are not aware of any theoretical treatment of this type of enzymic reaction, but a few groups have worked on polymer-modifying reactions in general (28, 29,30). In this discussion, we shall use the notation of Gonzalez et.al. (29).

The basic idea in polymer modification is that the reagent capable of transforming an unreacted into a reacted unit, will attack the units in the polymer at random. In our case, the reagent is the enzyme. The probability of a reaction occurring as a result of any particular attack, or the rate constant of the reaction, may be modified according to the environment of the particular unit. A simplified version of the model that we shall propose today may easily be explained by reference to Fig. 7: the rate of reaction for an unreacted unit with both neighbours of the same type will be set at k ("unmodified" reaction). If one of the nearest neighbours is a reacted unit (type 2 in Fig. 7), this will change the rate of reaction to ak (or ãk), and if both nearest neighbours have reacted (type 3 in Fig. 7) the rate of reaction will be changed to ck. The model was originally worked out in collaboration with my colleague Dr. Terence Painter, to check the theory of protection of neighbouring units as a result of the reaction of polysaccharides with periodate. In that case, a<1 and c<1.

From our point of view, the advantage of the model is that it is possible to explain a variation in composition of the product formed by introducing a variable a-value. With the epimerase, a variation in a may, for instance, be caused by a dependence of enzyme-substrate binding upon the calcium ion concentration. Both Platé et.al. (28) and Gonzales et. al. (29) have explicitly solved the differential equations for the density of reacted units at a time <u>t</u> after the start of the reaction, and it is thus in principle possible to calculate, at any degree of conversion, the number of sequences containing 1, 2, 3, --- reacted units, ($r_1(t)$, $r_2(t)$, $r_3(t)$ -). Similarly, a formula is available to calculate the density of unreacted units ($q_1(t)$, $q_2(t)$, $q_3(t)$ -). Even for this case, where only the two nearest-neighbours are of importance, the formulae are rather cumbersome for sequence lengths greater than 3.

<u>Fig. 7.</u> Rate constants for the reaction of different types of units in a partially reacted, linear polymer.

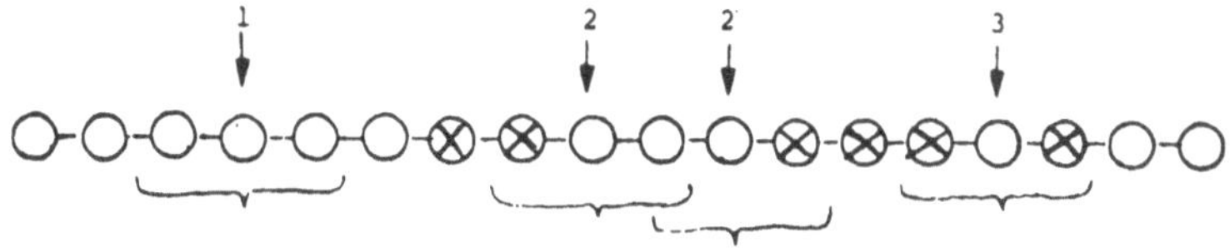

-O- unreacted unit, -⊗- reacted unit.

Unit of type 1: rate constant = <u>k</u>
Unit of type 2: " " = a<u>k</u> or ä<u>k</u>
Unit of type 3: " " = c<u>k</u>

For a=1, ä=0, c=0

$$r_1(t) = \tfrac{1}{2}(1-e^{-2kt})\exp[e^{-kt}-1]$$

$$r_\ell(t) = \exp[e^{-kt}-1]\left\{\frac{1}{\ell!}(1-e^{-kt})^{\ell} - \frac{1}{(\ell+1)!}(1-e^{-kt})^{\ell+1}\right\}$$

The data presented in Table 3 suggested that the enzyme had little, if any, effect on alginate with a high transition frequency (many MG-blocks). This indicates that the enzyme may not be capable of converting an M-unit into a G-unit when it occurrs between two pre-existing G-units. In the present model, this is equivalent to putting <u>c</u> equal to zero. The fact that the enzymic reaction appears to have an end point also points to complete protection of some M-units.

The introduction of c=0 leads to a substantial simplification of the formulae, and for a few special cases a simple, general formula may be arrived at. In Fig. 7 is shown the general formula for the case where reaction is permitted on one side of a reacted unit but not on the other, and also not between two reacted units. The rate of reaction is the same for the two types of units that are permitted to react, *i.e.* no cooperative effects occur. The number, or weight, fraction of all G-blocks making a significant contribution can now be calculated, and, once known, this result can easily be translated into the nmr-"language" by summing the contribution of each length of block to the various diads (MM, MG, etc.) and triads (GGG, GGM, MGG, MGM etc.). In Table 9 these calculated diad frequencies are given for comparison with those obtained by epimerization at low and high calcium-ion concentrations. Particularly for reaction at low calcium-ion concentrations, there is a conspicuous discrepancy between the experimental and calculated results, the GG-diad being far too low according to this non-cooperative model.

To increase the frequency of occurrence of the GG-diad it is necessary to enhance the rate of reaction in positions adjacent to G-units. In the model this corresponds to increasing the value of a. We have worked out a computer program for numerical integration of the differential equations for c=0, thus allowing calculations similar to those just described to be carried out for any value of a. The effect on the diad frequencies of increasing a from 1 to 100 is demonstrated in Fig. 8 for a polymer with F_G=0.36. The number of block transitions (MG and GM) decreases steadily and the polymer becomes more and more like a mixture of long M- and long G-blocks, that is, an effect in the direction we want.
To explore the possibility of explaining the epimerase reaction with this model, a number of experiments were carried out to obtain samples having different degrees of conversion.

The results obtained with samples epimerized at low calcium-ion concentrations are shown in Table 10 and compared with diads and triads predicted by the model, with a set at 50.

Table 10. Diad and triad frequencies for three samples epimerized at low calcium ion concentration compared to those predicted by the model (a=50).

Sample		F_M	F_G	F_{MM}	F_{MG}	F_{GM}	F_{GG}	F_{MMM}	F_{MMG}	F_{GMM}	F_{GMG}	F_{GGG}	F_{GGM}	F_{MGG}	F_{MGM}
I	Exp.	0.19	0.81	0.08	0.11	0.11	0.70	0.046	0.035	0.035	0.074	0.606	0.094	0.094	0.016*
	Theory	0.19	0.81	0.06	0.13	0.13	0.68	0.047	0.015	0.015	0.123	0.550	0.127	0.127	0.005
II	Exp.	0.32	0.68	0.21	0.11	0.11	0.57	0.17	0.04	0.04	0.07	0.474	0.096	0.096	0.014*
	Theory														
III	Exp.	0.51	0.49	0.40	0.11	0.11	0.38	0.339	0.065	0.065	0.041	0.315	0.065	0.065	0.045*
	Theory	0.51	0.49	0.40	0.11	0.11	0.38	0.34	0.060	0.060	0.045	0.28	0.095	0.095	0.015
IV	Exp.	0.65	0.35	0.55	0.10	0.10	0.25	0.473	0.076	0.076	0.026	0.200	0.050	0.050	0.050*
	Theory	0.65	0.35	0.55	0.10	0.10	0.25	0.490	0.065	0.065	0.025	0.175	0.075	0.075	0.015

*estimates of a lower accuracy

Sample I is epimerized for a ling time and represents at least an approach to an end point for the reaction under these conditions. The fact that 19 % M remains again supports the idea of protected M-units. Of the remaining M, more than 40 % exist as singlets, *i.e.* GMG triads. In comparing experimental and theoretical figures it is easy to see that the agreement now is very good compared to the situation with a=1. The most conspicuous trend seen in the Table is the constancy of F_{MG}, and consequently also F_{GM}, during the reaction. This is consequence of the high a-value used. Since we are only dealing with situations where c=0, coagulation of G-blocks cannot occur, and the number of blocks will consequently be determind by the number of G-units introduced by reaction 1 of Fig. 7. Since this reaction is suppressed by selection of a high a-value, it will make only a minor contribution during the last part of the reaction. This is demonstrated in Fig. 9, where diad frequencies are shown as a function of conversion.

Fig. 8. Diad frequencies as a function of the a-value for a polymer with 36 % guluronic acid (F_G=0.36).

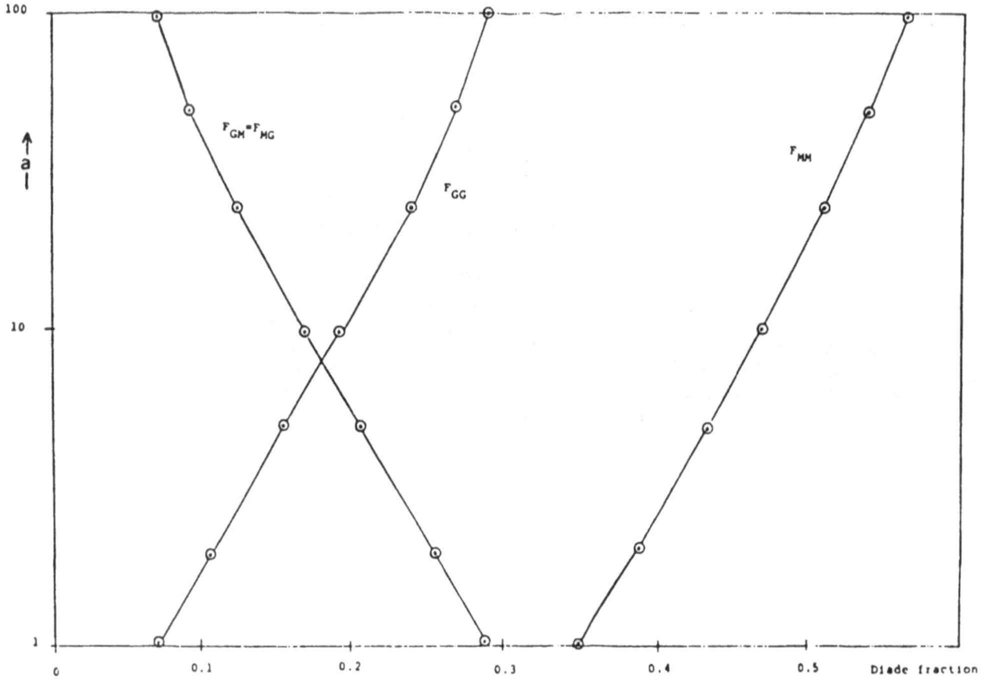

Deviations from the theoretical values are to be expected, since the requirements for constant conditions throughout cannot be fulfilled. Thus, as G-units are produced, calcium ions will be bound more strongly to the polymer, and a decrease in free calcium in solution is to be expected. The extent to which this influences the results is uncertain. Since the equivalent ratio of calcium to alginate is approximately 1:4, the influence will be most pronounced at the start of the experiment. It seems reasonable to conclude that the epimerase reaction under these conditions can be described with reasonable accuracy by this model, provided that a is given a high value. For high a values the interpretation is not unequivocal, however, and I will get back to this point later.

Fig. 9. Diad frequencies as a function of degree of conversion according to the theoretical model with a=50.

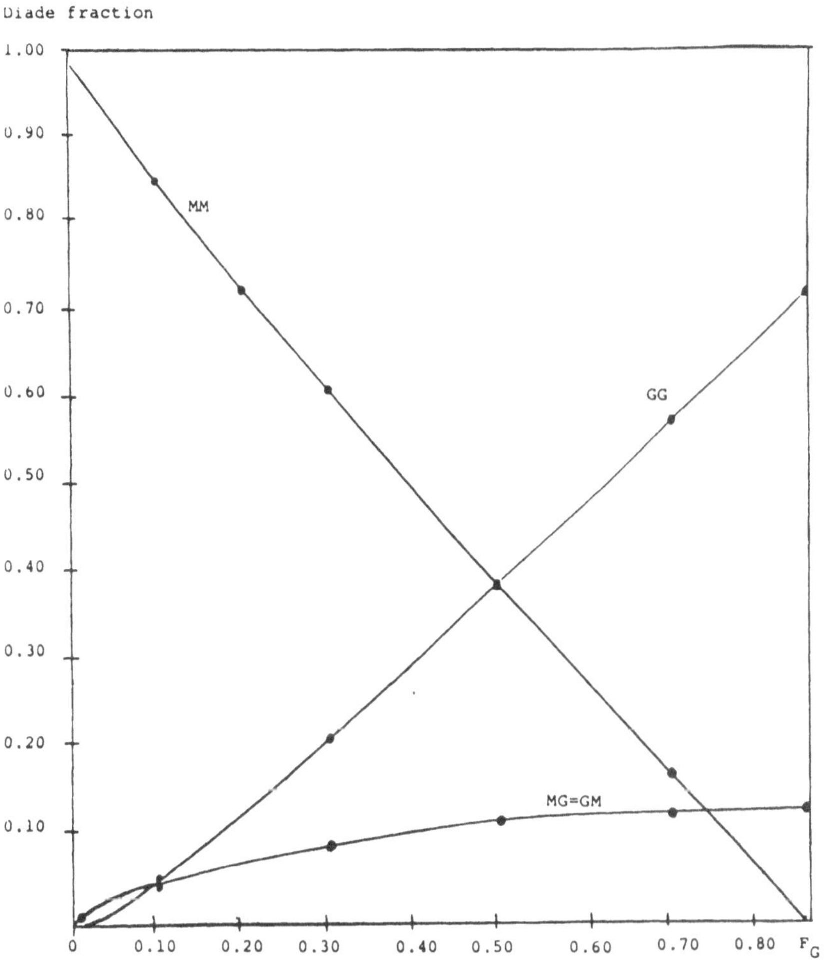

Table 11. Composition of products formed by the epimerase reaction at high calcium ion concentration (6.8 mM).

Sample		F_M	F_G	F_{MM}	F_{MG}	F_{GM}	F_{GG}	F_{MMM}	F_{MMG}	F_{GMM}	F_{GMG}	F_{GGG}	F_{GGM}	F_{MGG}	F_{MGM}
A	Exp.	0.58	0.42	0.40	0.19	0.21	0.21	0.34	0.06	0.06	0.13	0.17	0.04	0.04	0.17
	Theory, a=10	0.58	0.42	0.40	0.19	0.21	0.21	0.30	0.095	0.095	0.08	0.12	0.12	0.12	0.06
B	Exp.	0.64	0.36	0.48	0.16	0.16	0.20	0.40	0.08	0.08	0.08	0.19	0.01	0.01	0.15
	Theory a= 10	0.64	0.36	0.48	0.16	0.16	0.20	0.38	0.10	0.10	0.06	0.10	0.10	0.10	0.06

At high calcium-ion concentration, an experimental difficulty does not allow results to be obtained over the entire range of compositions. When a degree of conversion of between 25 and 45 % is reached, the product forms a loose gel, and the enzymic reaction stops. I shall therefore use only two independent results, representing the maximum conversion we can get at present, to demonstrate the composition of the product under these conditions. They are given in Table 11, sample A being the mannuronic-acid-rich alginate at half-normal concentration, sample B the calcium-soluble fraction of the above at normal alginate concentration (3.5 mM). Both samples show a most unexpected distribution of G-units, with F_{GGG} and F_{MGM} both high, and the other G-triads very low. This makes two confliting demands for the selection of an a value: if F_{MGM} is to be high, a will have to be low; and if F_{GGG} is to be high, a will have to be high. The theoretical values listed in the Table are for a medium value of a, and they demonstrate a perfect fit for diads but bery large deviations for triads, especially for the G-centered ones. It is thus quite clear that although this model may explain the results obtained at low calcium-ion concentration, the effect of increasing the calcium-ion concentration is not equivalent to decreasing the a-value.

Results like the ones obtained for G-triads in Table 11 are not compatible with the model, since they require high neighbour frequency (F_{GGG}) and at the same time high transition frequency of the MGM-type. The epimerase system thus seems

to violate one or both of the two fundamental precepts of this model:
a) One attacking agent b) Random attack of reagent on substrate units. In enzymic terminology, this would mean that we have either (a) more than one enzyme or (b) a single chain, or multiple attack mechanism. The response to variations in calcium-ion concentration among a number of different enzyme preparations reported by Ofstad et.al. (31) seems very difficult to explain in terms of only one enzyme. In our opinion, the available data fa our the idea of two enzymes of different binding specificity, one of these enzymes being controlled by the calcium-ion activity at the site of biosynthesis. This can only be verified by separation of the two entities and characterization of their response to calcium ions, a project under way in our laboratory.

To sum up, we consider the available evidence to suggest that alginate is produced primarily as a polymannuronate and then epimerized to form the characteristic binary copolymer. Since the sequence of guluronic acid units is the important factor determining the physical properties of the product, the way the particular sequences are introduced is of great interest. This is directly related to the question of how to describe a polymer of this type. Most probably the epimerase is not a single-enzyme, multi-chain system, but a more complex system where the local concentration of calcium ions is among the important factors in regulating the activity and course of reaction. The isolation and characterization of the epimerase, or epimerases, from the bacterium as well as from algae, will be necessary before we can reach any definite conclusion as to how they work.

Finally, a few remarks on the composition of alginate as it is isolated from the marine algae. As pointed out above, cytochemical studies have demonstrated that alginate is located in the cell wall and in the matrix (32). It is generally not possible to distinguish between these sources, and the isolated product of extraction will be a mixture of all alginate present in the sample. We have had the opportunity, through the cooperation of Professor Ralph Quatrano of Oregon State University, to investigate the composition of alginate from cell walls of *Fucus* zygotes. The composition of the alginate from 2-celled embryos of *F.disthicus* and from *F.vesiculosus*, 24 hours after fertilization is shown in Table 12.

Table 12. Composition of alginate from cell walls.

Alga	F_M	F_G	F_{MM}	F_{MG}	F_{GM}	F_{GG}
F.disthicus, embryo	0.29	0.71	0.27	0.02	0.02	0.69
F.vesiculosus, zygote	0.28	0.72	0.20	0.08	0.08	0.64
Ectocarpus	0.31	0.69	0.27	0.04	0.04	0.65

First of all, it is clear that this alginate is very rich in guluronic acid compared to the alginate extracted from a full-size plant (40-50 % guluronic acid). Secondly, the distribution of uronic acid units is very different from that of mature plants with a very low content of MG and GM-diads. The same tendency also clearly appears in alginate prepared from the filamentous, brown alga *Ectocarpus* after breaking the cells and removing all water- and acid-soluble substances. With this species, this should approximate to a cell-wall preparation. These alginates clearly approach the composition of long M-blocks and long G-blocks without the occurrence of the MG-block usually present in considerable amounts.

We would consider these results to support the assumption that the algae possess a control mechanism to manipulate the alginate composition according to the particular requirements of each type of cell and tissue. If differences are of the order af magnitude indicated here, also including the mannuronic-rich alginate present in reproductive tissue during most of the year (33), the heterogeneity of the total, extracted alginate will be too large to allow a meaningful description of the molecule. Combined studies of the epimerases and the composition of alginate from tissues as homogeneous as possible may prove to be the only way to give a complete description of the structure and biosynthesis of alginate.

Acknowledgement.

In addition to those already mentioned, I gratefully acknowledge the cooperation of Hans Grasdalen and Ragni Ofstad and the important contributions to this project by Professor Olav Smidsrød and the late Professor Arne Haug.

References.

1. Percival, E. and McDowell, R.H.: Chemistry and Enzymology of Marine Algal Polysaccharides, Academic Press 1967.
2. Stanford, E.E.C.: Chem News 47 (1883), 254.
3. Fischer, F.G., and Dörfel, H.: Physiol.Chem. 301, 224, (1955).
4. Smidsrød, O., Haug, A. and Whittington, S.G.: Acta Chem. Scand. 26, 2503 (1972).
5. Haug, A., Larsen, B. and Smidsrød, O.: Carbohydr. Res. 32, 217 (1974).
6. Haug, A., Larsen, B. and Smidsrød, O.: Acta Chem.Scand. 20, 183 (1966).
7. Haug, A., Larsen, B. and Smidsrød, O.: Acta Chem. Scand. 21, 691 (1967).
8. Larsen, B., Smidsrød, O., Haug, A., and Painter, T.: ibid, 23, 2375 (1969).
9. Fujibayashi, S., Habe, H. and Nisizawa, K.: J. Biochem (Tokyo) 67, 37 (1970).
10. Boyd, J. and Turvey, J.R.: Carbohydr. Res. 66, 187 (1978).
11. Min, K.H., Sasaki, S.F., Kashiwabara, Y., Umekawa M, and Nisizawa K.: J. Biochem. (Tokyo) 81, 555 (1977).
12. Grasdalen, H., Larsen, B., and Smidsrød, O.: Carbohydr. Res. 89, 179 (1981).

13. Grasdalen, H., Larsen, B. and Smidsrød, O., in A. Jensen and J. R. Steiner (Eds.) Proceedings of the Ninth International Seaweed Symposium, Scien. Press 1979, p. 309.

14. Smidsrød, O. and Haug, A.: Acta Chem. Scand. 26, 79 (1972).

15. Smidsrød, O. NTNF:s Institute of Marine Biochemistry, Report No. 34, 1973.

16. Lin, T.Y. and Hassid, W.Z.: J.Biol.Chem 241, 3282 (1966).

17. Lin, T.Y. and Hassid, W.Z.: ibid. 241, 5284 (1966).

18. Pindar, D.F. and Bucke, C.: Biochem J. 152, 617 (1975).

19. Gorin, P.A.J. and Spencer, J.F.T.: Can J. Chem. 44, 993 (1966).

20. Larsen, B. and Haug, A.: Carbohydr. Res. 17, 287 (1971).

21. Haug, A. and Larsen, B.: ibid, 17, 297 (1971).

22. Ofstad, R. and Larsen, B.: unpublished results.

23. Larsen, B. and Haug, A.: Carbohydr. Res. 20, 225 (1971).

24. Grasdalen, H. and Larsen B.: To be published.

25. Jacobson, I., Bäckström, G, Höök, M, Lindahl, U. Feingold, D.G., Malmström, A and Rodén, L.: J. Biol. Chem. 254, 2975 (1979).

26. Larsen, B. and Haug, A.: Proceedings of the 7th International Seaweed Symposium in Sapporo, Japan, 1971, University of Tokyo Press, 1973.

27. Madgwick, J., Haug. A. and Larsen, B.: Acta Chem. Scand. 27, 3592 (1973).

28. Platé, N.A., Litmanovich, A.D., Noah, O.V. Toom, A.L. and Vasilyen, N.B., J. Polymer Sci. 12, 2165 (1974).

29. Gonzales, J.J. and Kehr, K.W.: Macromolecules 4, 369 (1971)

30. Frensdorff, H.K.: Macromolecules 4, 369 (1971).

31. Ofstad R. and Larsen, B. Proc. Xth International Seaweed Symposium, Gothenburg 1980, p. 485 (1981).

32. Vreeland, V., J.Histochem. Cytochem. 20, 358 (1972).

33. Baardseth, E., Grasdalen, H. and Larsen B.: Proc. Xth International Seaweed Symposium, Gothenburg 1980, p. 373 (1981).

PHOTOBIOLOGY OF SEAWEEDS: ECOPHYSIOLOGICAL ASPECTS

K. Lüning

Biologische Anstalt Helgoland, Meeresstation, D-2192 Helgoland,
Federal Republic of Germany

Abstract: One of the tasks of photobiological research in seaweeds is to determine and characterize critical light levels quantitatively for the various photobiological reactions of seaweeds. From the data existing so far the following critical light levels can be derived. The absolute minimum for survival of perennial, multicellular algae is of the order of 6 Einsteins per m^2 per year, and for kelp 70 E m^{-2} $year^{-1}$. These levels correspond to 0.05-0.1 % of surface light, or to 0.5-0.1 %, respectively. Further examples for critical light levels are presented: minimum and optimum irradiances for growth, lethal irradiances for spores, compensation and saturating irradiances for photosynthesis. Besides its use as an energy source for photosynthesis and growth, light is also interpreted by the plant as an environmental signal in the processes of photomorphogenesis. During the last 10 years several photoperiodic, as well as other photomorphogenetic responses have been discovered among seaweeds, and a challenge exists in performing more experimental research especially in this field.

Introduction: A reappraisal after a century of seaweed photobiology

The photobiology of multicellular algae started almost exactly hundred years ago with Engelmann's work (21, 22) on the relationship between algal photosynthetic rates and light quality and with Berthold's investigation (2) of the flora of the Gulf of Naples and his conclusions about the factors controlling the vertical distribution of seaweeds and also about other aspects of their photobiology, e.g. phototropism (3). At a time when other investigators believed that the photosynthe-

Xth International Seaweed Symposium
Printed in Germany

sis of green plants would be strongest in the green part of the spectrum, or that the role of chlorophyll might just be to protect the photosynthetic apparatus from light injury, and that chlorophyll itself might be a product of photosynthesis (see historical review in reference 29), Engelmann, by use of his ingenious bacterial method, clearly showed that the action spectrum of photosynthesis follows the absorption spectrum, not only of chlorophyll in green algae, but also of the accessory pigments in brown and red algae. Engelmann had to fight hard to defend his findings, but using the words of Rabinowitch (61; p.1144), "one cannot but admire the unfailing correctness of his conclusions, obtained by means of an experimental method most investigators would hesitate to use even for qualitative, not to speak of quantitative, purposes". With more up-to-date methods of oxygen determination, Engelmann's main finding of the general parrallelism between the action and absorption spectra was confirmed for seaweeds by Levring (42) who measured the photosynthesis of various seaweeds by the Winkler method in glass bottles screened with colour glass filters and exposed in the sea, and by Haxo and Blinks (35) using a monochromator and an oxygen electrode. The conspicuous deviation of photosynthetic action from absorption in red and blue light (the "red drop") was first noticed in Haxo and Blinks' results (35) for red algae, and this observation was one of the important findings leading to the discovery of the Emerson enhancement effect (20) and subsequently to the identification of photosystems I and II. But the "red drop" phenomenon is already apparent from Engelmann's (22) photosynthetic action spectra for red and blue-green algae, although its importance was only realized 70 years later.

One of the by-products of Engelmann's fundamental investigation was his hypothesis that, due to their pigments, red algae are best, and green algae are worst, fitted to survive at greater depths (21; pp.23-25). Or, in other words, each algal group is dominant at those depths where its colour is complementary to the colour of the prevailing light. I do not want to go into detail about the struggle over this hypothesis (see Rabinowitch's review; reference 61), but just try to summarize our present view, as recently put forward by Dring (17) and Ramus (66), according to which Engelmann's hypothesis is only right in one respect: All algae that live at depth must find a way to fill the "green window" in the absorption spectrum of chlorophyll a. This has been achieved by different methods in different algal groups, and even the green algae can efficiently fill-up the "green window" by increasing their chlorophyll a and b content (62). If we calculate the photosynthetic rates of various algae for naturally occurring underwater spectral distributions at different water depths in coastal water types, as has been done by Dring (17), we find that red algae are chromatically best adapted not only at great, but also at shallow water depths, simply because green light, which they use so efficiently, always makes up a very high proportion of natural underwater light. In narrow-waveband green light, such as exists near the lower algal limit in coastal, "green" waters, the dominance of red algae (mainly crustaceous reds as the deepest multicellular forms) certainly supports Engelmann's hypothesis. However, in narrow-waveband blue light, at greater depths in subtropical and tropical "blue" waters, red algae are photosynthetically disadvantaged, since blue light is used very inefficiently. In spite of this, red algae (especially crustaceous forms), are still present here, together with deep-water representatives of the other algal groups. The fact that the crustaceous red algae are so successful near the lower limit of algal vegetation in blue as well as green waters, in spite of their bad photosynthetic performance in blue light, may be due to the slow rate of attrition, as suggested by Larkum, Drew and Crossett (41). Green and brown algae are, of course, very well fitted to use the blue light in deeper clear waters, as pointed out by Levring (43). Except for red algae of the floridean type in green light therefore, there is little agreement between the action spectrum of photosynthesis and the spectral distribution of underwater light at greater depths. This conclusion also applies to brown and green algae at shallow depths, at which red light is lacking already substantially. As Halldal (32) put it: "The brown and the green algae inhabiting (the shallow) region have two photosynthetic high activity bands, one in the blue to green region, another in the red; thus these algae are from an ecological point of view not particularly adjusted to the prevailing light." The main objection to Engelmann's hypothesis is that, as is particularly evident at depth in blue waters, the parallelism between the photosynthetic action spectrum and the underwater spectral distribution has little or no effect on the vertical distribution of benthic marine algae.

It is a striking coincidence in the limited history of seaweed photobiology that just when Engelmann was producing his photosynthetic action spectra, Berthold (2, 3) worked on the depth distributions of algae in the Gulf of Naples, and, only a little later, Oltmanns (56) published his profound observations on the effects of culture and environmental conditions on the performance of benthic algae. Both of these authors stressed that light quantity might be an important factor regulating the depth distribution of seaweeds. However, from the turn of the century until today, few hard data have been produced by phycologists relating to the quantitative light demands and light tolerances of marine algae. There have been several worthy attempts in this respect, e.g. by Harder, Ehrke or Montfort in the 1920's and 1930's on light demands for photosynthesis (see 29 for references), and by Biebl (4) on light resistance. However, these workers were largely hampered by the lack of adequate intruments for the control and measurement of light sources. Seaweeds soon were replaced by unicellular algae or isolated chloroplasts in the basic experiments on photosynthesis and as the seaweeds disappeared from the lime light of photosynthetic research, the specific ecophysiological problems posed by them disappeared also, and they received little attention from experimental photobiologists and phycologists - with exceptions, such as Halldal's work (31, 32) on photosynthetic action spectra, escpecially in the ultraviolet or Levring's investigation (43, 44) of photosynthetic performance of seaweeds in deep clear water. Today sophisticated optical and electronic equipment is available to control and measure light, and the phycologist has received valuable recommendations from physical oceanographers on how to measure light underwater (38, 72, 73). So he can now start to find out the minimum or optimum light levels that the seaweeds of the various depth zones require, and the light levels at which their photosynthesis and growth is inhibited. Some of the progress which has been achieved in recent years in this respect will be summarized in this paper.

Another aspect of the photobiology of seaweeds concerns light as an environmental signal in the processes of photomorphogenesis, rather than as an energy source in photosynthesis. After Garner and Allard's (26, 27) discovery of photoperiodism in higher plants, it was possible that seaweeds, many of which exhibit a characteristic seasonal pattern, would also be found which could recognize the season by measuring the length of the day, or, to put it more correctly, the length of the night. The first hints that photoperiodic reactions might in fact control algal life histories, came from Japanese investigators (11, 40) around 1960, and this lead to the first clear proof of a photoperiodic response in a seaweed species, namely in Porphyra tenera, by Dring (15, 16), and independently, by Rentschler (68). More such responses have been discovered since and these will be summarized, together with the effects of different spectral regions on morphological development in the last part of this paper. Regarding terminology and units, irradiance, measured in $W\ m^{-2}$ and photon flux density, measured in $\mu E\ m^{-2}s^{-1}$, will be used throughout this paper. For white fluorescent light and also for sunlight above water the following rule-of-thumb conversion factors (see reference 47) will be used to cope with the unit of illuminance (lux), which was mostly used in older work: $1\ W\ m^{-2} \sim 5\ \mu E\ m^{-2}s^{-1} \sim 250$ lux

A. Light levels at the lower limits of major algal zones

How much light reaches the lower limits of algal vegetation in the sea, and how much light do the more conspicuous large brown algae require for survival in the sublittoral ? At Helgoland we measured underwater light at two water depths in the blue, green and red range every 20 min, for one year (51). Three irradiance meters (photocells supplied with Schott glass filter combinations) were attached to an iron rack at 2.5 and at 3.5 m below M.L.W.S. The photocurrents were conducted by cable to a land station, where the data was stored on paper tape. The light (throughout in the following: photosynthetically active radiation, 400-700 nm) received at various biological boundaries within the sublittoral region was calculated from the data (Table 1). At the depth

Table 1

Integrals of photon flux density (400-700 nm) at different water depths (below M.L.W.S.) near Helgoland. (From reference 51).

Vegetation limit	Water depth (m)	Photon flux density ($E\ m^{-2}\ yr^{-1}$)	Percentage of surface irradiance (%)
Deepest forest (L. hyp.)	4	388	4.0
Deepest L. hyperborea	8	71	0.7
Deepest red algal crusts	15	6	0.05

where the deepest algae (crustaceous red algae) occur near Helgoland (15 m; see reference 44), 6 E m^{-2} $year^{-1}$ was received, corresponding to 0.05 % of surface irradiance. At 8 m depth, where the deepest Laminaria hyperborea plants are found, 71 E $m^{-2}year^{-1}$ or roughly 1 % of surface irradiance was recorded, and at the depth limit of the closed L. hyperborea forest (4 m), about 400 E m^{-2} $year^{-1}$ or 4 % of surface irradiance was measured. The only other study in which underwater light was recorded continuously throughout a year within the kelp zone, was performed near Igloolik (Canadian Arctic) by Chapman and Lindley (10) in order to find out if there was sufficient light for autotrophic growth of L. solidungula at a site which was only ice-free for about 2-3 months a year. At 8 m depth, where there was a rather dense kelp vegetation, 590 E $m^{-2}year^{-1}$ were recorded, and for the lower limit of individual kelp plants, near 20 m, the annual photon flux density calculated from occasional transmittance measurements was 49 E $m^{-2}year^{-1}$. This data compares well with the annual quantum budgets obtained for L. hyperborea near Helgoland. Furthermore, at both sites most of the year's light supply arrives during a short period of the year, the extreme case being, of course, the Arctic environment. Near Igloolik, 81 % of the yearly underwater light supply was received within the largely ice-free months, from August to early October (10). Near Helgoland, it is not ice which reduces underwater light to low levels for a great part of the year, but the high turbidity caused by sediment stirred up by the frequent storms between October and March.

90 % of the yearly underwater light supply arrives during the six months from April to September (33).

Wherever the lower vegetation limit is set by light and not by substrate or grazing activity we can expect the "annual quantum budgets" for the deepest kelps in different geographical regions to be of the same order of magnitude, and, perhaps an order of magnitude lower, the same should be true of the deepest multicellular algae. We can compare these values either by using absolute units, or by using the water depth at which the irradiance has been reduced to a certain percentage of its surface value. The latter procedure has the disadvantage that latitudinal differences in solar irradiance will enable algae in tropical regions to grow at somewhat greater light percentage depths than in temperate or polar regions, which receive about one-half or one-third, respectively, of the solar radiation in the tropics (28). Data for lower algal limits in regions whose water clarity was known in terms of Jerlov water type, show this generalization to be justified for temperate and subtropical regions, since the lower limit of algal vegetation generally occurs at 0.05-0.1 % of surface irradiance and the deepest kelp individuals occur at 0.7-1.4 %, or roughly 1 %, of surface irradiance (33). The relationship between the Jerlov water type and the depth at which a certain percentage of surface irradiance in the photosynthetically active range occurs, can be seen from Fig. 1, which is a modification of a diagram produced by Jerlov (37, 38), but extrapolated to the lower limits of benthic marine algae. It should be noted that the photosynthetically active range has been set here to 350-700 nm, and not to the 400-700 nm range which is commonly used in higher plant ecology. However, this difference in definition has little significance in coastal waters, in which UV is rapidly extinguished. The following examples will demonstrate how important the Jerlov water types are for the phycologist: At the French coast, near Roscoff, the Jerlov water type III or 1 prevails, according to measurements of Boutler et al. (7), and the deepest <u>Laminaria hyperborea</u> individuals occur at 25 m water depth. A lower depth limit of 25-30 m for <u>L. hyperborea</u> has been reported for many clear water sites along the European coasts, from France, Ireland, Scotland and Norway up to the Faroes. Near Helgoland

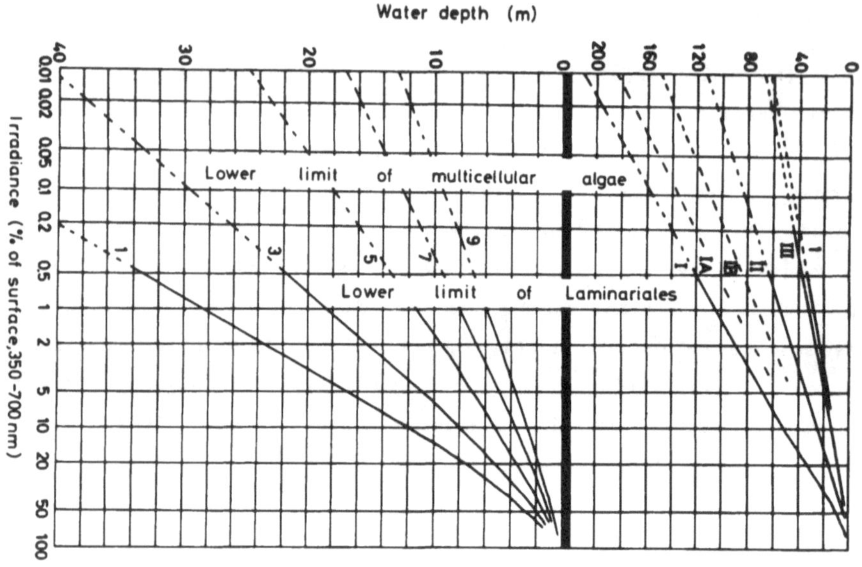

Fig. 1: Percentage of downward irradiance (350-700 nm) as function of water depth and Jerlov water types. Solid lines (range 100-0.5 %): according to Jerlov (37). Broken lines (range 0.5-0.01 %): projections of Jerlov's curves into lower percentage ranges. Critical levels of major importance are situated at 0.5 to 1 % of surface irradiance (lower limit of Laminariales) and at 0.05 to 0.1 % of surface irradiance (lower limit of multicellular marine algae). (From reference 51)

the water is much more turbid, Jerlov water type 7 (occasionally water type 5) prevails (51), and the lower limit of L. hyperborea occurs at 8 m water depth (45). At the other extreme, the depth limit of L. rodriguezii in the Mediterranean may be cited. Near Corsica Jerlov water type IA prevails (35), and 1 % of surface irradiance occurs, according to Fig. 1, at about 100 m depth. In fact, Fredj (25), from a submersible, observed the deepest individuals of L. rodriguezii at 95 m water depth. Similarly, we may expect to find the lower limit of algal vegetation, at 0.05-0.1 % of surface irradiance, at 30 and 50 m, respectively in Jerlov water types 3 and 1, corresponding to European clear water sites, or near 200 m in very clear oceanic waters (Jerlov water type I). Since these predictions agree closely with actual observations of the lower limit in these waters, it is tempting to reverse the whole procedure and predict the light climate from the "seaweed extinction depth" as suggested by Sears and Cooper (71). However, the available data on the light climate in places where seaweeds actually grow is at present too meagre to justify this approach, and more light measurements in the field are urgently required.

B. Critical light levels for growth and photosynthesis

In the course of his investigations of the distribution of benthic algae in the Gulf of Naples, Berthold (2) discovered that those deep-water algae that occurred at depths of 120 m or more, a small group of species such as Lithophyllum sp., Callithamnion sp. , Derbesia sp., or Phyllophora sp., were also found near the limit of algal growth in sea caves, and similar results have been obtained in more recent studies of sea cave vegetation (41, 55). As light quantity increased, either towards the entrance of caves, or towards the surface of the sea, Berthold found that the shade species were replaced in a characteristic sequence by many other species from all of the major algal groups, until in strong light there remained another small group of algae - the light-tolerant specialists consisting mostly of brown algae, but with some greens and reds. In late summer and autumn, even these largely disappeared, and Berthold recognised that this was probably due to excessive insolation. From the similarity of the distribution patterns in caves and in the open sea, Berthold concluded that light quantity was the main factor governing the vertical distribution of benthic algae. Although Berthold did not recognise that the changes in vegetation with depth in the sea could be due either to the decrease in light quantity or to the concurrent change in light quality, he was obviously right in regard to the vegetation pattern in sea caves which must be largely attributed to light quantity. Oltmanns (56) also recognised the importance of light quantity in regard to vertical algal distribution, after cultivating algal germlings from randomly distributed spores of several species in what we would call today a gradient of light quantity, although this experimental technique was not used again until

Halldal and French (33) introduced their light-temperature gradient plate. Oltmanns produced his gradient by use of sunlight passing a grey filter wedge, and he found that the germlings of the different species developed in neatly limited ranges of light quantity. He concluded: "For each species there must be a maximum, optimum and minimum of light quantity, and the further apart the minimum and maximum of a species lie, the greater will be its distribution range. As there are eurythermic and stenothermic organisms, there will be also euryphotic and stenophotic algae" (56; p.414).

The importance of light quantity for vertical algal distribution was stressed again by the work of Feldmann (23), who suggested that the sublittoral should be devided into an upper sub-zone ("étage infralittoral supérieur") which harbours the photophilous algae and a lower sub-zone ("étage infralittoral inférieur"), where the shade-loving algae dominate. This suggestion was supported by a symposium on zonation terminology (held in Genoa, 1957; summary in reference 59) with the revised names "infralittoral" for the zone of photophilous algae and marine phanerogams, and "circalittoral" for the zone of shade algae. The limit between these zones is thought to occur at 50-60 m at low latitudes and in clear water, or at 20 m at higher latitudes (58), and this roughly agrees with the depth at which Laminariales disappear in northern latitudes, i.e. at the 1 % surface light level (Fig.1).

Berthold and Oltmann's idea that the depth range occupied by each sublittoral algal species is mainly determined by the prevailing light quantity must be modified, to take account of other factors that we now know to be important, such as competition, as emphasized by Chapman (9). However, from a photobiological point of view, we also have to admit that we still lack quantitative data of the light requirements and light tolerance even for most of the common species. Only a rather fragmentary picture is available from recent measurements by various authors. This is presented in Table 2, but at least it shows the magnitudes which are to be expected.

The minimum light requirement for growth of laminarian gametophytes, as determined in white fluorescent light, lies in the order of

Table 2

Critical light levels for photosynthesis, growth and survival of benthic marine algae in artificial light (a-d) or sunlight (e-f). (For references see 47).

	Photon flux density ($\mu E\ m^{-2}s^{-1}$)	Irradiance ($W\ m^{-2}$)
(a) Minimum light requirement for growth		
Gametophytes of *Laminaria* spp.	0.25-0.5	0.05-0.1
(b) Saturating light levels for growth		
Germlings of deep-water red algae	4-9	0.7-1.7
Gametophytes of *Laminaria digitata*	8	1.5
Early sporophytes of *Laminaria* spp.	15	3
Adult thalli of *Codium fragile*	28	6
Adult thalli of *Laminaria saccharina*	70	14
Adult thalli of *Chondrus crispus*	94	19
Adult thalli of *Fucus* spp.	150	30
Adult thalli of *Pelvetia canaliculata*	250	50
(c) Compensating light levels for photosynthesis		
Sublittoral red algae	5-20	1-4
Laminaria spp.	5-15	1-3
Eulittoral green algae	5-25	1-5
Fucus serratus	10-20	2-4
(d) Saturating light levels for photosynthesis		
Deep-water red algae	60-70	12-14
Chondrus crispus, *Palmaria palmata*	180-210	36-42
Laminaria spp., *Macrocystis pyrifera*	150-220	30-44
Eulittoral red and green algae	400-500	80-100
Fucus spp.	500-600	100-120
(e) Lethal light levels		
Spores of *Laminaria* spp. (after 6-30 min)	360	70
(f) Inhibiting light levels for photosynthesis		
Adult fronds of *Laminaria hyperborea*	1250-1500	250-300

0.5 $\mu E\ m^{-2}s^{-1}$, which is equivalent to 25 lux in white light. Light-saturation of growth occurs for laminarian gametophytes and in germlings of deep-water red algae, at 5-10 uE $m^{-2}s^{-1}$. In adult, macroscopic thalli, growth is light-saturated at about one order of magnitude higher than in the few-celled, microscopic stages, probably because of the high proportion of non-photosynthetic tissue in the macroscopic stages. Furthermore, the growth of eulittoral species is not light-saturated until photon flux densities that are 2-3 times higher than for sublittoral algae. This is also evident from the saturating light levels for photosynthesis, which occur at about double the levels required for growth: i.e. 60-70 uE $m^{-2}s^{-1}$ for deep-water red algae, and 500-600 uE $m^{-2}s^{-1}$ in the eulittoral Fucus spp. This is analogous to the differences in level of photosynthetic light-saturation that are observed between shade and sun plants in terrestrial habitats (6). However, another typical trait of shade plants, namely a rather low rate of dark respiration, which results in a low compensation point and a photosynthetic surplus at rather low irradiances, is not apparent from the available data on compensation points in benthic marine algae. More investigations of this point are required, covering all of the dominant forms throughout the sublittoral. We also need more information on the levels of light and of ultraviolet radiation which inhibit photosynthesis, or even, at sufficiently high irradiances, become lethal. It is possible that differential sensivity towards high levels of light and ultraviolet is a major factor in algal distribution in the supralittoral, eulittoral, as well as in the upper sublittoral, but we have too little data at present (see Table 2). Spores of Laminaria spp. are killed within minutes, when exposed to irradiances which must be quite common in the uppermost part of the sublittoral, and the inhibition of photosynthesis at high irradiances is probably just as important for benthic algae growing near the water surface, as it is for phytoplankton (34). However, there has only been one recent investigation of this problem (in Laminaria spp.; reference 14) and it awaits further thorough analysis.

In addition to having lower compensation points and showing saturation of photosynthesis at rather low irradiances, shade plants are characterized by other traits, such as a high pigment content per unit weight, slow growth rates and a pronounced strategy to build up reserves before the onset of the light-poor season (6, 30, 67). The latter aspect is also seen in seaweeds, since the algal flora of the deeper sublittoral consists almost entirely of perennial forms. Regarding the pigment content of seaweeds, older investigations (for summary see reference 29) and also recent work (63-65) have shown that individuals of the same species growing in strong light have a lower chlorophyll a content and usually also exhibit a change in the ratio of accessory pigments to chlorophyll a. As pointed out by Dring (17), the phenotypic changes in pigment composition that have been observed to occur with depth (63-65, 69) are all qualitatively similar to those induced by decreasing light quantitity, either in the field (63-65, 69) or in the laboratory (8, 75), and, therefore, phenotypic or genotypic (i.e. between species) pigment changes with depth can be interpreted as responses to light quantity and not to light quality. We still lack experimental evidence, however, on whether the pigment composition of seaweeds will change in response to different light qualities, as well as to different levels of light quantity.

C. Photoperiodic and other photomorphogenetic reactions

Which of the steps in the development of numerous benthic marine algae are triggered by daylength, and which by other environmental factors, such as temperature or the seasonal pattern of nutrient concentration ? Only few species have been investigated in this respect. The photoperiodic responses discovered so far, which are all short day responses, are listed in Table 3. Most of these are absolute (qualitative) responses, which means, for example, that in short days the majority of the prostrate individuals of Scytosiphon lomentaria form erect thalli or the majority of Trailliella intricata individuals of Bonnemaisonia hamifera form tetrasporangia, whereas these reactions are totally blocked by long days. On the other hand, in a quantitative photoperiodic response the inapproppiate daylengths do not totally block the reaction. For instance, in the Conchocelis-stage of Porphyra tenera, short days dramatically increase the formation of conchosporangia, but a few sporangia are still being formed in long days. All of the responses listed in Table 3 are induced, if the daylength decreases below some critical value and, correspondingly, the night exceeds some critical length, but they are also completely (partially in the quantitative

Table 3

Photoperiodic short day reactions in benthic marine algae in which a night break is effective.
(A) = absolute (qualitative) response, (Q) = quantitative response.

Species	Enhanced reaction in short days	Source
Red algae:		
Porphyra tenera (Q)	formation of conchosporangia	15,16,67
Bangia fuscopurpurea (Q)	formation of conchosporangia	70
Asparagopsis armata (A)	formation of tetrasporangia	57,49
Bonnemaisonia hamifera (A)	formation of tetrasporangia	48
Calosiphonia vermicularis (A)	formation of tetrasporangia	52,53
Brown algae:		
Petalonia zosterifolia (A)	formation of erect thalli	48
Petalonia fascia (Q)	formation of erect thalli	48
Scytosiphon lomentaria (A)	formation of erect thalli	19,48
Green algae:		
Monostroma grevillei (A)	sporing of Codiolum-stage	48,49
Monostroma undulatum (A)	sporing of Codiolum-stage	49

cases) inhibited, if the long night is interrupted by a short pulse of light. The effectiveness of such a night-break is generally regarded as a diagnostic character of a genuine photoperiodic response. There have been a few reports, however, of benthic marine algae in which sporangium formation occurred only in short-day conditions, but a night-break was apparently ineffective , e.g. the Hymenoclonium-stage of the red alga Acrosymphyton purpuriferum (13) or Rhodochorton floridulum (76), although more recent work has established that a night-break is effective in the latter species, provided that it is combined with a 10 h rather than an 8 h photoperiod (Dring and West, unpublished). Ineffective night-breaks are not completely unknown in higher plant photoperiodism, but they have only been reported for long-day plants (74).

It may not be accidental that the photoperiodic responses reported so far have been found either in species which are characterized by a heteromorphic life cycle, such as in the red and green algae listed in Table 3, or in species like Scytosiphon lomentaria and Petalonia spp., in which there is a seasonal alternation between a prostrate system and erect, macroscopic thalli. The photoperiodic control ensures, in all of these plants, that the more or less cryptic phases are only induced to form the sporangia or the initials for the next macroscopic generation at the right time of the year. Or to put it the other way round, after the macroscopic phase has produced the cryptic phase, long days prevent the latter from immediately producing the macroscopic phase. For example, the Conchocelis-stage of Porphyra tenera, the Trailliella-stage of Bonnemaisonia hamifera, the prostrate system of Scytosiphon lomentaria and the Codiolum-stage of Monostroma grevillei, are all prevented by long days from initiating the macroscopic stage in summer, and each has to wait for the daylength to decrease to that critical value which is characteristic of the species, or, as we will see later on, of the particular geographical strain.

Photoperiodic control thus restricts the initiation of certain developmental stages to a particular season of the year, but this may be further restricted, if temperature acts as an additional limiting factor. In the Trailliella-phase of Bonnemaisonia hamifera tetrasporangia are formed in short days, with daylengths less than 12 h, only at temperatures around 15°C, and not at 12 or 20°C (49). This combination of conditions only leave open a rather small "window" during the year, which occurs in autumn at Helgoland (Fig.2) and at many other localities in the temperate region. The window is somewhat broader at coasts where the water temperature does not decrease so rapidly in late autumn, as, for instance in southern Japan (11) from where Bonnemaisonia hamifera possibly started its geographical expansion at the turn of the century (24, 54). Even here, however, the Trailliella-stage has only been found with tetrasporangia in the field between October and early winter, which is similar to the phenology reported for other localities. Thus, although the plant is exposed to inductive daylengths in spring, the water temperatures are apparently too low for tetrasporangium formation to occur, and in this way the species can discriminate between autumn and spring. The correct combination of short daylengths and relatively high temperatures is never achieved at some high latitudes, however, and in northern Scandinavia, for example, where the Trailliella-stage occurs without its gametophyte, it seems that the

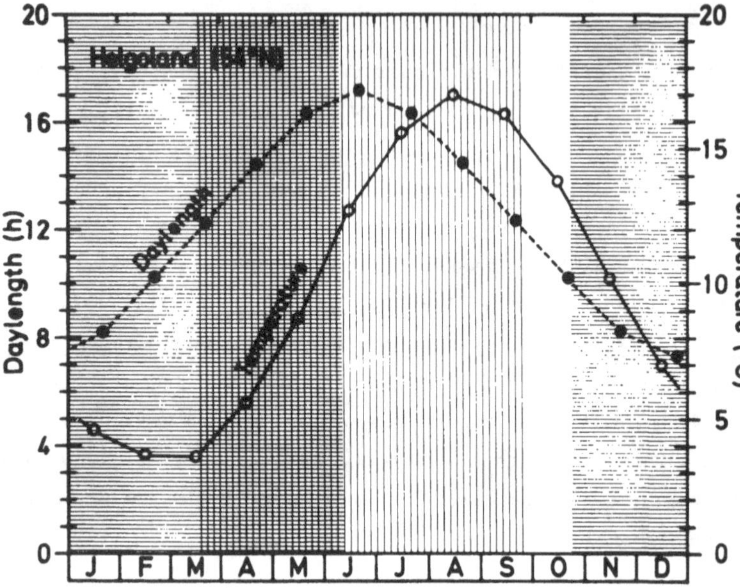

Fig. 2: Relationship between tetrasporangium formation in the Trailliella-stage of Bonnemaisonia hamifera and daylength or water temperature near Helgoland. Vertical lines indicate inhibition by daylength, horizontal lines by temperature. (From reference 49)

life-cycle is never completed. The photobiology of seaweeds may thus solve some of the riddles posed by their geographical distribution, as described in the case of Bonnemaisonia hamifera by McLachlan, Chen and Edelstein (54).

In Scytosiphon lomentaria the temperature range within which the short day response (i.e. formation of erect thalli from a prostrate system) occurs, is much wider than in Bonnemaisonia hamifera, although high temperatures block the formation of erect thalli completely, irrespective of daylength. Several geographical strains have been investigated (48), which are all genetically different in respect to critical daylength and inhibiting temperature. Almost no erect thalli are being formed at 15°C in a strain from Iceland, at 20°C in a strain from Helgoland, and at 23°C in a strain from Rovinj, in the Mediterranean. It appears, however, that, in all three localities, the formation of erect thalli is stopped in the summer by long daylengths, before the inhibiting temperatures are reached (Fig. 3). However, the water temperatures presented in Fig. 3 refer to the sublittoral, whereas Scytosiphon lomentaria occurs commonly in tide pools. Thus in late

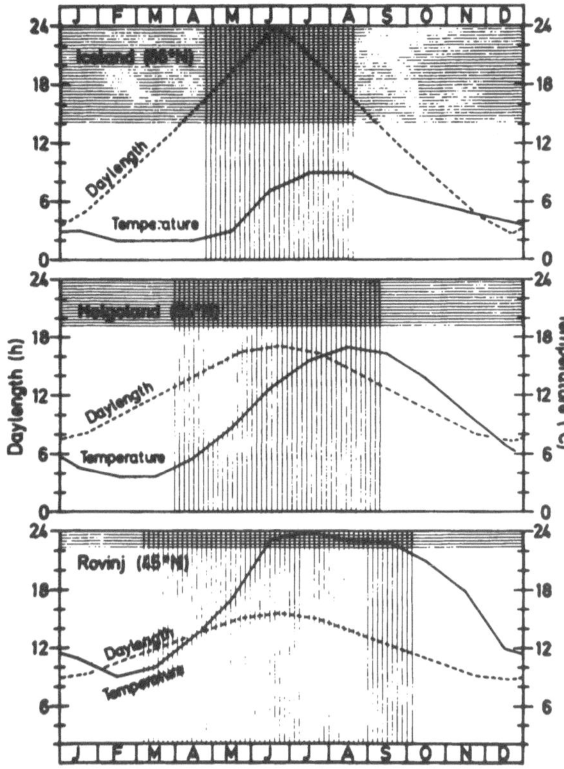

Fig. 3: Relationship between induction of erect thallus formation in three isolates of <u>Scytosiphon lomentaria</u> and daylength or temperature at three localities. Vertical lines indicate inhibition by daylength, horizontal lines by temperature. (From reference 49)

summer the species may be exposed to higher temperatures than indicated in Fig. 3, which could inhibit the production of new erect thalli after the daylength had fallen below the critical level (e.g. from late August onwards in Iceland).

As in many photoperiodic organisms, both plants and animals, the critical daylength increases towards the poles. In strains of <u>Scytosiphon lomentaria</u> from the Mediterranean, Helgoland, and Iceland, the daylengths which blocked erect thallus formation, were found to be 11.5, 13, and 16.5 h, respectively (Fig. 3). This pattern means that, near its southern limits of distribution, where water temperatures in summer may approach the lethal limit for the macroscopic thallus, the species is triggered to produce this phase in winter. At northern latitudes, on the other hand, where a restriction of the photoperiodic response to very short daylengths would result in very low light levels and the threat of ice, the critical daylength is longer and thalli can be produced in spring. However, continuous light also blocks erect thallus formation in the Icelandic strain, and this reaction, possibly combined with the temperature block mentioned above, prevents the start of a new generation of erect thalli in summer, which would run into the problem of how to survive the winter under ice. In this eulittoral or upper sublittoral species, the cryptic, prostrate system is obviously better adapted to survive in severe winter conditions. Scytosiphon lomentaria does not always exhibit this photoperiodic control of development. Non-photoperiodic strains have been isolated from Denmark (39), or Australia (12). The occurrence of photoperiodic and non-photoperiodic strains within a single species is not unknown. In some higher plants, the difference between a photoperiodic and a non-photoperiodic strain can be attributed to a single Mendelian gene, and many species of higher plants and animals contain ecotypes with very different photoperiodic behaviour, as well as non-photoperiodic strains (1, 74).

To date the search for photoperiodic responses in seaweeds has been mainly restricted to species which fit neatly into a petri dish. A substantial challenge exists, however, in finding the trigger for such dramatic seasonal events as the onset of new blade formation in Delesseria sanguinea or Laminaria hyperborea which regularly occurs around December. Photoperiodic control is clearly a strong possibility since the perennial red alga Constantinea subulifera produces new blades only in short days, and this response has all the hallmarks of a genuine photoperiodic reaction (60), and receptacle formation in Fucus distichus is also restricted to short days (5). It is obviously time that we enlarge the size of our culture vessels.

In addition to photoperiodic responses, a number of other photomorphogenetic reactions have been discovered in benthic marine algae. Laminarian gametophytes, for example, grow vegetatively in red or green light, but are induced to become fertile only after they have received a certain dose of blue light (50, 46). Germlings of Scytosiphon lomentaria form filaments in red or green light, but blue light induces the formation of both hairs and two-dimensional crusts (18). But research into the influence of light quality on the morphogenesis of seaweeds is still in its infancy, and this is especially true for the larger algae. And so, hundred years after the photobiology of seaweeds was given such a good start by the remarkable insights of Engelmann, Berthold and Oltmanns, we can say little more than: We now know about many more aspects of which we are ignorant.

References

1. Beck, S.D.: Insect Photoperiodism. Academic Press, London and New York 1968.
2. Berthold, G.: Pubbl. Staz. Zool. Napoli 3, 393-536 (1882).
3. Berthold, G.: Jb. wiss. Bot. 13, 570-717 (1882).
4. Biebl, R.: Protoplasma 46, 63-89 (1956).
5. Bird, N.L., McLachlan, J.: Phycologia 15, 79-84 (1976).
6. Björkman, O.: In: Photophysiology 8 (Ed. by A.C. Giese), pp. 1-63. Academic Press, New York 1973.
7. Boutler, J., Cabioch, L., Grall, J.-R.: Bull. Soc. Phycol. Fr. 19, 129-140 (1974).
8. Brown, T.E., Richardson, F.L.: J. Phycol. 4, 38-54 (1968).
9. Chapman, A.R.O.: Bot. Mar. 16, 80-82 (1973).
10. Chapman, A.R.O., Lindley, J.E.: Mar. Biol. 57, 1-5 (1980).
11. Chihara, M.: Sci. rep. Tokyo Kyoiku Daigaku 10, 121-154 (1961).
12. Clayton, M.N.: J. exp. mar. Biol. Ecol. 25, 187-198 (1976).
13. Cortel-Breeman, A.M., Ten Hopen, A.: Phycologia 17, 125-132 (1978).
14. Drew, E.A.: Brit. phyc. J. 9, 217-218 (1974).
15. Dring, M.J.: J. mar. biol. ass. U.K. 47, 501-510 (1967).
16. Dring, M.J.: Nature 215, 1411-1412 (1967).
17. Dring, M.J.: Limn. Oc. (In Press) (1981).
18. Dring, M.J., Lüning, K.: Z. Pflanzenphysiol. 75, 107-117 (1975).
19. Dring, M.J., Lüning, K.: Planta (Berl.) 125, 25-32 (1975).
20. Emerson, R.: Science 125, 746 (1957).
21. Engelmann, T.W.: Bot. Ztg. 41, 1-13 and 18-29 (1883).
22. Engelmann, T.W.: Bot. Ztg. 42, 81-110 (1884).
23. Feldmann, J.: Rev. Algol. 10, 1-139 (1937).
24. Feldmann, J.: Année Biol. 33, 49-56 (1956).

25. Fredj, G.: Bull. Inst. océanogr., Monaco 71 (1421), 1-42 (1972).

26. Garner, W.W., Allard, H.A.: J. agric. Res. 18, 553-606 (1920).

27. Garner, W.W., Allard, H.A.: J. agric. Res. 23, 871-920 (1923).

28. Geiger, R.: Die Atmosphäre der Erde. Karte 1: Jährliche Sonnenstrahlung. Perthes, Darmstadt 1965.

29. Gessner, F.: Hydrobotanik. Band I. VEB Deutscher Verlag der Wissenschaften, Berlin 1955.

30. Grime, J.P.: In: Light as an ecological factor (Ed. by R. Bainbridge, G.C. Evans, O. Rackham), pp. 187-207. Blackwell, Oxford 1966.

31. Halldal, P.: Physiol. Plant. 17, 414-421 (1964).

32. Halldal, P.: In: Optical Aspects of Oceanography (Ed.by N.G. Jerlov, E. Steeman Nielsen), pp. 344-360. Academic Press, London 1974.

33. Halldal, P., French, C.S.: Plant Physiol. 33, 249-252 (1958).

34. Harris, G.P.: Arch. Hydrobiol. Beih. Ergebn. Limnol. 10, 1-171 (1978).

35. Haxo, F.T., Blinks, L.R.: J. gen. Physiol. 33, 389-422 (1950).

36. Ivanoff, A.: Ann. Géophys. 13, 22-53 (1957).

37. Jerlov, N.G.: In: Marine Ecology. Vol. I. Part 1 (Ed. by O. Kinne), pp. 95-102. Wiley-Interscience, London 1970.

38. Jerlov, N.G.: Marine Optics. Elsevier, Amsterdam 1976.

39. Kristiansen, A., Pedersen, P.M.: Bot. Tidskr. 74, 31-56 (1979).

40. Kurogi, M.: Bull. Tohoku reg. Fish. Res. Lab. 15, 33-42 (1959).

41. Larkum, A.W.D., Drew, E.A., Crossett, R.N.: J. Ecol. 55, 361-371 (1967).

42. Levring, T.: Göteborg K. Vet.-och Vitt. samh. handl., Sjätte följden Ser. B. 5, 1-89 (1947).

43. Levring, T.: Bot. Mar. 11, 72-80 (1968).

44. Levring, T.: Proc. Int. Seaweed Symp. 6, 235-244 (1969).

45. Lüning, K.: Helgoländer wiss. Meeresunters. 21, 271-291 (1970).

46. Lüning, K.: J. Phycol. 16, 1-15 (1980).

47. Lüning, K.: In: The Biology of Seaweeds (Ed. by C.S. Lobban and M.J. Wynne), (In Press). Blackwell, Oxford 1981.

48. Lüning, K.: In: The Shore Environment, Vol.2: Ecosystems (Ed. by J.H. Price, D.E.G. Irvine, W.F. Farnham), (In Press). Academic Press, London and New York 1981.

49. Lüning, K.: Ber. Deutsch. Bot. Ges. (In Press) (1981).

50. Lüning, K., Dring, M.J.: Mar. Biol. 29, 195-200 (1975).

51. Lüning, K., Dring, M.J.: Helgoländer wiss. Meeresunters. 32, 403-424 (1979).

52. Mayhoub, H.: C. R. Acad. Sc. Paris 280, 2441-2443 (1975).

53. Mayhoub, H.: Bull. Soc. Phycol. Fr. 21, 48 (1976).

54. McLachlan, J., Chen, L.C.-M., Edelstein, T.: Proc. Int. Seaweed Symp. 6, 245-249 (1969).

55. Norton, T.A., Ebling, F.J., Kitching, J.A.: In: Fourth European Marine Biology Sympsosium (Ed. by D.J. Crisp), pp. 409-432. University Press, Cambridge 1971.

56. Oltmanns, F.: Jb. wiss. Bot. 23, 349-440 (1892).

57. Oza, R.M.: Bot. Mar. 20, 29-32 (1976).

58. Péres, J.M.: Oceanogr. mar. Biol. 5, 449-533 (1967).

59. Péres, J.M., Molinier, R.: Rec. Trav. mar. Endoume 13(22), 5-15 1957).

60. Powell, J.H.: The life history of a red alga, Constantinea. Ph.D. thesis, Univ. Microfilms Inc., Ann Arbor, Michigan 1964.

61. Rabinowitch, E.I.: Photosynthesis and Related Processes. Vol. II. Part 1. Interscience Publishers, New York 1951.

62. Rabinowitch, E.I.: Photosynthesis and Related Processes. Vol. II. Part 2. Interscience Publishers, New York 1956.

63. Ramus, J., Beale, S.I., Mauzerall, D., Howard, K.L.: Mar. Biol. 37, 223-229 (1976).

64. Ramus, J., Beale, S.I., Mauzerall, D.: Mar. Biol. 37, 231-238 (1976).

65. Ramus, F., Lemons, F., Zimmermann, C.: Mar. Biol. 42, 293-303 (1977).

66. Ramus, J.: In: The Biology of Seaweeds (Ed. by C.S. Lobban and M.J. Wynne), (In Press). Blackwell, Oxford 1981.

67. Raven, J.A., Smith, F.A., Glidewell, S.M.: New Phytol. 83, 299-309 (1979).

68. Rentschler, H.G.: Planta 76, 65-74 (1967).

69. Rhee, C., Briggs, W.R.: Bot. Gaz. 138, 123-128 (1977).

70. Richardson, N., Dixon, P.S.: Nature 218, 496-497 (1968).

71. Sears, J.R., Cooper, R.A.: Mar. Biol. 44, 309-314 (1978).

72. Tyler, J.E. (Ed.): Light in the Sea. Dowden, Hutschinson and Ross, Stroudsburg, Pennsylvania 1977.

73. Tyler, J.E., Smith, R.C.: Ocean Sciences 1, 1-103 (1970).

74. Vince-Prue, D.: Photoperiodism in Plants. McGraw Hill, London 1975.

75. Waaland, J.R., Waaland, S.D., Bates, G.: J. Phycol. 10, 193-199 (1974).

76. West, J.A.: In: Contribution to the Systematics of Benthic Marine Algae of the North Pacific (Ed. by I.A. Abbott, M. Kurogi), pp. 213-230. Jap. Soc. Phycology, Kobe 1972.

THE ROLE OF OLEFINIC HYDROCARBONS IN SEXUAL REPRODUCTION OF MARINE BROWN ALGAE

D.G. Müller

Fakultät für Biologie der Universität Konstanz
D-7750 Konstanz, Germany

A. Introduction

Phenomena connected with sexual reproduction are especially apparent in various marine algae. Processes such as fertilization of Fucus eggs (1) or the periodic maturation and liberation of oogonia and antheridia in Dictyota (2) have been described many decades ago. In those early days, first evidence was found that there is a pronounced interaction between eggs and spermatozoids or male and female isogametes (1, 3, 4). In many cases clouds of spermatozoids accumulate around an egg before plasmogamy takes place. Several workers have attempted to study such interactions. However, inherent problems, mainly the extremely small amounts of substances available, prevented significant progress. Recent improvements in analytical chemistry now made possible the analysis of several systems, including complete identification of the substances involved. In the following, I shall present the brown algal systems which have been successfully studied since 1968. In all cases, the biological part of the work was initiated by myself, including detection and collection of the substances, their isolation and respective bio-assays. Prof. Jaenicke of the University of Köln and his co-workers did the analytical chemistry and synthesis of all sexual hormones mentioned here. The substances discussed are termed hormones, because they fit a broad understanding of the term: they are secreted by specialized cells, transported through sea-water as a carrier medium and act upon

Xth International Seaweed Symposium
Printed in Germany

specific target cells.

B. The Organisms and Their Sexual Hormones

Ectocarpus siliculosus was the first species studied. It can be easily cultivated. Male and female gametophytes are morphologically identical and even the gametes cannot be distinguished by morphological characters: the species is isogamous. Nevertheless, there are physiological differences. Those gametes which are termed female have a strong tendency to settle down after a short motility period. After coming to rest and rounding up, a dramatic phenomenon occurs when male gametes are present: these male cells, which are perpetual swimmers, become excited and move in large numbers around female cells until a zygote is formed. It can be easily demonstrated by a suitable bio-assay that this response is caused by a secretion

Organism	*Ectocarpus siliculosus* *E. draparnaldioides*	*Cutleria multifida*	*Laminaria digitata* *L. saccharina* *L. hyperborea*	*Fucus serratus* *F. vesiculosus*
Structure	Ectocarpen 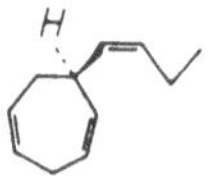 $C_{11}H_{16}$	Multifiden  $C_{11}H_{16}$	unknown $C_{11}H_{14}O$	Fucoserraten  C_8H_{12}
Function	attraction	attraction	release + attraction	attraction
Additional Substances	none	Aucanten Ectocarpen	Ectocarpen Pentadecane	none

Fig.1. Sexual hormones in Phaeophyceae and species in which they are produced.

coming from female cells after settling down. From now on, it took about two and a half years of continuous work to maintain and extract mass cultures of female gametophytes until finally the chemists were able to write down the structural formula (5). The substance was named ectocarpen (Fig.1).

A literature survey at this state revealed that other brown algae could be expected to possess similar hormonal interaction systems. Accordingly, Cutleria multifida was selected as the next system to be examined. Cutleria is a springtime annual species on the northern Mediterranean coast and can be cultivated in the laboratory. It has a classical type of heteromorphic alternation of generations. The sporophyte is a perennial crust, whereas the gametophytes are conspicuous and morphologically identical. The species is strongly anisogamous with large-size female zoids and spermatozoid-like male gametes. Female gametes, after they have come to rest, attract large numbers of male gametes in the same manner as seen in Ectocarpus. Substances secreted by female gametes were isolated and examined by gas chromatography. Three compounds were found. The most prominent fraction was identified as the attractant, whereas the two other compounds were much less effective (6). One of these is ectocarpen, the other was named aucanten and is a cyclic $C_{11}H_{16}$ olefin with a six-numbered ring. The Cutleria attractant was named multifiden, its structure is given in Fig.1 (7). The oldest literature report indicating chemical communication between egg and spermatozoids refers to the genus Fucus (1). Volatile secretions from eggs of Fucus serratus caused the attraction of spermatozoids. The respective compound was identified as a conjugated octatriene and given the name of fucoserraten (8).

The three compounds listed here are the only definitely known attractant molecules in brown algae to date. A fourth molecule, which in my opinion is the most interesting one encountered so far, is found in the genus Laminaria. A main feature of the order Laminariales is the alternation between a macroscopic sporophyte and a microscopic gametophyte stage. In dioecious species, male and female gametophytes can be grown separately, and under suitable conditions sexual organs are formed. Eggs are spontaneously released from the oogonia. In contrast, spermatozoids do not leave the antheridia, even if they are fully mature. Mass discharge of spermatozoids occurs when a drop of medium is added from a culture of ripe female plants with eggs (9). Extracts from such female mass cultures yield three major compounds: ectocarpen, n-pentadecane, and a third substance of the composition $C_{11}H_{14}O$. Bio-assays with this unknown compound show that it is responsible for the release of spermatozoids as well as for their subsequent attraction to the egg (10). This Laminaria system has been under study for more than two years. It is the most complicated system so far encountered in brown algae. The amount of active material is extremely small. Even with large-scale mass cultures no more than a few nanograms can be obtained per day. Instead, the chemists are using the fragmentary information on chemical structure available at present to synthesize possible model substances. Such synthetic candidates are then compared with the native compound using high-resolution glass capillary gas chromatography. It is hoped that this indirect approach will one day reveal the correct structure of the Laminaria hormone.

C. Specificity

Two species of the genus Fucus, F. serratus and F. vesiculosus, use the same attractant molecule (11). Therefore, spermatozoid attraction by the egg cannot be a basis for species specificity in this case. In fact, cross-combinations of living eggs and spermatozoids confirm this conclusion very clearly (12). Since for the taxonomist and the field phycologist the two species of Fucus are clearly distinct entities and do not noticeably interbreed, effective hybridization barriers must be postulated. Evidence for such species-specific factors has been presented by Bolwell et al. (13). They report that fertilization is based on associations between fucosyl- and mannosyl-containing ligands on the egg surface and specific carbohydrate-binding receptors on the sperm surface. Such interactions might possibly represent the basis for species-specific recognition.

The identity of the attractant in two species of the genus *Fucus* indicates that attraction between gametes is bound to a supraspecific level. In order to examine this phenomenon on a broader basis additional taxa in the genera *Ectocarpus* and *Laminaria* were examined.

Genus *Ectocarpus*: Ectocarpen was discovered as the male-attractant in *Ectocarpus siliculosus*. Another taxon in the genus is *E. draparnaldioides*. It represents the gametophytic phase of a form of *E. fasciculatus*. Gametes of *E. draparnaldioides* interact with complementary gametes of *E. siliculosus* in respect of attraction, although plasmogamy does not occur. Gas-chromatographic analysis of eluates from female gametes revealed that the attractant of *E. draparnaldioides* is ectocarpen (14).

Genus *Laminaria*: As mentioned above, eggs of *Laminaria digitata* secrete a substance with the composition $C_{11}H_{14}O$ which acts as a spermatozoid-releaser and -attractant. Although its chemical identity is not yet completely known, it can be recognized by its gas-chromatographic characteristics. Examination of eluates from *L. hyperborea* and *L. saccharina* clearly shows that these species form the same compound (14). Bio-assays confirm complete cross-interaction concerning release and attraction in all three species (9). Thus, at present, it seems that sexual hormones in brown algae are not species-specific. They are probably common to all species within one genus. Additional information is necessary in order to give more weight to this preliminary statement.

D. Phylogenetic Considerations

The Phaeophyceae are regarded as a monophyletic class. The chemical structure of the few sexual hormones identified to date nevertheless indicate some interesting relations to phylogeny (Fig.2). The order Ectocarpales is considered to be the most primitive. The genus Ectocarpus uses one single compound, ectocarpen, as its attractant. The approach mechanism perform-

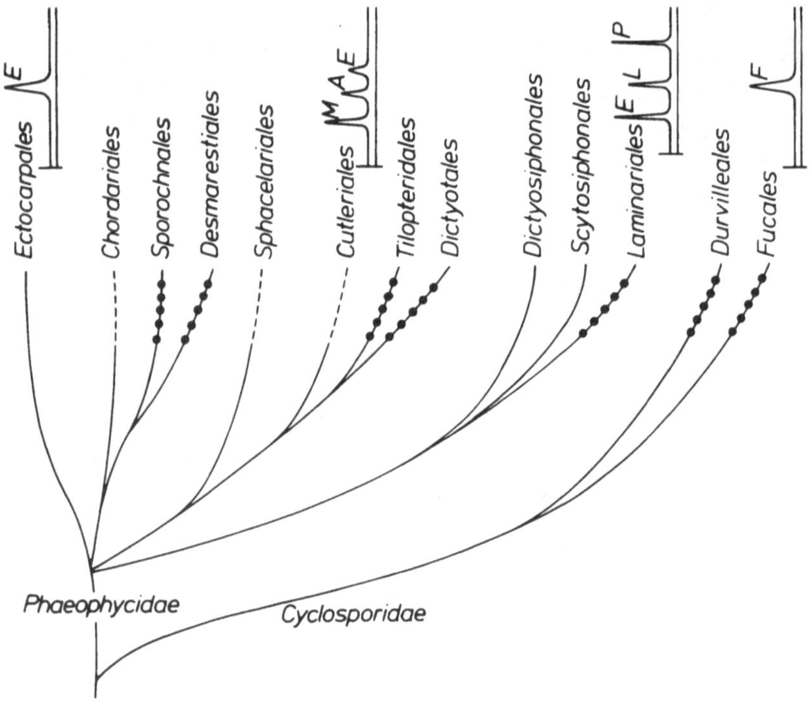

Fig.2. Phylogeny in the class Phaeophyceae, modified after Wynne and Loiseaux (20).
Schematic gas-chromatograms indicate substances found in eluates from female gametes. A: aucanten; F: fucoserraten; E: ectocarpen; M: multifiden; P: n-pentadecane; L: spermatozoid-releasing and attracting substance from Laminaria with unknown structure.
—— isogamy, ---- anisogamy, o-o-o-o- oogamy

ed by the male gametes is comparably primitive as will be shown later. The more advanced orders, Cutleriales with anisogamy and Laminariales with oogamy, possess in addition to their respective sexual hormones two accessory substances of unknown function. In both cases one of these is ectocarpen. Until a better explanation is found, the occurrence of ectocarpen may be taken as a phylogenetic reminiscence. In contrast, the genus Fucus, which belongs to a different line (Cyclosporidae), forms again one single compound which significantly differs in its chemical structure from the cyclic C_{11} compounds in Phaeophycidae. It is obvious that these thoughts are of a preliminary nature and more information is needed to establish well-founded relations. Members of the orders Dictyotales and Desmarestiales are at present being studied, and I would like to use this opportunity to ask colleagues for support. One of the critical points in studying sexual hormones in brown algae is to find suitable species that meet several essential requirements: the organism should be dioecious, and large amounts of female gametes must be available, either in the field or from laboratory cultures. Therefore, should you happen to come across materials that seem suitable for this type of work, I would be extremely anxious to start a joint effort and submit them to our analytical procedures.

E. Quantitative Studies

Soon after the structure of an interesting natural product is known, the substance is made available by laboratory synthesis in many cases. Then it is possible to study the response of male gametes to attractants on a quantitative basis. In such experiments, the attractive potential of known amounts of hormone dissolved in an inert solvent is measured. In Ectocarpus, Cutleria and Fucus the most powerful attraction is exhibited by the authentic molecules (15). Any alteration of the chemical structure reduces the attractive potential of the hormone

locality	locality / gametes ♀ \ ♂	1 Nap RB1	2 VF 01	3 Ros 6	4 Man 14b	5 Ber 42	6 WH 5d	6 Pen 9d	7 Bft 15b	8 Wil 18a	9 Tam 2b	10 PAr 27a	11 Vic 7	12 Nfl 31f	13 Cph 40-11
1	Nap DA2	+	+	+	+	+	∩	∩	+	+	+	+	+		+
2	VF 05	+	+	+	+	+	∩	∩	+	+	+	+	+	+	∩
3	Ros 3	+	+	+	+	+	∩	∩	+	+	+	+	+	+	
4	Man 3	+	+	+	+	+	∩	∩	+	+	+	+	+		
5	Ber 21	+	+	+	+	+	∩	∩	+	+	+	+			∩
6	WH 8b	∩	∩	∩	∩	∩	+	+	+	∩	∩	∩			+
	Pen 5	∩	∩	∩	∩	∩	+	+	+	+	+	+			+
7	Bft 18c	+	+	+	+	+	∩	∩	+	+	+	+	+		
8	Wil 11a	+	+	+	+	+	∩	∩	+	+	+	+	+		
9	Tam 18b	+	+	+	+	+	∩	∩	+	+	+	+	+		+
10	PAr 10n	∩	∩	∩	+	∩	∩	∩	+	+	+	+			
	PAr 12f	∩	∩	∩	+	∩	∩		+	+	+	+			
11	Vic 1	+	+	+	+	∩	∩	∩	+	+	+	+	+		
12	Nfl 31e		+												
13	Ska 6b	+	+	+	+	+		∩	+	+	+	+	+		
	Cph 40-1		+						+			+			+
14	Tro 2		+												

Fig.3. Results of cross-combinations with gametes of different sex.

+: plasmogamy observed. Semicircles: attraction, but plasmogamy not observed. Open fields: not examined. Results from Müller (17), supplemented with new data.

Collection data for new strains:
Ska6: Klagshamn, South Sweden, Sept. 1978;
Cph 40: Copenhagen, June 1979, from unilocular sporangium;
NFL 31: Kelly's Island, Newfoundland, Sept. 1977, from unilocular sporangium;
Tro 2: Tromsö, Norway, June 1978, from unilocular sporangium;
Vic 7: Sorrento, Victoria, Australia, May 1978, from unilocular sporangium.

(12). Attempts at measuring biological effects of substances that are almost insoluble in water are problematic. Detailed knowledge is necessary as regards partition characteristics between solvent phase and sea-water before absolute attractant concentrations in the medium can be estimated. Such information has been recently obtained for the first time by Boland et al. (16) for multifiden and a number of related molecules. In the case of multifiden, the threshold concentration for attraction is 3×10^{-7} M in the solvent. Taking into account the solubility characteristics of the molecule, the actual threshold concentration in the aqueous phase is about 6×10^{-9} M.

F. Biological Species Concept

The last two chapters will again deal with Ectocarpus siliculosus. It is well known that taxonomy in the genus Ectocarpus is complicated due to great variability of morphological characters. Thus, widely diverging opinions exist among taxonomists concerning species delimitation. Consideration of the biological species concept can possibly add new aspects to this problem. For this study, gametophytes from various geographic areas were compared. Successful plasmogamy was used as an indicator of biological relationship. The results (Fig.3) show that gametophyte populations in the entire North-Atlantic as well as South Australian waters are essentially cross-compatible and may, therefore, be considered as one syngen (17). The degree of morphological variability is remarkable, especially concerning branching patterns and gametangial shape. However, some unexpected exceptions emerged during this study. Two localities on the north-east American coast deviated from the general world-wide cross-compatibility pattern: the Woods Hole, Mass., area and Port Aransas, Texas. Ectocarpus from these places shows partial sexual isolation. The Woods Hole population is highly separated from almost all other populations. In Port Aransas, the male gametes are fertile with females from almost all places studied, but the female gametes are much more restrictive. They are non-compatible in all combinations except with their own local males, other North-American neighbours and males from the Isle of Man. Human olfactory sensation and gas-chromatographic analysis confirms that female gametes from all localities secrete ectocarpen. Failure of plasmogamy in the cases just described is possibly due to alterations in specific cellular recognition mechanisms such as those described by Bolwell et al. (13) in the genus Fucus. Microscopic examination shows that in addition to failure of plasmogamy the initial contact between the front flagellum of the male gamete and the surface of the female gamete does not occur. These phenomena may, therefore, be interpreted as a first step leading to the formation of a new species, starting with sexual isolation.

G. Kinetics of Sexual Approach in Ectocarpus

The failure of cellular contact during sexual reproduction in certain crosses of Ectocarpus siliculosus gametes offers the possibility to study in detail the approach of male gametes to the attractant source. In a first attempt dark-field video recordings were made and analyzed graphically (18). It became clear that male gametes of Ectocarpus do not perform a topotactic response, but find the attractant source in an indirect way. When swimming freely in the medium they move with maximum speed in straight lines. Upon contact with a boundary, they reduce their speed and tend to maintain the contact, now swimming in loop-like paths. In the presence of ectocarpen, this contact response is enforced dramatically, and the locomotion changes to circular paths, the diameter of which is inversely proportional to the attractant concentration. The combination of these behavioral components finally leads to arrival at the female cell and to circular or hairpin-like paths around it. The poor optical resolution of the video technique did

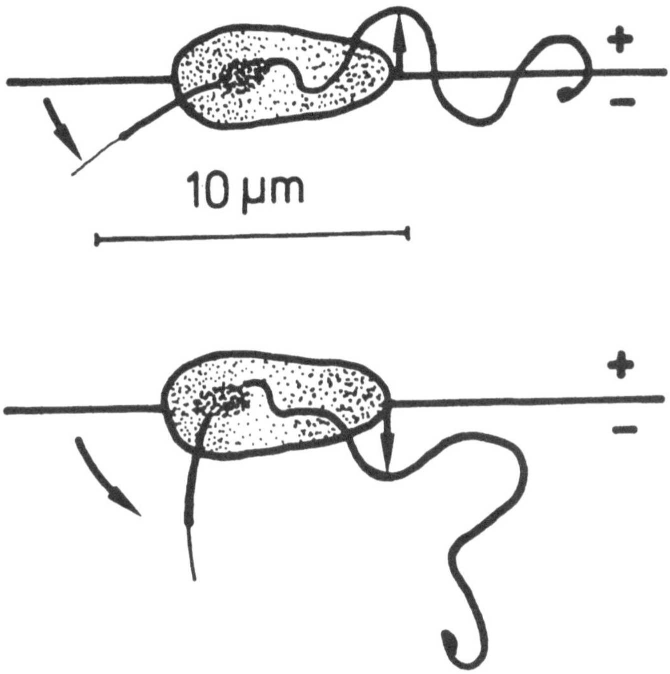

Fig.4. Male gametes of Ectocarpus siliculosus seen from the "ventral" side, which is oriented toward the cover-glass. Extreme positions of front and hind flagella reconstructed from high-speed micro-cinematography.

not permit to study the flagellar beat pattern associated with these locomotive reactions. Therefore, high-speed cinematography was used, which allowed to study flagellar movement with phase-contrast microscopy. Due to focusing problems, only cells moving in contact along the cover-glass were recorded. During contact with a boundary, exclusively one side of the cell is exposed to the surface: it is that side where the flagella emerge, which we call ventral (Fig.4). The front flagellum exhibits meander-like beats that start from the base and move upward at a frequency of about 30 Hz. Normally the hind flagellum is directed obliquely backwards. This unilateral position results in a persistent sideward locomotive component. When compensated by counteraction of the front flagellum, straight tracks result. Likewise, the front flagellum may act asymmetrically to enforce clockwise circular paths. The most striking feature are rapid beats of the hind flagellum. This structure is not flexible, it can only be bent as a whole. Within 10 msec, it deflects from its "normal" position to a right angle sideward or even obliquely forward. This beat of the hind flagellum results in a dramatic hairpin-like curvature of the track.

Male gametes perform all these flagellar beat patterns without chemical stimulation. The presence of ectocarpen, therefore, does not cause qualitative changes in locomotive behaviour of male gametes. The apparent effects induced by chemical

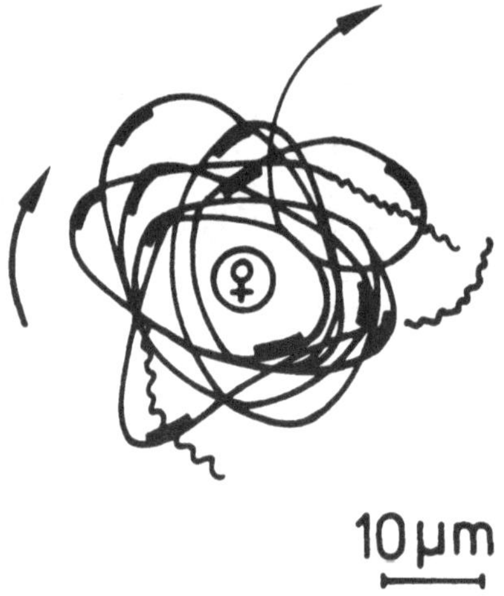

Fig.5. Ectocarpus siliculosus. Track of one male gamete around a female gamete. Reconstructed from high-speed cinematography. Sections with strong inclinations of the hind flagellum are emphasized.

stimulation result exclusively from quantitative changes. The most obvious effect induced by chemical stimulation is an increased frequency of rapid hind-flagellar deflections (19). This reaction seems to be the major component of small-diameter circles and hairpin-turns in the vicinity of a female cell (Fig.5). The cinematographic sequences being presented will illustrate this.

References

1. Thuret, G.: Ann. Sci. Nat. Bot. Sér. IV, 2, 197-214 (1854).
2. Williams, J.L.: Ann. Bot. (Lond.) 19, 531-560 (1905).
3. Berthold, G.: Mitt. Zool. Stat. Neapel 2, 401-413 (1881).
4. Kuckuck, P.: Wiss. Meeresunters. Abt. Helgoland N.F. 3, 95-116 (1899).
5. Müller, D.G., Jaenicke, L., Donike, M., Akintobi, T.: Science (Wash.) 171, 815-817 (1971).
6. Müller, D.G.: Biochem. Physiol. Pflanzen (BPP) 165, 212-215 (1974).
7. Jaenicke, L., Müller, D.G., Moore, R.E.: J. Amer. Chem. Soc. 96, 3324-3325 (1974).
8. Jaenicke, L., Seferiadis, K.: Chem. Ber. 108, 225-232 (1975).
9. Lüning, K., Müller, D.G.: Z. Pflanzenphysiol. 89, 333-341 (1978).
10. Müller, D.G., Gassmann, G., Lüning, K.: Nature (Lond.) 279, 430-431 (1979).
11. Müller, D.G., Gassmann, G.: Naturwissenschaften 65, 389 (1978).
12. Müller, D.G., Seferiadis, K.: Z. Pflanzenphysiol. 84, 85-94 (1977).
13. Bolwell, G.P., Callow, J.A., Callow, M.E., Evans, L.V.: J. Cell. Sci. 36, 19-30 (1979).
14. Müller, D.G., Gassmann, G.: Naturwissenschaften (1980) in press.
15. Müller, D.G.: Z. Pflanzenphysiol. 80, 120-130 (1976).
16. Boland, W. et al. (1980) manuscript in preparation.
17. Müller, D.G.: Phycologia 18, 312-318 (1979).

18. Müller, D.G.: Arch. Protistenk. 120, 371-377 (1978).
19. Geller, A., Müller, D.G.: J. exp. Biol. (1980) in press.
20. Wynne, M.J., Loiseaux, S. Phycologia 15, 435-452 (1976).

THE DOMESTICATION AND CULTIVATION OF CALIFORNIAN MACROALGAE

M. Neushul

Marine Science Institute, University of California, Santa Barbara, CA 93106, USA

Introduction

In order to devise a cultivation strategy that would allow one to plant, cultivate and harvest a macroalgal crop, one must define the plant requrements for light, space and nutrients and ensure that they are met. In attempting to do this at Santa Barbara, we have taken an ecological approach, by first studying the natural habitats where macroalgae occur, and paying particular attention to the patterns of plant distribution that occur in the sea. We have made diving surveys of kelp forests along the U. S. West coast, and have studied their macroscopic and microscopic features. We have grown plants in dish and tank culture in the laboratory. Recently we have experimentally planted in the sea (1). Our goal is to use ecological information as a basis for developing a new "maricultural ecology" that will be a basis for effective marine farming.

Studies of Algal Habitats

The Macrocystis forests of California were first explored by diving in the early 1950's, when scuba equipment became available. At that time it was my privilege to work with W. J. North, and to have the aid and guidance of the late E. Y. Dawson. Each new forest that we explored seemed to be unique and different from all others (2). Our preliminary reconnaissance and floristic work was augmented by the use of line transects, which provided quantitative and diagrammatic views of the habitats being studied

Xth International Seaweed Symposium
Printed in Germany

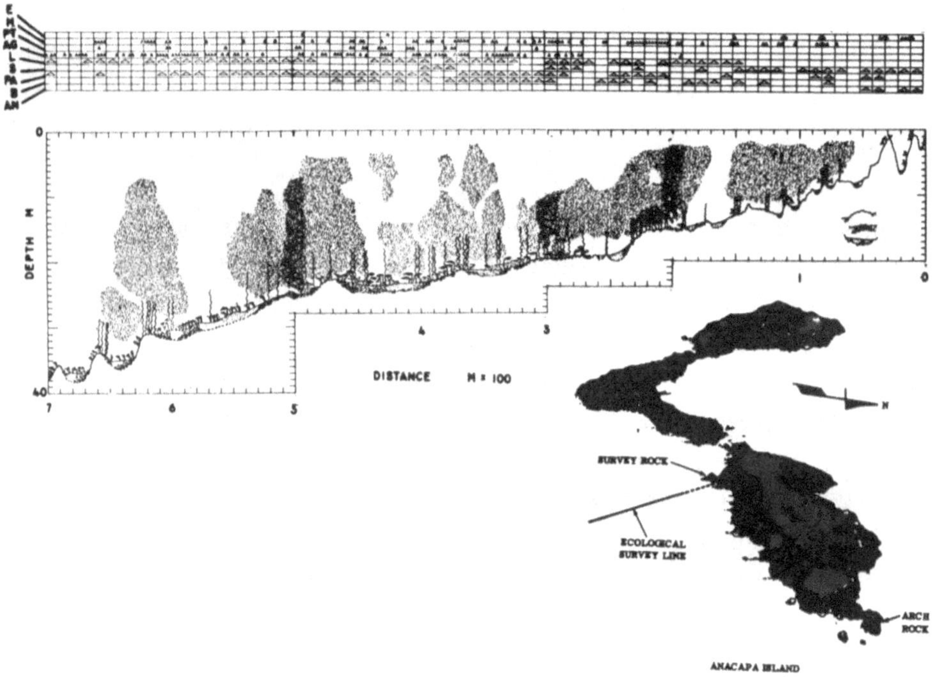

Figure 1. A line transect was established in a kelp forest off Anacapa Island, California, and ran for 700 M, from the intertidal down to a depth of 37 M (5). The distribution of Macrocystis in midwater was measured with a recording fathometer, and along the bottom by divers with tape recorders. Distribution patterns along the transect are shown for: Eisenia (E), Macrocystis (M), Pterygophora (P), Agarum (AG), and the animal genera Lytechinus (L), Strongylocentrotus (S), Pateria (P), Botruanthus (B) and Anthopleura (A).

(3). Tape recorders were adapted for use underwater, and a computer was used to process the data obtained (4). For the first time, in these early studies, the full range of plant distribution could be studied from the upper intertidal down to the lowest depth limits of vegetation. Also, by using microscopes, the small-scale micro-habitat features of the benthic turf could be studied.

A kelp forest off Anacapa Island, California, was studied in detail. A transect line was marked by an anchored cable, extending over 700 M in length along the sea floor, from the intertidal, down to a depth of 37 M (Fig. 1). Eighty species of algae were collected and identified from along the line. The patterns of plant distribution along this transect were determined, to a large measure, by the light that was available.

Other transects through kelp forests near La Jolla, California, were similar, with plants extending down to 53 M (3). In the more turbid waters of Puget Sound, the lower limit of vegetation was at 25 M (4). In both areas the greatest diversity of plant types was encountered in the lower intertidal and upper subtidal. As might be expected, the number of taxa decreased as the uppermost and lowermost extremes of vegetation were approached. If one collected only from the intertidal, almost half of the local flora would have been missed.

Doty (6) has shown that patterns of zonation in the intertidal can be related to the amount of time a substrate is submerged and exposed. Similar, striking patterns of zonation can also be found in the subtidal. For example, at Punta Eugenia in Baja California, a tropical flora and fauna is found in shallow water, while a luxurious kelp forest flourishes below the thermocline (2).

Marine ecologists have given considerable attention to the biological interactions that influence the structure and composition of intertidal and subtidal communities. Echinoderms, for example, have been shown to introduce pattern into a community through their grazing activity. Urchins can and do completely consume entire kelp forests. The Anacapa Island kelp forest was devastated by grazers in 1966 (Fig. 2). Another example of the effect grazing animals have in subtidal habitats is

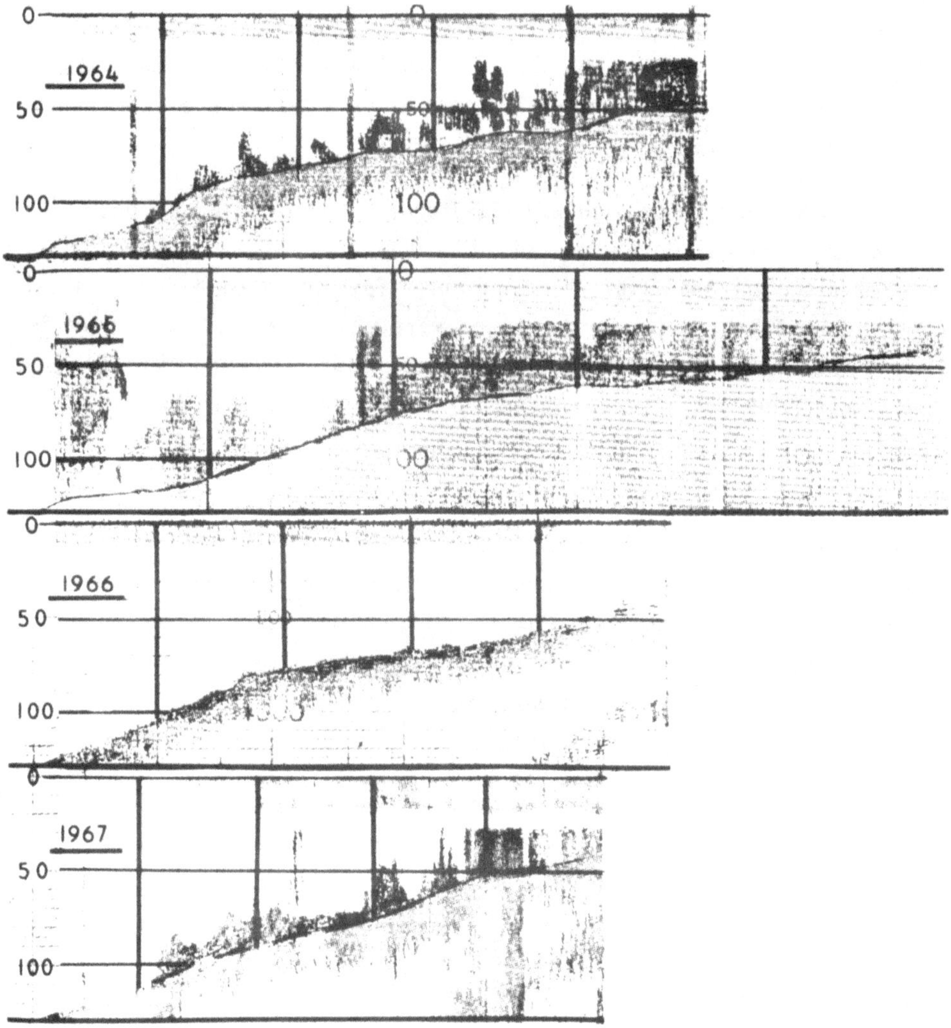

Figure 2. Sonar tracings were made with a recording fathometer along 500 M of the Anacapa transect line. These show the distribution of giant kelp along the line. Depth is given in feet, and the vertical lines are drawn 125 M apart to delimit equivalent segments of bottom for each fathometric trace, since the traces vary in length depending on the speed of the ship at the time they were run. Dates shown are Sept. 23, 1964; Aug. 20, 1965; Aug. 30, 1966 and Aug. 29, 1967. Cross-sectional areas of kelp shown are 2,190 M^2, 4,445 M^2, zero and 1,960 M^2 illustrating the dramatic changes that can occur in a kelp forest over relatively short periods of time.

provided by the un-planned "experiment" that involved an oil tanker that ran aground in a small bay in Baja California in 1957. It released oil that killed most of the animals in the bay. The algae were, for the most part, unaffected in the subtidal, and when freed from grazing pressure, reproduced very effectively. A prolific, dense growth of plants developed (7). This was very different from the open vegetation that had been found previously.

Algal reproduction in the sea normally involves the production of spores and their settlement and germination in the turf, or parvosilvosa, that occurs at the sea floor (8). Turf collected from the Anacapa Island transect contained about half (44) of the total number of species collected along the transect line. These were all less than 1 cm high at maturity. Only 13 of the species collected had individuals over 10 cm high. Since most of the species are microscopic, or nearly so, a diver cannot see them and study them directly.

In an attempt to learn more about this micro-habitat, we transplanted rocks from the transect line into laboratory tanks where they could be studied over long periods of time and held under controlled conditions. In this study, we initially collected 24 species from the rocks. Eleven species more were discovered after the rocks were held in the laboratory. Thus, the first collection missed about one third of the plant species present. These plants must have existed as microscopic spores or perennating fragments on the rock surfaces.

We have also used plastic fouling plates, placed on a sea-floor platform, to study algal micro-habitats. The plates were designed to be taken from the sea, studied microscopically in the laboratory and then returned to the sea. A "relocation" frame was made to fit over the plates, when they were surveyed, so specific germlings could be precisely re-located. By following populations of sporelings and young plants, as they grew and developed in a complex community of bryozoans, hydroids, and other animals, one could see how inhospitable the turf community can be.

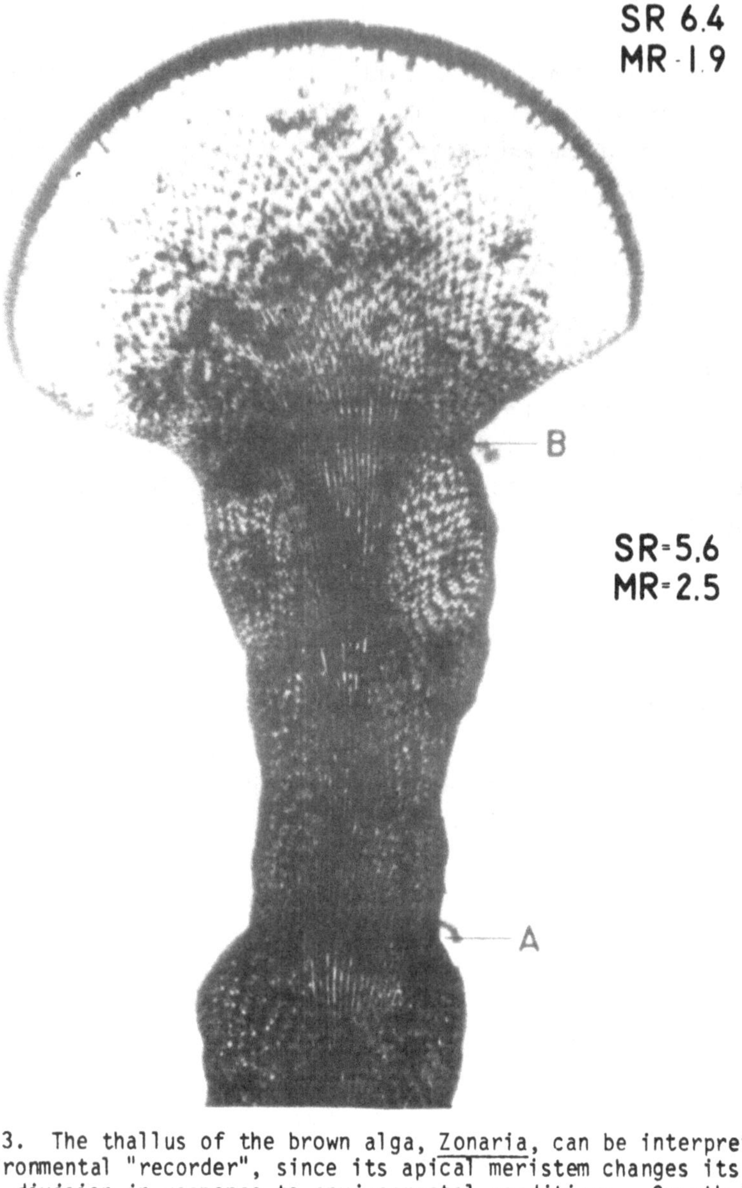

Figure 3. The thallus of the brown alga, Zonaria, can be interpreted as an environmental "recorder", since its apical meristem changes its pattern of cell division in response to environmental conditions. Growth between points A and B took place over a two week period, while the plant was in the sea. Growth from point B to the apical meristem took place over an eight-day period when the plant was kept in a laboratory splash tank in full sunlight. Good growing conditions resulted in a broadened, fan-like thallus, while poor growing conditions produced a narrower thallus. By counting the numbers of cells in the apex and the margins, Dahl (10) calculated a margin ratio (MR) and by determining the number of transverse versus longitudinal divisions in the apical cells he calculated a split ratio (SR). A 1 mm scale is shown.

Temporal Changes in Algal Habitats

Changes in the giant kelp forests of California have occured since aerial photograhs were first taken of them in the 1920's. Aerial photographs have also been used recently to document the seasonal changes in kelp density that occur each year. These changes can be attributed to variations in the degree to which upwelling and nutrient enrichment occur.

Changes with season in kelp forests are recorded in the shells of the abalone, Haliotis rufrescens. Olsen (9) has shown that the alternating pink and white bands, seen in the shells of these molluscs, are records of the yearly development cycle of the annual kelp, Nereocystis. Normally, the abalone consumes red algae and produces a pink shell, but when Nereocystis breaks loose and large amounts of this brown alga are available to the abalone, they consume it and produce white shell bands.

Changes over very short periods of time can also be "recorded" by organisms in the sea. Dahl (10) showed that the thallus of the brown alga Zonaria responds to different environmental conditions, by producing either a wide, or a narrow thallus (Fig. 3). Dahl measured a plant taken from the sea, and then returned it to its collection point for two weeks. He then collected it again and grew it in a laboratory splash tank for 8 days in full sunlight. The plant formed a broad blade under laboratory conditions, and a narrow one under conditions in the sea, thus acting as a "recorder" of growth conditions.

Algal Life Histories

Studies at Santa Barbara since 1963 have focused on accumulating fundamental data to document the interplay between the environment and the life forms and life histories of benthic algae. Major environmental determinants that influence life histories have been identified, and attempts have been made to manipulate life histories experimentally. By growing and observing plants in the laboratory, it is possible to calculate the chances of their reproductive success in the ocean. The mariculturalist would need also to calculate the probable rates of algal growth, for the phase or phases that would constitute a crop. He would need to have control of the life history in order to produce seed stock when it is needed.

Studies of the life history of Eucheuma by Polne (msc.) have shown that the plant can be controlled by holding germlings in the laboratory under conditions of light and temperature that induce dormancy. These can then be stimulated to grow when needed for outplanting.

The life histories of Macrocystis and Botryocladia can also be controlled in the laboratory. These two plants are discussed in detail here, since they have been the subjects of experimental manipulation both in the laboratory and in the sea. In order to gain an appreciation of how a life history may operate in nature, we grew plants in the laboratory to determine what their reproductive potential was. By comparing reproductive potential with what actually occurs in the sea, we can identify life-history phases that are potentially vulnerable, and the seasons when large losses of individuals are likely to occur.

The reproductive potential of Botryocladia

The reproductive potential of Botryocladia pseudodichotoma can be determined by following a hypothetical population of individuals, starting with a female, carpospore-producing plant, like that collected on July 17, 1977, from Naples Reef near Goleta, California. The plant, which was growing at a depth of 10 M, consisted of 53 vesicles, of which 40 were fertile bearing a total of 915 cystocarps. The cystocarps each produce about 1,385 spores per day. For 915 cystocarps there would be about 1.27 x 10^6 carpospores produced every day. If these all settled and grew into tetrasporophytic plants, and each of these became reproductively mature after forming a single vesicle (which our plants did in the laboratory), then a very large increase in individuals would occur. Tetrasporophytic vesicles were found to produce 76,404 tetrasporangia, each with four spores. Assuming that all of these were released and germinated, one would have a population of 3.88×10^{11} gametophytes produced per day per adult female.

The reproductive potential of *Macrocystis*

The reproductive potential of *Macrocystis* can be determined by starting with a single female gametophyte, which in the sea we assume produces only a single egg, which may be fertilized and grow into a single macroscopic sporophyte. At maturity the sporophylls of this plant would bear about 2.9×10^3 cm^2 of sporogenous sorus (11), which produce about 2.07×10^3 spores per cm^2 of sorus per minute (12). Therefore, one adult sporophyte would produce about 8.64×10^9 spores per day. If we assume that every spore that is released settles and produces a gametophyte, we will have the same number of gametophytes produced every day.

Losses from micro- and macro-habitats in the sea

If one female plant of *Botryocladia* can produce 1.94×10^{11} female progeny every day and one female gametophyte of *Macrocystis* can produce 4.32×10^9 female progeny every day, and the oceans are not filled with these plants, it must be because there are substantial losses of individuals where we assumed there would be none.

The process of spore settlement and attachment must be a crucial one for survival. Relatively little is known about this process, but it is known that algal spores can take as long as 15 hours to form their maximal adhesive bond with glass. Once attached, they can resist hydraulic dislodgment forces of nearly one hundred times their weight (13). Algal spores differ in their contents and sinking rates (14). Starting with a

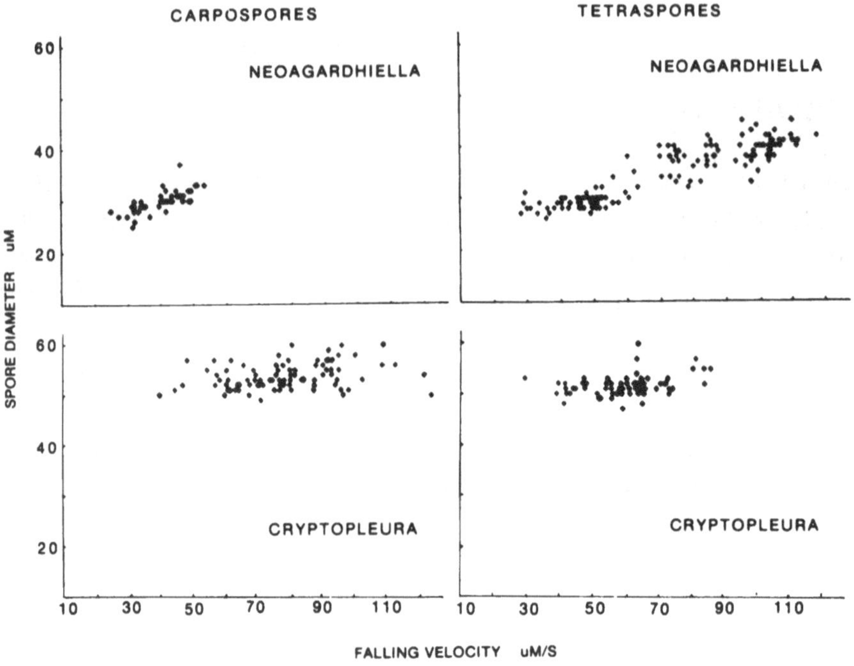

Figure 4. The carpospores and tetraspores of Neoagardhiella and Cryptopleura were studied in a horizontally-positioned microscope, equipped for video-taping. Here both the diameter and falling rate were determined for single spores. The scatter diagrams shown here illustrate the degree of diversity in spore size and falling velocity, both within a given species, and between the two genera studied. Both the size and the content (which influences density and hence sinking rate) varied from spore to spore (after Okuda and Neushul, msc.).

spore population from a single plant, we have shown that there are differences between spores in size and sinking rate (Okuda and Neushul, msc.) (Fig. 4). The implications of these differences are not yet appreciated but presumably they reflect adaptations to conditions that the spores are likely to experience when they reach the sea floor.

Once spores settle and begin to germinate, they can be studied more easily. Fouling plate studies from April, 1972, through March, 1973, showed that natural recruitment took place, but that along with this there was a continuous loss of sporelings. Pterosiphonia reproduced slowly, but had a higher survival rate than the other species. Hence, it was able to establish a population on the plates, while the other species were less well able to do so.

Losses of full-grown plants are more easily documented than those of sporelings. In Macrocystis forests, when plants are dislodged during storms, they frequently tangle with one another, and are cast up on the beach. This is a major cause of loss of adult individuals (16). Substantial loss is also attributable to grazing, as mentioned earlier.

Experimental outplantings of Botryocladia and Macrocystis

Experimental outplantings of laboratory-reared plants have been carried out in order to study the survival and growth of these plants in the sea under natural conditions. Plantings have been made in a number of different locations near Goleta.

Botryocladia germlings were grown from carpospores on $100 cm^2$ pieces of mat-surface plastic. By inverting the plastic in an aquarium tank after the spores had become attached, and by bubbling air up beneath it, water motion in the boundary layer near the spores was produced. Nearly 100% germination resulted. Syntagmatic germination occured as adjacent sporelings coalesced. Upright growths developed from the sporelings, but fewer than the number of original spores, by a factor of from 2 to 5. Single-spore germlings grew more slowly than those arising from coalesced

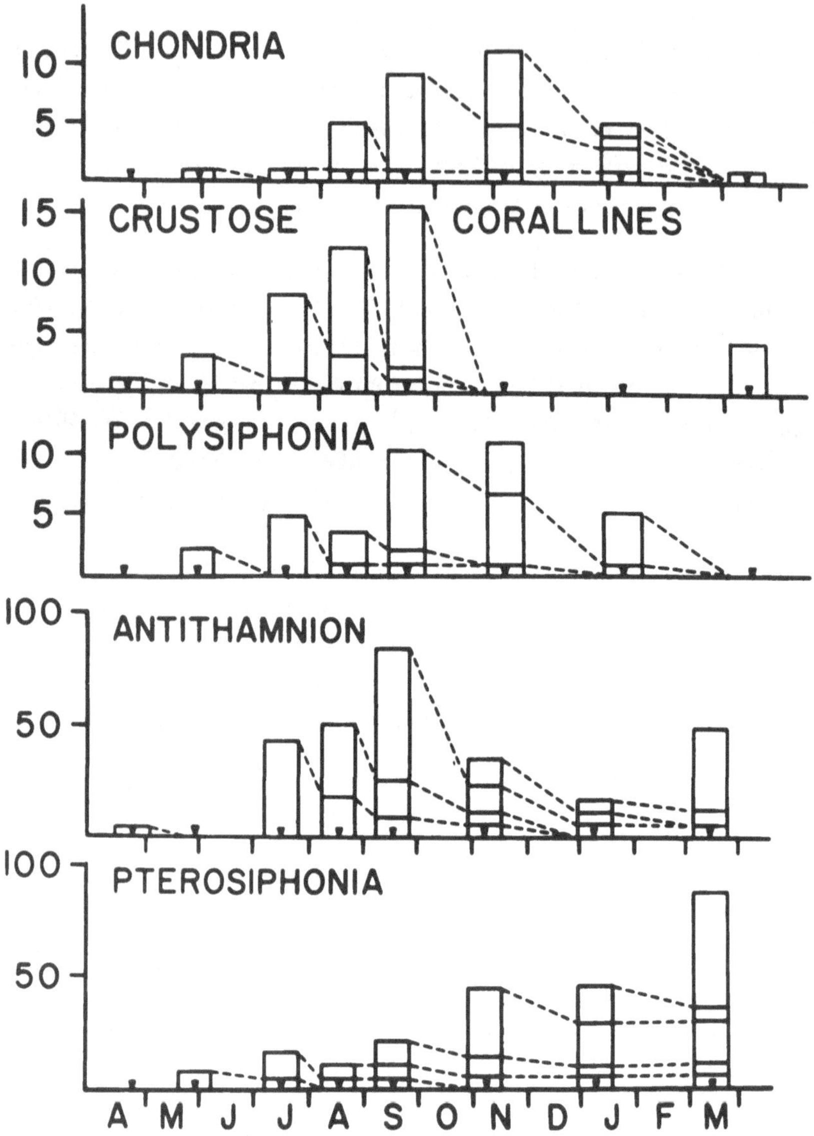

Figure 5. The results of a study of the survival of algal spores, once they settled, are seen in this study of fouling plates, placed in the sea, and studied from April 1972 to March 1973 (15). On eight occasions the plates were taken from the sea, studied microscopically in the laboratory, and replaced. The numbers of individuals counted are indicated on the left. The losses in each "month class" are indicated by the dotted lines that connect each bar. The whole bar represents the total number of individuals encountered in 160 cm^2 surveyed on five plates. The divisions in each bar indicate how many plants, at the time of each survey, were newly-settled, or survivors from previous "month classes." Pterosiphonia showed a lower recruitment rate than Antithamnion but, because it had a much lower loss rate, it persisted and built up a population of nearly 100 individuals after 12 months. None of the other plants were as successful.

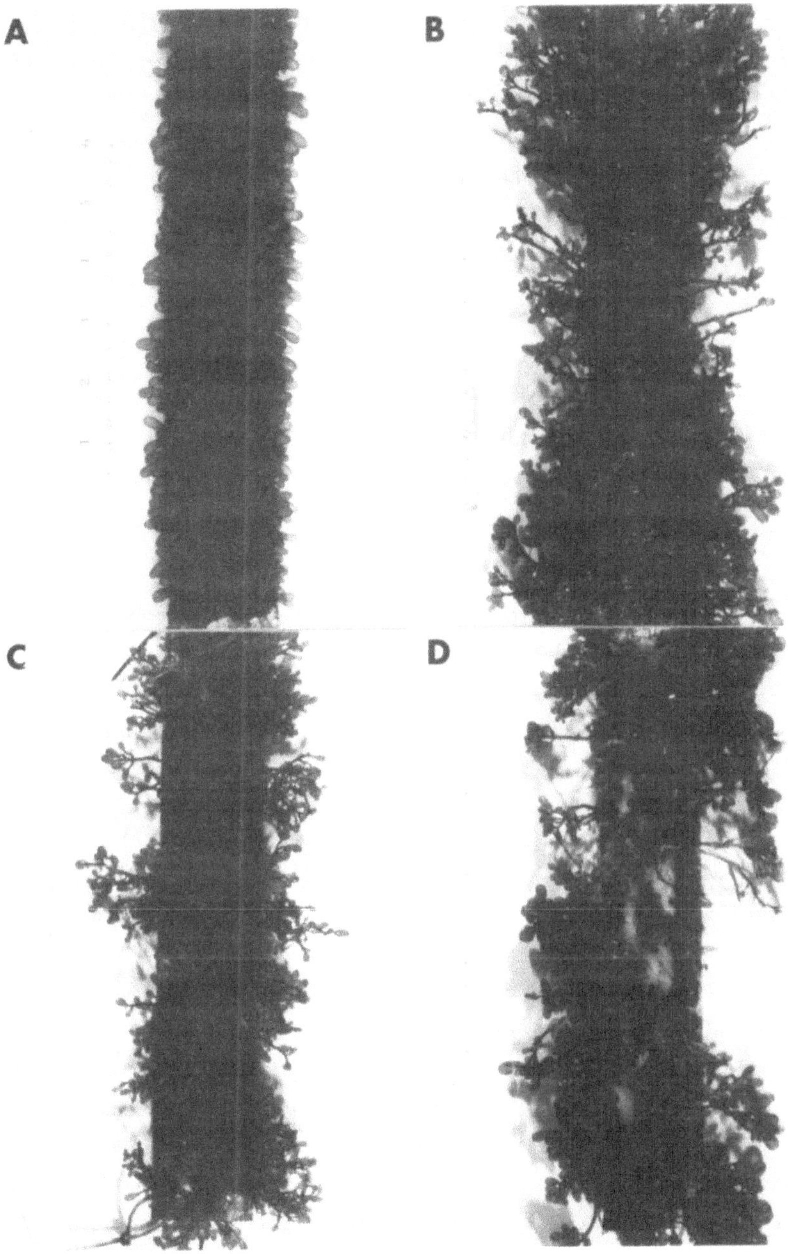

Figure 6. Botryocladia tetrasporophytes are shown after being produced in laboratory tank culture (A), just prior to being planted in the sea on May 12. The same sheet of plants are then shown in June (B), July (C) and October (D) after growth in the sea. It can be seen that the population was "thinned."

plants. The resulting tetrasporophytes became fertile in 8 months, growing at first at rates of 7% per day and later at 3-4% (wet weight). When tetraspores were formed, they were not released in detectable numbers but instead germinated in situ on the vescicles. Spores were released from senescent vesicles, suggesting that senescence and dehiscence may be involved in releasing fertile vesicles that might serve to distribute spores along the sea floor.

When plastic sheets with germlings attached were planted in the sea at Naples Reef, on May 12, 1978 (Fig.6), they supported a dense growth of plants. These were recovered and measured in June, July and October. Over this period of five months they were thinned. The total loss in weight from a sheet approached 50%. Ultimately, storm damage destroyed all the plantings.

Macrocystis sporophytes were raised from gametophytes in the laboratory and grown in a special greenhouse (17) until they had reached single-blade size and could be attached to floats for outplanting in the sea. Nine outplantings of 22 plants each were made at six sites in 1979 and 1980 over an 18 month period. These were both in, and adjacent to two kelp beds off Goleta Point. In total, 700 plants were put into the sea during this experiment. They were judged to have survived if they were still alive and growing after six months (Fig. 7).

Transplantings to sites in the kelp bed shaded by canopy were uniformly unsuccessful. The blades of the outplants became bleached and within two months sloughing increased and the plants died. Plant loss was not attributable to grazing by fish or invertebrates, but was probably due to low light intensity. The maximum losses at all sites occured during the stormy period of the year. Severe winter storms in 1979 destroyed all the outplantings at all of the sites. In some shallower sites, as much as 50 cm of sand covered the outplantings. The deeper outplanting sites (at 45 and 50 feet) seemed to benefit from cold, nutrient-rich water during the summer months. At these sites the nitrate nitrogen levels reached 2 to 3 ugm-at/l on occasion, while the shallower sites had nutrient levels of less than 1 ugm-at/l all of the time. The different rates of survival at shallow versus deep sites are probably due to these differences in

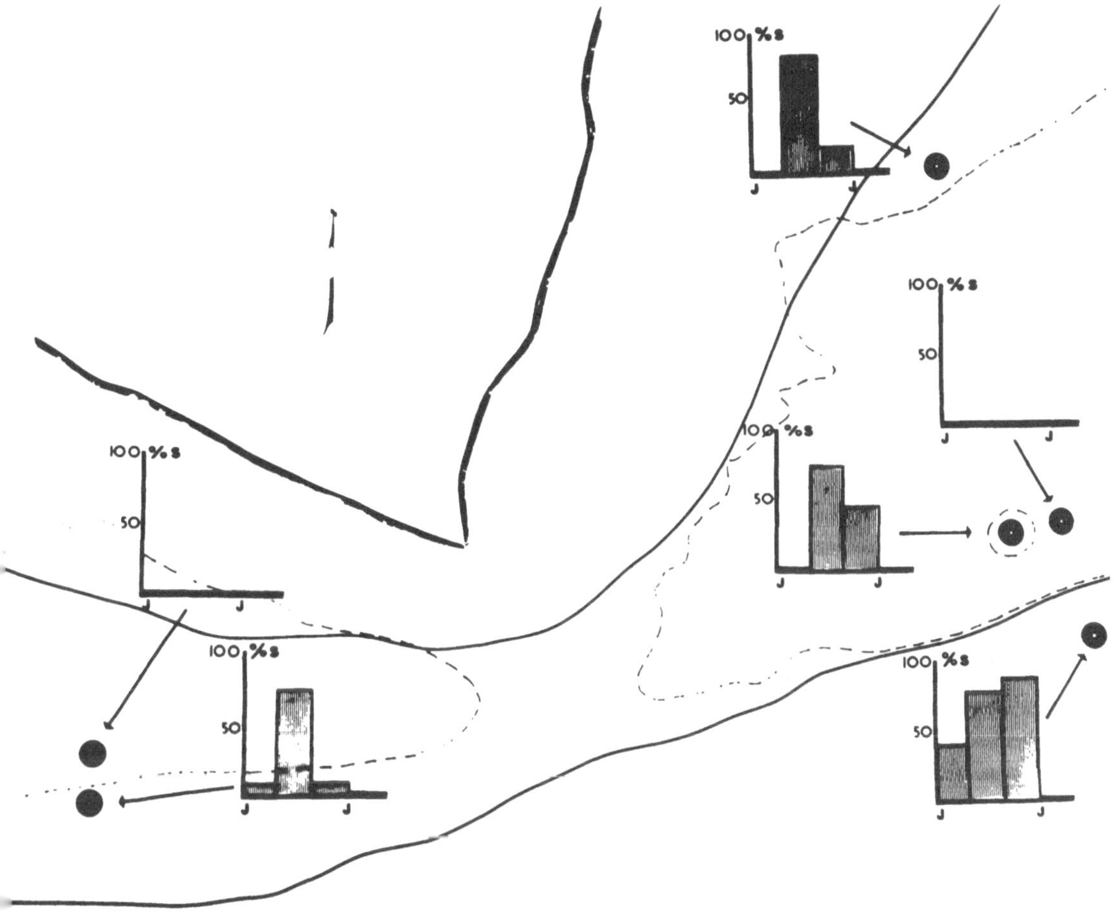

Figure 7. Macrocystis sporophytes were grown in the laboratory and then planted in the sea at the sites indicated, near Goleta Point. The proportion (%) of the originally outplanted plants that survived over (1) a stormy, (2) an upwelling, (3) a low nutrient and (4) the following stormy periods are indicated for each outplanting site. Survival under the kelp canopy did not occur, but when a clearing was made (canopy is denoted by a dotted line, the canopy was removed at the outplanting site encircled by a dotted line), the outplants grew well. Different survival rates at the shallower and deeper sites were presumably related to higher nutrient levels at the deeper sites.

nutrient levels.

Plant Adaptations

Exploratory studies of the natural patterns of plant distribution, of algal life histories, and of experimental outplantings in natural habitats, all suggest that the natural habitat is inhospitable to macroalgae which cannot adapt well for survival.

Ecologists have only recently begun to study the adaptive features of macroalgae. Searles (18) suggests that the tri-phasic life history of red algae is an adaptation which enables the plant to produce many diverse genotypes from a single sexual fusion event. Littler (19) and Littler and Littler (20) have studied a number of plants and suggest that thin, sheet-like forms and finely-branched plants have higher rates of photosynthesis than those with a thicker more elaborate thalli. They also point out that plants with a strong thallus would be adapted to resist wave action in nature and that special adaptations to resist grazing might operate in nature.

Neushul (21) advanced the hypothesis that macroalgal form results from a hydrodynamic adaptation to a specific habitat in either a boundary-layer, turf, surge or current zone (Fig 8). Studies of the hydrodynamic features of macro-algae have been undertaken at Santa Barbara by Charters and Anderson (22) who constructed a low-velocity water tunnel where macroalgae, and models of these plants could be tested. An examination of water-flow patterns around models is particularly informative (Fig. 9). Flow patterns of water over a rod, and across a plate are well known. The flow of water over the combination of a rod and plate, which simulates a stipe and blade, exhibits a novel pattern. Here vortex shedding from the rod, is modified by the plate to produce a flow pattern that "stirs" a rather thick layer of water adjacent to the plate. This stirring action of the water flow enhances gas exchange and nutrient uptake by a blade. The water flow pattern over a Macrocystis blade is very similar to that

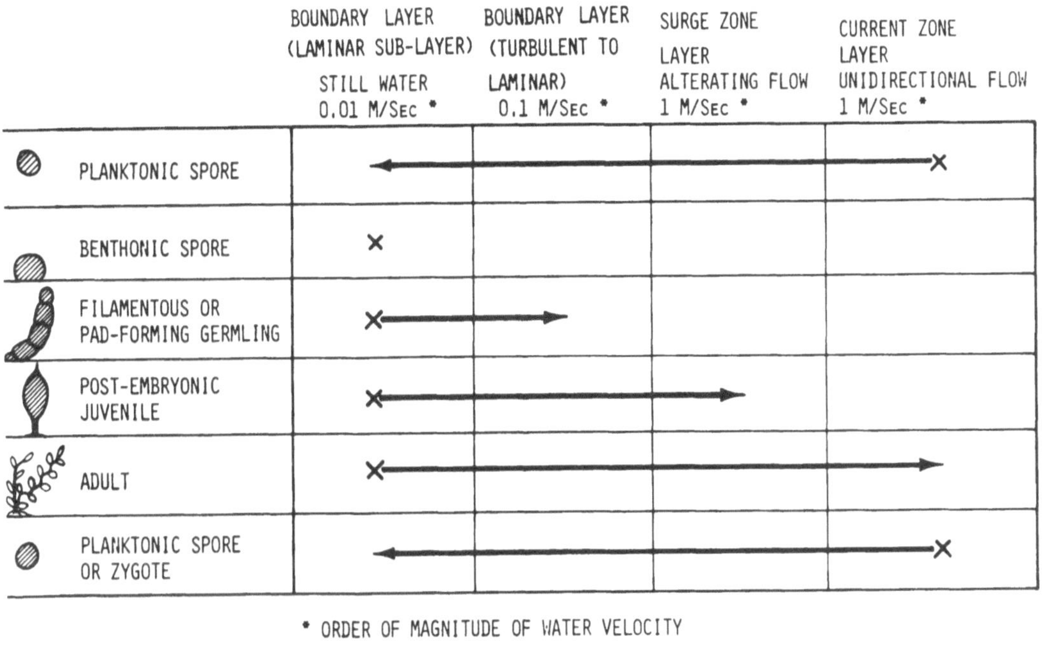

Figure 8. A relationship between form and function of algal life forms is suggested when life-history phases (on the left) are placed in the hydrodynamically-defined habitats (at the top) that these phases normally occupy (21). The spore would, at first, be planktonic, and then benthonic. Subsequent growth would produce plant forms characteristic of boundary layer, surge zone and current zone habitats. It has been suggested that the morphological types, diagrammed on the left, are adapted to function in these specific hydrodynamic habitats.

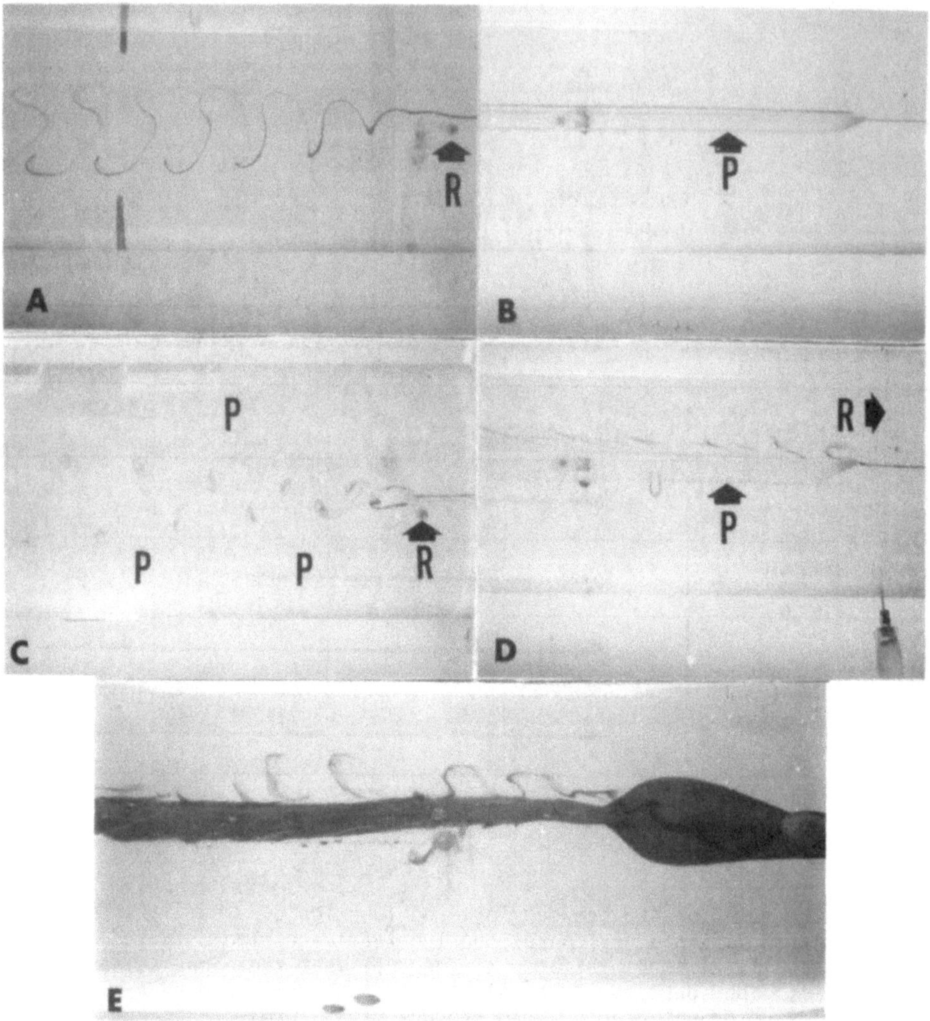

Figure 9. The use of both models and plants, in controlled-flow hydrodynamic experiments shown here, allows one to see the flow patterns around both (22). A dye-streamer is injected into a water tunnel upstream (to the right) of the test object. The manner in which the water is disturbed by the object can then be clearly seen as the water flows from right to left. A rod ("R" seen alone, from above in photo C) and a flat plate ("P" seen alone, from the side in photo B) act together in photos A (rod from above) and D (rod from side) to produce a "scrubbing-like" action over the plate. The plate alone (B) produces no such action, the water flow over it being entirely smooth and laminar. The rod alone (C) sheds a series of vortices. The rod and plate, which is a model for a stipe and blade of Macrocystis, are seen to produce scrubbing action in a relatively thick layer of water adjacent to the plate (D). A similar pattern is seen behind the vesicle of a Macrocystis blade, passing over the blade (E) (photos courtesy of A. C. Charters).

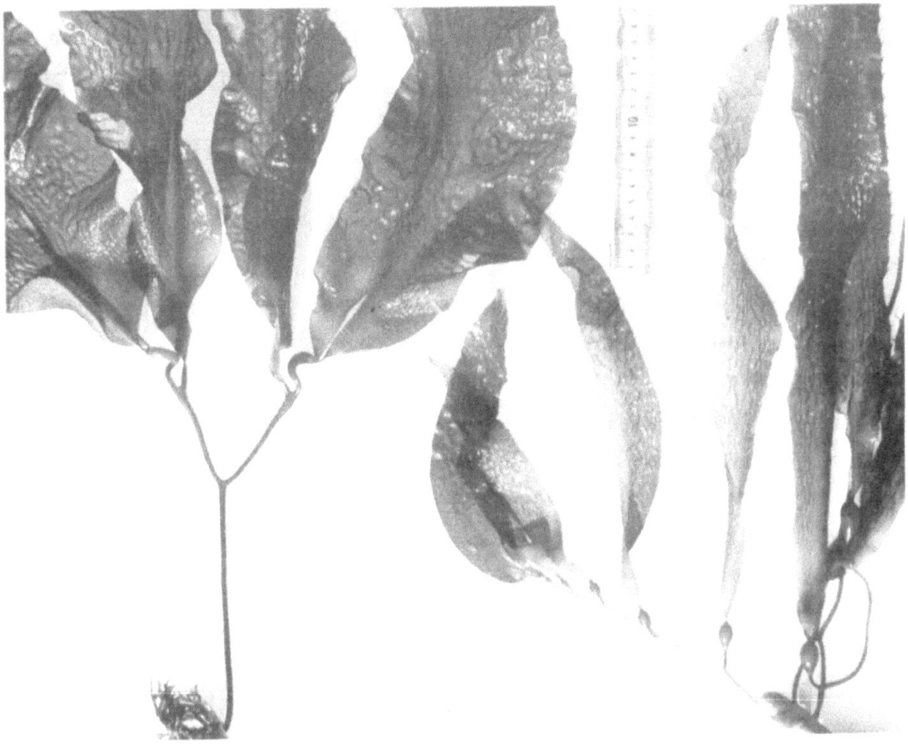

Figure 10. An intergeneric kelp hybrid between Macrocystis and Pelagophycus is shown (on the left) beside a Macrocystis frond (on the right). The hybrid is short and produces only two fronds which are terminated with coiled, inflated vesicles. The hybrid is not well-adapted for life in the surge zone to which its size confines it.

over the rod-and-plate model, suggesting that it may be hydrodynamically adapted to enhance nutrient uptake and gas exchange.

The adaptive value of a given morphological configuration is illustrated by studies of genetically uniform kelps planted in the sea, and by the fate of an outplanting of intergeneric hybrids. Crosses between Macrocystis and Pelagophycus were raised in the laboratory and then planted out into the sea. These hybrids are intermediate in morphology between the two parental genera, both of which are very large and adapted for survival in the current zone. The hybrids, in contrast, are only 1-2m long and produce only two coiled pneumatocysts (Fig. 10). When planted in the sea, they were exposed to surge action, and since they were unable to float up out of the surge zone they were abraded on the sea floor and ultimately destroyed. A plant that was held in the laboratory did reach maturity and formed sori (Fig. 11). Microscopic examination of the sori revealed that they did not contain mature sporangia. Also, attempts to collect spores from them were unsuccessful. Thus, the hybrids were not adapted for the water motion habitat that they were confined to, and were not able to reproduce. These features might, in fact, be considered positive by a mariculturalist interested in avoiding the problems that would arise if fertile kelps were introduced into an area where they did not naturally grow, and escaped from cultivation. The morphological mal-adaptation of the hybrid might be advantageous if the plants were to be grown on rope lines or other planting strucures removed from surge and the bottom.

Studies of genetically uniform kelps outplanted into the sea have been instructive. M. pyrifera, from San Diego, and M. angustifolia, from Goleta, were raised from gametophytic stocks, and outplanted on floats in the sea. Much to our surprize the local species was the least successful. It produced large holdfasts that grew over the floats and weighed them down so that abrasion damage destroyed the plants. In contrast, the smaller holdfasts of M. pyrifera did not do this, and the plants survived and grew well. In this case the method of planting selected for the species with the smaller holdfast. The mariculturalist can be instrumental in determing the success of a crop through his choice of

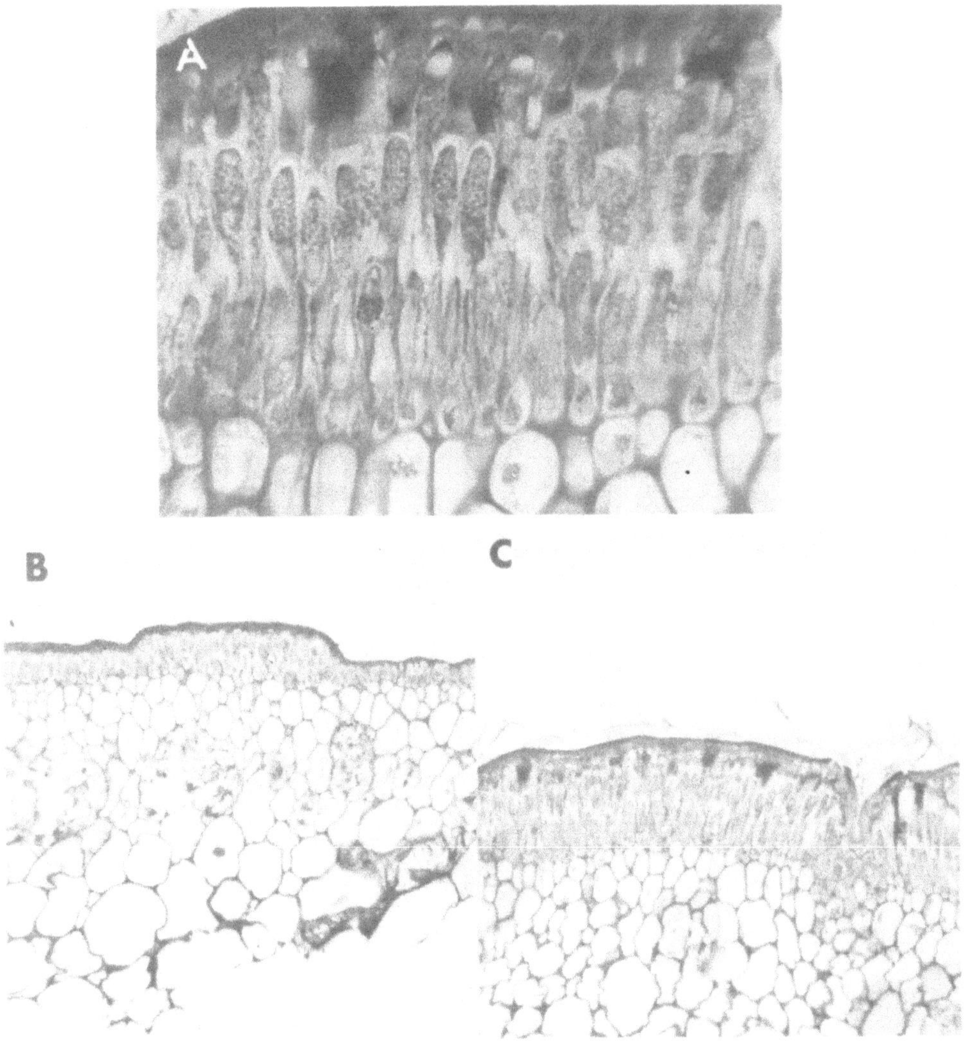

Figure 11. A sorus containing paraphyses and sporangia (A) was formed from meristoderm (B and C) on the blades of the intergeneric kelp hybrid between Macrocystis and Pelagophycus, but the sporangia did not produce spores. It was thus assumed that the hybrid was sterile.

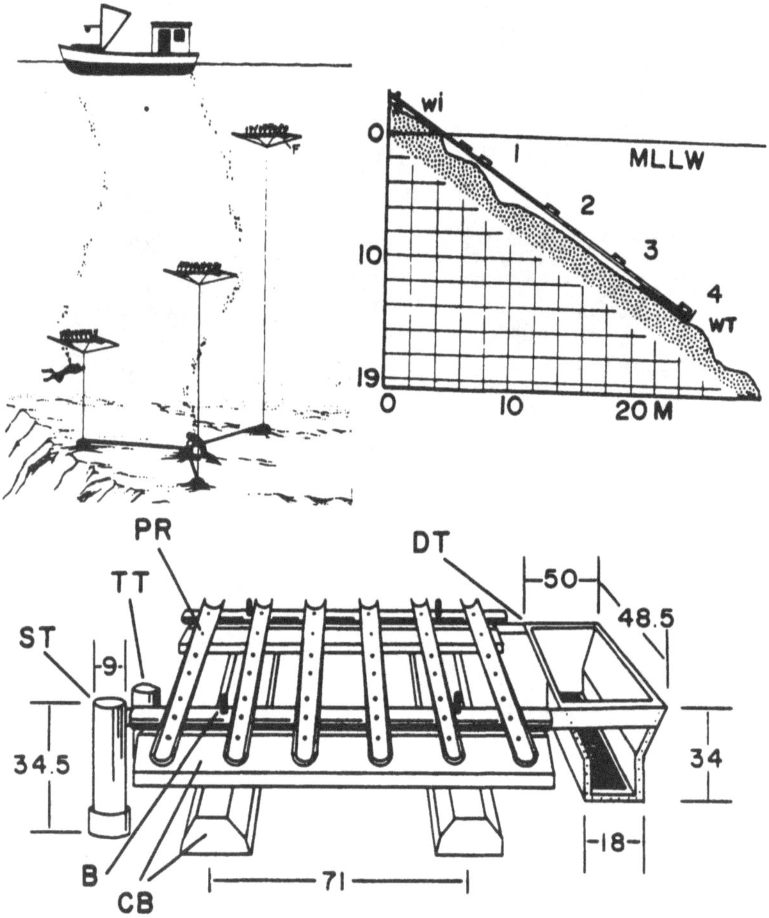

Figure 12. Experimental outplantings have been attempted using buoyant rafts that were untethered and allowed to float to the sea surface, where plants attached to them were detatched and photographed aboard a small boat (24). The plants were then re-attached and the rafts re-lowered into position. Another approach to depth-adjustment was to use rails anchored to a heavy weight, and positioned on a steep profile running down into the water (25). Four carts (1-4) ran on the rails and were pulled up and down by a cable attached to a winch (Wi) on the emergent end of the tracks. A third method was to use a platform on the sea floor composed of concrete blocks. A sediment trap (ST) and drift trap (DT) were attached. Concrete blocks (CB), corner bolts (B), a thermometer case (TT) and rails to which fouling plates were bolted (PR) are also illustrated. The dimensions given are in centimeters (15).

seedstock and method of planting.

Discussion

Field and laboratory studies at Santa Barbara indicate that the design of an optimum cultivation strategy begins with an understanding of the natural ecology of the plant to be grown. While some obviously detrimental factors, such as boundary-layer competition and micro-grazing, can be avoided simply by providing fresh substrate, other problems cannot be so easily circumvented. A systematic study of the major plant requirements for light, space and nutrients can be accomplished by laboratory and field tests, as Matsumoto (23) has done for Porphyra. This type of study, control of the life history of the plant, and an appreciation of local oceanographic conditions are all essential if maricultural success is to be attained.

One of the most challenging problems facing the mariculturalist is to provide adequate nutrients. An effective method of nutrient "irrigation" is essential where there are inadequate supplies in the sea. Chapman and Craigie (26) fertilized natural populations of Laminaria with sodium nitrate which resulted in rapid growth during the low-nutrient period of the year when growth is normally slow. Nutrient "irrigation" is presently used for Laminaria mariculture in China. Dr. W. J. North, in his studies of open ocean macroalgal cultivation, is artificially upwelling nutrient-rich seawater for his crops. If these methods can be commercially applied in extensive open-ocean farms, then new sources of food and energy will be at hand.

Acknowledgements

The progress made at Santa Barbara has been due, in large measure, to the contributions of Dr. A. C. Charters in algal hydrodynamics and Mr. D. Coon in algal growth. Dr. A. Gibor (tissue culture) and Mr. M. Anderson (algal culture and hydrodynamics) have made major contributions, as has V. Kuga who has grown many crops of sporophytes for our work. Mr. R. Fetter cultured the Botryocladia plants illustrated in this paper, while M. Polne, and Dr. W. N. Wheeler have contributed valuable insights into the ecological physiology of red and brown algae. Mr. J. Woessner has broadened our appreciation of phycological studies in the orient, and Dr. A. C. Mathieson has generously shared his expertise with us.

I would like to thank Dr. B. W. W. Harger and his co-workers for their contributions to our knowledge of macro-algal mariculture, and in particular the "resource-need" approach to cultivation. Major contributions to our program have also been made by visitors from abroad, who have included Drs. Luning, Norton, Arzzi, Okuda, Sanbonsuga and Rahimian.

The work discussed here has been supported by government and private-sector funds. The U. S. National Science Foundation has contributed through its Engineering, Oceanographic, and Applied Science Divisions (NSF- ENG 76-22720; OCE-76-24360; OCE-79-09241; PFR-7911715). The Department of Commerce, Sea Grant Program, has contributed to the above-mentioned kelp outplanting studies (USDC R/CZ-46). Support from the National Science Foundation and the Gas Research Institute, through the General Electric Company, has made it possible for me to attend these meetings, for which I am grateful.

References

1. Wheeler, W.N., Neushul, M., Harger, B.W.W.: Proc. Int. Seaweed Symp. 10, (in this volume) (in press).

2. Dawson, E.Y., Neushul, M., Wildman, R.D.: Pac. Nat. 1, 1-81 (1960).

3. Neushul, M.: Proc. 5th. Mar. Biol. Symp. Acta Universitatis Gothoburgensis. Botanica Gothoburgensa 3, 161-176 (1965).

4. Neushul, M: Ecology 48, 83-94 (1967).

5. Neushul, M., Clarke, W.D., Brown, D.W.: In: Philbrick, R.N. (ed.), Proc. Symp. Biol. Calif. Islands. Santa Barbara Botanic Garden: Santa Barbara, California. pp. 37-55 (1967).

6. Doty, M.S.: Ecology 27, 315-328 (1946).

7. North, W.J., Neushul, M., Clendenning, K.A.: Comm. Int. Explor. Sci. Mer Medit. Symp. Pollut. mar. par. Microorgan. Prod. Petrol, Monaco. pp. 335-354 (1964).

8. Neushul, M., Dahl, A.L.: Helgolander wiss. Meeresunters. 15, 480-488 (1967).

9. Olsen, D.: Biol. Bull. 134, 139-147 (1968).

10. Dahl, A.L.: Bot. Mar. 14, 76-112 (1971).

11. Neushul, M.: Amer. J. Bot. 50, 354-359 (1963).

12. Anderson, E.K., North, W.J.: Proc. Int. Seaweed Symp. 5, 73-86 (1966).

13. Charters, A.C., Neushul, M., Coon, D.: Limnol. and Oceanogr. 18, 884-896 (1973).

14. Coon, D., Neushul, M., Charters, A. C.: Proc. Int. Seaweed Symp. 7, 237-242 (1972).

15. Neushul, M., Foster, M.S., Coon, D.A., Woessner, J.W., Harger, B.W.W.: J. Phycol. 12, 397-408 (1976).

16. Rosenthal, R.J., Clarke, W.D., Dayton, P.K.: Fish. Bull. 72, 670-684 (1974).

17. Charters, A.C., Neushul, M.: Aquatic Bot. 6, 67-78 (1979).

18. Searles, R.B.: Amer. Naturalist 1154, 113-120 (1980).

19. Littler, M.M.: Bot. Mar. 22, 161-165 (1980).

20. Littler, M.M., Littler, D.S.: Amer. Naturalist (in press).

21. Neushul, M.: In: Abbott, I.A., Kurogi, M. (eds.), Contributions to the Systematics of Benthic Marine Algae of the North Pacific. Japanese Soc. Phycology: Kobe, Japan. pp. 47-74 (1972).

22. Charters, A.C., Anderson, S.M.: Journal of Hydronautics 14, 3-4 (1980).

23. Matsumoto, F.: Mem. Dept. Fish. Animal and Vet. Sci., Hiroshima University. J. Fac. Fish. Animal Husb. 2, 249-327 (in Japanese with English summary) (1959).

24. Neushul, M., Haxo, F.T.: Amer. J. Bot. 50, 349-353 (1963).

25. Neushul, M., Powell, J.H.: Ecology 45, 893-894 (1964).

26. Chapman, A.R.O., Craigie, J.S.: Mar. Biol. 40, 197-205 (1977).

MARINE PHYCOLOGY AND EXPLOITATION OF SEAWEEDS IN SOUTH AMERICA

E. C. de Oliveira F.

Dep. de Botânica, Universidade de S. Paulo
C. Postal 11461, S. Paulo (SP), Brasil

Introduction

The idea of this paper is to give an overall picture of the state of marine phycology in South America including a brief view of its history, present stage of development, and economic potential. It is an attempt to serve two purposes - to stimulate a stronger scientific and technical cooperation within the continent, and to serve as a sort of preparation for an eventual International Seaweed Symposium. Bolivia and Paraguay have no coast-line. Surinam and Guianas were not included. Main emphazis is put on Brazil, Chile and Argentina.

Historical Development

Although the South American Continent comprises an extensive coast line facing the Caribbean, the Atlantic, the Antarctic and the Pacific Ocean, and belonging to several independent countries (Figure 1), the devel - opment of marine phycology has several points in common and followed the same basic lines.

Analysing the development of the marine phycology in Brazil, Oliveira F. divided it in 3 phases which, with some time adjustments could be extended to the other countries. The first phase started in the beggin- ing of last century, extending up to 1950-60's. This phase is char - acterized by the publication of new taxa and lists of species, usually without descriptions nor illustrations, and incomplete location of the samples, such as "ad oras Brasiliae", for instance. It could also be

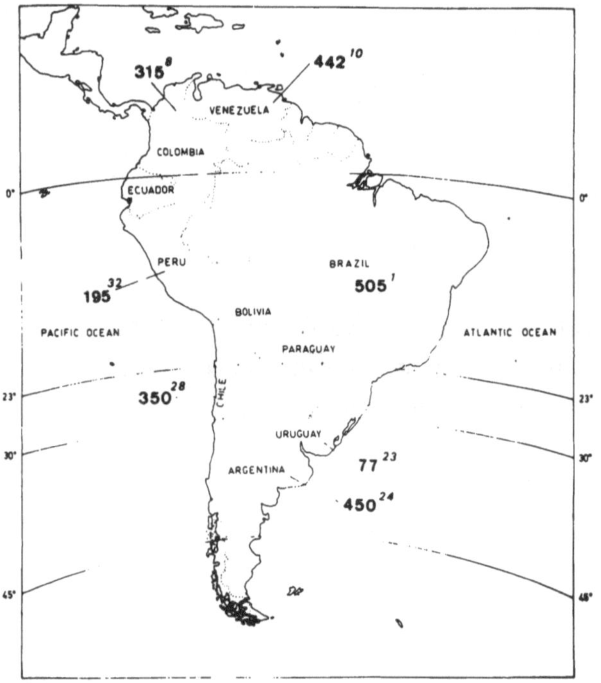

Figure 1. Political map of South America - bold face numerals indicate the number of species of seaweeds for each country (Cyanophyta excluded); italic numerals indicate the references for the authorities.

called "phase of the itinerant botanists" and was a consequence of the large expeditions from European countries: Challenger, Beagle, Vetor Pisani, La Bonite, Fregate Novara are some examples. The collections, scattered through several European herbaria, are usually poor. Important contributions were published by Bory de Saint Vincent, Montagne, Greville, Grunow, Martens, Dickie, Piccone and Moebius among others. More recently, near the end of this first phase and following the more critical and carefull papers of Europeans such as Skottsberg, De Toni and Levring, there was an increase in the number of North American phycologists such as Howe, Taylor and Dawson, just to mention some of the more prolific. The second phase started around the 1950's, a little before or latter in the different countries. It is characterized by the work of the first local phycologists. The research line is still floristic, in the line of the so called α taxonomy. However these studies were based on more abundant material, include a precise location of the samples, extensive descriptions and illustrations. The third phase beggan in the last decade and is just starting. It is characterized by a higher level taxonomy and by a diversification of research lines, such us ecology, population biology, life histories, citology and utilization. This phase is florishing more in Chile and Brazil and is a consequence of the increase in the number of local researchers, some specialized outside their countries, and probably also a reflection of the growing interest and concern about the importance of the marine algae. Of course the mentioned phases are artificial grouping of dominant study lines through the time and are not water tight, but overlap and interpenetrate.

Marine Phycology Today in South America

During a Symposium of Marine Algae, within the II Congresso Latino-Americano de Botânica held in Brasília, Brazil, in 1978, a group of phycologists representing Argentina (M. L. Mendoza), Brazil (M. C. Marino), Chile (B. Santelices), Colombia (G. Bula Meyer) and Peru (C. Acleto) was appointed to organize a survey of the phycologists in Latin America. This was published as "Ficologistas Latino-americanos" in 1979 by the "Sociedade Brasileira de Microbiologia". Although the survey is obviously incomplete it gives an idea of the phycologists working in each country. Brazil, Argentina and Chile congregate the larger groups, though active phycologists are also working in Peru, Colombia and Venezuela (Table 1).

A general idea about the state of marine phycology in the continent is given bellow for each country. For obvious reazons we do not list all the pertinent literature for each region, but just a few recent and key papers where the reader will find the necessary references.

Table 1. Phycology in South America (adapted from "Ficologistas Latino-americanos, Soc. Bras. Microbil., 1979). B-G. Blue-Green, R. Red, B. Brown , G. Green, P. Phytoplancton, S. Several.

	n. phycologists		main interest								
			Taxonomy-Biology						Ecol.	Expl.	Others
	fresh w.	marine	B-G	R	B	G	P	S			
ARGENTINA	8	5	1	2	1		4	2	1	1	1
BRAZIL	31	29				5	11	33	2	4	5
CHILE		6						3	1	1	
COLOMBIA		4						2	2		
ECUADOR		1								1	
PERU	1									1	
TOTAL	40	45									

Colombia

This is the only country with coasts to the Caribbean and the Pacific. At the Caribbean side the first collections were the ones of Linden and Schott in 1842-43, and 1857. Additional studies in the area were performed by Taylor (2), (3), Barriga-Bonilla et al. (4) and Kapraun (5). Acevedo (6) compiled all the species referred to Colombia by the previous authors. More recently R. Schnetter published some papers about the flora of the Caribbean coast (see (7) for additional references). Recent efforts towards the knowledge of the Caribbean flora of Colombia are being developed in Universidad Tecnologica del Magdalena by G. Bula Meyer, a local phycologist. The flora of the Caribbean coast has the same basic elements found in Venezuela and the West Indies,and amounts, up to now to 315 species (8). We have no recent informations about phycologists working in the Pacific coast of Colombia. This region was visited by the Allan Hancock Pacific Expedition and the algae studied by W. R. Taylor. The flora, both intertidal and subtidal, seems to be very poor as seen from the collections made by the Hancock Expedition (9). However, the lack of critical studies becomes obvious in a literature search.

Venezuela

The first papers about Venezuelan marine algae were published by Kunth (1825), J. Agardh (1847) and Murray (1888-89), (apud 10). In this century extense references to the marine algae of Venezuela were given by Taylor (2), Albornoz and Rios (11), Rios (12), Hammer and Gessner (13), Gessner and Hammer (14), Diaz-Piferrer (15), Lemus (16) and Taylor (17). This last paper of Taylor comprises a check-list of Venezuelan marine algae and was superseded only by the recent work of Ganesan (10). More details about the history of marine phycology in Venezuela and an up to date list of species and literature are given by Ganesan (10). The beggining of local phycology started in 1960's with the foundation of the Instituto Oceanográfico de la Universidad de Oriente, at Cumaná that is actually the more important centre of phycology in the country. With a few exceptions phycological work has been restricted to taxonomy. The number of species referred to the area is 442 (10), most of them occurring also in the West Indies and Brazil.

Brazil

The starting point of phycology in Brazil is certainly associated with the coming of Felix K. Rawitscher, a distinguished german physiologist, to the Botany Department at the University of S. Paulo. Rawitscher, not himself commited to phycology, was a student of Oltmanns and encouraged Aylthon B. Joly into this area. Joly got his training in phycology with W. R. Taylor, in Ann Arbor. He started publishing in phycology in 1950, however, most of his papers were prepared in the 60's, a decade also dedicated to the training of several students. The Botany Department, at the University of S. Paulo, was a nucleus of phycological training not only for marine but also for fresh water phycology, attended by Brazilians and by other Latin-Americans students, as well. Today, although S. Paulo is still the larger phycological centre in Brazil there are phycologists, trained in Brazil, working in several points of the Brazilian coast in local Universities and research institutions. The number of marine research stations has being increasing and there are three phycologists from abroad working permanently in Brazil. The main approach is taxonomic but there is a trend towards diversification of the research lines. More details about the development of marine phycology in Brazil can be found in Oliveira F. (18), (1). The last mentioned paper presents a critical list of species recognizing 505 good species (Cyanophyta and Diatoms not included), (see Table 1). Reasons are given to consider 132 records as dubious or of uncertain occurrence in Brazil. Although dividing the Brazilian coast line in four regions, Oliveira F. (1) recognized that there is just one flora in the whole area, that is the Caribbean. This flora presents a gradual impoverishment towards the south, till Uruguay.

Uruguay

The coast line of Uruguay is small and a significant part of it is not suitable to algal growth due to water turbidity and lack of substrate. Even so the flora can be considered poor and was little known till recently. The first paper to deal specifically with the flora of Uruguay is the one of Howe (19). Another usefull papers are the ones of Taylor (3), (20), Cunillera and Oliveira F. (21), Coll and Oliveira F. (22). The last one comprising a list of all the taxa referred to the region. The thesis of Coll (23) is the more complete work and refers to 77 species for the area. Oliveira F. (1) considers the Uruguayan flora as a depauperate extension of the Caribbean one with just a few representative of cold water species from the south. A more detailed history of phycology in Uruguay and a complete list of references is given by Coll (23).

Argentina

The majority of papers published in the last century and the first decades of this century dealt with the flora of southern Argentina, specially Tierra del Fuego, Falkland Islands (Malvinas) and Antarctic. Phycological studies by local researchers are recent and started after 1960. There are three main centres of phycological research in the country. 1. CIBIMA (Centro de Investigacion de Biologia Marina) in Buenos Ayres with research stations in Puerto Deseado and in Ushuaia. 2. Museo Argentino de Ciencias

Naturales Bernardino Rivadavia, in Buenos Aires. 3. CNP (Centro Nacional Patagonico), in Puerto Madryn. The first one congregate the larger number of phycologists. The main approach is taxonomic, existing a clear division of research fields between the phycologists, each one studying a particular group of seaweeds. However other lines of research such as ecology and the chemistry of algal products have also been pursued (24). The benthic flora, comprises 450 species (Cyanophyta and Diatoms excluded) as referred to by Kuhnemann (25). In opposition to the situation in other Southern American countries the marine Cyanophyta are well known in Argentina thanks to the studies of D. R. Halperin (cf. Mendoza, 1978, for references). More details about the phycological development in Argentina is given by Mendoza (24), who also gives a complete list of papers on Argentinian seaweeds. The flora is typically cold water with a relatively small percentage of warm water elements, restricted to the northern part of the coast. The richest flora is to be found in Patagonia and Tierra del Fuego. The subject is fully discussed by Kuhnemann (26).

Chile

After the period of the large expeditions and the studies of visitor scientists,marine phycology in Chile seems to have started with the the stablishment of the Marine Station of Montemar (Estacion de Biologia Marina de Montemar) in 1945. The first papers by Chilean phycologists were published in 1948 by A. H. Llana. These were followed by the papers of H. Etcheverry in the 1950's and 1960's. However it was in the 1970's that marine phycology flourished in Chile with the several papers of K. Alveal, D. H. Kim , B. Santelices and their colaborators. This last decade also saw a diversification of the research lines. The three main centres of phycological research in Chile are in Valparaiso (Estacion de Biologia Marina de Montemar), Santiago (Universidad Catolica) and Concepcion (Universidad de Concepcion). The development of marine phycology in Chile can be seen in detail in Etcheverry (27) and Santelices (28). The flora of continental Chile has about 350 species (28), though this number will certainly increase with further studies. In a recent publication Santelices and Abbott (29) found 40 new references of marine algae to Chile, what gives an idea of what can be expected in the near future. The Chilean flora is typically temperate, with a large number of subantarctic and endemic species whose diversity decreases towards the north. The flora of the oceanic islands, on the other hand have a large number of warm waters species. Good references to literature on Chilean algae are given in Etcheverry (27), Levring (30), Kim (31) and Santelices (28). A phytogeographic characterization of the temperate coast of the Pacific was presented by Santelices (54).

Peru

The first authors to study the flora of Peru were J. Agardh, material collected by Humboldt and Bomplant in 1802, Piccone (1886), material collected by the Vettor Pisani Expedition,and De Toni (1903), (apud 32). One of the more important studies about the flora of Peru was published by Howe (33). Other important works are the ones of Collins (34), Taylor (35), Dawson et al. (36) and Acosta Polo (37) where other references can be found. Although Peru is within the tropical zone it does not show a characteristic tropical flora (9). The flora amounts to 195 species (38). Most of the papers have a taxonomical approach, though publications concerning economical aspects were also published (32), (37), (39). The main centre of phycological studies in Peru is at Universidad Nacional Mayor de San Marcos, Lima.

Ecuador

There is not much phycological work going on in Ecuador to my knowledge. In the Catalogue of South American Phycologists only one name appears, M. A. Mayor, in Guayaquil. The collections made by the Allan Hancock Expedition were studied by Taylor (9). It seems that the continental Ecuador received very little attention as compared to the "Archipielago de Colon" (Galapagos Islands). The nature of the flora is tropical and mangroves, with their characteristic bostrichietum are a noticeably feature of the area (9). The continental flora seems to be an extension of the northern tropical areas and though appearing poor is certanly insufficiently known as is the Colombian Pacific coast.

Exploitation of Seaweeds in South America

Seaweeds have been exploited mainly in Chile, Argentina and Brazil (Table 2) although some exploitation have also been carried on in Peru and Uruguay. A general view of the situation was given by Michanek (40), who

Table 2. Seaweed production in South America (tons of dry weight)

	1971	1972	1973	1974	1975	1976	1977	1978	1979
Argentina (40), (56)									
Rhodophyceae			21400	1546	1381	1317	1519	2372	1514
Phaeophyceae			1700		123	50	198	186	163
Brazil (45)									
Rhodophyceae	110	215	1810	2239	837	384	222	207	287
Chile (43)									
Rhodophyceae	4817	4807	2484	6894	10625				
Phaeophyceae					210				
Peru (39)									
Rhodophyceae	102		13	145	11	118			
Total	5029	5022	27407	10824	13187	1869			

points out that a total of 23 and 30 thousands tons, wet weight, represent harvest estimations for 1971-1973 for red algae in the Atlantic and Pacific coasts, respectively. For brown algae the values are 75 and 1 thousands tons respectively. A more recent view is given below with references for more detailed informations.

Alginophytes

The higher potentials for alginate production are certainly in the extense belts of *Macrocystis* growing in the cold waters of Argentina, Chile and southern Peru. However the stocks of *Lessonia* and *Durvillaea* should also be considered. For Argentina Kühnemann (41) refers a density of 8-15 kg/m^2, fresh weight, of *Macrocystis*, without however giving the total area. More informations about the ecology of *Macrocystis* belts in Argentina are given by Barrales and Lobban (1975). Jensen estimated a potential of 15-20 thousand tons of *Macrocystis* floor and 1-1.5 thousand tons of alginates in Argentina (41). In Chile exploitation of *Lessonia* and *Macrocystis* is recent and production of alginate is small in comparison with agar. However the potential is large and exploitation seems to be growing (43). In Peru the species more important belongs to *Macrocystis*, *Lessonia* and *Eisenia* (32), (39). However no data is available about the stocks and exploitation of these plants. In the lower latitudes of the South American Atlantic potential alginophytes are the species of *Sargassum* which are specially abundant in the SE coast of Brazil (44), (45), although a mixture of Dictyotaceae could also be considered in some places. It is worthy mentioning here the *Laminaria* belts that grows in deep water in the SE coast of Brazil, which seems to occur in considerable amounts (45), (46). A similar situation occur in Venezuela (54). However they are certainly not as important as the ones found in the cold water area of the continent.

Agarophytes and Agaroidophytes

The more important resource for agar production in the continent are the *Gracilaria* spp. as pointed out to Brazil (45), Argentina (41), Chile (43), Venezuela (55) and Peru (39). Species of *Gracilaria* have been processed for agar in Chile, Argentina and Brazil, and a significant amount of raw material has been exported to Japan and European countries. The genus is represented by about two dozen species in the Tropical Atlantic (1), (3), and by one or a few species in the more temperate waters (47), (48), (32). The available production data are given in Table 2. Other possible sources of good agar are the Gelidiaceae, specially *Pterocladia capillacea*, in Brazil and Uruguay (45), (23). However the natural stocks are not large enough to support a permanent exploitation. In Chile *Gelidium* sp. has being exploited but the amount is also small and decreased from 110 tons in 1965 to 8 tons in 1975 (43). Other species of Gelidiaceae occur in Peru and other countries but the stocks are probably small.

Of the agaroidophytes it is worthy mentioning Hypnea spp., specially H. musciformis in the warm Atlantic (45),(54), Gigartina chamissoi and Iridaea spp. in the Pacific (28), (43),in addition to other species of value, but occurring in small quantities, such as Eucheuma spp., for instance . Gigartina skkotsbergii was also considered as a promising species in Argentina (41).

Other Uses

Besides its use in the manufacture of cement,calcium carbonate is widely used in the Brazilian agricultural land to neutralize the acid soils. There are very extense belts of corallinaceous algae along the Brazilian platform at depth of 20-40 m in the NE littoral (49), (50). The surface sediments contain more than 75% of calcium carbonate (51). Assuming a thickness of 5 m (51), the shelf reserve is 2×10^{11} tons, that is more than 50 times the estimated land reserve (Dep. Nac. Prod. Min., 1972). Besides the encrusting corallines there are also extensive deposits of Halimeda segments in the NE coast of Brazil containing 93% of calcium carbonate (51).

In what concerns the use of seaweeds as food, fodder and manure the utilization is small in the continent though there are some references in the literature. The use of seaweed as food is restricted to Porphyra spp. in Brazil, consumed only by the oriental immigrants (45). The consumption of Durvillaea, Gigartina and Porphyra seems to be more significant in Chile (52), (43) and Peru (32), (37).

Discussion

The brief historical view of phycology development in South America was intended to show: i.that a significant improvement of phycological knowledge in the area was attained only recently, after the work of the first resident researchers; ii. that marine phycology is a young science in the continent. This is probably not peculiar to Phycology but seems to be the general situation in most fields of marine sciences, for over 60% of the actual researchers pursued a course of study in the last decade, between 1971-1978 (53). In consequence there is much pioneer work to be done in several points of the continent although some areas could be considered quite well known in terms of its flora. The last five years have seen an uprise of new approaches, specially in Brazil and Chile, developing simultaneously with the taxonomical studies. The reazons of this diversification is not clear. It could be due to the belief of more progressive workers that one can produce good pieces of research in higher level taxonomy, physiology, citology and other aspects of biology without having to give all the intermediate steps, i.e, the belief that α taxonomy is not a pre-requisite for Ω taxonomy. Or, on the other hand, it could represent just an effort of the local researchers to follow more modern and fashionable approaches in order to have their work accepted for publication in the periodicals of the developed countries. In other words, to become recognised by the international scientific society. This implies in further troubles, including the writing or their work in a foreign language, and is a point that should be carefully analysed in terms of its real contribution to the development of the local science.

However there is no doubt that marine phycology is experimenting at this time a burst of progress in some Southern American countries. The foundation of the Sociedade Brasileira de Ficologia (Brazilian Phycological Society) this year was an important phycological event in Brazil, with possibly good reflections in all the continent. It will, hopefully improve the communication among the Brazilian phycologists enabling a better coordination of the phycological activities in the country, without rigidly directing it. We hope also to acquire a momentum to be heard by the government agencies that control the production of seaweeds in Brazil, for the establishment of sound policies towards a rational exploitation.

Seaweed production in South America is in this moment mainly concentrated in Argentina, Chile, Brazil and Peru (Table 2). The main resources exploited today is *Gracilaria*. However expansion of production from the natural stocks is somewhat limited. In Argentina although the local production of agar reachs about 300 tons per year, the growing fields are restricted to a small area in Chubut (25). In Chile, with a production of almost 10 thousands tonns (dry weight) of *Gracilaria* in 1975 (43) good grows are limited to more sheltered areas (K. Alveal V., personal communication). In Brazil the natural populations are declining (45) perhaps due to overharvesting , with a consequent increase in the price (Figure 2). The optimistic data given by Michanek (40) for Brazil are misleading and based on incorrect

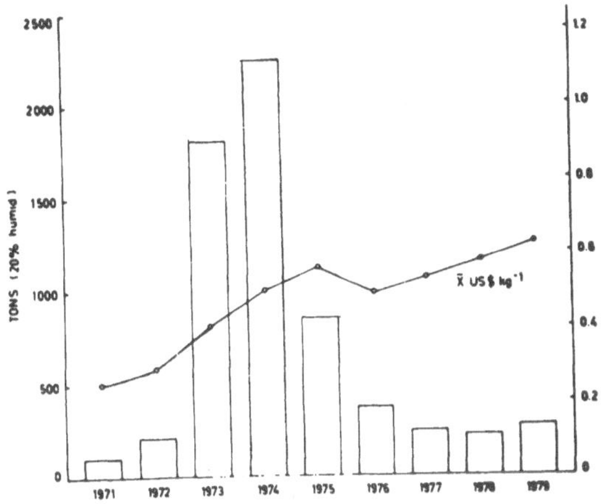

Figure 2. Seaweed production in Brazil. The scale on the right represents the mean price of exportation in US dollars per kg of dry weight. The seaweeds exported are *Gracilaria* sp. and *Hypnea musciformis*.

informations. Other red algae with potential for increasing exploitation are *Hypnea musciformis* in Brazil (45), *Gigartina skkotsbergii* in Argentina (41) and *Iridaea* spp. and *Gigartina chamissoi* in Chile (43). The extensive deposits of crustose corallines and *Halimeda* in Brazil remain intact (49). Exploitation of Gelidiaceae seems to be possible only through cultivation since the natural stocks are not large enough and recuperate slowly.

References

1. Oliveira F., E.C. de: Algas marinhas bentonicas do Brasil. Univ. S. Paulo, Thesis. S. Paulo 1977.
2. Taylor, W.R.: Allan Hancock Atlantic Exped. 2, 1-193 (1942).
3. Taylor, W.R.: Marine algae of the eastern tropical and subtropical coasts of the Americas. Univ. Michigan Press, Ann Arbor 1960.
4. Barriga-Bonilla, E., Hernandez C., J., Jaramillo T., I., Jaramillo M., R., Mora O., L.E., Pinto E., P., Ruiz C. M.: La Isla de San Andrés . Univ. Nac. Colombia, Bogota (1969).
5. Kapraun, D.F.: Carib. J. Sci., 12, 199-203 (1972).
6. Acevedo R., G. de J.: Rev. Acad. Colomb. Cienc. Exact. Fis. Nat. 13, 237-257 (1968).
7. Schneter, R.: Bibliot. Phycol. 24. J. Cramer, Vaduz 1976.
8. Bula M., G.: An. Inst. Inv. Mar. Punta Betin 9, 45-71 (1977).
9. Taylor, W.R.: Allan Hancock Pacific Exped. 12, 1-528 (1945).
10. Ganesan, E.K.: Reseña historica, catalogo, bibliografia y distribucion de las macro-algas bentonicas y fanerogamas marinas citadas para Venezuela. Univ. de Oriente, Thesis. Cumaná 1978.
11. Albornoz, O., Rios, N. de: Lagena 8, 3-12 (1965).
12. Rios, N. de: Lagena 8, 41-50 (1965).
13. Hammer, L., Gessner, F.: Bol. Inst. Oceanogr. Univ. Oriente 6, 186-256 (1967).
14. Gessner, F., Hammer, L.: Int. Rev. ges. Hydrobiol. 52, 657-692 (1967).
15. Diaz-Piferrer, M.: Carib. J. Sci. 7, 1-13 (1967).
16. Lemus, A.J.: Bol. Inst. Oceanogr. Univ. Oriente 13, 23-46 (1974).
17. Taylor, W.R.: Bol. Soc. Ven. Ciencias Nat. 22, 71-101 (1976).
18. Oliveira F., E.C. de: Bolm Fac. Fil. Ciencias Univ. S. Paulo, (bot.) 26, 1-277 (1969).
19. Howe, M.A.: Bull. Torrey bot. Club 57, 605-610 (1931).
20. Taylor, W.R.: Pap. Mich. Acad. Sci. 24, 127-164 (1939).
21. Cunillera, F.J.F.C., Oliveira F., E.C. de: An. 25 Congr. Nac. Botanica, Mossoró, (RN), 135-139 (1977).
22. Coll, J., Oliveira F., E.C. de: Bot. Marina 19, 191-196 (1976).
23. Coll, J.: Algas marinhas bentonicas do Uruguai. Univ. S. Paulo, Thesis, S. Paulo 1976.
24. Mendoza, M.L.: Contr. Tecn. CIBIMA, B. Aires 32, 1-27 (1978).

25. Kühnemann, O.: Memorias del Seminario sobre Ecologia bentónica y sedimentación de la plataforma continental del atlántico sur. 94-111, UNESCO, Montevideo 1979.

26. Kühnemann, O.: Physis 31, 117-142, 295-325 (1972).

27. Etcheverry D., H.: Rev. Biol. mar. 7, 63-182 (1958).

28. Santelices, B.: Actas I Symp. Algas Mar. Chilenas, (B. Santelices, ed.), 1-34, Santiago 1979.

29. Santelices, B., Abbott, I.A.: Phycologia 17, 213-222 (1978).

30. Levring, T.: Lunds Univ. Arsskr., N.F., Avd. 2, 56(10), 1-185 (1960).

31. Kim, D.H.: Gayana 1, 1-82 (1971).

32. Acleto O., C.: Univ. Nac. Mayor S. Marcos, Museo Hist. Nat. J. Prado , Lima, 5, 1-85 (1971).

33. Howe, M.A.: Mem. Torrey Bot. Club 15, 1-185 (1914).

34. Collins, F.S.: Rhodora 17, 89-96 (1915).

35. Taylor, W.R.: Pap. Michigan Acad. Sci. Arts Let. 31, 57-90 (1947).

36. Dawson, E.Y.: Nova Hedw. 13, 1-111

37. Acosta P., J.: Univ. Nac. Mayor S. Marcos, Museo Hist. Nat. J. Prado , Lima 7, 1-9 (1977).

38. Acleto O., C.: II Congr. Latinoamericano Botânica (Resumos), Brasília, 154 (1978).

39. Acleto O., C.: Explotacion de algas marinas en el Peru. Rev. Latino - amer. Ficologia 1(in press).

40. Michanek, G.: FAO Fisheries Tech. Paper 138, 1-127 (1975).

41. Kühnemann, O.: Contr. Tecn. CIBIMA, B. Aires 5, 1-35 (1970).

42. Barrales, H.L., Lobban, C.S.: J. Ecol. 63, 657-677 (1975).

43. Joyce, L., Santelices, B.: Biol. Pesq. Chile 10, 3-26 (1978).

44. Oliveira F., E.C. de, Paula, E.J. de: Proc. 9 Int. Seaweed Sympos. , S. Barbara, (eds. Jensen, A., Stein, J.R.), Science Press. Princeton (1979).

45. Oliveira F., E.C. de: A exploração de algas marinhas no Brasil: situação atual e perspectivas futuras. Rev. Latinoamer. Ficologia 1 (in press).

46. Oliveira F., E.C. de, Quege, N.: Inst. Pesq. Tecn., Pesq. Desenv., S. Paulo, 1, 1-18 (1978).

47. Halperin, D.R. de, Boraso, A.L.: Contr. Tecn. CIBIMA, B. Aires 8, 1-152 (1971).

48. Kim, D.H.: Bot. Marina 13, 140-162 (1970).

49. Coutinho, P.N.; Atas VII Simp. Geol., Fortaleza, 59-65 (1975).

50. Kempf, M.: Mar. Biol. 5, 213-224 (1970).

51. Milliman, J.D., Amaral, C.A.B.: Soc. Bras. Geol., An. XXVIII Congr., P. Alegre, 335-344 (1974).

52. Etcheverry D., H.: Rev. Biol. Mar. 8, 153-174 (1958).

53. UNESCO, Directorio de Ciencias del Mar de America del Sur. Oficina Regional de Ciencia y Tecnologia para Amer. Latina y el Caribe, Montevideo 1979.

54. Santelices, B.: Phycologia 19, 1-12 (1980).

55. Diaz-Piferrer, M.: El Farol, Caracas 222, 18-22 (1967).

56. Meyer, A.: Personal communication based on "Ministerio de Economia - Argentina", for years 1974-1979 (projection).

REFLECTIONS ON THE INTERRELATIONSHIPS BETWEEN RED SEAWEED SOURCE CHEMISTRY AND USES

D. J. Stancioff

Marine Colloids Division, FMC Corporation
P. O. Box 308, Rockland, Maine 04843, USA

Introduction

Over twenty years ago, Stoloff and Silva (1), tentatively concluded that it was possible to attribute taxonomic significance to the polysaccharides of the red seaweeds. Some years later, for the purpose of improving seaweed utilization, Stoloff (2), proposed a classification system that was actually based on polysaccharide type rather than on traditional criteria. His reasoning was that if color could be used to separate various classes of plants it was equally appropriate to use polysaccharide structure for species differentiation. Since that time we have learned a lot more about the chemistry of these polysaccharides (3-6), more species have been examined (7, 8), and we have learned that in the Gigartinaceae the tetrasporophytes and the gametophytes contain different polysaccharides (7-9), a fact of fundamental significance. We are therefore not quite as sanguine about proposing a new classification, however, we should not reject Stoloff's proposals but rather integrate them into our present system, which should profit by having a new dimension added to it.

My own interest in classification grew, in part, out of frustration with mislabeled samples, problems of identification and the failure of improperly labeled raw material to perform as expected. Basically, however, my interest stems from the strong conviction that there are logical patterns in nature that we need to understand if they are to be of any benefit to

Xth International Seaweed Symposium
Printed in Germany

us. Besides, I believe that Stoloff was on the right track.

A Different Viewpoint

At this point I do not wish to propose a new classification system but rather I would like to explore how our point of view as users of seaweeds, as engineers and chemists and marketers might contribute to algal systematics.

In industry we practice our own brand of classification based on certain characteristics that are useful to us. Thus, it is not surprising that we should seek a correlation between plant source and polysaccharide structure, properties and applications. However, we also look at other, less apparent properties. For example, ease of processing is very important to us so we are very conscious of how well the plant material breaks down, how fast the polysaccharide is extracted. Consequently, we know that certain groups of plants are more easily extracted than others. We are also interested in plant color, odor, yield, ease of recovery, or any other property that may be important in identification and processing of our raw materials.

All these attributes are catalogued formally or informally and our seaweed raw materials are classified according to how well they perform in each required function.

It is not that we see things any differently from the biologist. We simply observe different properties because they matter more to us; yet these properties may be just as useful at showing relationships between organisms as any properties the biologist may have chosen to study.

Color. Color as a means of identification is naturally very important since the primary seaweed class groupings are based upon it.

Usually, the plants that we see are commercial shipments that have been dried and are partly bleached. The colors of the rehydrated plants are therefore quite different from those in the living algae. This can, however, be turned to our advantage. In the fresh plant chlorophyll often masks the red and blue pigments which makes many red seaweeds look brownish. In the partially bleached weed the chlorophyll is almost gone, and the reds and blues may be considerably attenuated.

The different shades that the plants go through during bleaching are characteristic for each species even though there may be some variations due to season, location or exposure. Sometimes a color may even be common to a whole genus. For example, all Hypnea species that we have seen go from deep brownish purple through various shades of wine red, bright red and orange to yellow, whereas Iridaea species never go through red or orange, but rather, tend towards dull reddish purple or sometimes purplish blue through pinkish brown, almost always with a touch of brown remaining to dull the blue purple color.

Chondrus colors are similar to those of Iridaea with a little brown to dull the purplish or pinkish tints. However, purplish blue of Iridaea is rare in Chondrus.

The magnifoliate Californian Gigartinas, G. exasperata, G. harveyana, G. corymbifera all go through a deep red color with little purple in it, whereas the finely branched, terete types such as G. acicularis, G. pistillata, G. tenella, G. serrata, G. canaliculata tend towards purplish pink. In contrast the Mastocarpus group of Gigartina shows much more red.

An otherwise unrelated group that has much in common in color as well as other features, and that might perhaps be considered as a sort of bridge between Iridaea and Chondrus is what we might call the greenish-dull-purple group comprising I. boryana, I. cornucopiae, C. ocellatus and the distinct broad forms of C. crispus found on the North Western coast of

Spain and in Portugal.

Some of these color differences may have some ecological significance such as an adaptation to intertidal or subtidal conditions, however, I believe that there may be more to it than that, and that by comparing colors qualitatively and quantitatively we may be able to add one more dimension to our understanding of the relationships between the various species.

Structural Properties. The production of seaweed polysaccharides begins with an aqueous extraction. Even the way in which the plants break down during the extraction process gives an indication of their type.

Agarophytes whether they are Gelidiaceae or Gracilariaceae are notoriously difficult to break down whereas the carrageenan bearing plants, particularly the Gigartinaceae, disintegrate much more readily but, for some curious reason, the tetrasporophytes break down more slowly than the gametophytes. The former actually seem to stiffen when first immersed in boiling water as if they contained some albumin-like protein that coagulates with heat, whilst the latter immediately begin to disintegrate along their tips or edges.

Hypnea species are fairly difficult to extract and so are most of the Phyllophoraceae. Among the Solieriaceae, those plants that are more closely related to *Agardhiella* are most difficult to extract than the ones that are closer to *Eucheuma spinosum* even though all contain very much the same type of polysaccharide.

Odor. Seaweeds, like any vegetable, also have their own distinctive odors depending on whether they are wet or dry, fresh or rehydrated, heated at high pH or low pH. In our laboratory, it is not unusual to hear someone say, "This smells like an agar weed" or "This stinks like *Grateloupia*." The Gigartinaceae have a sweetish odor when rehydrated. Unfortunately we cannot put this subjective information to any use at this time.

Polysaccharide Structure. A first generalization we can make is that, in all the red seaweeds tested so far, the major water soluble polysaccharides are galactans that have a linear structure with alternating α-1→3 and β-1→4 linkages.

The 1→3 linked sugar is apparently always D-galactose.

The 1→4 linked sugar may be either D-galactose or 3,6-anhydro-D-galactose or L-galactose or 3,6-anhydro-L-galactose. If the sugar is L- then the galactan belongs to the agar group; if it is D- it belongs to the carrageenan group.

In both the carrageenan or the agar structures the backbone chain may be masked by the sulfate ester groups. Methyl ethers may mask the agar chain. Without going further into detail we can list the following:

Carrageenan types: Lambda-, Kappa-, Iota-, Furcellaran.
Agar types: Agaropectin, Agarose, Porphyran.

All orders so far examined contain at least some species with polysaccharides of the agar type. In the Bangiales we find porphyran and the Nemaliales are well represented with a large number of species of *Gelidium* and *Pterocladia* all containing agarose and agaropectin. In the Crytonemiales only *Gloiopeltis* contains an unusual 6-sulfated agarose. All *Gracilarias* and *Corallopsis* in the Gigartinales contain agarose and agaropectin while in the Ceramiales the only three species examined contain agarose.

A carrageenan of the furcellaran type is found in two species of Cryptonemiales *Tichocarpus crinitus* (10), and *Endocladia muricata* (6).

In the Gigartinales the more evolved families apparently contain only carrageenan. There is only one exception - *Ahn-*

feltia plicata, which contains agarose.

It is difficult to make any generalizations about the Solieriaceae. They all seem to contain polysaccharides with either furcellaran, *kappa*, or *iota* structures, with all three represented in *Eucheuma* (11).

The Phyllophoraceae and Gigartinaceae contain *kappa*-type carrageenan - but in general the polysaccharide is a hybrid between *iota*- and *kappa*.

A generalization that we can safely make about the polysaccharides of the Gigartinaceae, is that the tetrasporophyte contains only a nongelling polymer that may have a high 6-sulfate content (*lambda*-carrageenan) or a low 6-sulfate content (*xi*-carrageenan).

Recent results have confirmed our previous work (6) that the high 6-sulfate content is typical of *Chondrus* while the lowest sulfate is found in *Gigartina*, particularly in the type species, *G. pistillata*. *Iridaea* and *Rhodoglossum* fall in between. However, there are several *Gigartina* species that overlap with *Iridaea*, *G. volans* being very close to *Chondrus* in this respect.

A most convincing proof of the similarities of *Chondrus* and *Iridaea lambda*-carrageenan is that anti-*Chondrus* antibodies also react against *Iridaea* but not against *G. californica* (12) which contains a *xi*-polysaccharide.

Properties. We have now learned enough about the correlations between chemistry and properties of the polysaccharides of the red seaweeds to be able to predict properties fairly well if we know the chemical structure and to guess at the chemistry if we know the properties.

Nevertheless there are some subtle differences in properties that we cannot predict from chemical analysis. These subtleties we believe are due to variations in sulfate distribution on the molecules. For example, we can make a mixture of iota-carrageenan from Eucheuma spinosum and kappa-carrageenan from E. cottonii that will give the same analytical results as a sample of G. stellata extract. The gel properties in milk or water will however be quite different. A seaweed species can therefore be classified according to the properties of its polysaccharides in milk, water, or any other system. Such a classification has just as much validity as a chemical one and may, like an immunological test, be far more sensitive.

Now let's take an example: From the point of view of the user G. volans and G. leptorhynchos are not Gigartina but Chondrus. First of all the polysaccharides are more like those in Chondrus. The kappa fractions are strong gelling in milk and water and the lambda fractions show a high 6-sulfate content particularly in G. volans.

Colors of both weeds are different from most Gigartinas tending towards purplish-brown. Both plants show a distinct tendency towards dichotomous branching, although, in G. leptorhynchos this is hidden behind all the proliferating pinnules on the surface of the flattened main branches. In both plants the tetrasporangial sori are buried in the blade or main branches, although in G. leptorhynchos they may also be found on the pinnules.

Finally, both have loose or, in the case of G. leptorhynchos, only slightly developed enveloping tissue around the cystocarp (13).

If we consider all the above we conclude that both species fall somewhere between Chondrus and the type species for Gigartina, G. pistillata. Still they seem to be closer to the former than the latter.

Conclusion

A classification system based on only one or two characteristics, whether morphological or chemical, is an over-simplification. The greater the number of features that we can measure and compare the greater the validity of our system.

We have seen that even within a single species as, for example, *Gigartina pistillata* there are two very different polysaccharides depending on whether the plants are gametophytes or sporophytes. We find the same pattern in other members of the Gigartinaceae family. Since our classification system contains two polysaccharides it is actually strengthened. But, we also find that sporophytes in the *Gigartina pistillata* have a lighter color and that they break down differently in hot water indicating a different cell wall structure. These features are also repeated in other Gigartinaceae which gives us additional dimension to our system and greater confidence in it.

In sum, the processor and the chemist see other features of the seaweeds than the botanist and may classify them differently. Nevertheless, such attributes as good filtration properties, or good extract color, or gel properties and casein binding can and no doubt will also make their contribution to plant systematics.

References

1. Stoloff, L., Silva, P.C.: Econ. Bot. 11, 327-330 (1957).
2. Stoloff, L.: Econ. Bot. 16, No. 2, 86-94 (1962).
3. Rees, D.A.: Adv. Carbohydrate Chem. Biochem. 24, 267-332 (1969).
4. Yarotsky, S.V., Shashkov, A.S., Usov. A.I.: Bioorg. Khim. 3, 1135-1137 (1977).

5. Bhattacharjee, S.S., Yaphe, W., Hamer, G.K.: Carbohydr. Res. 60, C1-C3 (1978).

6. Stancioff, D.J., Stanley, N.F.: Proceedings Sixth International Seaweed Symp., Santiago de Compostela (1968).

7. Chen, L.C-M., McLachan, J., Neish, A.C., Shacklock, P.F.: J. mar. biol. Ass. U.K. 53, 11-16 (1973).

8. Peats, S.: Proceedings Tenth International Seaweed Symposium, Goteborg (1980).

9. Parsons, M.J., Pickmere, S.E., Bailey, R.W.: New Zealand Journal of Botany 15, 589-95 (1977).

10. Usov, A.T., Yarotsky, S.V., Shashkov, N.D., Biopolymers 19, 977-990 (1980).

11. Doty, M.S., Santos, G.A.: Aquatic Bot. 4, 143-149 (1978).

12. Evelegh, M.T., Vollmer, C.M., McCandless, E.L.: J. Phycol. 1489 (1978).

13. Kim, D.H.: A Study of the Development of Cystocarps and Tetrasporangial Sori in Gigartinaceae. Nova Hedwigia XXVII, (1976).

MARINE PHYCOCULTURE IN CHINA

C. K. Tseng
Institute of Oceanology, Academia Sinica, Qingdao, China

Farming of land plants, specifically the seed plants, has a long history of many thousand years. With the continual development of agronomy - the science and art of producing land crop plants - people are able to produce more and more, better and better plant products per unit area of cultivated land, and now the earth is able to feed some four billion people.

Farming of sea plants, specifically the algae, has a very short history, perhaps three hundred years. Actually, algal farming on a truly scientific basis started only about twenty years ago. The late start of scientific algal farming is due to several reasons. Man is a land animal and depends on plants growing on land for food, drugs etc. To the primitive man the sea was something very dreadful and mysterious, and plants growing there were nothing but weeds. It certainly took a lot of trial and error before man was able to appreciate some of these sea plants for food, as drugs, etc. One of the sea plants which has long attracted man's interest was the purple laver, *Porphyra* spp., called *nori* by the Japanese and *zicai* by the Chinese. The Orientals appreciate the *Porphyra* so much that natural production on the beaches was unable to meet the demand of the people. Undoubtedly the ancient Orientals many years ago must have tried hard to produce the much appreciated *Porphyra* without success until about three hundred years ago in Japan and two hundred years ago in China, the Japanese *hibi*-collector method and the Chinese rock-cleaning method were devised, which greatly promoted the *Porphyra* production. However, it was not until the late fifties and early sixties that the *Porphyra* cultivation began to be put on truly scientific bases. Another

Xth International Seaweed Symposium
Printed in Germany

sea plant greatly appreciated by the Orientals is the Laminaria, especially L. japonica, the Japanese kelp. This seaweed has been appreciated by the Chinese people for at least fifteen hundred years, but it was not until 1952, three years after the founding of the People's Republic of China, that the Laminaria was successfully produced by artificial cultivation on floating rafts. These two cases show that, even with the present-day knowledge of modern sciences, it is still not such an easy matter to domesticate a wild marine plant into a real crop plant, which will grow at the time and in the place designed by man and do better than their wild relatives. In fact, so far there are only a few marine plants which can be regarded as truly marine crop plants, perhaps three more beside the above mentioned Laminaria and Porphyra.

1. Phycoculture defined

The difficulties involved in the domestication of the seaweeds, transforming from wild plants into crop plants are manifold. The most fundamental one is that in the farming of seaweeds we are dealing with a very different kind of plants than in the farming of land plants which are exclusively seed plants. First of all, there is a difference between the reproductive units of these two kinds of plants. The land crop plants reproduce by seeds which are multicellular, macroscopic and generally preserved by drying and eventually sown into the ground. The marine crop plants belonging to the algae reproduce microscopic unicellular spores which cannot be preserved by drying or stored in any other way. The spore-liberating algae should be brought as close to the desired substrate as possible, so that the liberated spores will have the necessary place to attach to and grow. Secondly, there is the difference between the growth media of the two kinds of plants. The land crop plants grow on soil and in the air, whereas the marine crop plants grow in the sea, attached to some sort of substrates. In the case of the land crop plants, the air is relatively gentle stationary with respect to the source of light, the energy for effecting photosynthesis. In the case of the marine crop plants, the sea is in constant vigorous motion with perpetual pounding waves and ever-changing tidal levels, and the substrates to which the seaweeds attach must be able to withstand the ceaseless water motion and remain stable enough to ensure remaining in a definite level to receive adequate light for photosynthesis. All these and other differences will naturally lead to differences in the technical methods of cultivation. It is, therefore, necessary to introduce a special term to designate the cultivation of seaweeds for commercial purpose. Currently three terms are being applied for this purpose: marine aquaculture, mariculture and marine agronomy. All these terms have the same limitation, and that is, they technically include in one way or another the commercial cultivation of marine seed plants. It is not impossible that eventually such marine seed plants as the mangroves and the eelgrass may be subjected to commercial cultivation. In view of the fundamental difference between the seed plants and the algae, therefore, I suggest that for the commercial cultivation of the algae in general, the term phycoculture should be applied to designate a new type of agriculture employing the algae, -- the primitive photosynthetic, spore-bearing thallophytic plants in the commercial cultivation activities. Thus, phycoculture may be defined as the science and art of commercial cultivation of the algae. It follows that for the commercial cultivation of seaweeds the term marine phycoculture will have to be applied.

2. Phycoculture of Laminaria japonica

For perhaps two thousand years, the Japanese kelp Laminaria japonica vernacularly known in China as haidai (meaning "sea ribbon"), has served as a subsidiary food as well as an anti-goiter remedy. In the Chinese literature in the 5th and 6th century haidai, then known as kunbu, was imported from Korea and Japan. In the present century, old China imported large quantities of this kelp from Japan. In the twenties, probably in 1927, Laminaria japonica was accidentally introduced from Japan to Dalian, Liaoning Province on the northern coast of the Yellow Sea. Since then the Laminaria becomes a naturalized member of the Chinese marine flora. Its distribution is, however, restricted for many years to Dalian, where some quantities of the haidai were harvested annually, the largest annual output being about 60 tons.

The first few years after the founding of the People's Republic of China in 1949 China continued to import large quantities of the haidai from Japan to satisfy the need of the people. In 1951 the marine phycologists and mariculturists in Qingdao initiated a program working on the biology of the Laminaria and methods of cultivation. In the fifties, three crucial methods concerning the artificial cultivation of the Laminaria had been devised, namely, (1) the floating raft method of cultivation, (2) the method of culturing summer sporelings in low temperature, and (3) the method of fertilizer application in the sea. With the help of the first two methods and on the basis of the studies on the growth and development of the Laminaria sporophyte in relation to the temperature factor, experiments were successfully carried out on the southward transplantation of the haidai. These four achievements laid the foundation for the development of the Laminaria aquacultural industry (8, 9).

In the sixties great effects had been exerted on the expansion of the haidai aquacultural industry, on improvement of the existing methods of this mariculture and on basic research and application of the results in the cultivation. In this period the following achievements had been made: a) The natural population of the Laminaria proved to be genetically mixed in nature, and three new strains bred (3). b) Improvements in the fertilizer application methods made (21). c) A method of tip-cutting manipulation to increase production devised (25).

In the seventies the Laminaria cultivation industry had been further expanded and the following achievements had been made: (1) strains high in production and in iodine contents bred (7); (2) the etiology of two kinds of diseases was clarified and preventive measures suggested (27).

The short history of China's Laminaria cultivation industry shows that it has been undergoing a continual expansion for the last thirty years. The annual production has increased enormously from the trial production of 62 tons fresh haidai (about 12 tons dry) in 1952 to about 275,000 tons dry haidai in 1979, in which about 15,000 hectares of marine farms were engaged in the commercial cultivation of this Laminaria (9). Some sketches of the more important achievements of the Laminaria cultivation are presented in the following.

1) The floating raft method of artificial cultivation. This method is characterized by three important processes. The first is the spore collecting process which is equivalent to the seed-sowing in land crop farming. The Laminaria frond with abundant mature sporangial sori is subjected to partial drying in the air and then placed in a small body of sea water in a container. The liberated zoospores will soon attach themselves to the substrates, the seeding cords, which are placed in the water beforehand. The second is the setting-up of the floating raft for the cultivation with the seeding cords hanging beneath or between bamboo segments, which are connected with each other in series resembling a raft floating on the sea and attached to the sea bottom by coarse ropes. The superiority of the raft method lies in

maintaining the cultivated kelp plants in the selected water level. This is one of the keys to the success of our kelp cultivation. This is because at our Yellow Sea and Pohai coasts the sea water is quite turbid and the tidal range quite large, and with the raft method of cultivation maintenance of the seaweed at the desired depth is ensured for the best photosynthetic activities. The third is the sporeling transplantation, which is equivalent to the rice-shoots transplantation in the rice farming. This is a crucial process ensuring appropriate density for the growth of the Laminaria fronds. In the spore collecting the zoospores attach to the cords at random and in great density, and if the resulting sporelings are allowed to grow in this manner, the growth would be too dense, and the resulting plants would be small and thin. It is therefore prerequisite that when the sporelings reach 10-15 cm in height, they are detached from the seeding cords and twisted on coarser cultivation ropes. The length of the cultivation ropes and the number of sporelings per rope are different in different places, and in the Qingdao region the rope is somewhat less than 2 meter long, on which more than 30 sporelings are positioned (8, 9, 28; Fig. 1).

Fig. 1. Floating raft method of Laminaria cultivation.

Fig. 2. Laminaria japonica harvesting.

2) Method of low temperature culture of summer sporelings. In the early stage of the studies on the Laminaria, investigations were made on the biology of this kelp, especially on the growth and development of the gametophytes in relation to some environmental factors, such as temperature, light intensity and light period (hours of illumination per day). In 1954 sufficient information about these conditions had been accumulated. For instance, it was experimentally proved that the optimal temperature for growth of the gametophyte was about 15°C, and that for their development into sporophytes about 10°C, whereas at 20°C or above, the development of the gametophytes to sporophytes was inhibited. On the basis of the results obtained, a method for large scale culture of gametophytes was devised, the entire process of the growth and development of gametophytes being completed in a rather short period, generally 10-12 days (4).

In the first few years of the application of the floating raft method of cultivation, spore collecting took place in autumn (generally in the middle of October in the Qingdao region); the sporelings thus produced is called autumn sporelings. Soon after the seeding cords were placed in the sea, the spores of various algae such as *Ectocarpus*, *Enteromorpha*, *Licmophora* etc., in a short time adhered to the cords and within a few days grew into macroscopic thalli, densely covering the seeding cords, while the *Laminaria* spores thereon remained in the microscopic gametophyte stage. Coverage of the cords by the undesired weeds seriously prevented the gametophytes to receive sufficient light, hence greatly retarding their growth. Under these circumstances, the sporophytes sometimes failed to appear entirely, and when they did appear, they would grow to the desirable size for the transplantation process sometimes in January-February of the next year.

A better understanding of the growth and development of the gametophytes and of the growth of the autumn sporelings under the natural conditions led us to believe that a possible means of overcoming the threats of the weeds to the sporelings would be to collect spores and grow the sporelings in early summer in artificially cooled culture solution, instead of in middle autumn as in the usual case. In middle autumn when the sea water temperature drops down to about 20°C or lower, favorable to the growth of *Laminaria* sporophytes, the sporelings which have appeared, would be able to grow in the sea much faster than the contaminating weeds the growth of which will be suppressed. In 1954, on the basis of the above assumption, an experiment was set up, the results of which entirely confirmed our assumption, thus materializing the object of the experiment. Since these sporelings are cultivated in early summer, they are known as "summer sporelings", the summering being taken place in artificially cooled culture rooms, the water temperature generally maintained at 8-10°C. Therefore this mehtod is known as "low-temperature culture of summer sporelings" or simply "summer sporelings culture method" (8, 9, 19).

Since the introduction of the summer sporelings culture method, it has been well received by the workers and technicians of the kelp cultivation industry. After the extention of the method, some improvements have been made by the workers and technicians. Experiences of these years show that application of the method not only solves the problem of the weed-growths, but also greatly enhances the production by 30-50%, since the summer sporelings have an advantage of two to three months' growth. This method has an additional advantage of improving the laboring condition of the workers, since the sporeling transplantation process takes place earlier in November to December, when the sea water temperature is about 10-15°C, whereas in the old autumn sporeling method, this process is effected in January to February of the next year, when the sea water temperature is lowest in the whole year, generally about 1-2°C. This method also saves a lot of cultivation materials, thus greatly lowering the cost of the production. The introduction of this method greatly facilitates extending the *Laminaria* cultivation industry to the coastal regions south of the Changjiang (Yangtze) River. Otherwise the problem of sporeling supply would be an unsurmountable difficulty.

3) *Fertilizer application in the sea*. Up to 1955, commercial cultivation of the *haidai* was limited to a few bays in the three cities Dalian, Yantai and Qingdao. For some years we had tried in vain to grow tha *haidai* in the outer sea region of Qingdao. Chemical analyses of the sea water showed that in these places the nitrate-nitrogen content was very low, only about 2-3 mg/m^3, at most not over 5 mg/m^3 (total nitrogen content not over 50 mg/m^3), while in the inner bays where the *haidai* was cultivated commercially, the nitrate-nitrogen content was about 20 mg/m^3 (total nitrogen content over 200 mg/m^3). It was therefore suspected that the failure of the *Laminaria* to grow in the outer sea region was due to low nitrogen content, and that fertilizer appli-

cation would solve the problem of extending the _Laminaria_-cultivation to the outer sea region. Fertilizing the sea is, however, a difficult problem. Unlike the stationary soil, sea water is in constant motion; to make fertilizer application in the sea successful, there must be means of minimizing the losses due to water movement and concurrently assuring that the _Laminaria_ would receive the necessary fertilizer. After a series of experiments we have successfully employed porous clay bottles as tools for fertilizer application; the porosity of the clay bottles allows the diffusion of the fertilizer, which become available to the _haidai_ in their vicinity. The results of the experiments showed that by employing this method of fertilizer application, it was possible to cultivate the _Laminaria_ commercially also in the outer sea regions. Since clay bottles were employed in this method it is known as "clay bottle fertilizer method". Extension of this method was welcome by the cultivation workers and technicians of the industry, and in the following four to five years the method was popularly accepted by the industry, and various improvements on the method were made (1, 8, 9, 20, 21).

The clay bottle method of fertilizer application has, however, several defects. The clay bottles employed were rather clumsy, their cost quite high, much labor involved in the hanging of the bottles, and the addition of fertilizer, and the fertilizer effect undesirably low. Therefore various improvements had been made by the workers, for instance, the employment of small bottles, or even plastic bags with a few very minute punched holes. Further practical experiences showed that with large farms, hundreds of acres in area, simply spraying the fertilizer, solution resulted in saving both labor and fertilizer. It gives better growth and the loss of fertilizer is minimal. With the exception of special tasks or small areas of cultivation, the clay bottle fertilizer application method is no longer employed, and application of the fertilizer solution by spraying is generally adopted. For the production of a ton dry kelp about 100-150 kg of ammonium sulfate is consumed. The fertilizer effect is therefore still quite low, and it is estimated that the _haidai_ absorbes only about 1/6 of the fertilizer applied. It is necessary to find out means of enhancing the fertilizer effect to save the fertilizer and consequently some further studies have been made.

So far nitrogen fertilizer has been the only fertilizer employed since our Yellow Sea coastal regions are particularly deficient in nitrogen. It is therefore necessary to have a systematic knowledge of the absorption of the nitrogen fertilizer by the _Laminaria_. Experiments were accordingly conducted and nitrogen-starved fronds were used as experimental material. The results showed that these fronds would absorb a definite quantity of nitrate-N or ammonium-N to supplement the lack of these ions in the plant body. It was further found that the absorption was closely related to the concentration of the solution used, the duration of the soaking, the pH of the solution and the degree of N-starvation of the plant. The results also showed that under normal condition, the nitrogen absorption gradiant of the _Laminaria_ is quite high, and the plant can absorb the nitrogen compounds for its normal growth only when their concentration is high enough, generally at the level of 20 mg/m^3 nitrate-N (23).

Our experiments also showed that the _Laminaria_, having absorbed sufficiently large quantities of the nitrogen fertilizer, will continue its normal growth within a definite time. For instance, soaking the kelp in sea water solution of sodium nitrate of 80-100 gm/m^3 will let the plant absorb sufficient nitrogen fertilizer for maintaining its need of nitrogen for 5-7 days. The so-called "fertilizer application by soaking method" was proposed on the above basis. This is proved to be suitable for the cultivation of juvenile sporelings only a few cm long. With fronds of several meters, evidently this method is not applicable unless some sort of special mechanical devices is created. On the basis of the same principle was also proposed the so-called "intermittent fertilizer application method" (26), which is, in brief, a method of fertilizer application by rotation. This is also a method of saving fertilizer. Another measure for the saving of fertilizer is to stop fertilizer application in the latter stages of growth of the Laminaria, since experimental results showed that about one month or longer before the harvest of the kelp, stopping fertilizer application does not influence the yield. However, some believe that the quality of the resulting product will be somewhat inferior. At present the above measures of saving fertilizer are practiced by some aquaculture farms, and are not yet universally accepted.

4) Southward transplantation of commercial cultivation of the Laminaria. China's coastal regions of the East China Sea south of the Changjiang (Yangtze) River are very fertile. According to our investigation in 1956, the nitrate-nitrogen content of the sea water in the Zhoushan region of the Zhejiang Province was 88-227 mg/m^3 and that in the Xiapu region of the Fujian Province 88-123 mg/m^3, respectively 11 times and 6 times more than that of the fertile region in the Qingdao region. Cultivation of the Laminaria at these two provinces, if possible, would not need any application of fertilizer, thus decreasing greatly the cost of production, and broaden the area of cultivation. The turbidness of the sea water in these regions will not be a serious problem for the commercial cultivation of the haidai, since with the floating rafts method the water level of cultivation may be regulated at will. The difficulty is undoubtedly the relatively long period of the high sea

water temperature there. Therefore an experiment was conducted on the growth of the Laminaria in relation to water temperature. The results of the experiment showed that, although the optimal temperature for the Laminaria growth was 5-10°C, growth was still good enough at 13°C, and even at as high as 20°C, a Laminaria frond of 1-2 m was still able to effect some growth. An investigation on the sea water temperature characteristics of the Zhejiang and Fujian provinces led us to believe that the period of low temperature was sufficient for the haidai to grow to commercial standard in these provinces south of the Changjiang River (8, 9, 22, 24).

On the basis of the above experiment and investigations, we conducted an experiment in 1956 on the cultivation of the Laminaria at Gouqi Island, Zhejiang Province with the cooperation with the Yellow Sea Institute of Fisheries and the Zhejiang Institute of Fisheries. The results of the experiment fully confirmed our postulation that haidai of commercial standard can be cultivated in the region south of the Changjiang River, and laid the ground work for the large scale southward extension of the Laminaria cultivation industry to the Eastern China Sea coastal provinces, the haidai production of which now accounts for about one third of China's total production (18).

5) Genetical studies and breeding of new strains. Genetical studies showed that natural populations of the Laminaria now under cultivation in China are genetically mixed in nature and are considered to possess a high level of hybridity. Hence, they are good original materials for selection and breeding works. It has also been pointed out that the harmful effect of continual inbreeding is relative and conditional; in some inbred lines there are harmful effects while in others no harmful effects are found, which eventually form strains of certain good characters. On the ba-

sis of the basic principles of segregation of inbreeding hybrids and directional selection, our genetical workers employed methods of continual inbreeding and selection for the breeding work. In the inbreeding process, single frond of the *Laminaria* is used to collect the spores instead of using several fronds as in the large scale production of the sporelings in the industry. In this way, the gametophytes and the resulting sporophytes all come from one single kelp frond (3). In the selection process, selection was made with mature *Laminaria* fronds with certain special desirable characters, such as broader and thicker frond, better adaption to higher water temperature, high production, higher iodine contents etc.

In the sixties, three strains of the *Laminaria* were bred, one with broad frond, one with long frond and one with thick frond. Experiments showed that such morphological characters as frond-length, frond-width, frond-thickness and stipe-length are all quantitative inheritable characters, the development of which are controlled by polygenes as well as various environmental factors. The heritability of frond-length was found to be about 0.5.

In the seventies the *haidai* mariculture industry was in need of strains with both high production and high iodine content. To China the *Laminaria* not only provides our people with a subsidiary food but also serves as the raw material of our algin industry, and in the production of the algin, it is further required that the by-products, including the iodine, must be recovered. Thus, high iodine content has also become an object of our breeding work. In 1970 our phycological workers conducted large scale selection work in the nine cultivation regions of five coastal provinces and employing the half-frond method they selected 19 plants with high production and high iodine contents. Breeding was carried out by repeatedly inbreeding and selection, and to the offsprings of a single frond, x-radiation was effected. After a few years of cultivation and examination of the genetical characters, two strains with high production and high iodine content were bred, which answered better to the demands of the industry. Experiments showed that the iodine content of the Laminaria is also a character of quantitative inheritance, just as the frond-length. The two strains are now under cultivation on large areas in some of the haidai mariculture regions and have won the approval of the haidai cultivation farms.

6) Pathogenic diseases in Laminaria cultivation. In recent years two kinds of pathogenic diseases in Laminaria cultivation have been recognized and well documented. One of these is the malformation disease of Laminaria summer sporelings in green house culture in artificially cooled water of about 10°C. The diseases sporelings are characterized by oogonia and abnormal, malformed sporophytes, which consequently died and dropped from the seeding nursery cords. After a series of experiments in the laboratory and in one of the sporeling cultivating stations, the conclusion has been reached that the cause of the malformation is the presence of hydrogen sulfide in the culture solution. Experiments showed that hydrogen sulfide in a concentration of 0.50 ml/l inhibited 96% of germination, 0.13 ml/l delayed 50% gametophyte formation, 0.005 ml/l caused shrinkage of the egg, and the half toxic dose of the oogonium was about 0.018 ml/l. The organisms involved were found to be sulfate-reducing and sulfide-producing saprophytic bacteria such as Micrococcus, which were found abundantly in the black rust of the cast-iron pipes of the water circulation system and in the rotted spots of the Laminaria fronds, as well as in the marine mud in the vicinity of the station. The etiology of the disease have been clarified, preventive measures have been taken by separating the sporeling cultivation system from the mature sporophyte

cultivation system and by sterilizing the sporeling cultivation water system with bleaching powder before the seeding process.

The other well documented disease is the frond twist disease of the cultivated Laminaria on the floating rafts. The diseased have abnormally twisted fronds with greatly swollen stipes and shortened rhizoidal holdfasts. The contagious and biotic nature of the disease had been experimentally proved, and the latent period found to be quite long, 60-70 days. Electron microscope examination revealed presence of numerous polymorphic mycoplasm-like organisms, mostly coccoid and some ovoid, dumb-bell and amoeboid in shape. Experiments also showed that the antibiotics of the tetracyclines were effective in controlling the disease.

7) Transport and accumulation of nutrient matter and tip-cutting method. From the results of the frond-cutting experiments, the condition of the accumulation of nutrient matter, the photosynthetic intensity and the analyses of the chemical constituents of the different parts of the fronds it is evident that the nutrient matter is transported from the upper parts downward. The accumulation of the nutrient matter is greatest in the basal part and less near the apex. Experiment employing P^{32} as trace element also showed that P moves from the apex to the basal part of the frond through the trumpet tissues, then to the stipe and finally to the holdfast. There is no indication whatsoever of a movement of P in the reverse direction. The rate of the movement observed was about 12 cm in an hour.

On the basis of the downward movement of nutrient matter in the Laminaria fronds, and the phenomenon of natural casting of the distal parts of the fronds, we have proposed the "tip-cutting method" to increase production. By this method the distal parts of the *Laminaria* , sometimes as much as one third of the intire frond, are cut off at certain time interval to improve the lighting condition of the fronds and the water movement in the *Laminaria* farms. Such a measure will improve the conditions of the growth of the *Laminaria* and enhance the quality of the product. The cut-off distal parts of the frond will be cast off anyway under the natural condition; these parts are not good for food but are good raw material for the algin industry. By taking such measures, the increase of production generally amounts to about 15%.

3. Phycoculture of *Porphyra*

The species of *Porphyra* are popularly known in China as *zicai* (meaning purple vegetable), and have long been appreciated by our coastal people as delicious food. China has a long history of utilization and cultivation of the *Porphyra*. The earliest records of the occurrence of *zicai* and its food value appeared in books published in the years 533-544. In the Sung Dynasty (A. D. 960-1279) *zicai* had already become local goods from the Haitan Island of the Fujian Province selected for presenting to the emperor annually. This means that about one thousand years ago the Chinese had already regarded the *Porphyra* as a delicacy. About 200 years ago the simple "rock-cleaning" method of cultivation (10) was devised in which the *zicai* rocks were mechanically cleaned of the attached seaweeds and barnacles in autumn, just before the mass liberation of the *Porphyra* spores. The cleaned rocks would then provide the space required for the unknown spores to attach and grow. The whole process is very simple, and people have to depend upon nature's mercy to give them abundant spores. The condition is entirely similar to the Japanese "bamboo-*hibi* planting" method. Up to the early fifties the source of the spores had been a mystery to the *Porphyra* aquaculturists and the phycologists as well.

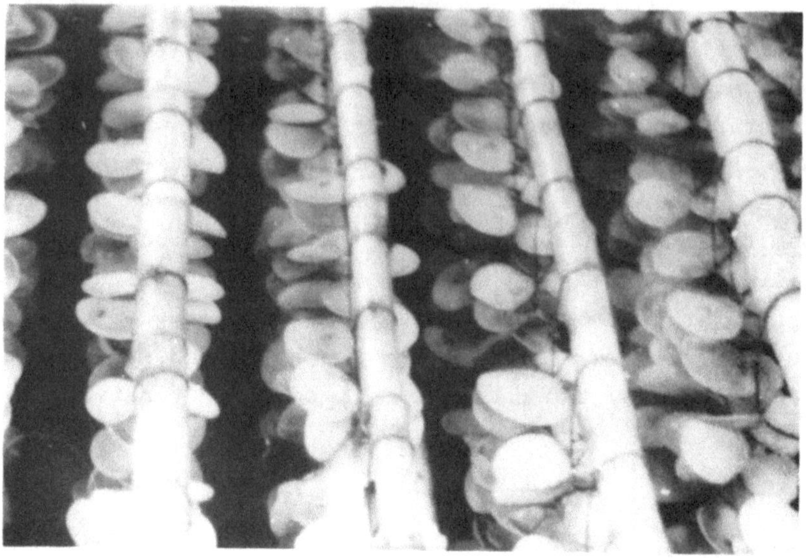

Fig. 3. Meretrix shells in tanks for cultivation of Conchocelis-phase.

Fig. 4. Porphyra cultivation on net.

In the early fifties Drew's discovery of the Conchocelis as a stage in the life history of Porphyra stimulated both the Japanese and the Chinese phycologists to delve into studies of the Porphyra life history to search for the "missing link". The problem has finally been solved by both the Japanese and the Chinese phycologists (11-15), who independent of each other, discovered the missing link, the conchospore, a term which we coined in 1955 (12, 13) and which is now popularly accepted. For the next few years, efforts had been exerted on studies on the Conchocelis stage of the life history, particularly on large scale culture. The shell of Meretrix meretrix was found to be an excellent substrate for culture and a more or less standardized method for the Conchocelis culture was introduced.

In 1958-1959 the Institute of Oceanology and the Dalian Aquaculture Station conducted a cooperative experiment on the semiartificial spore collecting cultivation of Porphyra yezoensis in Dalian, and shells with mature Conchocelis were hung beneath the cultivation nets for supplying with conchospores (16). The unit production was 433 g/m^2 nets, 3-4 times higher than the control. This was the first Porphyra cultivation experiment in China in which the newly discovered conchospores were involved. In 1960-1961 we carried on an experiment in Qingdao on the Porphyra cultivation in which the conchospores were collected directly on the cultivation nets. The unit production was 473 g/m^2 nets, 6.4 times higher than the control. The Japanese fixed pillar method of cultivation was adopted in this experiment. The number of shells with the Conchocelis used in the all-artificial spores collecting cultivation experiment were less than half of that previously used in the semi-artificial spores collecting cultivation experiment in Dalian (6, 9). On the bases of the above experiments similar cultivation experiments on the Fujian species Porphyra haitanensis were conducted in Fujian Province. The results were very encouraging, and a new type of cultivation raft was employed which floated on sea surface at high tide and set down on the ground at low tide. This was the prototype of the semi-floating cultivation raft, now widely adopted in China's Porphyra cultivation in the intertidal regions.

In 1964 the Ministry of Fisheries sponsored a cooperative project to promote *Porphyra* cultivation in Fujian Province, which is the "home" of the "rock-cleaning method" and has a long tradition of producing *zicai* for commercial purpose. Several institutes joined this project, including the Institute of Oceanology. The species selected was the local *Porphyra haitanensis* which is a fast-growing plant well known for its very much elongated thallus in the natural habitat. In 1959 we had collected a plant of this species, measuring 9.3 meters in length. Our experiences with the cultivation of the northern *P. yezoensis* involving the indoor culture of the *Conchocelis*, the collection of the conchospores on the cultivation nets, and the cultivation in the intertidal regions proved to be of great help, although the species involved has ecological characteristics quite different from that of the northern species. These experiences together with the semi-floating raft method of cultivation, originated in the Fujian Province, had made cultivation of *Porphyra haitanensis* a success (9).

In the meantime, cultivation of *P. yezoensis* in North China had also made some progress. A small-scale experiment applying the semi-floating raft method of cultivation, originally devised for *P. haitanensis*, has yielded good results to the northern species, showing the superiority of this method for intertidal cultivation. The experimental production reached a high record of 1090 g/m^2, 2.5 times more than the production of the former fixed-pillar method. Since 1970 the semi-floating cultivation method has been widely adopted by mariculturists for the intertidal cultivation of *P. yezoensis* in north China (6, 9).

In the year 1979-1980 the total area of zicai cultivation in China was 3,400 hectares and the production 7,200 tons dry weight, the product being in the form of square sheets, round sheets or loose thalli. The southern species Porphyra haitanensis accounted for about 90%, the northern species P. yezoensis for about 10% of the total production. In Guangdong Province experimental production employed P. guangdongensis and in Taiwan Province P. sp. The production of both is still very small (9).

On the basis of results obtained from experimental studies of the biology of the Porphyra, the following methods of commercial cultivation have been devised:

1) Method of mass culture of the Conchocelis. Phycocultural methods to be widely accepted by the Chinese mariculturists have to be economical and easy to manipulate. A number of experiments have been conducted on the growth and development of the Conchocelis to find out the best conditions for the growth of the vegetative stage and for the production of the conchospores. On the basis of the results obtained, methods of mass culture were devised. In the following the method for the culture of the Conchocelis of Porphyra yezoensis, now popularily accepted by the mariculturists, will be discribed and comments on that of P. haitanensis made.

   For cultivating of the Conchocclis phase of P. yezoensis in Qingdao, the carpospores are collected in early and middle May and are seeded on shells of the mollusc Meretrix meretrix at a density of 200-300 spores per cm^2. The culture takes place in large, elongated, shallow concrete tanks containing seawater, 20-30 cm in depth. The seawater used should be previously subjected to sedimentation in the dark for at least three days, and fertilized with nitrogenous and phosphate nutrients to contain 0.1 mg-atom N (KNO_3) and 0.01 mg-atom P (KH_2PO_4). Water temperature is not controlled and is subject to fluctuations with the changes in air temperature. Light intensity is partially controlled by shading cloths under the glass roof. Plane-type culture has been adopted, and the shells with the Conchocelis are arranged orderly one by one with the concave surface (seeded with conchospores) facing upwards. The shells are cleaned with soft brushes once every two weeks. When the pH of the seawater rises to about 8.5, the water should be changed. The water should be gently stirred a few times a day (5, 6, 9).

In order to produce the maximal number of conchospores per unit area, measures have been taken to control the light intensity and light period according to the requirement of the different stages of growth and development. Three stages are generally recognized. The first is from early May to early July, during which the water temperature gradually rises to 22-23°C. This is the period for the active vegetative growth of the Conchocelis. In this period the light intensity is controlled at about 3,000 lux in the middle of the day. The second stage is from early July to late August or early September, in which the water temperature gradually rises from 22-23°C to 26-27°C, sometimes in the middle of August, and then gradually decreases. This is the critical period for conchosporangial formation, and success or failure to obtain a good crop of the conchospores depends on correct handling of the light factor. The principle is to enhance the accumulation of the necessary reserves for sporangial formation but to delay the actual formation of the sporangia before the peak of the high temperature. After the peak and with the temperature dropping, formation of the sporangia should be encouraged. Therefore in early July or slightly later, the maximum light intensity should be lowered to about 1,500 lux. Then after middle August with the temperature falling, the maximum light intensity should be further reduced to 750 lux. In this way, formation of maximal quantity of sporangia will be ensured. The third stage is from late August or early September to late September when the water temperature gradually drops to about 23°C. This is the period of conchospore formation. The light intensity is held at 750 lux and the light period reduced to 8-10 hours per day by artificially darkening the culture rooms by the shading screens. Under such measures, conchospores appear in late September, and mass discharge of these spores generally takes place in early and middle October. By adopting the above measures, a shell of about 40 cm^2 may discharge as many as 20 million conchospores in the spore-collecting period. To provide conchospores for the cultivation nets of a hectare Porphyra farm 5-7 m^2 of shells will be needed (5, 6, 9).

In the mass culture the Conchocelis stage of Porphyra haitanensis in Fujian Province, carpospores collecting takes place in February to March and the seeded shells are tied in pairs in strings, each with 4-8 pairs, and hung from poles in tanks with culture solution about 25 cm in depth. The other details of the methods of culture are similar except that the water temperature is much higher (to 31°C) and the high water temperature period (higher than 25°C) is longer in the culture (9).

2) Method of seeding conchospores on cultivation nets. In the case of Porphyra yezoensis, before the seeding process, repeated examination of the culture are made to see the progress of the spore formation. When the number of spores reaches fifty thousand per shell and day, the culture is ready for the seeding process, and the cultivation nets, made of synthetic fibers, are placed in the tanks on the shells with the Conchocelis. The light intensity on the surface of the culture tanks should be regulated to 1,500-3,000 lux or more, since it has been shown experimentally that adherence and germination of the conchospores require

strong light. The conchospores are non-motile and thus agitation of the water in the tank is necessary, for which a jet pump is usually employed. It has been shown experimentally that the largest mass discharge of the conchospores takes place in teh early morning before 10 a. m. Therefore the seeding process is done at that time. When spore discharge reaches one hundred thousand per shell, the formal seeding can be started. When the daily discharge reaches one million spores per shell per day, the seeding process can be shortened; a few minutes are enough for a sufficient number of spores to attach to the cultivation net. A density of 3-30 spores per square mm fully satisfies the requirements of the seeding process, and the nets may be taken out. Usually for each tank of shells, spore-collecting can be conducted 1-4 times, sometimes as much as 6 times a day, each time for a few minutes to few hours depending on the number of the spores adhered (5, 6, 9).

In the case of *Porphyra haitanensis*, the shells with mature *Conchocelis* have to go through a stimulating process otherwise the mass discharge of the spores will not be obtained. Ordinary method for the stimulation is to put the shells into string bags, and hang them in the sea overnight, bringing, the usual practice is to spread spore "solution" to the nets on the sea, and the nets are densely constricted and hung at a bamboo frame floating on the surface of seawater. Early in the morning the shells with the stimulated *Conchocelis* are put in a container with seawater and the spores will soon be discharged forming spore solution (9).

3) Method of cultivation of leafy thalli in the sea. In the late fifties to early sixties, the fixed pillar type of cultivation, popular in Japan, was adopted for the experimental cultivation of *Porphyra yezoensis* in the intertidal region in North China. The production was merely 100-200

g/m^2, not promising enough to meet the requirement of the commercial cultivation in China. The low transparency of the seawater (1-3 m) and the wide tidal range (4-6 m) in North China undoubtedly accounted for the low production. Later, the semi-floating raft method of cultivation originated in Fujian Province, has been adopted and the production per square meter has reached 500 g or more. This method has the advantage of the rafts remaining on the sea surface even at high tide, thus giving the plant an opportunity to receive sufficient light for photosynthesis. In the Qingdao region the spore collecting takes place in early to middle October when the surface seawater temperature drops down to about 20°C. Cultivation of the Porphyra also takes place in the near shore region not exposed during the low tide. In that case, the floating raft method is employed. In places where the total nitrogen concentration of the seawater is less than 50 mg/m^3, fertilization with nitrogenous nutrients has to be employed. Small boats with pumps are generally used for spreading the fertilizers in solution. Floating raft cultivated Porphyra often has weed problem and devises are made to raise up to the nets above the sea surface for sunning which, if handled correctly, will kill the weeds but not the Porphyra. On the floating rafts, as a rule, the young sporelings are not as abundant and do not grow as well as those on the semi-floating rafts in the intertidal region. The Porphyra is an intertidal plant and needs some hours of exposure daily to effect better growth. Therefore, one of the accepted practices is to let the seeded rafts remain in the intertidal region for some time and after the conchospores have germinated and growths appeared, the rafts are moved to the sea and tied to the supports of the semi-floating rafts (5, 6, 9).

In the case of Porphyra haitanensis in the Fujian and Zhejiang provinces, the semi-floating raft method is generally adopted for its cultivation. Investigation showed that the favourable tidal level for cultivating this Porphyra is at the lower littoral region, quite different from its natural habitat, quite high in the tidal range. This is because in its natural habitat, it grows in very exposed places affected by strong wave action, resulting in ceaseless splashing of the higher littoral rocks by waves (9).

4. Other types of marine phycoculture in China

Commercial cultivation of the Laminaria and the Porphyra represents the two principle types of marine phycoculture in China. Besides, Undaria pinnatifida, Eucheuma gelatinae and Gracilaria verrucosa, called respectively qundaicai, qiongzhi and jiangli in China are also commercially cultivated (9).

Undaria is cultivated by the same raft method as the Laminaria and is often mix-planted with the latter on the same floating raft in Qingdao and Dalian. It has a shorter growing season, maturing much earlier than the Laminaria and is harvested in March, thus not interfering with the growth of the haidai which is harvested in June to July. The annual production amounts to only a few hundred tons (9).

Eucheuma gelatinae is a raw material for extracting carrageenan and has been commercially cultivated in the Qionghai and Wenchang districts of the Hainan Island since 1960. At first, Eucheuma was cultivated by sending divers to insert cuttings of the seaweed in sublittoral reefs. Since 1974 new cultivation methods have been devised. For the "seed", thalli are selected that are thick, sturdy and without epiphytes. The selected thalli are then divided from the bases of the branches, about 40 branch systems from each kg of the thalli. These "seeds" are tied to branches of dead corals by means of rubber rings and scattered on selected sea bottoms in a density of about 75,000 branch systems per hectare. Arrangement of the "seed" thalli in order on the sea bottom is made by divers. The same lot of the Eucheuma farm is harvested once a year. Because of the limited area of the coral reefs at Hainan, only about 200 hectares of the reefs are farmed and about 300 tons of the dry seaweed are produced annually (9).

Gracilaria verrucosa is an agarophyte and is produced naturally in large quantities in the southern provinces. In the late fifties to early sixties, commercial cultivation was carried out in Guangdong Province by inserting branches of the seaweed in the splits of bamboo sticks about 15 cm long, which were then planted in littoral farms. Later, cultivation of the seaweed was carried out by scattering juvenile Gracilaria of more than 5-6 cm in length, either gathered from the natural habitat or cultured from spores in littoral farms sheltered from strong wave action. In recent years the net cultivation method has been employed with some success. In one cultivation season of about three months, the production may reach two to three tons in dry weight. To date, however, the commercial cultivation is still on a small scale in Guangdong and the principal source of the Gracilaria as an industrial raw material comes from the natural habitat (9).

In Taiwan Province, cultivation of Gracilaria verrucosa was initiated in 1967. The industry has been expanding rapidly during the last years, and both monoculture of Gracilaria alone or polyculture with added milkfish, shrimp and crabs have been practice with success. In 1977, Gracilaria ponds occupied 223 hectares of land, producing 6,804 tons. The producing areas are located mostly in the Pingdong (Pingtung) and Tainan districts in south Taiwan. Gracilaria is uniformly planted in the bottom of the ponds using cuttings or the torn sections as the "seeds". They are tied to bamboo sticks or covered with old fishing nets to avoid drifting. About four to six thousand kg of fresh Gracilaria are planted in a monoculture pond of one hectare and the production averages 13 tons per hectare. While monoculture Gracilaria pond produces more seaweed per unit area, the polyculture pond with milkfish, shrimps and crabs in addition to Gracilaria yields better profit (2, 9).

A short review of the present conditions of the marine phycoculture in China shows that China's young phycoculture industry, although still a very small one in comparison with the other established industries, has firmly established itself in China and is still full of vitality. The Laminaria cultivation industry appears to have become mature and approached its limit of development. But the Porphyra cultivation industry, and expecially the Gracilaria cultivation industry and the Eucheuma cultivation industry are still very young and in their early stage of development.

References

1. Cheng, T. H.: Econ. Bot. 23(3), 215-236 (1969).
2. Chueh, C., Chen, C.: Seaweed economics in Taiwan (Manuscript) (1978).
3. Fang, T. C., Wu, C. Y. et al.: Scientia Sin. 12(7), 1011-1018 (1963).
4. Ren, G. Z. In: Tseng, C. K., Wu, C. Y.: Manual of Cultivation of Haidai (Laminaria japonica), pp. 72-95, Science Press, Beijing 1962. (in Chinese)
5. Section of Experimental Phyco-ecology, Institute of Oceanology, Academia Sinica: Scientia Sin. 19(2), 253-259 (1976).
6. Section ef Experimental Phyco-ecology and Section of Systematic Phycology, Institute of Oceanology, Academia Sinica: Cultivation of Tiaoban-zicai (Porphyra yezoensis Ueda), Science Press, Beijing 1978. (in Chinese)
7. Section of Seaweed Genetics and Breeding, Institute of Oceanology, Academia Sinica and Qingdao Institute of Marine Fisheries: Scientia Sin. 19(2), 243-252 (1976).

8. Tseng, C. K. In: Tseng, C. K., Wu, C. Y.: Manual of Cultivation of Haidai (Laminaria japonica), pp. 99-112, Science Press, Beijing 1962. (in Chinese)

9. Tseng, C. K. In: Lobban, C., Wynne, M. J.: The Biology of the Seaweeds. (in press)

10. Tseng, C. K., Chang, C. F.: Shengwuxue Tongbao 3, 29-33 (1956). (in Chinese)

11. Tseng, C. K., Chang, T. J.: Act. Bot. Sin. 3(3), 287-302 (1954)*

12. Tseng, C. K., Chang, T. J.: Act. Bot. Sin. 4(1), 27-46 (1955a)*

13. Tseng, C. K., Chang, T. J.: Scientia Sin. 4(3), 375-398 (1955b)

14. Tseng, C. K., Chang, T. J.: Act. Bot. Sin. 5(1), 33-48 (1956)*

15. Tseng, C. K., Chang, T. J. In: Collected Papers of the 5th Meeting of the Western Pacific Fisheries Research Commission, pp. 34-38, Science Press, Beijing 1962. (in Chinese)

16. Tseng, C. K., Chang, T. J., Li, J. J.: Kexue Tongbao 1959(5), 169-171 (1959). (in Chinese)

17. Tseng, C. K., Chang, T. J., Zhao, R. Y.: Kexue Tongbao (1959(5), 171 (1959). (in Chinese)

18. Tseng, C. K., Liu, T. G. et al.: Stud. Mar. Sin. 3, 102-118 (1953).

19. Tseng, C. K., Sun, K. Y., Wu, C. Y.: Acta Bot. Sin. 4(3), 255-264 (1955a)*.

20. Tseng, C. K., Sun, K. Y., Wu, C. Y.: Acta Bot. Sin. 4(4), 375-392 (1955b)*.

21. Tseng, C. K., Wu, C. Y.: In: Collected Papers of the 9th Meeting of the Western Pacific Fisheries Research Commission, pp. 37-40, Science Press, Beijing 1966. (in Chinese)

22. Tseng, C. K., Wu, C. Y., Sun, K. Y.: Acta Bot. Sin. 6(2), 103-130 (1957).

23. Tseng, C. K., Wu, C. Y. et al. In: Collected Papers of the 5th Meeting of the Western Pacific Fisheries Research Commission, pp. 21-32, Science Press, Beijing 1962 (in Chinese)

24. Wu, C. Y. In: Tseng, C. K., Wu, C. Y.: Manual of Cultivation of Haidai (Laminaria japonica), pp. 34-71, Science Press, Beijing 1962. (in Chinese)

25. Wu, C. Y. In: Tseng, C. K., Wu, C. Y.: Manual of Cultivation of Haidai (Laminaria japonica), pp. 275-287, Science Press, Beijing 1962. (in Chinese)

26. Wu, C. Y., Tseng, C. K. et al.: Kexue Tongbao 1959(24), 829-830 (1959).

27. Wu, C. Y., Kao, N. S. et al.: Oceanol. et Limnol. Sin. 10(3), 238-250 (1979)*.

28. Zhang, J. C. In: Tseng, C. K., Wu, C. Y.: Manual of Cultivation of Haidai (*Laminaria japonica*), pp. 166-186, Science Press, Beijing 1962. (in Chinese)

---

* Papers in Chinese with English abstracts or summaries.

Part II

O R D I N A R Y   L E C T U R E S

1. Distribution, Morphology, Taxonomy

LIFE HISTORY OF GYMNOGONGRUS GRIFFITHSIAE (TURNER) MARTIUS (PHYLLOPHORACEAE, GIGARTINALES) *

Marilza Cordeiro-Marino
Secção de Ficologia, Instituto de Botânica, Caixa Postal, 4005, 01000- São Paulo, SP, BRASIL.

Arturo Candia Poza
Departamento de Biología y Tecnologia del Mar, Pontifícia Universidade Catolica, Casilla 127, Talcahuano, CHILE

INTRODUCTION

Many members of Phyllophoraceae differ in life history patterns in Gigartinales. Life history is an important information in such group of economic exploitation for carrageenan extraction. In *Gymnogongrus*, problems in life history are pointed by Dixon (1) who grouped species producing gametangia and an internal carposporophyte, and species forming an external tetrasporophyte from gametangia.

Comments on culture studies on *Gymnogongrus* (2) emphasized the existence of two types of life history. In the first, there is alternation of two heteromorphic generations (an erect gametophyte alternating with a crustose tetrasporophyte); in the second, the tetrasporangial phase develops onto a gametophyte. Tetrasporangial nemathecium of *G. griffithsiae* in field plants was called tetrasporoblast (3), although previous named otherwise (4,5).

Tetrasporoblast (3) was renamed carpostetraspororangial outgrowths (6), because plants showing tetrasporangia, immersed cystocarps and normal carposporangia, and with spermatangia were not known. This gap is being filled in the present contributions.

In Brazil, *G. griffithsiae* is the only member of Phyllophoraceae, found from Rio de Janeiro to Rio Grande do Sul (7).

---

* Research partly supported by CNPq Grant nr. 2222.0547/77.

Xth International Seaweed Symposium
Printed in Germany

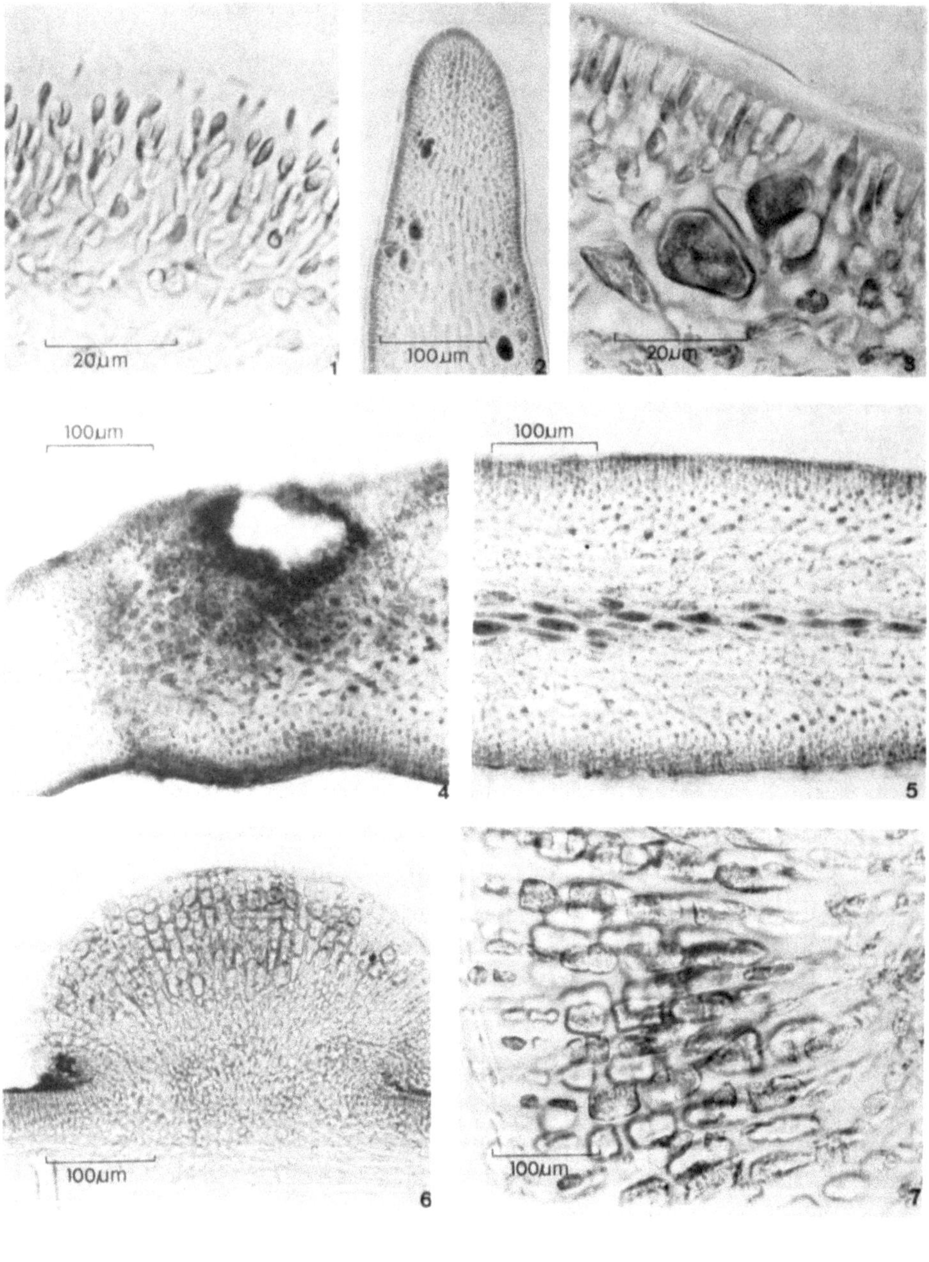

Fig. 1-7. Field plants, structural details. 1. Spermatangia. 2. Apex with carpogonial branches. 3. Carpogonial branch. 4-5. Carposporophyte, gonimoblast filaments and enlarged medullary cells. 6. Nemathecium, rows of tetrasporangial cells. 7. Ripe tetrasporangia.

MATERIAL AND METHODS

Material was collected in States of São Paulo, from São Sebastião, Itanhaém and Ilha do Cardoso, and in Santa Catarina, from Itapema, Porto Belo and Garopaba. Two series of unialgal cultures, obtained in Provasoli's enriched seawater medium (West, pers. comm.), were kept in growth chambers. One, from vegetative apices (collection from Praia Cibratel, Itanhaém, Jan. 9, 1978) and kept at 10°C, $16:\overline{8}$; 20°C, $12:\overline{12}$; 1800-2000 lux in Instituto de Biología, Universidad de Concepción, Chile. The other from tetraspores (collection from Praia do Bairro de São Francisco, São Sebastião, on Nov. 14, 1978) and kept at 20°C, $10:\overline{14}$, 23°C, $14:\overline{10}$; 1500-2000 lux, in Instituto de Botânica de São Paulo, Brazil.

Studies were made with both hand- and freezing microtome-sections from either alive or 4% formalin seawater fixed material.

OBSERVATIONS

Field plants are monoicious and grow intertidally on rocks, sometimes at the sand limit. The tetrasporophyte develops onto the gametophyte which forms nemathecia at the thallus surface, sometimes entirely surrounded by them. Spermatangia are grouped at the apex of the thallus, in the outermost layer of the cortex; each spermatangial mother cell yields one or two colorless spermatia (Fig. 1). Carpogonial branches, also grouped at the apical portion (Fig. 2), are formed by a large supporting (auxiliary) cell bearing a three-celled carpogonial branch (Fig. 3). Both supporting cell and the first carpogonial branch cell often carry one or two either simple or branched sterile filaments. Fusion of the carpogonium and auxiliary cells has never been observed. Carposporophyte develops from the supporting cell of a single procarps, but many supporting cells may participate in the formation of a carposporophyte. Gonimoblast filaments develop from the supporting cell and usually grow inward through the medulla. Medullary cells, both near or far the supporting cell are gradually enlarged and show dense contents. Pit connections appear between enlarged medullary cells and the

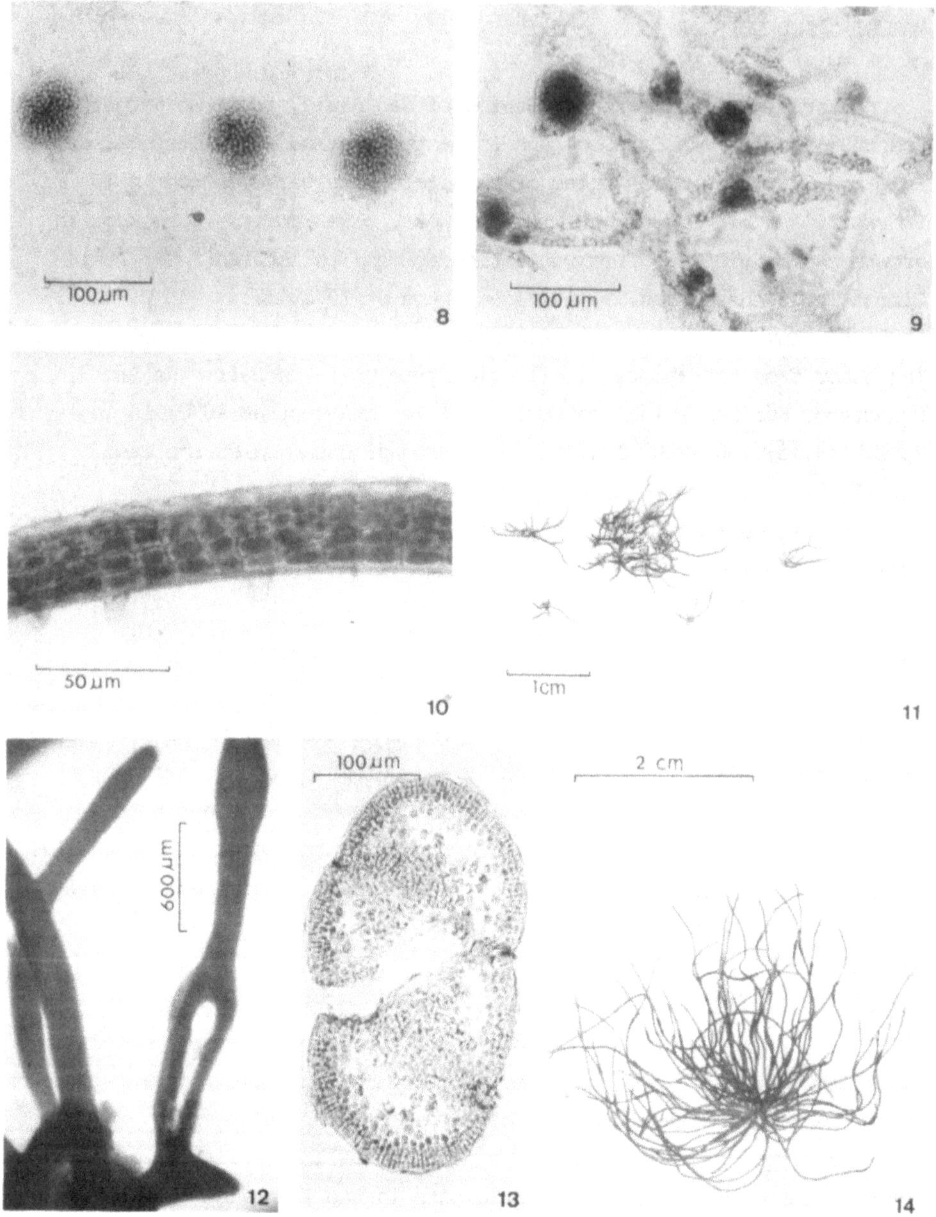

Fig. 8-14. Cultured plants. 8. Mediate discal type. 9. Filamentous mediate type. 10. Polystromatic crust, cross section. 11. Two months old thalli. 12. Four months old thalli, inflate portion, peripheral cracks. 13 Carposporophyte. 14. Eleven months old thalli.

gonimoblast filaments. Carposporangia are formed from both gonimoblast filaments and medullary cells (Fig. 4-5). Carpospores germinate inside the gametophyte wall and reach the thallus surface by gonimoblast filament growth or by irregular and localized crack on the thallus surface. Nemathecium shows a monostromatic hypothallus and a polystromatic perithallus wich erect filaments and secondary pit connections between adjacent cells. Perithallic filaments, either simple or dichotomous, show intercalary cells, each giving rise to a tetrasporangial mother cell (Fig. 6). Tetrasporangia are in rows, with 4-5 successive cells of a single filaments ending by 1-2 sterile terminal cells (Fig. 7). Ripe tetrasporangia are globular or ellipsoidal, 13.8x9.0µm, and cruciately divided.

In culture, 2-4 days after inoculation the tetraspores are divided into two cells and follow either one of two developmental patters (8): mediate discal type (Fig. 8) and filamentous mediate type (Fig. 9), both giving rise to polystromatic crusts, 84-124µm in diameter, composed by a basal layer of the retangular cells; these produce short rizoids in the lower part and simple or ramified filaments tightly compacted in the upper part (Fig. 10). One month old crusts form multiaxial upright thalli near the center. Several thalli are produced from a single crust. Two months old thalli are terete, some dichotomous and 4mm in lenght, while others remain undivided (Fig. 11). Three months old thalli produce procarps and spermatia. Four months old thalli show inflate portions and sometimes peripheral cracks (Fig. 12). The inflate portions are carposporophytes (Fig. 13). After 6-7 months cultured plants reach 2cm in lenght and produce nemathecia. Eleven months old thalli exhibit numerous branches resembling plants in nature, some with irregular ramifications, others mainly dichotomously branched (Fig. 14). Growth was well succeded under all culture conditions but much better at 23°C, 14:10. Cultured plants resemble field specimens entirely.

DISCUSSION

Tetraspore development in G. griffithsiae in nature remains unknow. In culture, the patters is similar to what has been shown for G. furcellatus, Gigartina teedii (Roth) Lamouroux, and G. chamissoi (C.

Agardh) J. Agardh (Candia unpublished). Spermatia formation and position in the thallus are the same for all other Gigartinaceae members. These features and others already pointed out for Phyllophoraceae and Gigartinaceae (9) reinforce previous suggestions of two closely related families with a common ancestral (10, 11). Chemin (4) and Gregory (5) described the nemathecia of G. griffithsiae as composed by intra and extramatrical tissues developed from a large auxiliary cell. Schotter (3) called the structure tetrasporoblastic and stated that the early stages of nemathecial formation are similar to the carposporangial phase; only later the major part of the gonimoblast filaments grow outwards the thallus, instead of inward, to form a wart on the thallus surface. Magne (12) inferred that fertilized carpogonium develops into a reduced, parasite tetrasporophyte in Phyllophora and Gymnogongrus of Phyllophoraceae. He also commented on tetrasporoblastic Phyllophoraceae with carposporophyte becoming very much reduced, represented only by the zigote that develops into a carposporophyte including the carpospore which, in its turn, grows as a reduced tetrasporophyte. This apparently digenetic group was considered in the reality trigenetic (12).

In the present study, field and culture plants of G. griffithsiae are monoicious and show carposporophyte very similar of Ahnfeltia concinna J. Agardh immersed in the thallus (13). The tetrasporophyte develops onto the gametophyte, with structure similar to the nemathecia of G. crenulatus (14). In G. griffithsiae the nemathecium shows a structure very similar to the independent, crustose tetrasporophyte of G. furcellatus (Candia, unpublished). Thefore, from the structural details the life history of G. griffithsiae is trigenetic, represented by a monoicious, independent, macroscopic gametophyte that carries an internal carposporophyte immersed in the thallus plus an external tetrasporophyte forming a wart on the thallus surface.

SUMMARY

Studies of field and culture plants of G. griffithsiae shows details of spermatium formation, carpogonial branch, carposporophyte, and tetrasporophyte development. A trigenetic life history is proposed, based on a monoicious, independent macroscopic gametophyte that carries an internal carposporophyte and a external reduced tetrasporophyte (nemathecium). Structural data support the relationships between Phyllophoraceae and Gigartinaceae.

REFERENCES

1. Dixon, P.S.: Biology of the Rhodophyta. Otto Koeltz Science Publ., Koenigstein 1977 (reprint).
2. Masuda, M., De Cew, T.C., West, J.A.: Jap. J. Phycol., 27, 63-73 (1979).
3. Schotter, G.: Bull. Inst. Oceanogr., Monaco, 64 (1383), 1-99 (1968).
4. Chemin, E.: Soc. bot. Fr., 80 (-10), 755-770 (1933).
5. Gregory, B.D.: Journ. Linn. Soc. London, 49 (331), 531-551 (1934).
6. Dixon, P.S., Irvine, L.M.: Seaweeds of the British Isles Part. 1. Introduction, Nemaliales, Gigartinales. British Museum (Nat. Hist.), Londres 1977.
7. Oliveira Filho, E.C.: Algas marinhas bentônicas do Brasil. Depto Bot. USP, São Paulo 1977.
8. Inoh, S.: Saiso no hassei (Development of marine algae spore), Tokyo 1947.
9. Kim, D.H.: Nova Hedwigia, 27, 1-146 (1976).
10. Kylin, H.: Lunds Univ. Arsskr. N.F. Avd 2, 28 (8), 1-88 (1932).
11. De Cew, T.C., West, J.A.: J. Phycol. 13, 16 (1977 suppl.).
12. Magne, F.: Soc. bot. Fr. Mémoires, 247-268 (1972) .
13. Magruder, W.H.: Phycologia, 16 (2), 197-203 (19 77).
14. Ardré, F.: Rev. algol. nov. sér., 3 (2), 151-176 (1978).

TAXONOMIC AND BIOLOGICAL STUDIES ON SPECIES OF IRIDAEA BORY IN CENTRAL CHILE

H. Etcheverry, G. Collantes and V. Ríos

Department of Biology, Faculty of Mathematics and Natural Sciences, University of Chile, Valparaíso, Chile

Introduction

The intention of this study is to simplify the systematics of the genus Iridaea Bory and of some species in Central Chile, in order to clarify their status as species both their macro and microstructure were determined together with ecological observations of Iridaea boryana, I. laminarioides, I. caespitipes, I. crispata, I. ciliata and I. undulosa, necessary for the purpose of this investigation. The observations were carried out during two years (1978-1980) along the rocky coast of Central Chile, especially Montemar (Valparaíso). Over 1.000 herborized and fresh specimens obtained in localities such as Valparaíso, Gulf of Concepción, Bahía Tic-Toc (Chiloé), various zones of Magallanes and in the Chilean Antarctic were reexamined and studied.

List of Stations

Valparaíso:	Horcón	32°30'	Lat.	S.
	Ventana	32°47'	"	"
	Montemar	32°57'	"	"
Pichilemu(Pta. Lobos)		34°23'	"	"
Talcahuano:	San Vicente	36°34'	"	"
	Tumbes	36°33'	"	"
Golfo Concepción:	Dichato	36°32'	"	"
	Desembocadura Bio-Bio	36°49'	"	"
Ancud:	Bahía Ancud	41°51'	"	"
Chiloé:	Bahía Tic-Toc	43°40'	"	"

Xth International Seaweed Symposium
Printed in Germany

Punta Arenas:	Agua Fresca	53°10' Lat. S.
	Fuerte Bulnes	53°40' " "
Bahía Orange:		55°31' " "
Antártica Chilena:	Isla Decepción	63°04' " "
	I. Greenwish	63° " "

Lastly, unialgal cultures of *Iridaea boryana* and *I. laminarioides* were made for studying the discoid phase up to the formation of small young fronds.

We disagree with Kim (1976) and other authors who include *Iridaea* species in the *Gigartina* genus.

Observations and Results

The *Iridaea* genus can be characterized according to the following description. Membranous thallus with one or more erect fronds growing from adhesive discoid holdfast. The frond are multiaxial cladoms, with short cylindrical stipe, broad, single and entire or deeply clefted with a sharp apex and rounded base. The histological structure presents a filamentous medulla and cortex composed of laterally adjoined anticlinal filaments forming the epidermis in the external layer. Heterothallic Tetrasporic plant with embedded sori and tetraspores cruciately divided. The cystocarpic plants with globose cystocarps deeply embedded and distributed along the surface of the frond with or without pericarp. A three celled carpogonial branche. The supporting cell functioning as the auxiliary cell and producing gonimoblastic filaments. Fig. 1-5.

Habitat: Intertidal and subtidal in rocky coast. *I. laminarioides* occurs in supralitoral zones in open coasts, especially in Montemar.

The following elements are essential for determining the species. Morphological, size and shape of frond. Form of both, the stipe and the apophysis, presence of papillae on the surface of the frond and lateral emergences. Histological, number and dispositions of cells of anticlinal filaments.

The form and distribution of the cystocarp and tetrasporangial Sori- Ecological, nature of the substratum and vertical distribution.

Other features which some authors consider important for defining species, such as location of the cystocarps and tetraspores in the frond (fairly close to the surface or external), are not considered, since we think that the position is the result of the growth of algae. Likewise the presence of absorbent filaments among the carpospores. We share Abbott's opinion who thinks that such characteristics can be only carefully used.

The artificial culture of I. boryana, was obtained from cystocarpic and tetrasporic plants, following the same technique employed for Porphyra columbina Mont. (Etcheverry et al. 1977). The fixation of the reproductive elements occurred between the first and third day. The growth and diameter of the carpospores was carefully controlled. After suffering several mitosis the carpospores produced pluricelular crustose discus 50-70 μ wide. The process continued with the formation of straight papillae, when the discus stopped growing. Small fronds with a multiaxial organization corresponding to the tetrasporic plants evolved from the papillae. The cultures are six months old and the formation of tetraspores was not observed. Fig. 6-8.

As conclusion, the type cycle of Iridaea laminarioides Bory is trigenetic.

The electron microscopic observations were carried out according to the same technique employed by Etcheverry for the study of "The Conchocelis Phase of Porphyra columbina (1979). Unfortunately due to the nature of the discus the micrographs do not give any further information but the general features of the cellular constituents corresponds to the Rhodophyta.

We shall examine the species, particularly, groups of species: or complex.

1. Iridaea boryana (Set. et Gard.) Skottsb., I. laminarioides Bory and I. caespitipes (Set. et Gard.) Skottsb., which share the same fundamental characteristics, i.e. entire frond, narrowed at the end, cuneate, shaped or oval lanceolate with smooth edges and apophysis canaliculate. I. caespitipes numerous fronds sprout from a common base. The frond thickness is 400-500 μ with loose medulla and a cortex with rather large anticlinal filaments with an average of 8-12 cells each; tetrasporic sori are placed in the centre of the frond or they are close to the surface. The cystocarps which are in the centre of the cortex show pericarpium proprium.

   I. boryana was stablished on the base of young specimens; the second I. laminarioides on adult specimens. I. caespitipes differs from both. I. boryana and I. laminarioides since it has a stoloniferous holdfast. Numerous fronds sprout from the base just as I. boryana which may well be an ecological character since this species grows in brealing waters and the jointed frond allow a better resistance to the waves.

   According to the priority assigned to the creation of these species. I. laminarioides Bory, Coquille p. 105 pl. 11 f. 1 A, B, C would became the type and I. boryana and I. caespitipes, would remain as synonyms.

2. I. crispata, I. ciliata and I. undulosa constitute another complex, as Dr. Levring says (1960) "It is impossible to separate these three species", all of them have been included in the Gigartiridaea Section for their fimbriated apophysis, cuneated to reniform frond base, single emergences (I. undulosa), ramified emergences (I. ciliata); on the other hand I. crispata has a cordate base with emergences. These characteristics prove to be inacurate since, transitional forms are to be found when examining a large number of specimens from the same locality. The three species occur from Valparaíso to Magallanes and in the Falkland

Islands (I. undulosa). The prevailing characteristic of the adult plant is the presence of simple papillae which eventually turn into compound papillae. Fig. 9-11.

We propose priority to I. crispata Bory 1826, keeping I. ciliata, I. undulosa and I. augustinae as synonyms. I. ciliata should be used only for those species occurring in Magallanes and the Falkland Islands.

The frond in I. crispata is rather thin 400 µ thick with a loose medulla. The cortex is composed of anticlinal filaments, 4-9 cells long. The external filaments are oblong shaped, rather long, while the internal ones are globose shaped.

3. I. membranosa J. Ag. 1872, 1879 p. 10 is maintened as valid species. The specimens of I. membranosa were collected in Pichilemu (Pta. Lobos). No specimens have been found in Valparaíso, although this species is supposed to occurs in this zone. We think there must be an error of locality as it may have been confused with Grateloupia cutleriae. Dr. Levring (1960) describes specimens for Canal Chacao in Chiloé.

The largest collected specimens have a thin lanceolated, flat frond with proliferations in the edges, which sprout from a common base. It is a distinctive feature. The medulla has a loose reticule and a cortex with 5-6 rows of anticlinal globe-shaped filaments. Cystocarps fill the medulla.

Conclusions

Only the following species of Iridaea occur in Central Chile, according to the previous analysis and discussion of both results and observations.

1. Iridaea laminarioides Bory 1827, which includes I. boryana (Set. et Gard.) Skottsb. and I. caespitipes (Set. et Gard.) Skottsb., as

synonyms.

2. *I. crispata* Bory 1826, Synonyms: *I. ciliata* Kutz., *I. undulosa* Bory and *I. augustinae*.

3. *I. membranosa* J. Ag. 1872.

According to the observations of a species of *Petrocelis* carried out in Montemar and on the basis of cultures we think that it may well be that it is part of the life cycle of an *Iridaea* species.

The ultrastructure of *I. laminarioides* in young plants is similar to that observed by other authors (Dodge 1977, Hara 1972) namely the morphology of cells and chloroplast which is present, with the characteristics of the *Polysiphonia* type (Hara 1972) without pyrenoid.

The ultrastructure of the papillae which stem out from the growing discus reveal a multiaxial organization which is a characteristic of the genus Iridaea.

As a general conclusion we thing that a systematic approach should be biological and that the largest possible number of parameters should be employed in order to determine a species according to its importance and presence, thus it is rather riskfull to simplify or increase the number of species of a given genus on the base of inacurate characteristics.

References

1. Abbott, I.A.: Syesis. 4: 51-72, 25 figs. (1971).

2. Abbott, I.A.: Field Studies which Evaluate Criteria Used in Separating Species of Iridaea (Rhodophyta). In Abbott, I.A. and Kurogi, M. (Ed.). Contributions to the sistematics of benthic marine algae of the North Pacific, pp. 253-264. (1972).

3. Agardh, J.G.: Species genera et ordines algarum vol. 3 (1). Epicrisis systematies floridearum. Leipzig pp. I-VII 1-724. (1876).

4. Alveal, K.: Rev. Biol. Mar. 14 (3): 85-119. (1971).

5. Bory de Saint-Vincent, J.B.: Cryptogamie. In Duperry, L.I. (Ed.). Voyage autour du monde sur La Coquille, pendant........ 1822-1825. Botanique Paris 301 pp., 38 pls. (1828).

6. Bory de Saint-Vincent, J.B.: Iridaea Dict. Class. Hist. Nat. 9: 15-16 (1826).

7. Chen, L.C.M., Edelstein, T., Mc.Lachlan, J.: Phycologia 13 (4): 287-294, 14 figs., 1 table (1974).

8. Dion, p., Delépine, R.: Rev. Algol. N.S.T. XIV (4): 327-341. (1979).

9. Dodge, J.D.: The Fine Structure of Algal Cells. XII + 261 pp. (1973).

10. Etcheverry D. et G. Collantes S. Cultivo artificial del luche, Porphyra columbina (Montagne 1845) Rhodophyta, Bangiaceae. Rev. Biol. Mar. Dep. Oceanol. Univ. Chile 16 (2): 195-202 (1977).

11. Fralick, J.E., Cole, K.: Syesis 6: 271-272, 2 figs. (1973).

12. Edelstein, R., Chen, L.C.M., Mc. Lachlan, J.: Phycologia 13 (2): 99-107, 22 figs. (1974).

13. Hara, J.: Procced. VIIth Int. Seaw. Symp. pp. 153-158 (1972).

14. Kim, D.H.: Nova Hedwigia XXVII 1: 94, figs. 1-237 (1976).

15. Kylin, H.: Lunds Univ. Orskr. N.F. Avd. 2, 28 (8): 1-88 (1932).

16. Kylin, H.: Lunds XV + 673 pp., 458 figs. (1956).

17. Levring, T.: Lunds Univ. Arskr N.F. Avd. 2, 56 (10): 1-83 (1960).

18. Mikami, H.: Sc. Papers Inst. Alg. Res., Fac. Sci. Hokk. Univ. 5 (2): 181-285 (1965).

19. Montagne, C.: Algas. In C. Gay (Ed.). Historia física y política de Chile. Botánica 8 pp. 351-355; 334-340 (1852).

20. Norris, R.E., Kim, D.H.: Development of thalli in some Gigartinaceae. In Abbot et al. Contributions to the systematics of benthic Mar. Alg. of the North P.C. pp. 265-279 (1972).

21. Searles, K.B.: Univ. Calif. Publ. Bot. 43: 1-86, 33 figs., 12 pl. (1968).

22. Sheath, R.G., Hellebust, J.A., Sawa, T.: Phycologia 18 (1): 1-12 (1979).

23. Setchell, W.A., Gardner, N.L.: Nat. Acad. Sci. Proc. 22 (8): 469-473 (1936).

24. Setchell, W.A., Gardner, N.L.: Univ. Calif. Publ. Bot. 19 (6): 195-244, pls. 23-29 (1937).

25. Skottsberg, C.: K. Sv. Vet. Akad. Handl. ser. 3 (ie ser 5) 19 (4) 92 pp., 7 figs., 3 pls. (1941).

26. West, J.A. et al.: Journ. Phycol. 14: 416-426 (1978).

PLATE I.

Figure	1.	Habit of *Iridaea laminarioides* Bory.
"	2.	T.S. Sterile frond.
"	3.	T.S. Tetrasporic frond.
"	4.	T.S. Frond with cystocarps.
"	5.	Carpogonial branch.

PLATE II. Growth of crustose phase in *Iridaea laminarioides* Bory.

Figure	6.	Liberation of tetraspores.
"	7.	Formation of the discus.
"	8.	Discus after six months of growth.

PLATE III. Details of the frond base and of the papillae.

Figure	9.	*Iridaea undulosa* (Margins and apophysis).
"	10.	*Iridaea crispata* (Lobulate frond and superficial papillae.
"	11.	*Iridaea crispata* (Cuneate base and papillae).

Pl. I.

Fig. 1

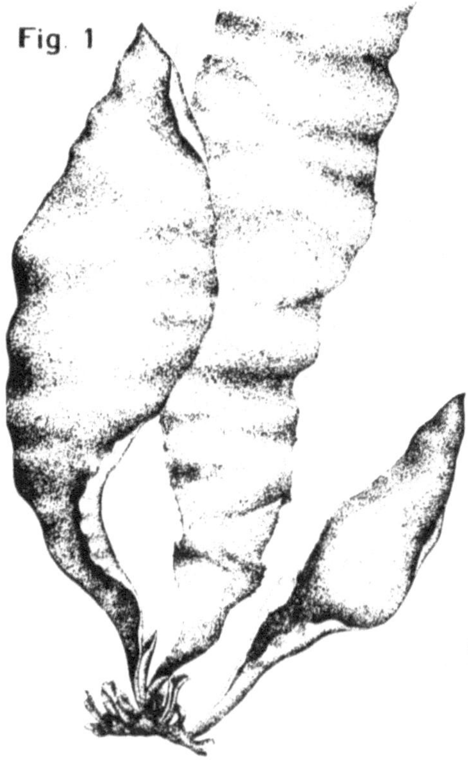

Fig. 2

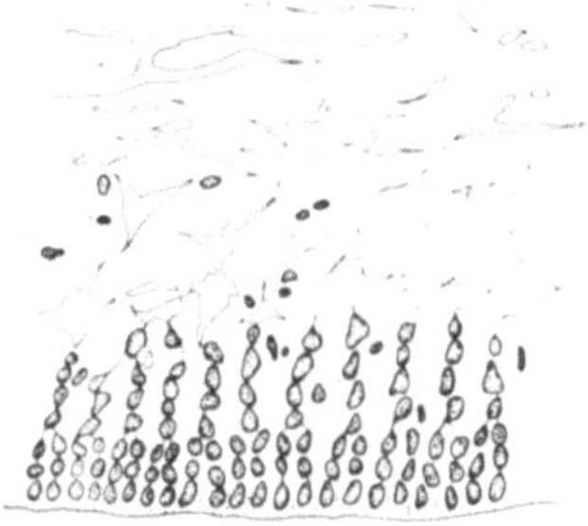

Fig. 3

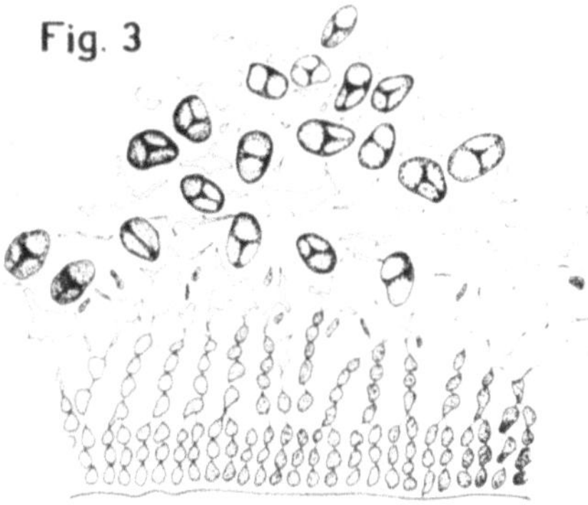

Fig. 4

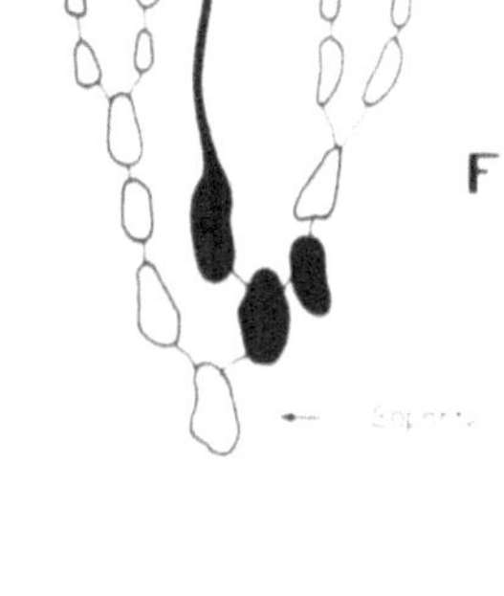

Fig. 5

Pl. II

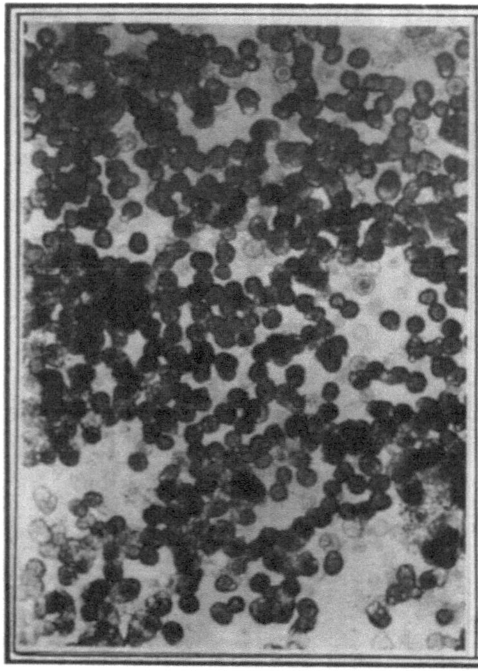

Fig. 6

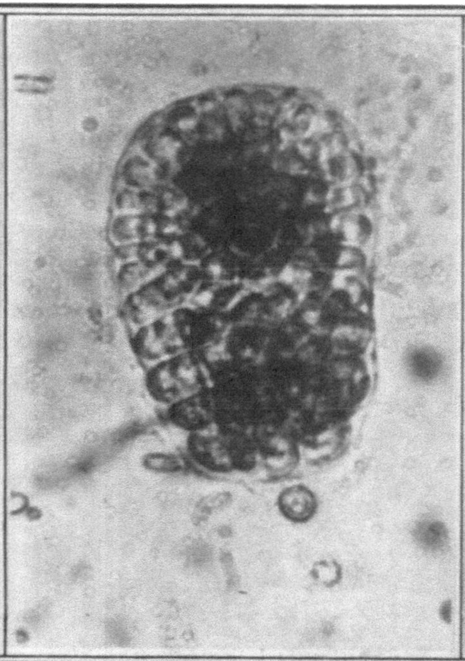

Fig. 7

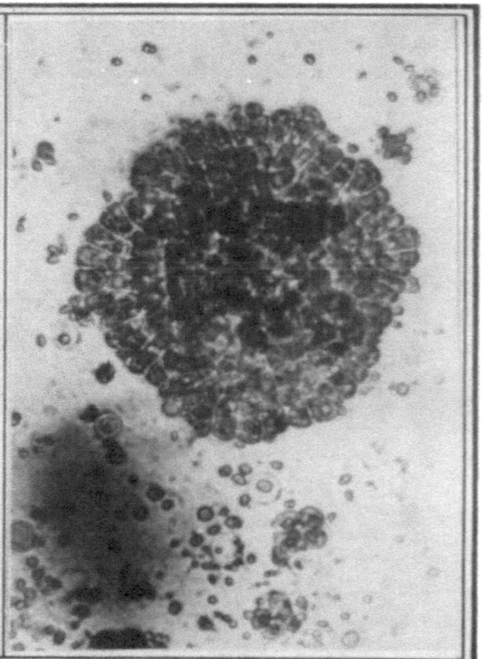

Fig. 8

Pl. III.

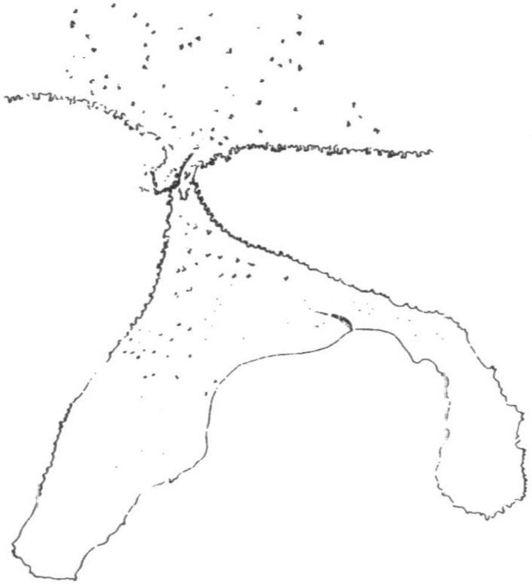

Fig. 9

Fig. 10

Fig. 11

MORPHOLOGICAL STUDIES ON AGARDHIELLA SCHMITZ (GIGARTINALES, RHODOPHYTA) FROM VENEZUELA

E.K. Ganesan

Departamento de Biología Marina, Instituto Oceanográfico, Universidad de Oriente, Cumaná, Venezuela

Introduction

Among the marine algal collections of eastern Venezuela, a red alga with many features of Agardhiella tenera (J.Ag.) Schmitz (sensu 12) attracted my attention, since it was shown recently (13, 14) that two taxa (Neoagardhiella baileyi Kützing) Wynne & Taylor and Solieria tenera (J.Ag.) Wynne & Taylor) with different thallus and cystocarpic structure have been passing under this biomial in the warm and cold waters of the western Atlantic ocean. It has also been pointed out recently (9, see also 3) that Neoagardhiella Wynne & Taylor is a superfluous name for Agardhiella Schmitz. Two other species of Neoagardhiella i.e., N. ramosissima (Harvey) Wynne & Taylor (2, 14) from Brazil and the Caribbean Sea and N. gaudichaudii (Montagne) Abbott (1) from California are currently known. A study of the Venezuelan material showed that it belonged to N. baileyi, which is currently called Agardhiella subulata (C. Ag.) Kraft & Wynne (3). For comparative purposes, material of A. subulata from Woods Hole and N. gaudichaudii from California was also studied.

Material and Methods

Venezuelan material of A. subulata was collected by me from Los Bordones beach in July-August, 1976, near the Institute of Oceanography, Cumaná, Venezuela from cast ashore material.

Xth International Seaweed Symposium
Printed in Germany

Prof. G.F. Papenfuss kindly provided material of A. subulata from Woods Hole collected in July 1960 and N. gaudichaudii was provided by Dr. I.A. Abbott (no. 6361) and by Dr. A. Nonomura collected by him from Mussel Pt., Pacific Grove, California on 14 April 1973. All the collections were preserved in 4-5 % seawater formalin. Methods as given in the papers (1, 5) were followed.

Results

Agardhiella subulata from Venezuela: Plants subtidal, dull pink to greenish, up to 20 cm high, base discoid, main axis very slender at the base (Figs. 1, 2), 0.5 mm broad and 2-4 (5) mm above, terete, delicate and soft throughout, branching irregular alternate, radial or dominantly unilateral, branches constricted at their bases and with acute apices, longitudinal medullary filaments few in number of broad (15-18 um) and narrow filaments (3-7 um) with interconnecting and rhizoidal filaments derived from the cortex; medulla occupying ca 50 % of thallus width in most parts, in basal parts longitudinal filaments scarcely evident, medulla filled with filaments running from cortex to cortex; tetrasporangia zonate (Fig. 3), 50 um long and 30 um broad; carpogonial branches generally 3-celled (Fig. 17), rarely 4 (Fig. 5) or 5-celled (Fig. 6); hypogynous cells commonly project towards one side (Figs. 5, 17) auxiliary cells with 3 large subsidiary cells (Fig. 16); cysto carps markedly projecting or half-immersed, 1-2 (rarely 3) per uni area; in sections with a prominent central sterile tissue (Fig. 4); carposporangia in discrete groups, each group separated by sterile filaments.

Agardhiella subulata from Woods Hole: In several anatomical and reproductive details, my observations on this material agreed well with the earlier studies (1, 4, 8, 14).

Neoagardhiella gaudichaudii from California: A transection 10

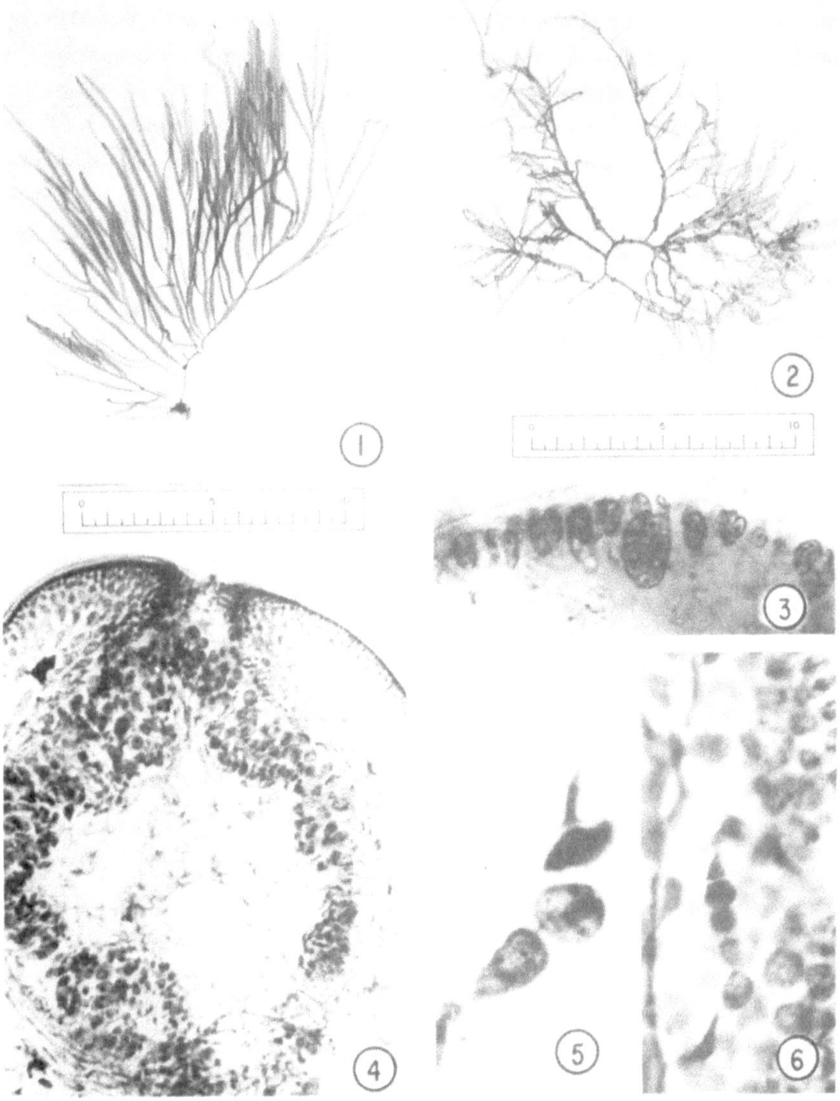

Figs. 1-6. Agardhiella subulata from Venezuela. Fig. 1. Habit of a tetrasporic plant. Fig. 2. Habit of a cystocarpic plant. Fig. 3. Outer cortex of a tetrasporic plant in section showing a zonate tetrasporangium X 388. Fig. 4. T.S. of a cystocarp. Note the central sterile tissue and carposporangia in discrete pockets X 140. Fig. 5. A four-celled carpogonial branch. Note the densely stained and protruding hypogynous cell X 688. Fig. 6. A five-celled carpogonial branch arising from one of the medullary filaments X 450.

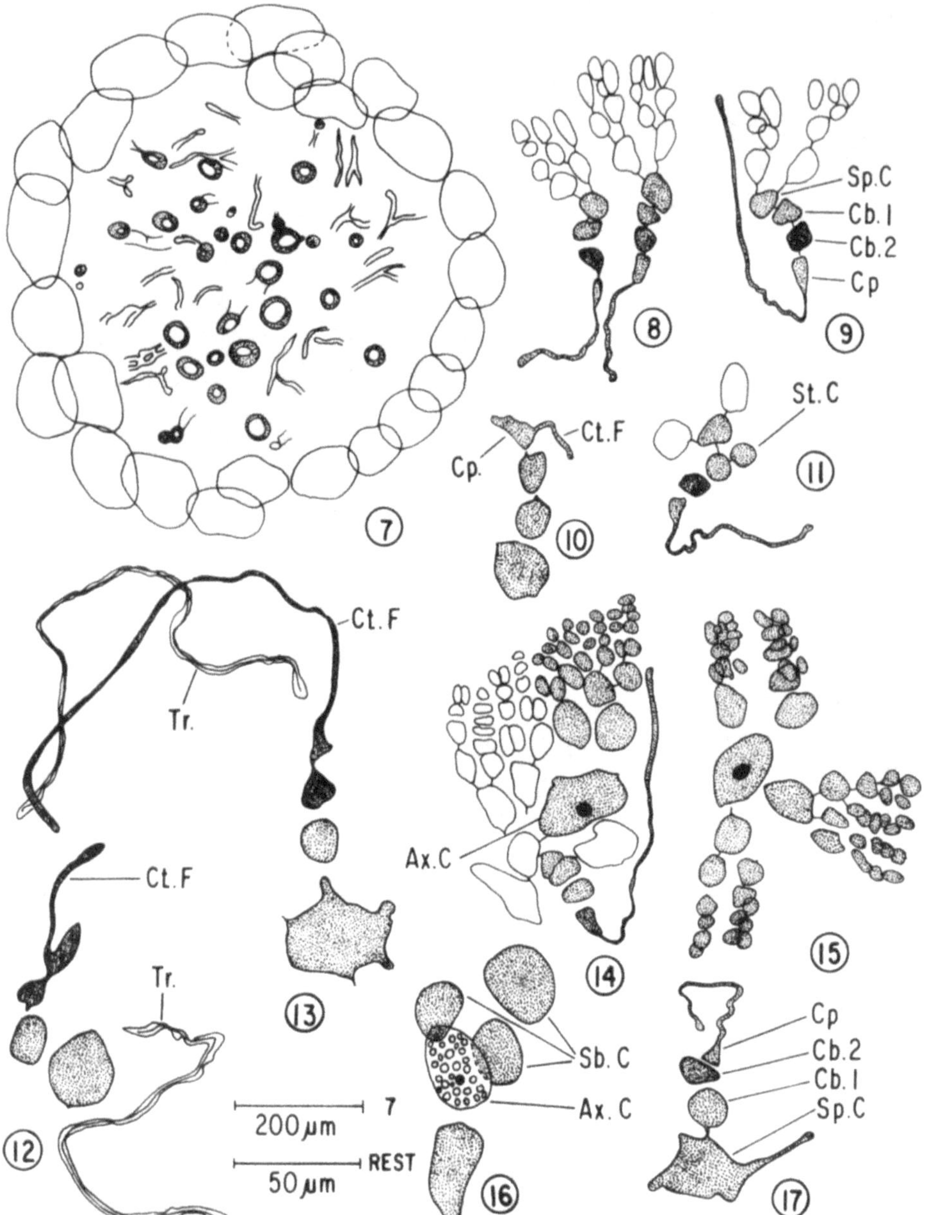

Figs. 7-15. N. gaudichaudii from California. Fig. 7. T.S. of thallus 10 mm below apex. Figs. 8,9, 11. Carpogonial branches. Figs. 10,12,13. Origin of connecting filaments. Figs. 14, 15. Aux. cells. (Fig. 7, Nonomura material, rest Abbott 6361. Figs. 16-17. Aux. cell & Carp. br. of A. subulata from Venezuela.

mm below apex of a branch is illustrated in Fig. 7. Compared to A. subulata, medullary filaments are far more numerous and thicker-walled. Carpognoial branches (Figs. 8, 9, 11) do not have protruding hypogynous cells and auxiliary cells bear 2-4 many celled laterals as the same size as adjacent cells (Fig. 14). Formation of connecting filaments are illustrated in figs 10, 12, 13. Cystocarp structure is the same as described in a recent study (1).

Discussion

Excepting for the smaller size of tetrasporangia, in other features, the Venezuelan plants agreed well with the earlier studies (1, 4, 8, 14) on A. subulata (=N.baileyi). A. subulata is reported here as a new addition to the Venezuelan marine algal flora. Although the Pacific coast N. gaudichaudii has long been considered as conspecific with the Atlantic A. subulata (see 14), a recent critical study (1) had shown for the first time certain significant differences in vegetative and cystocarp structure to recognize them as 2 species. My observations on the Caribbean A. subulata and the Californian N. gaudichaudii confirm this view point. Besides, the present study revealed certain differences in the structure of the carpogonial branches and auxiliary cells. In N. gaudichaudii hypogynous cells of the carpogonial branches do not project towards one side and the auxiliary cells lack 3 large subsidiary cells (see also 11 and compare with 4, 8, 14). Since recent authors (3, 9) are inclined to restore the generic name Agardhiella Schmitz rather than Neoagardhiella Wynne & Taylor, the following new combination is proposed for N. gaudichaudii.
Agardhiella gaudichaudii (Montagne) comb. nov.
Basionym. Gigartina gaudichaudii Montagne (1842, p. 255); Montagne (1846, p. 69, pl. 143, fig. 1).

Acknowledgements

Grateful thanks are expressed to the Consejo de Investigación, Universidad de Oriente, Cumaná, Venezuela for the financial assistance to participate in the Seaweed Symposium and to Prof. G.F. Papenfuss, Drs. I.A. Abbott and A. Nonomura for providing material from Woods Hole and California. The author is also thankful to Mr. A. Pernia and Mr. E. Hernandez for their help in the preparation of the photographs and drawings respectively.

References

1. Abbott, I.A.: J. Phycol. 14, 48-53 (1978).
2. Joly, A.B., Ferreira, M.M., Pinheiro-Vieirera, F., Yoneshigue-Braga, Y.: Arq. Est. Biol. Mar. Univ. Fed. Ceará 8, 177-181 (1968).
3. Kraft, G.T., Wynne, M.J.: Phycologia 18, 325-329 (1979).
4. Kylin, H.: Acta Univ. Lund. 24, 1-127 (1928).
5. Min-Thein, U., Womersley, H.B.S.: Aust. J. bot. 24, 1-166, (1976).
6. Montagne, J.F.C.: Ann. Sci. Nat. sér. 2, 18, 241-282, (1842).
7. Montagne, J.F.C.: Phyceae. In Gaudichaud. C. Voyage Autour du Monde, Bertrand, Paris. 1-112, (1846).
8. Osterhout, W.J.V.: Ann. Bot. 10, 403-427 (1896).
9. Papenfuss, G.F., Ganesan, E.K.: J. Phycol. (suppl.) 14, 34 (1978).
10. Smith, G.M.: Marine Algae of the Monterey Peninsula, California. Stanford Univ. Press, Stanford, California 1944
11. Smith, G.M.: Cryptogamic Botany, vol. 1. Algae & Fungi, Second edition, McGraw-Hill, New York 1955
12. Taylor, W.R.: Marine Algae of the Eastern Tropical and Subtropical Coasts of the Americas, Univ. Michigan Press, Ann Arbor, Michigan 1960
13. Taylor, W.R., Rhyne, C.F.: Smithsonian Contrib. Bot. 3, 1-16 (1970)
14. Wynne, M.J., Taylor, W.R.: Hydrobiologia. 43, 93-107 (1973).

ON DEEP-SEA RED ALGAL VEGETATIONS IN THE JAPAN SEA, AND A NEW NEMASTOMACEOUS ALGA, PREDAEOPSIS JAPONICA N. GEN. N. SP.

M. Kajimura
Marine Biological Station, Shimane University, Saigo, Oki-gun, 685 Japan

Introduction

Deep-sea marine algal vegetations of 181 species belonging to 105 genera in 47 families have been studied by the author with a steel dredge designed by himself at the depth ranging from 15m to 60m in the southwestern part of the Japan Sea in the past 14 years since 1966. Among them, 120 species belonging to 77 genera in 24 families are red algae including five new species and one new genus.

Results

Deep-sea red algal vegetations at the depth ranging from 15m to 60m: Scinaia moniliformis, Scinaia tokidai sp. nov., Scinaia sp., Scinaia sp., Scinaia sp., Actinotrichia fragilis, Galaxaura falcata, Ptilonia okadai, Dolisea fimbriata, Bonnemaisonia hamifera, Asparagopsis taxiformis, Gelidium pusillum, Gelidium pusillum f. foliaceum, Gelidium tenue, Desmia japonica, Desmia hornemanni, Peyssonnelia caulifera, Lithophyllum okamurai, Lithothamnium erubescens, Marginisporum aberrans, Amphiroa dilatata, Alatocladia modesta, Corallina pilulifera, Jania decussato-dichotoma, Carpopeltis angusta, Carpopeltis articulata, Callophyllis adhaerens, Callophyllis adnata, Callophyllis sp., Kallymenia perforata, Kallymenia sessilis, Kallymenia stipitata, Predaea tokidai sp. nov., Predaeopsis japonica

Xth International Seaweed Symposium
Printed in Germany

gen. et sp. nov., Halarachnion latissimum, Meristotheca papulosa, Meristotheca sp., Plocamium leptophyllum var. flexuosum, Plocamium oviforme, Plocamium telfairiae, Hypnea sp., Phacelocarpus japonicus, Sarcodia ceylanica, Trematocarpus pygmaeus, Gracilaria sublittoralis, Tylotus lichenoides, Stenogramme interrupta, Gigartina tenella, Gloioderma iyoensis, Gloioderma sp., Chrysymenia grandis, Cryptarachne polyglandulosa, Coelarthrum albertisii, Rhodymenia intricata, Rhodymenia sp., Lomentaria catenata, Champia bifida, Champia expansa, Champia japonica, Champia parvula, Binghamiella californica, Spermothamnion codicola, Callithamnion sp., Antithamnion nipponicum, Hollenbergia sp., Platythamnion yezoense, Platythamnion sp., Gymnothamnion okiense sp. nov., Acrothamnion pulchellum, Crouania attenuata, Delesseriopsis elegans, Wrangelia japonica, Griffithsia coacta, Griffithsia corallina, Griffithsia japonica, Plumariella minima sp. nov., Spyridia filamentosa, Ceramium boydenii, Ceramium cimbricum, Ceramium gracillimum, Campylaephora crassa, Centroceras clavulatum, Centroceras distichum, Microcladia dentata, Hemineura schmitziana, Branchioglossum ciliatum, Hypoglossum barbatum, Hypoglossum geminatum, Hypoglossum serratifolium, Erythroglossum minimum, Erythroglossum repens, Delesseria violacea, Schizoseris minima, Acrosorium flabellatum, Acrosorium polyneurum, Acrosorium uncinatum, Acrosorium yendoi, Martensia denticulata, Heterosiphonia japonica, Heterosiphonia pulchra, Dasyopsis plumosa, Dasya sp., Polysiphonia elongella, Polysiphonia tokidae, Chondria expansa, Chondria intricata, Chondria ryukyuensis, Chondria stolonifera, Laurencia brongniartii, Laurencia hamata, Laurencia intermedia, Laurencia pinnata, Pterosiphonia pennata, Symphyocladia marchantioides, Symphyocladia pennata, Herposiphonia fissidentoides, Herposiphonia subdisticha, Herposiphonia tenella, Euzoniella flaccida, Euzoniella ocellata, Leveillea jungermannioides.

Predaeopsis japonica gen. et sp. nov. Thalli mucous, foliaceous, erect, lobed, 1-2.5cm high, 3cm wide, pink to deep pink in color, thickened below, with a wedge-shaped short stipe at the base, margin smooth, surface with delicate wrinkles. Medulla consisting of slender, loosely disposed, infrequently branched filaments. Cortical layer consisting of filaments branched dichotomously 3-6 times, outer cortical cells linear-oblong. Rhizoidal filaments arising from inner cortical cells. Chromatophore irregularly band-shaped. Plant dioecious. Carpogonial branch 2-celled, formed on an inner cortical cell. Carpogonial cell elongate and bullet-shaped, hypogynous cell elongate. Auxiliary cell intercalary on a cortical filament. Many nutritive cells formed from the cells adjacent to the auxiliary cell. Several connecting filaments arising from the lower portion of the fertilized, enlarged, spherical carpogonium, branching in irregularly dichotomous manner, fusing with auxiliary cell at their terminals. Gonimoblast initiated at the terminal portion of the connecting filament, developing outward. Mature cystocarp with 2 subspherical to pear-shaped gonimolobes maturing in sequence, all gonimoblast cells becoming carposporangia, with no pericarp. Bisporangial mother cells formed terminally, 1-4 in number, on cortical filaments of the female gametophyte, divided unevenly by a horizontal septum. Germination of the bispore belonging to the diprotocellular type (1). Spermatangial mother cells borne on cortical filaments terminally or subterminally, 1-4 in number, each producing 1-3 spermatangia. This new species different from P. weldii comb. nov. (2) having 3-celled carpogonial branch and gonimoblast initiated from auxiliary cell terminally. This new genus distinguished from Predaea (3, 4, 5, 6, 7, 8, 9, 10) in having bisporangia on the female gametophyte.

References

1. Inoh, S.: Embryological Study of Marine Algae, Hokuryukan, Tokyo 1947
2. Kraft, G., Abbott, A.: J. Phycol. 7, 194-202 (1971).

1

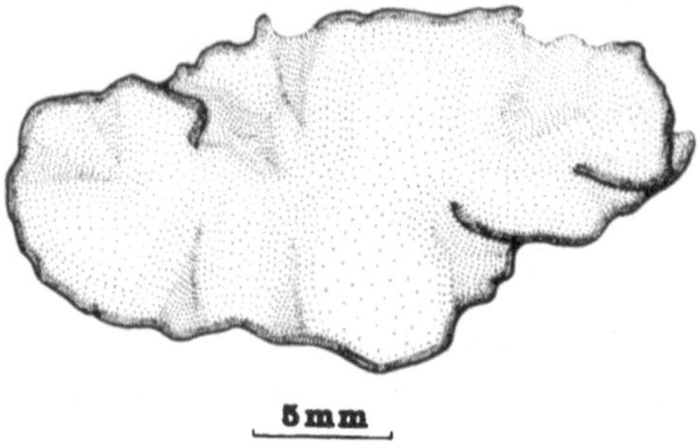

2

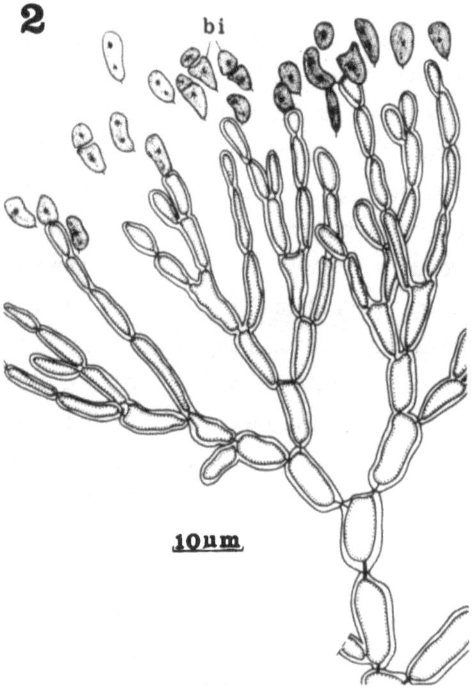

Figs. 1-2. *Predaeopsis japonica* gen. et sp. nov.
1. The holotype female specimen (OS 9836) collected by the author at the depth of 25m off Tsudo, the Oki Islands, Shimane Prefecture on April 18, 1979. 2. The bisporangia (bi) formed terminally on the cortical filaments of a female gametophyte.

3. Setchell, W. A., Gardner, N. L.: Proc. Calif. Acad. Sci. 4th Ser. 19, 109-215 (1930).
4. De Toni, G.: Noterelle de nomenclatura algologica. VII. Primo elenco di Floridee omonime, Brescia 1936
5. Feldmann, J.: Bull. Soc. Bot. Fr. 89, 104-113 (1942).
6. Ercegović, A.: Acta Adriatica 4, 27-39 (1949).
7. Børgesen, F.: Dansk Bot. Arkiv 14, 1-8 (1950).
8. Dawson, E. Y.: Pac. Nat. 1, 31-52 (1960).
9. Kraft, G. T., John, D. M.: Br. phycol. J. 11, 331-344 (1976).
10. Yoshida, T.: Jap. J. Phycol. 28, 69-74 (1980).

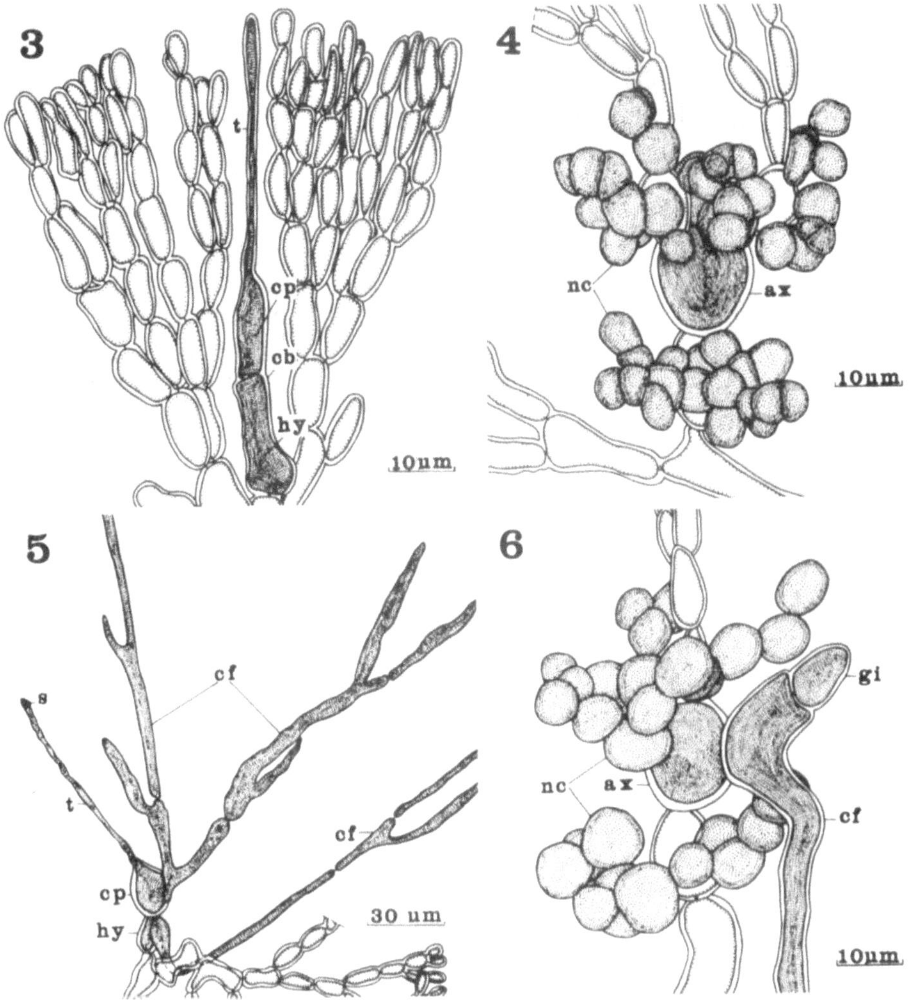

Figs. 3-6. Predaeopsis japonica gen. et sp. nov.
3. A carpogonial branch (cb) consisting of an elongate hypogynous cell (hy) and a bullet-shaped carpogonium (cp) with a trichogyne (t). 4. The intercalary auxiliary cell (ax) on a cortical filament with the adjacent nutritive cells (nc). 5. Two dichotomously branched connecting filaments (cf) arising from the lower portion of the fertilized, enlarged, spherical carpogonium (cp) positioned anterior to the hypogynous cell (hy) and with a trichogyne (t) to which a spermatium (s) attached terminally. 6. A gonimoblast initial (gi) arising from the terminal of the connecting filament (cf) fusing with an auxiliary cell (ax) accompanied with the adjacent nutritive cells (nc).

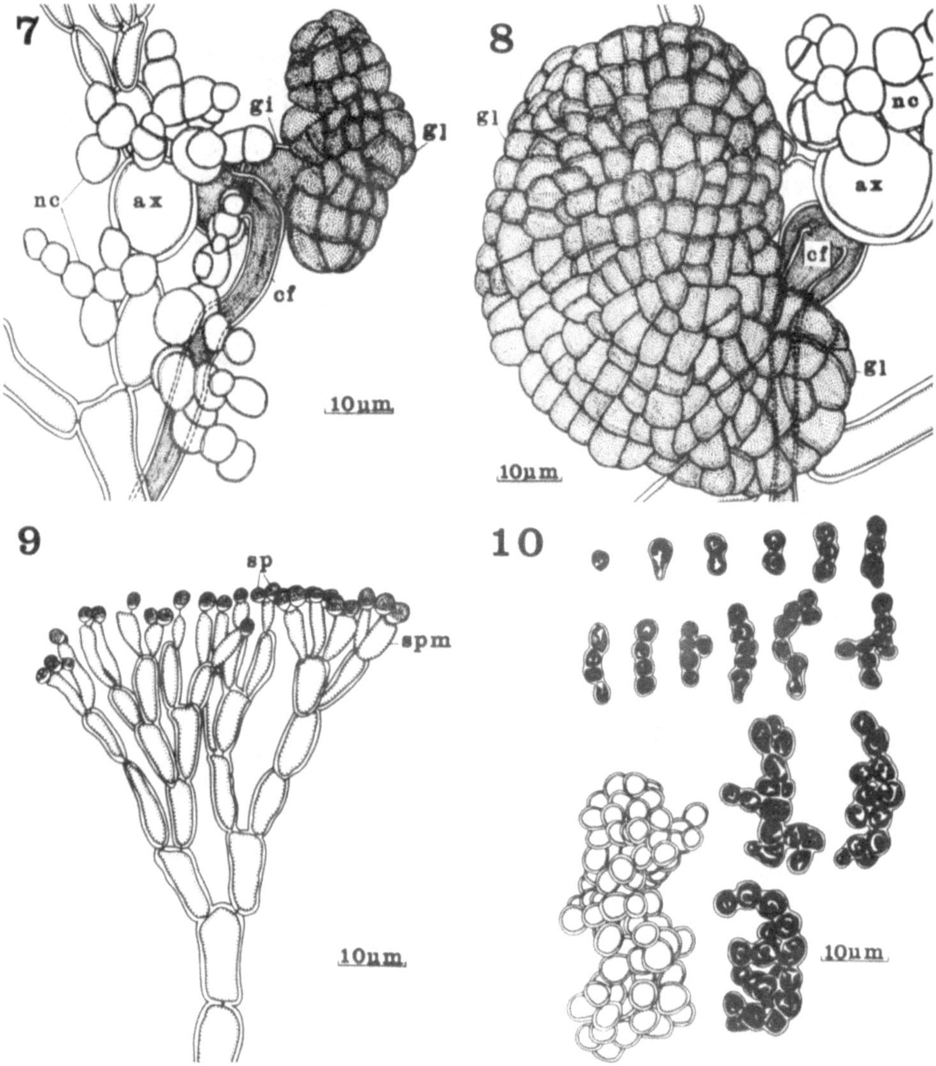

Figs. 7-10. Predaeopsis japonica gen. et sp. nov.
7. A young cystocarp with a gonimolobe (gl) and a gonimoblast initial (gi) originated terminally from the connecting filament (cf) fusing with the colorless auxiliary cell (ax), and the colorless nutritive cells (nc). 8. A mature cystocarp with two gonimolobes (gl), and a connecting filament (cf) fusing terminally with the colorless auxiliary cell (ax) accompanied with the adjacent colorless nutritive cells (nc). 9. Spermatangia (sp) produced terminally from the elongate spermatangial mother cell (spm). 10. Germination of the bispore belonging to so-called diprotocellular type.

ON THE IDENTITY OF LAURENCIA PINNATIFIDA FROM SCANDINAVIAN COASTS

Francis MAGNE

Biologie végétale marine, 7 Quai St Bernard, 75230 PARIS Cedex 05

SUMMARY: As studies have shown, from a long time two species are commonly confused under the name *Laurencia pinnatifida* in material from west-european coasts: the true *L. pinnatifida* (Hudson) Lamouroux and *L. platycephala* Kützing. The two are seemingly very near but they can be separated by precise characters.

In the scandinavian form of *Laurencia* till now related to *L. pinnatifida*, most of the characters are identical with the ones of *L. platycephala*.

From the beginning of the last century, a confused situation exists about *Laurencia pinnatifida*. The indications contained in the classical works on systematics and determination for this species are frequently ambiguous and conflicting.

A study of the *Laurencias* from the region of Roscoff has shown that two species, well distinct but very near in appearance, have been confused in fact.

One is the genuine *Laurencia pinnatifida* (Hudson) Lamouroux described (as *Fucus pinnatifidus*) by HUDSON; the corresponding type is, according to Dr. P. S. Dixon (1979), a specimen in the Petiver's herbarium. The other is *Laurencia platycephala* Kützing, described in 1865 and remained secret from this date; the corresponding type is the specimen n° 941.99-71 in the

Xth International Seaweed Symposium
Printed in Germany

	L. PINNATIFIDA (Hudson) Lamouroux	L. PLATYCEPHALA Kützing	L. HYBRIDA (DC) Lenorm. ex. Duby
color in living state	brownish red yellowish	ponceau red yellowish	violaceous greenish
attachment system	a disc and creeping prostrate axes	a disc	a disc
male organs	lateral in narrow and deep pocket	terminal, with hemispherical cavity	terminal, with hemispherical cavity
wall thickenings in medullary cells	+	-	-
emersion	well tolerated	not tolerated	well tolerated

Herb. Lugd. Bat. (Rijksherbarium, Leiden).

The characters of the two species have been studied in a recent publication (MAGNE, 1980); the more striking ones are shown in the table (and these of Laurencia hybrida (DC) Lenormand, in some means near, are added on).

So, it is appearing that the form from the danish and scandinavian coasts, that was studied by KYLIN (1923, 1928, and Rosenvinge (1923), in fact is away. Its characters, as demonstrated by the above authors and confirmated along in this study using specimens coming from the Kristineberg region, are these of L. platycephala: a disc without creeping axes at the base, terminal male organs with hemispherical cavity, no wall thickenings, living constantly immersed.

Nevertheless, it is not still possible to assert without restriction that this form is identical to *L. platycephala*. Indeed a discrepancy is existing about a character not still considered here but important, the formation of the tetrasporocysts. According to KYLIN (1923), these organs are in an adaxial position in the scandinavian form, but they are in an abaxial position in the examined *L. platycephala* specimens.

A new study of the tetrasporophytes of the scandinavian form is consequently to be done. If it reveals abaxial tetrasporocysts, against KYLIN's opinion, it will be one more argument to consider we are dealing with *L. platycephala*. If, on the contrary, it gives confirmation of adaxial tetrasporocysts, then we must think the scandinavian form is not corresponding to *L. pinnatifida* nor *L. platycephala* but to a third species to be identified or to be described.

The author is greatly indebted to Dr. P. S. DIXON for his kind aid in locating the type specimen of *L. pinnatifida*.

References

DIXON, P. S., 1979 - (Personnal communication).

KYLIN, H., 1923 - Svensk. vet. Akad. Handl., 63 (11): 1-139.

KYLIN, H., 1928 - Lunds Univ. Arsskr., N. F., 2, 24 (4): 1-127.

MAGNE, F., 1980 - Cah. Biol. mar., 21 (2): 227-237.

ROSENVINGE, L. K., 1923-24 - D. Kgl. Danske Vid. Selsk. Skrift Skrift., Naturvidensk., 7 (3): 287-485.

SEXUAL REPRODUCTION IN THE PALMARIACEAE

J. P. van der Meer

Atlantic Research Laboratory
National Research Council of Canada, 1411 Oxford St.,
Halifax, N. S., Canada, B3H 3Z1

Introduction

Species of Palmaria and Halosaccion have long interested phycologists because of the mystery that surrounded their life histories. Female plants appeared to be lacking although males were common for at least some species. The mystery was solved recently when microscopic females and an active sexual cycle were discovered (van der Meer & Chen, 1979; van der Meer & Todd, 1980).

Guiry (1974) removed Palmaria, Halosaccion, and Leptosarca from the Rhodymeniaceae and placed them in a new family, Palmariaceae. Subsequently this family was transferred to a completely new order, Palmariales (Guiry & Irvine 1974, Parke & Dixon 1976.) The new order was not universally welcomed as orders have traditionally been described on the basis of the female reproductive cells and the development of the carposporophyte and neither of these were known for the genera in question. The present paper describes sexual reproduction in P. palmata and H. ramentaceum, giving details about carpogonia, their relationship to neighbouring cells and the early post-fertilization development of the diploid phase.

[1]NRCC No. 18254

XIth International Seaweed Symposium
Printed in Germany

Methods and Materials

Conditions used to culture plants have been published (van der Meer & Chen 1979, van der Meer & Todd 1980). Recognition of fertilized carpogonia and diploid cell clusters was facilitated by mating mutant green females with wild-type males, a cross which yields red diploid tissue on a green haploid background. Whole plants were mounted for microscopic examination, some preparations being squashed in 0.5% HCl and stained with 2% aniline blue to study pit connections.

Results

Carpogonia of P. palmata and H. ramentaceum were very similar (Fig. 1-6, Table 1). The most striking difference was the position of carpogonia on the female discs and the absence of carpogonia on small upright shoots of H. ramentaceum. Size of carpogonia was influenced by age, younger ones being about 10-15 μm, older ones about 20-25 μm. Large carpogonia were less pigmented and in H. ramentaceum very large, colourless carpogonia in excess of 50 μm were encountered (Figs. 7,8). Cells associated with the carpogonia were unremarkable and closely resembled other cells of the thallus. There was no consistent carpogonial branch, and cells supporting the carpogonia were extremely variable. No intercalary carpogonia were observed.

A week after crossing, some of the carpogonia on the green females turned red indicating that the wild-type paternal genome was being expressed in the zygote (Figs. 9-11). For H. ramentaceum the red pigmentation was often very light because the carpogonia were located near the margin of the disc where all cells were pale. In both species all subsequent development of the diploid phase originated from the fertilized carpogonia. There were no auxiliary cells and no

Table 1. Characteristics of Carpogonia

Character	*P. palmata*	*H. ramentaceum*
Age when carpogonia first appear	4 days	5 days
Number of carpogonia at 8-12 weeks	5-20 (approx.)	50-200 (approx.)
Size of carpogonial cell	10-20 μm mode: 14 μm	12-25 μm mode: 16 μm
Size of trichogynes	3.0-4.5 μm wide; up to 375 μm long. some branched	3.5-5.0 μm wide; up to 360 μm long. some branched
Shape of carpogonial cell	mostly obovoid to spherical (variable)	mostly ovate (variable)
Location on older females (8-12 weeks)	near center of disc; young upright fronds	near margin of disc
Chloroplasts	present	present
Number of nuclei	1	1
Carpogonial branch	no constant features	no constant features
Auxiliary cell	none	none

cell fusions were observed during the early post-fertilization stages. For *P. palmata* the first division of the zygote was transverse, yielding two cells approximately equal in size, the "upper" one bearing the trichogyne (Fig. 12,13). The next division was often in the lower cell, producing a three- cell cluster (Figs. 14,15). After the first two divisions interpretation became difficult. In one instance the bottom cell had clearly developed into a five-cell branch before any further division of the upper cell (Fig. 16), but in other cases it appeared both cells had divided to yield a four-cell cluster. Some of the variability was caused by differences in the position of the zygote on the female plant. Those nearer the margin of the disc soon began to form elongated disc cells whereas those on young shoots never formed a disc. In *H. ramentaceum* the first division of the zygote was also transverse (Fig. 17). It appeared that both cells contributed to the formation of the diploid disc which always developed before any upright fronds were differentiated. Large elongated disc cells formed after only a few divisions, and were readily recognizable even when unpigmented (Fig. 18).

Discussion

The taxonomic position of *P. palmata* and *H. ramentaceum* can be determined with greater confidence now that information about sexual reproduction is available. The complete absence of a generative auxiliary cell in these species supports their removal from the Rhodymeniales. The question now becomes whether they should be placed in the Nemaliales because they lack an auxiliary cell or retained in a separate order, Palmariales, as suggested by Guiry & Irvine (1976).

Merging the Palmariaceae with the Nemaliales causes no difficulties as the latter already consists of a varied assemblage of taxa, some of which also have diphasic life

---

Figures: Carpogonia (c), of *P. palmata* (1-4) and of *H. ramentaceum* (5-8); Post-fertilization stages in *P. palmata* (9-16) and in *H. ramentaceum* (17-18). Diploid cells are shaded. Scale in Fig. 1 applies to all figures. For details see text.

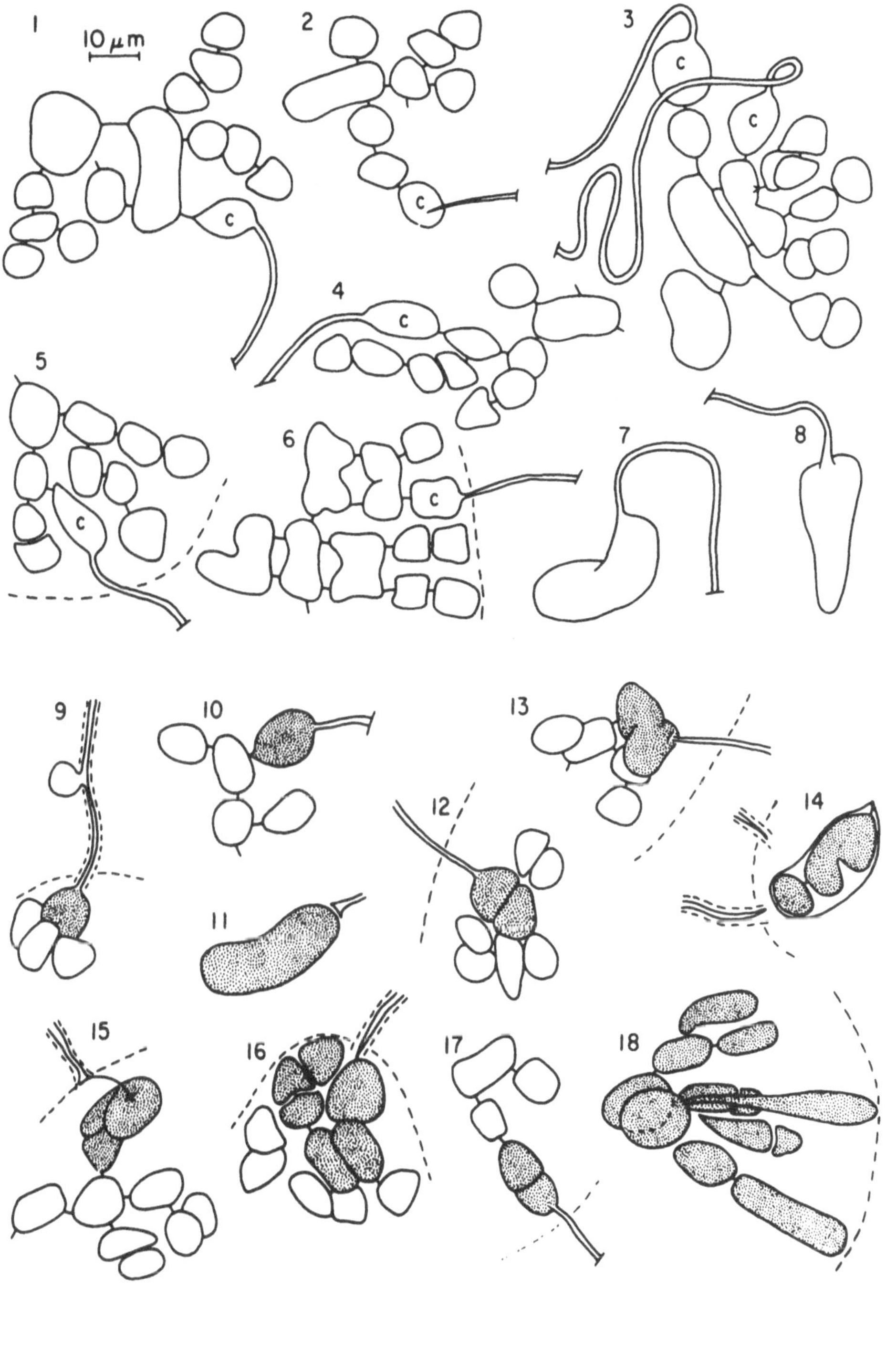
1
10 µm
c
2
3
4
5
6
7
8
9
10
11
12
13
14
15
16
17
18

histories (eg. Rhodochorton purpureum (Lightf.) West 1969). However, taxa presently in the Nemaliales have already generated much controversy and it would seem an unstable place at best to assign the Palmariaceae. Retaining the Palmariales appears to be more constructive. There is nothing to suggest that the diphasic life histories of P. palmata and H. ramentaceum have been derived from a triphasic pattern, and it is very possible that these taxa have originated from ancestors that never evolved carposporophytes. They appear to have adopted a highly protective reproductive strategy which makes zygote amplification via carpospores unnecessary. With the new information on sexuality, the Palmariales can be described as an order which completely lacks carposporophytes, and used to house the Palmariaceae and related taxa.

References

Guiry, M. D.: J. mar. biol. Ass. U. K. 54: 509-528 (1974).

Guiry, M. D., Irvine, D. E. G.: Proceedings of the Eighth International Seaweed Symposium Bangor, (1974) (In press).

Parke, M., Dixon, P. S.: J. mar. biol. Ass. U. K. 56: 527-594 (1976).

van der Meer, J. P., Chen L. C-M.: Can J. Bot. 57: 2452-2459 (1979).

van der Meer, J. P., Todd, E. R.: Can. J. Bot. 58: 1250-1256 (1980).

West, J. A.: J. Phycol. 5: 12-21 (1969).

EGG LIBERATION IN SOME JAPANESE SARGASSACEAE (PHAEOPHYCEAE)

T. Okuda
Department of Fisheries, Kyushu University 46-04, Fukuoka 812, Japan

Introduction

Japanese species in Sargassaceae have at least two aspects. First; some species used to be harvested for human food, and Hizikia, one of them, is still now popular as a sea food. One village, famous for the longevity of its inhabitants, is on the sea, and they often eat seaweeds raw or slightly processed. Second; these plants form underwater forests where living marine resources are fostered. In Japan, Marine Ranching Plan has been undertaken to build up or expand pertinent environment for such marine lives. Although kelps are elements of such underwater forests and have been served as food for abalones and other animals, species in Sargassaceae are dominant elements in northern Kyushu, western part of Japan. This study was partly supported by the subsidy of the Board of Fisheries intending to propagate sargassaceous algae (MRP55-I-6-1).

Since I have been interested in the rhythmic discharge of reproductive cells in marine algae, I started observations on the egg liberation in order to obtain enough fertilized eggs, which were supposed to settle on various substrate and to grow further. Three questions in obtaining fertilized eggs are as follows. 1) Is the egg liberation rhythmic or not? 2) Does it occur during the spring tides or the neap tides? 3) Are there any time lags between individuals or species?

Xth International Seaweed Symposium
Printed in Germany

Method

Eggs are liberated zonately in each receptacle beginning at its basal part. As these liberated eggs remain on the surface of receptacles with mucilage for only one day in some species, it is essential to make daily observations on the egg liberation. I picked up 4 species for this purpose because they were easy of access in the rough seas. Some plants were marked with vinyl spaghetti and were checked every day at their habitat. In making close observations on each receptacle, some branches were separated and kept underwater by a boat, and some receptacles were marked with vinyl spaghetti or sewing thread.

Sargassum horneri, a textbook species in Japan, is one of the most feasible species with its big receptacles. Female ones generally measure 25-30 mm in length, 5 mm in diameter, and it is not rare to find ones longer than 40 mm. This annual alga grows best in Sargassaceae and one incomplete thallus was reported to reach 8.4 m. Other 3 species, S. thunbergii, S. muticum, Hizikia fusiformis were also chosen.

Results

1) Sargassum horneri. As of now, more than 45 plants of this alga were marked including missing ones. Table 1 shows the dates of egg liberation in 18 plants. Plant 1 liberated eggs on May 13, one day before the last quarter of the moon. The second liberation occurred on May 17, 4 days after the first liberation. The third liberation occurred on 21st, one day before the new moon. As this table shows, liberation occurred during both the neap tides and the spring tides. Intervals vary from 2 to 8 days.

   Figure 1 shows all the intervals between successive egg liberation found in 28 plants. Liberation often occurred at 7-

Table 1. Dates of egg liberation in Sargassum horneri.

				◐			May, 1974					●			
	11	12	13	14	15	16	17	18	19	20	21	22	23	24	25
1			+				+				+				+
2			+						+						
3		+								+			+		
4		+							+						+
5		+							+						
6		+							+						
7			+								+				
8			+								+		+		
9		+							+			+			
10			+				+					+			
11		+					+					+			
12		+							+			+			
13			+						+			+			
14		+							+						+
15		+							+						
16		+							+						
17		+							+						
18	+								+						+

or 6-day intervals. There are some longer or shorter intervals, but since such different intervals are shared even by one individual plant, a certain interval length does not seem to be indigenous to one particular plant. If there are some groups of receptacles on one plant and liberate eggs alternately, the result will be the same as those shown in Table 1 and Fig. 1. So, I followed egg liberation on each receptacle.

One receptacle of 19 mm long liberated eggs at its middle area on May 30 (Table 2). The eggs remained following two days. On June 3, next liberation took place proceeding up-

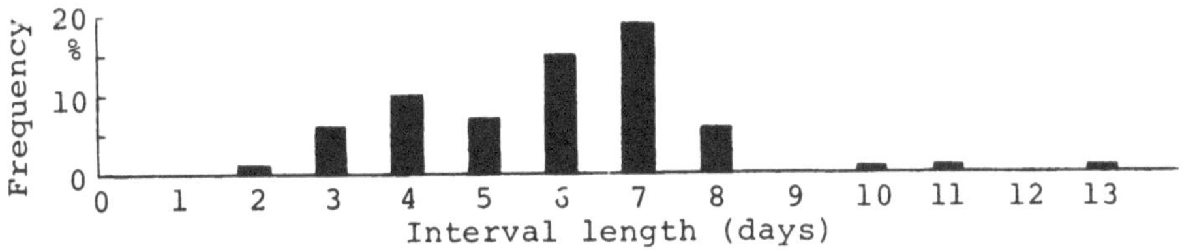

Fig. 1. Frequencies of the intervals between successive egg liberation in 28 individual plants of Sargassum horneri.

ward. Eggs remained through 5th, and the last liberation occurred on 7th at the distal end of this receptacle. All those on the same branch followed this one. In only a few cases, receptacles of another plant liberated eggs on two consecutive days, as is shown at the bottom in Table 2.

Table 2 shows that the egg liberation is simultaneous in one plant, and also that most of these receptacles did not stop growing. The growth rate of 118 receptacles from 17 plants averages 0.23 mm a day. The highest mean value in one branch is 0.41 mm a day.

2) *Sargassum thunbergii*. This common alga forms large beds at lowtide mark. During the spring tides, they are exposed to the air when the tide is low, but they are hardly exposed during the neap tides. As this alga has small receptacles, I failed to tag enough receptacles for following each one. I marked 17 plants that liberated eggs on May 28, 1974.

Table 2. Egg liberation in each receptacle on three individual plants of *Sargassum horneri*, 1979.

May 30	31	J. 1	2	3	4	6	7
19(12-14)	+	+	-	20(14-17)	+	-	20(17-20)
21(10-13)	+	+	-	22(13-20)	+	-	23(20-23)
22(13-16)	+	+	-	22(16-20)	+	-	24(23-24)
22(15-19)	+	+	-	24(19-21)	+	-	25(21-24)
23(15-17)	+	+	-	25(17-22)	+	-	25(23-25)
Jun. 7	**8**	**9**	**10**	**11**	**12**	**13**	**14**
30(0-7)	+	+	-	-	-	33(7-27)	+
30(0-11)	+	+	-	-	-	31(11-25)	+
31(0-13)	+	+	-	-	-	33(13-27)	+
33(0-15)	+	+	-	-	-	34(15-29)	+
44(0-19)	+	+	-	-	-	45(19-38)	+
May 19	**20**	**21**	**22**	**23**	**24**	**25**	**26**
18(0-5)	19(0-9)	+	+	-	-	-	21(9-13)
21(0-9)	22(0-17)	+	+	-	-	-	drop
24(0-10)	24(0-13)	+	+	-	-	-	24(13-20)
25(0-6)	26(0-12)	+	+	-	-	-	27(12-20)
25(0-14)	25(0-20)	+	+	-	-	-	25(20-22)

+, eggs on the surface; -, no eggs on the surface. (mm)

But intervals between successive liberation vary. If I had not marked each individual plant, only observing from the top of the breakwater, liberation should appear at 7- to 9-day intervals. The intervals between successive egg liberation found in 33 plants are shown in Fig. 2. In this alga eggs remain only one day.

3) Sargassum muticum. This species is one of a few monoecious algae in mainland Japan. It does not grow exposed areas. It often grows with S. thunbergii, but the upper limit of S. muticum is lower than the former species. Intervals between successive egg liberation are shown in Fig. 2. Eggs remain only one day.

4) Hizikia fusiformis. This alga is one of edible species in Japan, growing at a little higher level than S. thunbergii. This species seems to be more simultaneous than the former species in liberating eggs. Figure 2 shows the intervals between successive egg liberation found in 66 plants. Eggs remain only one day. Receptacles are about 10 mm long at the beginning, and I could follow only 3 receptacles in one individual. The growth rate averages 0.27 mm a day.

Consideration

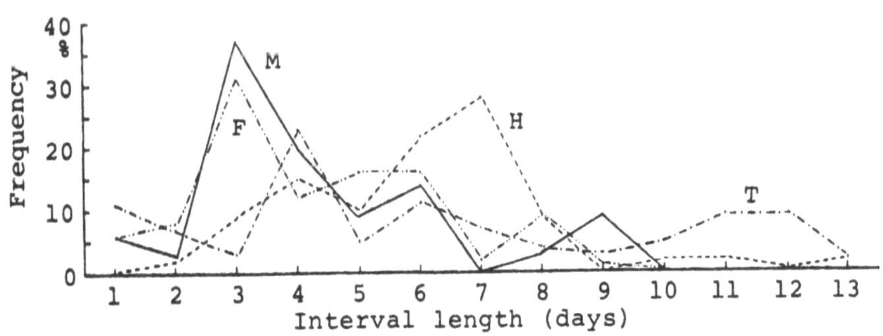

Fig. 2. Frequencies of the intervals between successive egg liberation in 4 species of Sargassaceae. F, Hizikia fusiformis; H, S. horneri; M, S. muticum; T, S. thunbergii.

Egg liberation in this group was first reported by Tahara (1) from a standpoint of rhythmicity. He happened to observe the egg liberation during the spring tides, so, making the second observation, he corrected his previous report. According to his second report (2), they do not always liberate eggs at semilunar intervals. After his report, however, eggs have generally been obtained during the spring tides. So, it seems to be widely accepted that the egg liberation occurs during the spring tides. In his report on Pelvetia, Subrahmanyan (3) has endorced this view. As is reported here, there is little possibility of semilunar nature harbored in receptacles.

Suto (4) reported that Hizikia liberated eggs every 7-8 days, which does not completely agree with my result. As he observed this alga near Tokyo, central part of Japan, indigenous character in plants seems to vary depending on localities. It is desirable to conduct observations at many localities as possible.

References

1. Tahara, M.: Bot. Mag. Tokyo 23, 151-153 (1909).
2. Tahara, M.: J. Coll. Sci. Tokyo Imp. Univ. 32, Art. 9, 1-13 (1913).
3. Subrahmanyan, R.: J. Indian Bot. Soc. 36, 373-395 (1957).
4. Suto, S.: Bull. Japan. Soc. Sci. Fish. 17, 9-12 (1951).

PORTERINEMA FLUVIATILE AS A STAGE IN THE LIFE HISTORY OF **SORAPION KJELLMANII** (FUCOPHYCEAE, RALFSIACEAE).

P. M. Pedersen

University of Copenhagen, Institut for Sporeplanter, Ø. Farimagsgade 2D, DK-1353 Copenhagen K., Denmark.

Introduction

Sorapion kjellmanii (Wille) Rosenvinge is a relatively little known crustose brown alga reported from rather scattered localities in the North Atlantic. It is characterized by the terminal, pear-shaped unilocular sporangia not surrounded by paraphyses. In the vegetative condition the plants may be recognized by the occasional apical branching of the upright filaments and the presence of a single disc-shaped chloroplast per cell. The species is normally reported with only unilocular sporangia.

Porterinema fluviatile (Porter) Wærn is also a relatively little known brown alga. The quadripartite plurilocular sporangia, each loculus with a separate opening, are characteristic for fertile monostromatic discs. If the plants develop into crustose thalli before fertility, the sporangia are formed terminally on the upright filaments in crown-like structures. Unilocular sporangia are seldom reported in this species.

Material and Methods

Sterile material of **Sorapion kjellmanii** was collected in shallow water on the eastern side of the island of Saltholm, the Sound, Denmark, on 19 August 1976 by Aase Kristiansen. The plants grew epilithically together with crusts of **Ralfsia**. A crude culture was e-

Xth International Seaweed Symposium
Printed in Germany

stablished, and later fragments were isolated by pipetting. The resulting unialgal cultures were reared in Provasoli's ES (5) with an initial addition of germanium dioxide to prevent growth of diatoms. Philips fluorescent tubes, warm white, 29, were used as light source. The culture conditions were as follows: 1) 5 $\pm$ 2.5°C, 30 ‰ S, 2.5 klx, 8 h light/16 h dark (SD); 2) 4° C, 30 ‰ S, 1.5 klx, 16 h light/8 h dark (LD); 3) 15°C, 30 ‰ S, 1.9 klx, LD.

Results

Under culture conditions 1 and 2 the plants grow very well vegetatively and develop into almost black crustose thalli. These plants agree with those found in nature in their morphology, showing frequent apical branching (Fig. 1). Each cell contains one disc-shaped chloroplast placed in the distal part of the cell. Hairs sometimes occur. The plants were cultivated continuously under conditions 1 and 2 since August 1976 and late January 1978, respectively, and under condition 3 through three months without formation of reproductive structures.

Sterile plants reared continuously under condition 1 through nine months were transferred to condition 2. As hope for occurence of

Fig. 1. Part of sterile **Sorapion kjellmanii** grown under condition 1, showing initial branchings of the upright filaments. 0 indicates broken filament. Camera lucida drawing of squash preparation.

reproductive structures in this species had almost been given up the plants were left here for 11 months. After this long period, however, germlings were observed in the culture. Pear-shaped unilocular sporangia occurred terminally on the upright filaments of the parent generation (Fig. 2), and terminal true hairs with a distinct basal sheath (Fig. 2, arrowhead) had developed abundantly on these old plants. the swarmers from the unilocular sporangia rounded up and formed embryospores which germinated unipolarly leaving the embryospore and part of the germination tube empty (Fig. 3). The germlings developed into new crustose plants (Fig. 4) and these plants also carried true, basally sheathed hairs (Fig. 5, arrowhead). After about six months in culture under condition 2 the new generation of crustose plants began forming sporangia. These developed terminally on the upright filaments often in crown-like structures (Fig. 6, arrows). The loculi of these peculiar plurilocular sporangia each liberated a single swarmer leaving the sporangia walls which soon collapsed (Figs 7, 8). Copulation between swarmers from the plurilocular sporangia was never observed. Instead, the swarmers rounded up and formed embryospores which showed the same germination patterns as the embryospores formed of the swarmers from the unilocular sporangia. Under condition 2 the germlings developed into new crustose plants with plurilocular sporangia.

Discussion

With regard to all known vegetative and reproductive characteristics, one chloroplast per cell, apical branching of the upright filaments, and the terminal pear-shaped unilocular sporangia, the plants from nature and later the plants in unialgal cultures are in agreement with **Sorapion kjellmanii**. As the plants with plurilocular sporangia can only be identified as **Porterinema fluviatile** due to the peculiar plurilocular sporangia I consider it experimentally proved that **Sorapion kjellmanii** and **Porterinema fluviatile** are

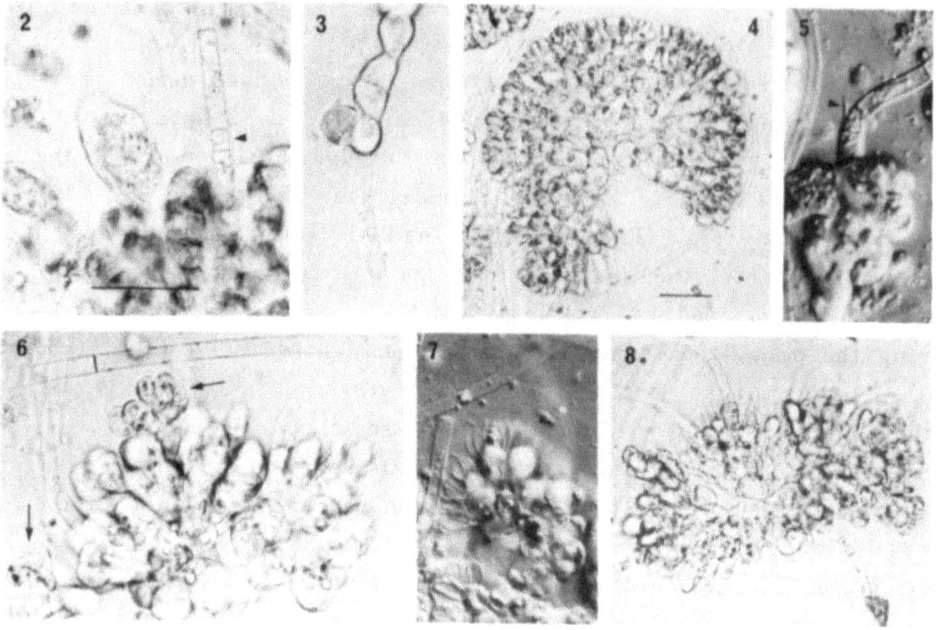

Figs 2-8. **Sorapion kjellmanii - Porterinema fluviatile** in culture. Fig. 2. Crustose plant with terminal, pear-shaped unilocular sporangia and a hair with a basal sheath (arrowhead). Fig. 3. Germling formed by germination of a swarmer from a unilocular sporangium. Note that the embryospore and a part of the germination tube is empty. Fig. 4. Crustose plant with initial development of terminal plurilocular sporangia (the **Porterinema**-stage). Fig. 5. Interference micrograph showing the basally sheathed hair (arrowhead). Fig. 6. Plant with plurilocular sporangia in crown-like structures (arrows). Fig. 7. Interference micrograph of empty plurilocular sporangia. Fig. 8. Plant with mostly empty plurilocular sporangia. Scales = 30 µm; large scale, Fig. 2, 3, 5, 6, 7; small scale, Fig. 4, 8.

the same species with unilocular and plurilocular sporangia, respectively. Whether the shift in reproductive structures is a result of a shift in the ploidy level is not known. Copulation between swarmers from the plurilocular sporangia has, however, never been observed, neither in culture (9) nor in nature (10).

In **Sorapion kjellmanii** plurilocular sporangia have been reported only in a phenology scheme (8), and no special attention has been paid to these structures, so that it is unknown whether these are **Porterinema**-like or not.

Fig. 9. Known distribution of **Sorapion kjellmanii** ●, (●➚ type locality Novaja-Semlja), **Sorapion simulans** Ⓢ, and **Porterinema fluviatile** ▲, ◪ unpublished record.

Porterinema fluviatile is normally reported with plurilocular spogia, in a few cases, however, also with unilocular sporangia (9, 10), even on the same plant (10). Wærn (10) has clearly intimated the connection between the two plants by saying that the unilocular sporangia of **Porterinema fluviatile** resemble those of **Sorapion simulans**, a species probably conspecific with **Sorapion kjellmanii** (7). It has also been suggested that **Sorapion simulans** is identical with **Ralfsia spongiocarpa** (1). The present study does not support the last suggestion.

The geographic distributions of **Sorapion** and **Porterinema** show good coincidence, as shown in Fig. 9. This may also be taken in support of their identity. **Porterinema fluviatile** was long believed to be restricted to brackish water (10), but it has later been found in several marine habitats (2, 6) and culture studies have shown that it is euryhaline (9).

The occurrence of hairs varies in the cultures, maybe due to variation in the nutrient content of the medium. Hairs are not mentioned in the diagnosis of **Sorapion kjellmanii** but have been reported later (3, 4). None of previous observers have mentioned a basal sheath. In **Porterinema fluviatile** hairs have been illustrated with or without a sheath (cf. 10, Fig 63j, m) so the observation of them is probably difficult, maybe depending on the age of the hair.

The present study leaves uncertainty about re-formation of plants with unilocular sporangia. In spite of this I have no doubt that the two plants are identical so that according to the rule of priority, both must be referred to as **Sorapion kjellmanii.**

References

1. Fletcher, R. L.: 8th Int. Seaweed Symp. (abstract) (1974).
2. Jaasund, E.: Bot Gothoburg. **4**, 1-174 (1965).
3. Kristiansen, Aa.: Bot. Tidsskr. **72**, 203-226 (1978).
4. Lund, S.: Meddr Grønland **156**(1), 1-247 (1959).
5. Nielsen, R.: Bot. Tidsskr. **67**, 245-269 (1972).
6. Pedersen, P.M.: Meddr Grønland **199**(3), 1-80 (1976).
7. Rosenvinge, L. K.: Meddr Grønland **20**, 1-125 (1898).
8. Sears, J. R., Wilce, R. T.: Ecol. Monogr. **45**, 337-365 (1975).
9. Wilce, R. T., Webber, E. E., Sears, J. R.: Mar. Biol. **5**, 119-135 (1970).
10. Wærn, M.: Acta phytogeogr. Suec. **30**, 1-298 (1952).

EFFECTS OF ENVIRONMENTAL FACTORS ON THE DIURNAL PERIODICITY OF TETRASPORES OF SOME GIGARTINALES (RHODOPHYTA)

M. Umamaheswara Rao and G. Subbarangaiah
Department of Botany, Andhra University, Waltair, India.

Introduction

Information available on the diurnal periodicity in spore release and on factors influencing the diurnal rhytm in the members of Gigartinales is very fragmentary. The present paper gives some information on the diurnal rhytms of spore release in *Gracilaria corticata* J. Ag., *Gracilaria textorii* (Sur.) J. Ag., *Gracilariopsis sjoestedtii* (Kylin) Dawson and *Hypnea valentiae* (Turn.) Mont. and on the effects of desiccation, salinity, light intensity and temperature on the diurnal periodicity in tetraspore shedding.

Materials and Methods

Fruiting plants of the four Gigartinales members, collected from the Visakhapatnam coast, were used for spore liberation experiments. Though tetrasporophytes were used for all the experiments, some observations were made with Carposporophytes to know the diurnal pattern in the shedding of carpospores. As described earlier (1) spore liberation experiments were conducted by placing tetrasporic or cystocarpic fronds in petridishes containing sterile seawater. These experiments were commenced in the evening at 6 PM and the spores liberated into the petridishes at 4 hr intervals i.e. at 10 PM, 2 AM, 10 AM, 2 PM and 6 PM were counted. To test the effect of desiccation fronds selected were first blotted with cloth; exposed to air in the laboratory for 0 (control), 15, 30, 45 and 60 min. and

Xth International Seaweed Symposium
Printed in Germany

then transferred to petridishes containing seawater for the estimation of spore output. Effect of salinity was tested using seawaters of 20, 30, 40 and 50‰ salinity. Desiccation and salinity experiments were conducted at room temperature (31±2°C) near a light source of 1500 lux to provide sufficient light during day time of the experiments from 10 AM to 6 PM. To test the influence of light intensity on the diurnal rhytm experiments were carried out at 0,1500 and 2000 lux. The respons of five different temperatures 15°C, 20°C, 25°C, 30°C and 35°C, was studied with G. corticata and H. valentiae by keeping petridishes in a temperature controlled dard incubator. Spore counting method was described in earlier papers (1, 2). Mean values of 10 experiments conducted with G. corticata and H. valentiae and 5 experiments with G. sjoestedtii and G. textorii are plotted in Figs. 1 to 5.

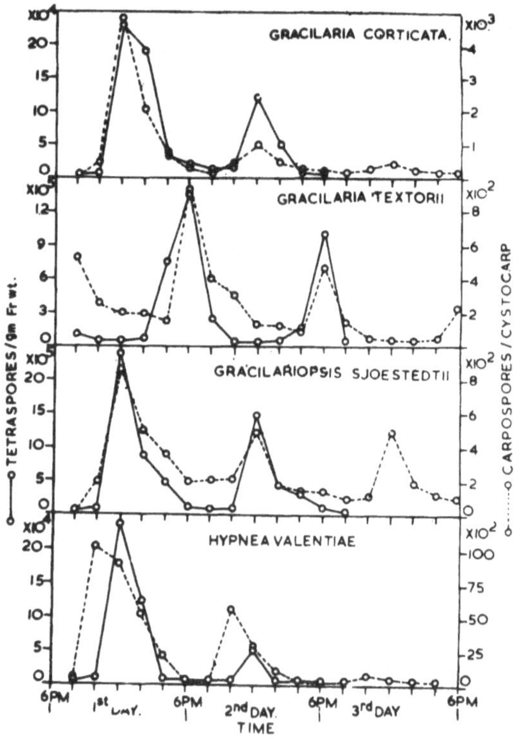

Fig. 1. Diurnal periodicity in the liberation of tetraspores and carpospores.

Observations

Fig. 1 shows the quantity of tetraspores and carpospores liberated at different times of the day and in different days in C. corticata, G. textorii, G. sjoestedtii and H. valentiae. In G. corticata and G. sjoestedtii peak output of tetra and carpospores was seen in the early morning hours between 2 AM and 6 AM. In H. valentiae similar diurnal rhytm was observed in tetrasporophytes, but peak shedding of carpospores was found during night time between 10 PM and 2 AM. In G. textorii maximum shedding of tetra and carpospores was observed in the evening hours from 2 PM to 6 PM. Data collected for 2 or 3 days (Fig. 1) clearly show that similar circadian rhytm exists in the spore release of the Gigartinales investigated.

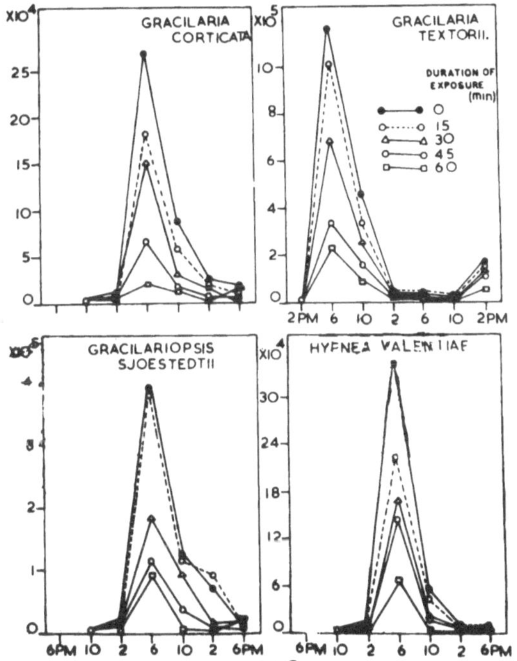

Fig. 2. Effect of desiccation on the diurnal rhythm in spore shedding.

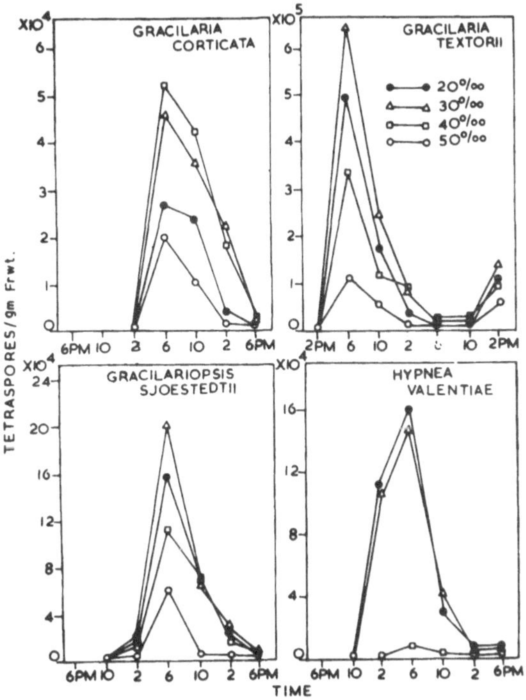

Fig. 3. Influence of salinity on the periodicity in spore shedding.

As shown in Fig. 2, the diurnal periodicity curves obtained for tetraspores did not vary in submerged conditions (D exposure) and in fronds exposed to air for 15 to 60 min. However, the spore output at different times of the day decreased from 0 to 60 min. exposure (Fig. 2). In the four salinities tested, similar diurnal fluctuations were seen (Fig. 3) and depending upon the salinity tolerance limits of the four algae, peak output was seen in *G. corticata* at 40‰, in *G. textorii* at 30‰ and in *H. valentiae* at 20‰ (Fig. 3). The time of peak liberation of spores was not affected either in the dark or in the two light intensities used (Fig. 4), though spore output decreased with increasing light intensity.

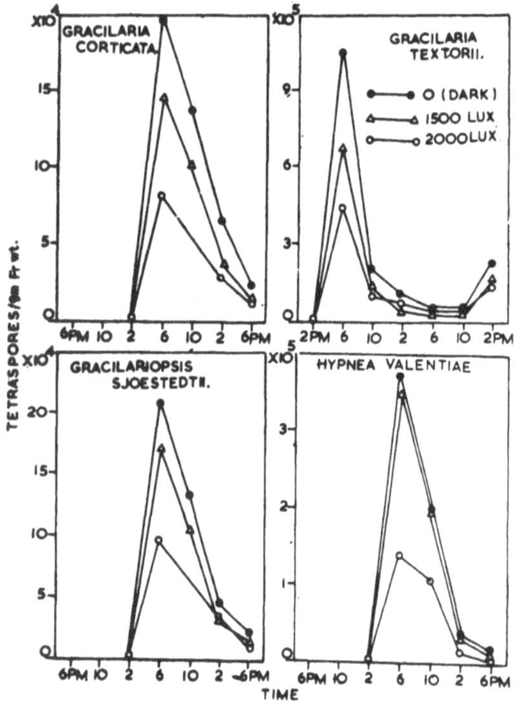

Fig. 4. Effect of light intensity on the diurnal changes in spore shedding.

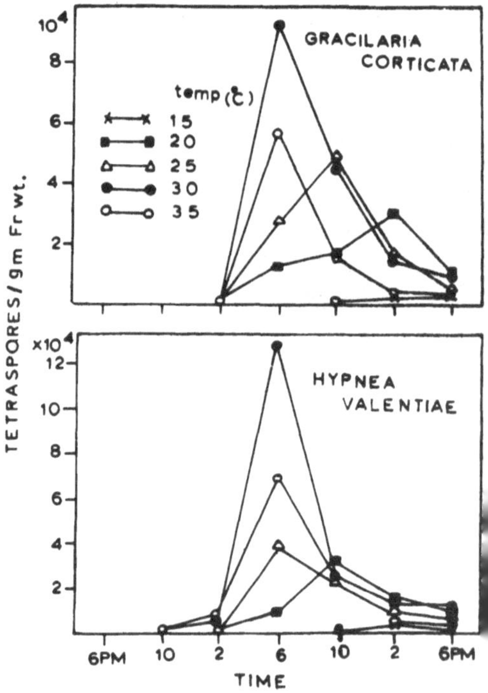

Fig. 5. Effect of temperature on the diurnal periodicity in spore release.

The pattern of the diurnal periodicity curves altered with temperature (Fig. 5). The normal periodicity with maximum spore shedding between 2 AM and 6 AM was observed at 30°C and 35°C in G. corticata and between 25°C and 35°C in H. valentiae. In temperatures below 30°C in G. corticata and below 25°C in H. valentiae the time of peak shedding varied markedly (Fig. 5). For instance in G. corticata peak number of spores was obtained after 4 hrs between 6 AM and 10 AM at 25°C, after 8 hrs between 10 AM and 2 PM at 20°C and after 12 hrs from 2 PM to 6 PM at 15°C. These changes observed in G. corticata and H. valentiae (Fig. 5) suggest that low temperatures delay and alter the diurnal pattern in spore shedding.

Discussion

In the present study a definite circadian rhythm was observed in spore release among the members of Gigartinales investigated (Fig. 1-5). The time of peak shedding also varied in the four algae. In G. corticata, G. sjoestedtii and H. valentiae peak output of tetraspores was seen from 2 AM to 6 AM (Fig. 1) as reported in G. corticata growing at Mandapam (3). The diurnal rhythm observed in G. textorii with an evening peak from 2 PM to 6 PM (Fig. 1) was similar to that reported in Gelidiella acerosa (4). Scanty information available in the literature indicates that the pattern of diurnal curve alters in relation to desiccation (5) and alternating periods of light and darkness (5, 6). The present findings clearlyshow that desiccation of fronds, salinity and continuous dark or light at different intensities have no effect on the diurnal periodicity observed in G. corticata, G. textorii, G. sjoestedtii and H. valentiae (Figs. 2-4). But the temperature of the seawater controlled the peak output of spores in a day (Fig. 5). The diurnal rhythm with an early morning peak from 2 AM to 6 AM was observed when seawater temperature in the experimental sets varied from 30 to 35°C for G. corticata and from 25 to 35°C for H. valentiae. In contrast to this, day time peak was observed between $15^{o}C$ and $20^{o}C$ or $25^{o}C$ (Fig. 5). In fronds treated for short periods at $40^{o}C$, peak liberation was observed 4 hrs earlier (night peak) between 10 PM and 2 AM (unpublished), further confirming the relationship between the diurnal rhythm and seawater temperature. In Gelidium amansii the time of peak shedding varied depending upon the seasonal changes in seawater temperature (7, 8). The experimental evidence gathered in the present study agrees with the results obtained on G. amansii.

Acknowledgements

The senior author expresses his sincere thanks to UNESCO, Paris for providing funds to attend the Symposium and to Prof. T. Levring, Marine Botanical Institute, Göteborg, for his help and encouragement.

References

1. Umamaheswara Rao, M. and Subbarangaiah, G.: Proc. Symp. Coastal Aquaculture, (1980 - In Press).
2. Umamaheswara Rao, M.: Proc. Symp. Marine algae of the Indian Ocean region (1979 - In Press).
3. Umamaheswara Rao, M.: J. Exp. Mar. Biol. Ecol. 21, 91-98 (1976).
4. Umamaheswara Rao, M.: Bot. Marina 17, 204-207 (1974).
5. Matsui, T. J.: Shimonoseki. Univ. Fish. 17, 185-231 (1969).
6. Sagromsky, H.: Pubbl. Staz. Zool. Napoli. 32, 29-40 (1961).
7. Katada, M., Matsui, T., Nakatuskhara, T., Kojo, T. and Miura, A.: Bull. Jap. Soc. Sci. Fish. 19, 471-473 (1953).
8. Katada, M.: Shimonoseki Coll. Fish. 5, 1-87 (1955).

THE MARINE SEAWEED OF CHILE'S TENTH REGION
(Valdivia, Osorno, Llanquihue and Chiloé)

R. Westermeier
Instituto de Botánica, Universidad Austral de Chile,
Valdivia - Chile

Introduction

Studies of the marine macroscopic algae of Chile were carried out first by foreign expeditions. Due to the fact the areas studied at this were restricted to the surrondings of important harbers or regions easy to reach. The knowledge of the marine algal comunities along the coast of Chile was rather desproporcionate. Nevertheless numerous are the papers publisched concerning this subject, among those (1, 2), (3), (4, 5) and (6, 7) must be mentioned. In accordance, studies of the litoral algae of the tenth region of Chile are only few. Algae from Corral, Seno de Reloncaví and the Bay and Gulf of Ancud were sampled by the Lund University Chile Expeditions 1948-49 and identified beside other algae by (8). Later studies are carried out by (9), (10) and (11, 12), were results from differents zones of this region are reported.

Therefore the primary object of this study is to inform about the marine macroscopic algae identified until now in the tenth region of Chile.

Methods

Sampling was carried out between the years 1976-79 in 4 provinces of the tenth region of Chile, between 39°24', 43°30' Lat. S., and 73°13', 73°45' Long. W., that correspond from

Xth International Seaweed Symposium
Printed in Germany

north to south to Valdivia, Osorno, Llanquihue and Chiloé. (Fig.1). The specimens were collected at ebb-tide along longitudinal and vertical transect from the upper to the lower litoral zone in differents habitats. Specimens originating from the sublitoral were obtained by diving, dragging and collecting the stranded algae.

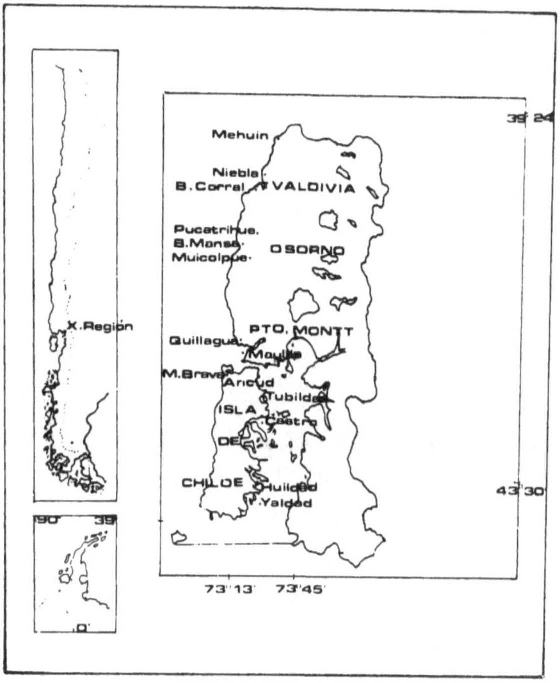

Fig.1: Map Tenth Region (South Chile) Showing the localities of sampling.

The identified algae became part of the herbarium of the Institute of Botany of the University Austral of Chile. Some of the specimens were conserved in seawater with 5% Formol.

Results

The total amount of algae reached 118 species, of which 77 species (65.3%) belong to the Rhodophyta and these became the most abundant group, followed by the Chlorophyta with 21 species (17.8%) and then by the Phaeophyta with 19 species (16.1 %). The Cyanophyta were presented with 1 species only (0.8%).

In the following the species are listed in alphabetical order Uptill now only 66 species were found in the 4 investigated areas and marked by the following letters, V:Valdivia, O:Osorno, L:Llanquihue and C:Chiloé.

Chlorophyta

1. Blidingia minima(Naeg.)Kylin (V)(C)
2. Bryopsis plumosa(Huds.)C.Ag.
3. Bryopsis rosae Gaud. (L)
4. Briopsis sp. (O)
5. Chaetomorpha aerea(Dillw.)Kütz. (V)(C)
6. Chaetomorpha linum(Müll.)Kütz.
7. Cladophora sp.
8. Cladophora sp.
9. Codium dimorphum Sved. (V)(C)
10. Codium fragile(Sur.)Hariot (O)(L)
11. Enteromorpha bulbosa(Suhr.)Kütz. (L)(C)
12. Enteromorpha clathrata(Roth)Grev. (L)(C)
13. Enteromorpha compressa(L.)Grev.
14. Enteromorpha intestinalis(L.)Link.
15. Enteromorpha prolifera(Müll.)J.Ag.
16. Monostroma sp. (V)(O)(L)
17. Rhizoclonium riparium(Roth)Harv.
18. Rhizoclonium tortuosum(Dillw.)Kütz.
19. Ulva lactuca L.
20. Ulva (E.)linza L.
21. Ulva sp.

Cyanophyta

1. Lyngbia confervoides C.Ag.

Phaeophyta

1. Adenocistis utricularis(Bory)Skotts.
2. Chordaria magellanica Kylin (V)
3. Colpomenia sinuosa(Roth)Derb.et Sol. (V)
4. Desmarestia ligulata(Lightf.)Lamour
5. Desmarestia munda Setchell & Gardner
6. Desmarestia sp. (L)
7. Durvillea antarctica(Cham.)Hariot
8. Ectocarpus confervoides(Roth)Kjellm.
9. Geminocarpus geminatus(Hook.f.et Harv.)Skottsb. (V)
10. Glossophora kunthii(C.Ag.)J.Ag. (L)
11. Halopteris hordacea(Harv.)Sauv. (V)(O)
12. Lessonia fuscesens Bory (V)(O)(C)
13. Lessonia nigrescens Bory
14. Macrocystis pyrifera(L.)J.Ag.
15. Myriogloia chilensis Levre (C)
16. Petalonia fascia(Müll.)Küntze (V)
17. Ralfsia sp. (V)(O)(C)
18. Scytosiphon lomentaria(Lyngb.)Endl.
19. Scytothamnus australis Hook.et Harv. (V)(C)

Rhodophyta

1. Ahnfeltia durvillaei(Bory)J.Ag.
2. Ahnfeltia setacea(Kütz.)Schmitz
3. Ahnfeltia sp. (V)
4. Anthithamnion flaccidum(Hook.f.et Harv)De Toni (V)(C)
5. Anthithamnion sp. (L)

6. Ballia callitricha(C.Ag.)Mont.
7. Ballia sp. (O)
8. Bossea orbigniana?(Dcne)Manza (V)(C)
9. Bostrichia sp. (L)
10. Bostrichia sp.
11. Callithamnion montagnei? Hook.f.et Harv.
12. Callithamnion sp. (V)(O)(L)
13. Callophyllis sp.
14. Callophyllis variegata(Bory)Kütz.
15. Centroceras clavulatum(C.Ag.)Mont.
16. Ceramium pacificum(Collins)Kylin (L)
17. Ceramium rubrum(Huds.)C.Ag.
18. Ceramium sp. (L)
19. Chaetangium fastigiatum(Bory)J.Ag.
20. Chondrus canaliculatus(C.Ag.)Grev.
21. Corallina chilensis Dcne
22. Corallina officinalis? L.
23. Dasyopsis sp.
24. Delesseria sp.
25. Delesseria sp.
26. Gelidium lingulatum J.Ag.
27. Gelidium pusillum(Stackh.)Le Jolis
28. Gelidium sp.
29. Gelidium sp. (V)
30. Gigartina chamissoi(C.Ag.)J.Ag. (L)(C)
31. Gigartina skottsbergii Setch.et Gardn. (V)(C)
32. Gracilaria verrucosa(Hudson)Papenfuss (V)(L)(C)
33. Grateloupia doryophora(Mont.)Howe
34. Grateloupia sp.
35. Griffithsia sp.
36. Gymnogongrus disciplinalis?(Bory)J.Ag.
37. Gymnogongrus furcellatus(C.Ag.)J.Ag.
38. Gymnogongrus sp. (V)
39. Herposiphonia sp.
40. Heterosiphonia merenia Falk. (V)
41. Heterosiphonia sp.
42. Heterosiphonia wurdemanni(Bailey)Falk. (V)(O)
43. Hildenbrantia le cannellieri Hariot
44. Hymenena laciniata?(Hook.f.et Harv.)Kylin (V)(L)(C)
45. Hymenena sp. (V)(C)
46. Iridaea boryana(Setch.et Gardn.)Skottsb.
47. Iridaea ciliata Kütz.
48. Iridaea dichotoma Hook.f.et Harv. (L)(C)
49. Iridaea laminarioides Bory
50. Iridaea sp. (O)(L)(C)
51. Iridaea sp.
52. Laurencia sp.
53. Lithophyllum sp.
54. Lithophyllum sp. (V)(C)
55. Lithophyllum sp. (C)
56. Lithothamnion sp.
57. Lithothamnion sp. (C)
58. Lithothamnion sp. (C)

59. Myriogramme croziere(Hook.f.et Harv.)Kylin (V)(L)(C)
60. Myriogramme livida?(Hook.f.et Harv.)Kylin (V)(C)
61. Myriogramme multinervis(Hook.f.et Harv.)Kylin
62. Phycodrys quercifolia(Bory)Skottsb.
63. Plocamium cartilagineum(L.)Dixon
64. Polysiphonia abscissa Hook.f.et Harv.
65. Polysiphonia sp.
66. Polysiphonia sp. (L)
67. Porphyra columbina Mont.
68. Porphyra umbilicalis(L.)C.Ag.
69. Porphyra sp. (V)
70. Porphyra sp. (L)
71. Prionitis lanceolata Harvey
72. Prionitis lyalli Harvey
73. Pterosiphonia dendroidea(Mont.)Falk.
74. Pterosiphonia sp.
75. Pugetia chilensis?(J.Ag.)Kylin
76. Schizoceris sp. (V)(C)
77. Rhodymenia skottsbergii Setch.et Gardn. (V)(C)

Discussion

Comparing the results of this study with those of (13, 14) in Quinteros and Valparaíso and (15)in Valparaíso and Magallanes (table 1), an increase of species from Quinteros to Magallanes is observed. The major number of species reported from Magallanes might possibily be effected by the high frecuency of sampling in done this region but also the species diversity is proposed higher in cold than in temperate water. Along the coast of Chile,(16)reported that the number of species decreased towards Ecuador; this result would correspond with the result analysed here. Although in the provinces that form tenth region of Chile there is not an increase in the number of species from the north to the south. The values in number and percentage of species are very close to each other.

Comparing these results with those of (8) it is pointed out that about 48% of the algae identified in the present investigation were cited by the author mentioned above, but it must be taken into consideration that, for example in Llanquihue, the collections were made in differents areas.

The presence of H.merenia in this list is particularly important, beccause it is collected for the first time in Chile. Its area of distribution was limited to the subantarctic islands (11).

Table 1 : Comparison of algae found along the coast of Chile.

Areas	CHLOROPHYTA		PHAEOPHYTA		RHODOPHYTA	
	Nr.	%	Nr.	%	Nr.	%
Quinteros	10	20.4	10	20.4	29	59.2
Valparaíso	14	18.2	14	18.2	49	63.6
X[th] región	21	17.8	19	16.1	77	65.3
Magallanes	25	16.0	20	12.8	111	71.2

References

1. Montagne, P.C.: Ann.Soc. Nat. Bot. 3, 347-356 (1835).
2. Montagne, P.C.: In C. Gay Botan. 8, 228-393 (1852).
3. Svedelius, N.: In Nordenskjöld 3, 238-316 (1900).
4. Skottsberg, C.:Kung. Sv. Vet. Ak. Han. 11, 1-56 (1921).
5. Skottsberg, C.:Kung. Sv. Vet. Ak. Han. 4, 1-92 (1941).
6. Levring, T.: In C.Skottsberg (Ed) 2, 601-670 (1941).
7. Levring, T.: Bot.Tradgord 15, 181-186 (1943).
8. Levring, T.: Ed. Lunds. Univ. Arssk. 10, 1-85 (1960).
9. Oliva, M.: Tésis Grado, Univ. Austral-Chile 1-62 (1967).
10. Alveal, K., Romo, H.: Gayana, Misc. 7, 3-28 (1977).
11. Westermeier, R., Ramírez, C.: Med. Amb. 2, 44-49 (1978).
12. Westermeier, R.: Rev. Biol. Pesq. En Prensa (1979).
13. Navas, E.: Rev. Universitaria 95-120 (1965-66).
14. Romo, H., Alveal, K.: Gayana, Misc. 6, 1-41 (1977).
15. Alveal, K., Romo, H., Valenzuela, J.: Biol. Marina, 1, 1-29 (1973).
16. Santelices, B.: J. Phycol. 13, 1-60 (1977).

2. Ecology

ON THE VEGETATION DYNAMICS OF GRACILARIA SPEC. IN PLAYA CHANGA, COQUIMBO, CHILE

Hans J. Black and Erika Fonck O'B.
Centro de Investigaciones Submarinas, Universidad del Norte,
Casilla 117, Coquimbo, Chile

Introduction

The investigations presented in this paper touch two important aspects. The first concerns the present state of marine botany in Chile as it has been described by various authors in recent years (6, 7, 8). The second concerns the economical role, the genus Gracilaria plays in Chile, especially in the region of Coquimbo (3, 5, 9). Thus this study contributes to the knowledge of the vegetation dynamics of Gracilaria populations in a defined area and continnues investigations recently begun to describe the basic ecofactors interaction in the biology og Gracilaria (9).

This presentation is using the term "vegetation dynamics" to mean the general dynamics of vegetation structure in an ecosystem and and its description with the aid of the measurable properties of this structure and their changes in time and space. Local components of structure are described in their short and long term interactions (2). The necessity of long term studies (9) also relates to the more theoretical problems encountered in a general analysis of the role of vegetation in benthic ecosystems.

As the general status of research on the genus Gracilaria for Chile has been recently summarized (9), it is only necessary here to point out again the complexity of problems in marine botany in the applied field.

Xth International Seaweed Symposium
Printed in Germany

Material and Methods

The methods follow in general those applied before in a neighbouring bay and deal with the same type of Gracilaria there described (9).

From June 1978 to February 1980 in Playa Changa off Coquimbo (29°57'S, 71°20'W) a sampling area of ca. 130 ha was defined. Along 11 profiles perpendicular to the coast line and 150 meters apart, 13 stations at 60 m intervals were sampled monthly, using a 1 m^2 frame. SCUBA divers collected the plant material inside the frame. Fresh weight was determined to an accuracy of $\pm$ 25 g. Biomass was expressed in kg m^{-2} fresh weight, using ranges for interpretation as presented in fig. 2. Before weighing the samples were cleaned from impurities and alga of errant or epiphytic caracter were removed, together with the zoobenthic part of the Gracilaria community.

The water movement was described using the lighthouse observations off Faro Punta Tortuga on the westernmost end of Coquimbo Peninsula, noted following key 26 (1). The observations cited here cover six hour periods of wind generated swell stronger than key no. 3. Durations of 24 and more hours are plotted seperately.

For sediment caracterization directed samples were subjected to standart geological methods so allowing determination of grain size and degree of sorting. This sampling was carried out once in June 1979.

Examination for the presence of cystocarpic plants was done during the period form March 1979 to February 1980.

To observe growing capacity, 60 thallus sections each 8 cm in length were burried in sand. The cuttings were weighed every 30 days over a 180 day experimental period.

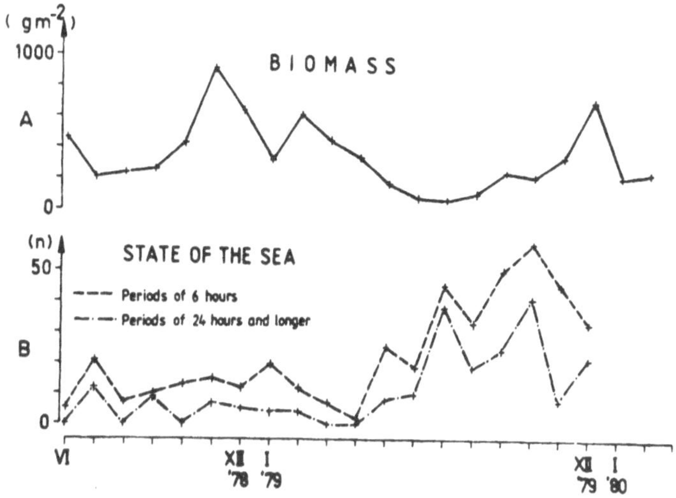

Fig. 1: Biomass of Gracilaria spec. and state of the sea

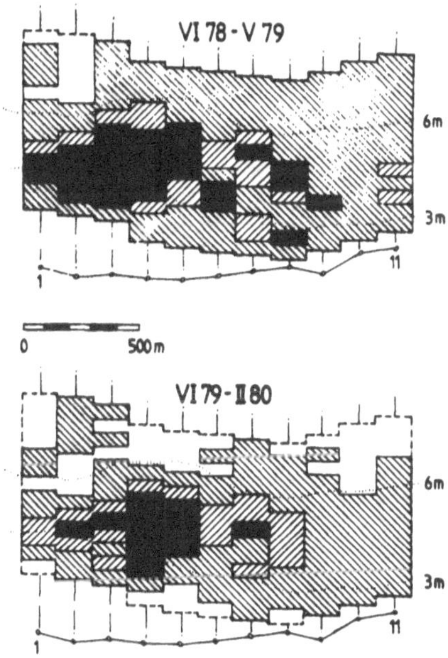

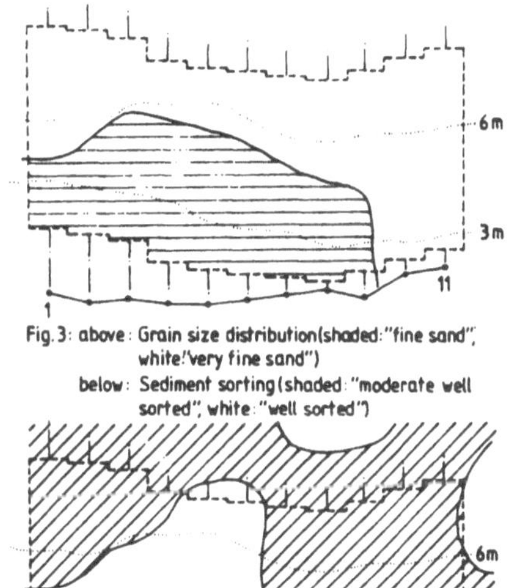

Fig. 3: above: Grain size distribution (shaded: "fine sand", white: "very fine sand")
below: Sediment sorting (shaded: "moderate well sorted", white: "well sorted")

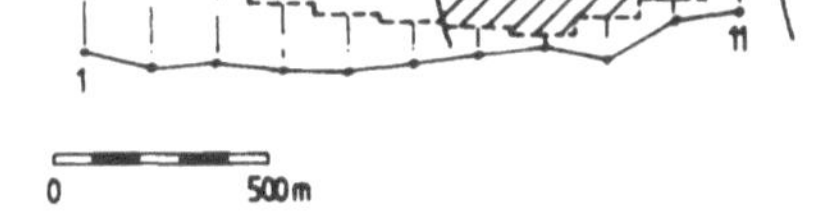

Fig. 2: Distribution of biomass: means of fresh weight in Gracilaria spec.

0 g m-2 · 1-500 g m-2 · 500-1000 g m-2 · > 1000 g m-2

Results

The investigation area off Playa Changa is covered by a <u>Gracilaria</u> meadow similar to that described for Bahía La Herradura (9). It contains only negliable quantities of epiphytes.

The biomass shows a seasonal change (fig.1). The time of investigation covers two saisons. In both saisons appear maxima in summer, in 1978 in November and in 1979 in December. The first maximum is higher than the second one. The winter minimun in 1978 shows higher values than the depression in 1979. As the second saison has not been sampled over the full duration, the decrease of the total biomass cannot be stated accurately, but is estimated to be of a remarkable order.

Changes in biomass are not correlated to the frequency of wind generated swells (9). In both winters a decrease due to reduced growth had allready taken place, before swells of long duration appeared. Also biomass starts to increase during spring and early summer, while swells of long duration still persist transporting large quantities of algae towards the coast.

Also the distribution in space, namely depth, shows changes. A major zone of high density exists between profiles "1" and "5", in 2.0 to 5.0 meters depths. In this zone about 90% of the biomass are concentrated. The smaller accumulation between profiles "6" and "9" is reduced to nearly zero towards the end of the second saison, the reduction taking place stepwise. The area without vegetation enlarges during the second season. An expanded area of low density vegetation was also observed at this time. Profiles "1", "2" and "3" were effected most at depths less than 3.0 m (fig.2). Grain size analysis shows a sediment distribu-

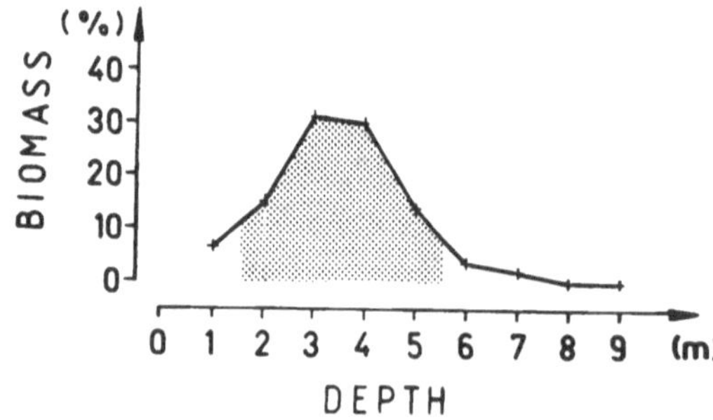

Fig. 4: Distribution of biomass in depth (VI 78 - II 80)

tion in which two types of sediments play a role: "fine sand" (3.9°Ø) and "very fine sand" (2.02°Ø). All biomass values of 1 kg m^{-2} and more were found in the fine sand area. In the second season, all high biomass values except the one in profile "7" were located in the area of well sorted sands. The decrease of biomass took place in the field of moderately well sorted sands, being composed of fine grains. The sediment types are well separated (fig.3).

The distribution pattern during winter months does not differ from that of the summer months, with respect to the depth, the lower limit of vegetation being between 8.0 and 9.0 meters.

The number of samples containing cystocarpic plants was very low: five in April 1979, four in May 1979 and one in August, October and December 1979 and in February 1980. These cystocarpic plants occured in small isolated groups at depths between 3.0 and 6.0 meters, except the one in December 1979 which was at 2.0 meters. There is no relationship between density of vegetation and presence of cystocarpic plants.

The buried plants showed no growth until they were uncovered, when they displayed the same pattern of growth as decribed for Gracilaria verrucosa (5)

Conclusions

As only very little cystocarpic material was present, the repopulation of Gracilaria vegetation by carpospores seems not to have importance in the maintanance of the meadows in this region. This is confirmed by earlier findings (5). The vegetative growth of thallus fragments plays the major role, which points out the importance of the substrate. Investigations in the early stages of growth of newly settled thallus fragments are thus beconing important.

The dynamics of the sediments enter as a main complex of factors the dynamics of vegetation in both, biomass and repopulation. As bioturbation is influencing the sedimentation process intensivly, further investigations should take this factor into account.

The capacity to grow new ramifications seems to be similar to that described for Gracilaria verrucusa (4,5) and does not include appearances to be reported.

The seasonal changes in biomass suggest defendence of being dependant on temperature and light regimes, as shown in laboratory experiments in the recent years (9). The lower maximun in biomass in the second cycle suggests, the meadow is in regression. This might be due to the relatively higher frequency of wind generated swell in the second season, resulting in the thinning out of vegetation throughout all the period. Mayor growth phase and destruction by water movement are not directly related but seem to interact in a highly complex manner, which is to be looked at in connection with sedimentation. As extraction of algae by man auments during periods of swell, the natural loss is impossible to be estimated. The finding of no change being present in the distribution pattern favours interpretation of the decrease being the expression of long time periodical changes. There is connection in the sedimentology of the substrate and the distribution pattern which needs further investigation. Maximum growth expected in shallow water is not achieved in this region because the effects of braking swell and extraction.

References

(1) Armada de Chile: C. de Met., p 36, (1961)
(2) Black, H.J.: Diss. Universitaet Kiel, (1978)
(3) Fonck, E.: U. del Norte, CIS, Informe, (1980)
(4) Jones, W.E.: J.mar.biol. Ass. UK, Vol. 38, 47-56, (1959)
(5) Kim, D.H.: Botanica Marina, Vol. XIII, 140-162, (1970)
(6) Lopehandia, J.: Actas I Symp. Algas Mar.Chil., 201-219 (1979)
(7) Romo, H. et al.: dito, 155-163, (1979)
(8) Santelices, B.: dito, 1 - 34 , (1979)
(9) Santelices, B. and Fonck, E.: dito, 166-200, (1979)

THE PHYTOBENTHOS OF A MEDITERRANEAN HARBOUR, PORT-VENDRES

C.-F. Boudouresque, J. Marcot-Coqueugniot and H. Nédélec
Laboratory of Botany (Marine phytobenthos and Pollution),
University of Luminy, 13288 Marseilles cedex 2, France.

Port-Vendres is a little French town of 6000 inhabitants, 15km north of the Spanish border, whose trading turnover is quite large : 150 000 to 200 000 tons yr^{-1}. In addition, it is a busy fishing and pleasure port. Most of the waste water is evacuated outside the harbour, but some drain pipes and secondary sewage outlets emerge in the pleasure dock. With the exception of the period from June to September, we quote relatively moderate pollutant contents; cationic detergents are more abundant than anionic ones; the organic carbon contents are higher near the bottom than near the surface; salinity is slightly lower than in the open sea (36-38°/₀₀) but can drop to 17 °/₀₀ after storms (1).

Material and Methods

Sampling was made by collecting the whole settlement together with its substratum. The laboratory-sorting concerned the following taxa : Bangiophyceae, Florideophyceae, Phaeophyceae, Chlorophyceae and Bryopsidophyceae. Phytosociological parameters have been defined in a previous work (2).

Qualitative minimal area, established by mean of species area curves (3) and Sørensen index area curves (4) (index values >.75) is about 160 cm^2 in the top 20 cm and about 100 cm^2 at 6m depth (Table I). Quantitative minimal area, defined as the .60 value on the Czekanowski index area curve (5), is 100 cm^2 whatever the depth.

Xth International Seaweed Symposium
Printed in Germany

Quadrat area (cm2)	Number of data	Sorensen-index mean	Czekanowski-index mean
1	24	0.454	0.373
4	24	0.583	0.468
25	24	0.463	0.639
50	24	0.569	0.745
100	24	0.629	0.773
150	12	0.693	0.823
225	6	0.709	0.854
450	1	0.699	0.889

Table I : Sørensen and Czekanowski indices between contiguous quadrats (Pleasure dock, site C, sample RB-122, -6m)

Diversity and Equitability

160 species and varieties were found in the whole port area : 17 Bangiophyceae, 71 Florideophyceae, 30 Phaeophyceae, 17 Bryopsidophyceae and 25 Chlorophyceae. In the summer, the number of species near the surface is in the range of 40 to 55 by sample. This value drops slightly to between 30 and 40 (i) with depth, (ii) towards the mouth (sites E and F) and (iii) towards the head (site A) of the port.

At a given site, the Shannon index and the Equitability are relatively homogeneous from point to point at the same season. In contrast, we note deep changes from year to year; it is however premature to ascert that these changes are coherently orientated (Table II). As a rule, Equitability values are in the range or very slightly lower than the open sea values (6).

Density of Reproduction

Reproduction density usually ranges between 1 and 3; these values correspond well to the mean values of most of the Mediterranean phytocenoses (6); possible seasonal changes are not obvious; the heavy summer drop in reproduction with depth seems to be the only clear gradient.

Date	Sample number	Shannon index	Equitability
July 1974	RB-102	4.11	0.77
July 1977	RB-113	3.33	0.61
	RB-114	3.40	0.63
	RB-115	3.02	0.56
	RB-116	3.44	0.62
	RB-117	3.37	0.62
August 1978	RB-119	2.72	0.48
	RB-120	2.55	0.44
	RB-121	2.56	0.44

Table II : Changes of Shannon index and Equitability at a same site (C, - 0.2 m) from year to year.

Dominance of Main Taxa

Qualitatively, i.e. regarding the number of species present, the dominant classes are, in decreasing order, Florideophyceae, Chlorophyceae and Bryopsidophyceae; within the Florideophyceae, Ceramiales and Cryptonemiales are largely dominant; in contrast, Nemalionales, Gelidiales, Bonnemaisoniales, Rhodymeniales and Gigartinales are almost completely lacking. The qualitative dominance of Florideophyceae increases with depth (Table III).

Quantitatively, more conspicuous fluctuations are to be noted near the surface; The Phaeophyceae are dominant from February to May (*Dictyota dichotoma*, *Ectocarpus confervoides*, *Sphacelaria furcigera*, *Cutleria adspersa*, *"Aglaozonia parvula"*, *Giffordia sp.*), then disappear more or less completely to be replaced by Chlorophyceae (*Ulva rigida*), Bryopsidophyceae (*Cladophora sp. plur.*) or Florideophyceae . Near the sea-bottom, the dominance of Florideophyceae (*Pseudolithophyllum lobatum*, *Gelidium melanoideum*, *Ceramium sp. plur.*, *Antithamnionella sarniense*) is tremendous throughout the year (73 to 96%).

Recolonization

Denudation experiments were carried out on the wharf wall. We wanted to know whether, as has often been assumed (7), harbour

Sample number	RB-124	RB-126	RB-128	RB-130	RB-119	RB-120	RB-113	RB-114	RB-122	RB-112
Date	Feb.79	Feb.79	Apr.79	May 79	Aug.78	Aug.78	Jul.77	Jul.77	Apr.79	Aug.76
Depth (m)	-0.2	-0.2	-0.2	-0.2	-0.2	-0.2	-0.2	-0.2	-6.0	-6.0
Bangiophyceae	15.4 +	13.2 +	11.6 +	11.1 +	11.5 +	13.2 0.2	18.6 6.7	14.6 8.6	3.3 +	3.3 +
Florideophyceae	28.2 34.3	31.6 31.8	23.3 22.4	35.6 16.2	40.4 27.2	37.7 33.9	25.6 34.0	29.3 27.9	55.0 72.8	56.7 95.8
Phaeophyceae	18.0 46.1	13.2 31.5	25.6 45.4	17.8 63.7	13.5 21.9	9.4 23.8	9.3 0.2	9.8 0.4	16.7 15.9	6.7 2.0
Chlorophyceae	20.5 2.1	23.7 11.9	23.3 3.9	17.8 2.8	15.4 36.5	17.0 37.9	20.9 3.0	22.0 3.9	11.7 4.6	10.0 0.2
Bryopsidophyceae	18.0 17.5	18.4 24.8	16.3 28.2	17.8 17.3	19.2 14.3	22.6 4.3	25.6 56.1	24.4 59.3	13.3 6.8	23.3 2.1

Table III : Dominance of main taxa in some examples of samples. In each cell of the table : above left, the qualitative dominance; bottom right, the quantitative dominance as a function of the percentage cover.

settlements are simply juvenile settlements halted in growth before climax, or perpetually regenerating after repeated damages. Three months after denudation, short life-history Phaeophyceae (Ephemerophyceae, 8) are largely dominant : *Ectocarpus confervoides, Giffordia granulosa, Giffordia sp.* Qualitative and quantitative similarities (Sørensen and Czekanowski indices) between the regenerating and control areas increase very quickly (Fig. 1). The regenerating area can hardly be distinguished from the control area 8 1/2 months after denudation, although not identical especially from the quantitative point of view : as a matter of fact, the encrusting *Lithothamnium* stratum is far from being reconstituted.

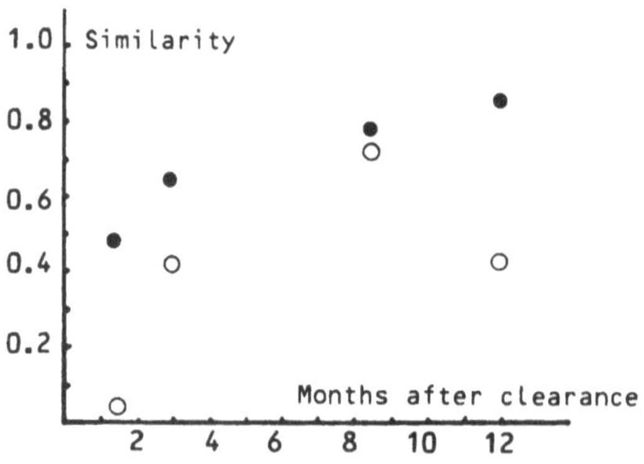

Fig. 1 : Sørensen and Czekanowski indices between regenerating and control settlements. Solid circles : Sorensen index; open circles : Czekanowski index (Site C, 0.2 m depth).

A comparison between Port-Vendres results and the recolonization rates in Mediterranean open sea settlements (9, 10) does not seem to present significant differences.

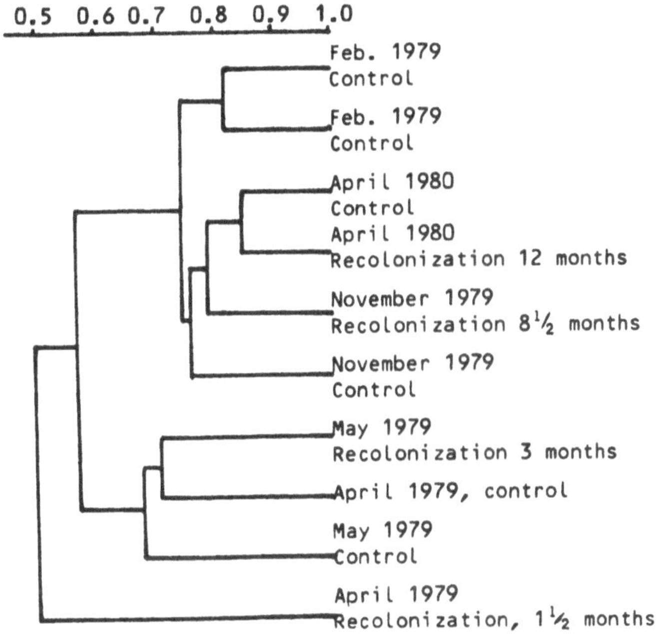

Fig. 2 : Dendrogram of regenerating and control settlements, based on data on presence/absence of species. (Site C, 0.2 m depth).

Phytosociology

Three phytocenoses seem to emerge from our results in Port-Vendres harbour : (i) the pleasure dock shallow phytocenosis, relatively rich in species, with a relatively high total covering, Shannon index and Equitability, and a normal reproduction density. (ii) The pleasure dock deep settlement, with a decrease of the above mentioned indices and parameters, and with an especially heavy drop in reproduction density; it may possibly be regarded as a reduced facies of the *Platythamnio-Compsothamnietum* phytocenosis (11). (iii) In the less polluted mouth of the port, the algal settlement is relatively poor in species, with a low Shannon index and is badly characterized; it is clear that we are faced with a hydrologically instable area : according to the predominant current, this area is alternately bathed by pure off-shore inflowing or polluted outflowing water.

Conclusions

Many of our results are in noway in agreement with generally accepted data concerning pollution impact on Mediterranean phytobenthos. In the most heavily and constantly polluted area, a relatively rich and well characterized shallow phytocenosis is to be found. This phytocenosis is *extremely miniaturized* , even more so than the other Mediterranean ones. The recolonization rate *invalidates the hypothesis that this settlement consists simply in stages of recolonization by opportunistic species*. It seems obvious that the Port-Vendres phytobenthos *reacts better to a moderate, but relatively constant pollution, than to a weaker but intermittent pollution* (12,13).

References

1. Casanobas, B.: Hydrologie du port de Port-Vendres. Etude préliminaire. Mémoire I.U.T. Perpignan, Fr., 61 p. (1979).
2. Boudouresque, C.F., Cinelli, F.: Pubbl. Staz. zool. Napoli, Ital., 40(2), 433-459 (1976).
3. Boudouresque, C.F., Belsher, T.: Cah. Biol. mar., Fr., 20, 259-269(1979).
4. Sørensen, T.: K. dansk. Vidensk. Selsk. Biol. Skrift,5,4,3-16,34 (1948).
5. Goodall, D.W.: Ordination of plant communities, Whittaker ed., Junk publ. The Hague, 99-149 (1978).
6. Panayotidis, P.: Comm. intern. Explor. sci. Médit., 27 (in press).
7. Littler, M.M., Murray, S.N.: Mar. Biol., Germ.,30, 277-291 (1975).
8. Feldmann, J.: Rev. algol., Fr., 10 (1-4), 1-340 (1938).
9. Boudouresque, C.F.: Helgol. wiss. Meeresunters, Germ.,24,202-218 (1973).
10. Huvé, P.: Recherches sur la genèse de quelques peuplements algaux marins de la roche littorale dans la région de Marseille, Thèse Doct. Etat Univ. Paris, 479 p., 65 pl. (1970).
11. Boudouresque,C.F., Belsher, T., Marcot-Coqueugniot, J.: Trav. sci. Parc nation. Port-Cros, Fr., 3, 89-120 (1977).
12. Belsher, T., Boudouresque, C.F.: Atti Tav. rot. internazion. "La Biologia marina per la difesa e per la produttività del mare", Livorno 1974, Stem-Mucchi impr. Modena, Ital, 215-260(1976).
13. Fresi, E.,Mazzella, L.,Cinelli,F.:Mem.Biol.mar.Oceanogr.,6,135-140(1976).

BIOGEOGRAPHY AND ECOLOGY OF THE SICILY CHANNEL

I. The algae of the Banks

F. Cinelli

Stazione Zoologica di Napoli, Laboratorio di Ecologia del Benthos, I-80077 Ischia Porto, Naples, Italy

Introduction

J.Feldmann, in his classic work "Recherches sur la Végétation de la Méditerranée - La côte des Albères", says: "si nos connaissances sur l'écologie des algue marines présentent encore beaucoup de lacunes, celles sur leur chorologie sont encore plus incompletes". This situation, forty years later on is improved and especially in the recent past, the literature on the Mediterranean marine benthic flora is being enriched mostly in the fields of the biogeography and ecology (1, 2, 3, 4). The Sicily Channel is one such area in which, with the exception of some recent studies, study of the benthic ecology and biogeography has been largely neglected. This work presents the results of two periods of investigations on the Sicily Channel to characterize the benthic phytogeography of the Banks and to identify the ecological requirements of the algal flora.

Material and methods

The material was collected during two oceanographic cruises of

Xth International Seaweed Symposium
Printed in Germany

the R/V Bannock (C.N.R.) in September 1975 and August 1976. The samples were collected from 15 stations at depths between 20 and 55 meters by SCUBA and preserved with 4% seawater-formol. For all sample a surface of 400 cm^2 was completely scraped clear of all algae. The localities sampled were the following: *Terrible Bank, Talbot Bank, Old Pantelleria Bank, Adventure Bank, Nerita Bank*. For a detailed morphological and geographical description of the area see 4 and 5.

Biogeographic characterization

A total of 180 species of macroalgae (*Chlorophyceae, Phaeophyceae* and *Rhodophyceae*) has been collected. They fall into four major biogeographic groups: 1). *Endemic*; 2). *Atlantic* (boreal, temperate and tropical); 3). *Indo-Pacific*; 4). *Cosmopolitan*, as reported in Table I. The percentage of species in the four groups is given for the Sicily Channel as well as comparative values for the Adriatic Sea and the Straits of Messina according to Giaccone (6) and Giaccone and Rizzi Longo (7).

TABLE I

	Region	Adriatic	Str.Messina	S.Channel
Groups	*Endemic*	46,02%	26,50%	33,90%
	Atlantic	41,47%	57,40%	53,50%
	Indo-Pacific	3,03%	2,00%	4,20%
	Cosmopolitan	12,61%	13,90%	10,50%

1). *Endemic*. About 61 species (33,90%) in the Sicily Channel are endemic to the Mediterranean. The number of endemics is relatively high. Giaccone and Rizzi Longo (7) for the flora of the Straits of Messina report 26,50% endemic elements and suggest that this value is due to the presence of a higher number of *Rhodophyceae* (as in the Sicily Channel) in comparison with other parts of the Mediterranean. Boudouresque (8), for the sub-biocoenoses occupying the shady and comparatively calm biotopes of the circalittoral, states that from the phytogeographical point of view its flora is mainly "mediterranean" with a conspicous number of endemics. According to the available information (9,10) it is possible to separate the endemic elements in two categories: *paleoendemic* and *neoendemic*. In our samples the only element of the first category is represented by *Laminaria rodriguezii* Bornet, a very peculiar species entirely different from all laminarians of Atlantic origin. Among the neoendemic species, the phylogeny of the genus *Cystoseira* is very interesting. Giaccone and Bruni (11) believe Mediterranean *Cystoseira* to have originated from one or few old Atlantic immigrants. According to these authors the species of *Cystoseira* found in the banks should have the following phylogenetic origin:*C. ercegovici* f. *latiramosa* Giaccone from the *C. discors-C. abrotanifolia* complex; *C. spinosa* Sauv.; *C. platyyclada* Sauv.; and *C. zosteroides* (Turner) C.Ag. from the *C.spinifera-C.opuntioides* complex.

2). *Atlantic*. The boreal and temperate Atlantic group is the richest one and includes 6 species of *Chlorophyceae*,19 *Phaeophyceae* and 53 *Rhodophyceae*, that is 43% of the total flora;thus,the Mediterranean represents a sort of refuge area and the southern limit of their diffusion. As for the tropical-atlantic group, which represent the 10,50% of the total (19 elements) it is composed of species common to the Mediterranean in the vicinity of the Straits of Gibraltar and nearby areas (Maroc,Mauritania,southern Spain and Portugal) and to the Atlantic in the Caribbean region, Some of these species have a Mediterranean origin (as *Codium bursa* (L.) C. Ag.) and only later colonized the Atlantic waters.

3). *Indo-Pacific*. This group represents only the 4,20% (7 elements) of the total flora in the Sicily Channel. As in the entire western-Mediterranean the species of Indo-Pacific origin are rather rare, the number being further reduced on the Banks due to the strong influence of north African surface current and particularly so for the species entering the Mediterranean within the Lessepsian migration (12). The only example of a Lessepsian immigrant is represented by *Lophocladia lallemandii* (Mont.) Schz. restricted mostly to the eastern part of the Aegean Sea as most Lessepsian immigrants.

4). *Cosmopolitan*. The cosmopolitan group represents 10,50% of the total and includes 6 species of *Phaeophyceae* and 13 of *Rhodophyceae*.

Structure and ecology of the algal population

The algal population of the Sicily Channel can be arranged into ecological groups and supergroups, according to Boudouresque et al. (13), as in Table II.

TABLE II

Ecological groups and supergroups	n. of Species	%
SI (sciaphilic infralittoral or *Acrosorikon*)	14	7,7%
SIC (sciaphilic infr.-circalitt. or *Rhodymenikon*)	12	6,6%
SC (s iaphilic calm condition or *Peyssonnelikon*)	48	26,6%
SCI (sciaph. calm infralitt. or *Aglaothamnikon*)	9	5,0%
SCIT (sciaph. calm infral. tolerant or *Udoteikon*)	6	3,3%
CC+CCT (circalitt. and circ. tol. or *Dasyopsikon* +		

Rhodriguezzelikon)	22	12,2%
ISR (infralitoral rock or *Corallinikon)*	11	6,1%
SRh (rheophilic or *Phyllarikon)*	15	8,3%
PhIsl (photophilic infralit. or *Mega-Cystoserikon)*	26	14,4%
SSBsl (superficial sciaph. exposed or *Mega-Schotterikon)*	8	4,4%
HP (phanerogam beds or *Posidonikon)*	5	2,7%
ETN (thionitrophic or *Ulvikon)*	4	2,2%

Table II shows that the majority of the flora is composed of species with sciaphilic requirements. The importance of the photophilic groups (Phsl) is certainly consistent with the extreme transparency of the water in this region. On the basis of information obtained from previous work in the Western Mediterranean, Cinelli et al. (4) stated that in the association of the sciaphilic algae growing under comparatively calm conditions (ASC) it is possible to recognise three sub-associations: ASCpi (calm infralittoral), ASCcc (calm circalittoral) and ASCr (rheophilic circalittoral). In the Sicily Channel, in which the bottom currents predominate both in the infralittoral and in the circalittoral , it is possible on the basis of this new data to recognize only one type of sub-association, the ASCr, characterized by a variable number of rheophilic species which form the canopy layer. The underlying stratum includes both photophilic and sciaphilic species in the infralittoral (groups PhIsl and SI), and mostly sciaphilic species (groups SC and CC+CCT) in the circalittoral.

Conclusions

From the phytogeographic point of view, the Sicily Channel appears to be a sort of boundary between the eastern and western Mediterranean. In this region the endemics are very important despite the predominance of the atlantic components in which a conspicous number of atlantic tropical species was found. The opinion of Giaccone (6) and Boudouresque (8) that the infralittoral zone represents the principal centre of formation and diffusion of the endemic species of the Mediterranean is confirmed. From the ecological stand point, the flora, living under the constant influence of the bottom currents is arranged in two different strata both in the infralittoral and in the circalittoral zone. The canopy layer, composed of rheophilic species acts as a sort of barrier to depress the water movement, reducing the impact of hydrodynamic factors and thus permitting the development of a considerable number of sciaphilic species.

References

1. Giaccone, G., Sortino, M., Solazzi, A., Tolomio, C. Lavori dell'Ist.Bot. e del Giard. Col. di Palermo 25,103-119 (1973).
2. Giaccone, G., Scammacca, B., Cinelli, F., Sartoni, G., Furnari, G., Giorn. Bot. Ital. 106, 211-229 (1972).
3. Giaccone, G., Sortino, M., Lavori dell'Ist. e del Giard. Col. di Palermo 25, 166-183 (1974).
4. Cinelli, F., Fresi, E., Mazzella, L., Ponticelli, M.P.,Giorn Bot. Ital. 113, 173-188 (1979).
5. Colantoni, P., Giornale di Geologia 40 (1), 181-207 (1975).
6. Giaccone, G., Rizzi Longo, L., Mem. Biol. Mar.Ocean. 6 (3), 69-123 (1976).
7. Giaccone, G., Annuario W.W.F. Italia 6 (19), 1-118 (1978).
8. Boudouresque, C.F., Bull.Mus.Hist.nat.Marseille 33,147-225 (1973)
9. Feldmann, J., Thèse Sci.nat. Rouen (1937).
10. Giaccone, G., Mem. Biol.Mar.Ocean. 2 (2), 37-59 (1972).
11. Giaccone, G., Bruni, A., Atti, Ist.veneto Sci.Lett.Arti. 131, 59-103 (1973).
12. Por, F.D.,Lessepsian migration,Springer-Verlag, Berlin-Heidelberg-New York 1978
13. Boudouresque, C.F., Belsher, T., Marcot-Coquegniot, J.:Trav. sci. Parc nation. Port-Cros 3, 89-120 (1977).

THE ECOLOGICAL IMPORTANCE OF KELP-LIKE HOLDFASTS AS A HABITAT OF INVERTEBRATES IN CENTRAL CHILE.

II. Factors affecting community organization.

J. Cancino(*) and B. Santelices

Laboratorio de Zoología, Pontificia Universidad Católica de Chile, Casilla 114-D, Santiago, Chile.

(*) Present address: Department of Zoology, University College of North Wales, Bangor, Gwynedd LL57 2UW, U.K.

Introduction

A critical evaluation of the non-productive, ecological roles played by brown algal holdfasts as habitats of invertebrates is not found in the literature. Most pertinent studies are either taxonomic treatments (1,2,3,4) with only occasional mention to aspects of population dynamics and faunal community organization, or compare the structure of these communities over a wide range of environmental conditions (5,6,7,8,9,10). Holdfasts normally provide refuge from wave impact and predation (2,3,4,5,6,7,11) and function as settlement areas and nursery grounds as is the case for some filamentous algae (12,13). However no attention has been paid to the patterns of holdfast colonization by invertebrates and to the effects of holdfast morphology in structuring such communities. This report is the second part of a study aimed at understanding the ecological roles played by brown algal holdfasts in the intertidal rocky habitat of Central Chile. It will analyse holdfast attributes important in structuring the faunal community and will characterize holdfast colonization patterns. A detailed analysis of the structure of this community has been given elsewhere (14).

Materials and methods

A total of 31 holdfasts of <u>Lessonia nigrescens</u> Bory and 22 of <u>Durvillea antarctica</u> (Chamisso) Hariot were collected between Jan-

Xth International Seaweed Symposium
Printed in Germany

Fig. 1. Schematic representation of chambers and channels in holdfasts of L. nigrescens (A) and D. antarctica (B).

uary 1976 and March 1978 in three localities of Central Chile (Horcón, Valparaíso and Las Cruces). Both algae are typically from the infralittoral fringe (15) on exposed rocky shores. The holdfasts of these 2 species differ in morphology (Fig.1). Usually only one chamber exists in D. antarctica and numerous in L. nigrescens.

In the laboratory, the larger macro-invertebrates (≥1mm) were removed from the holdfasts. Then the area of water interchange (defined as the sum of the area of all holes≥5mm on the holdfast surface) and the internal volume of the holdfast (free of algal tissue) were measured. Each holdfast was then dissected under a microscope to remove the smaller invertebrates. Holdfasts heavier than 0.5k were subsampled; four randomly selected, radial sections examined from each holdfast. The animals so gathered were sorted by species, counted and wet weighted. Biomass data of the 43 invertebrate taxa found were used to calculate diversity indexes (16,17) and association values among holdfasts (18). Samples were grouped following the linkage method (19).

Results and discussion

Volume of internal cavity is positively correlated with holdfast weight (Figs.2A,2D) and invertebrate biomass (Figs.2B,2E), which is expected, since chambers and channels in the holdfasts are excavated by the animals. However, the largest amounts of biomass were found

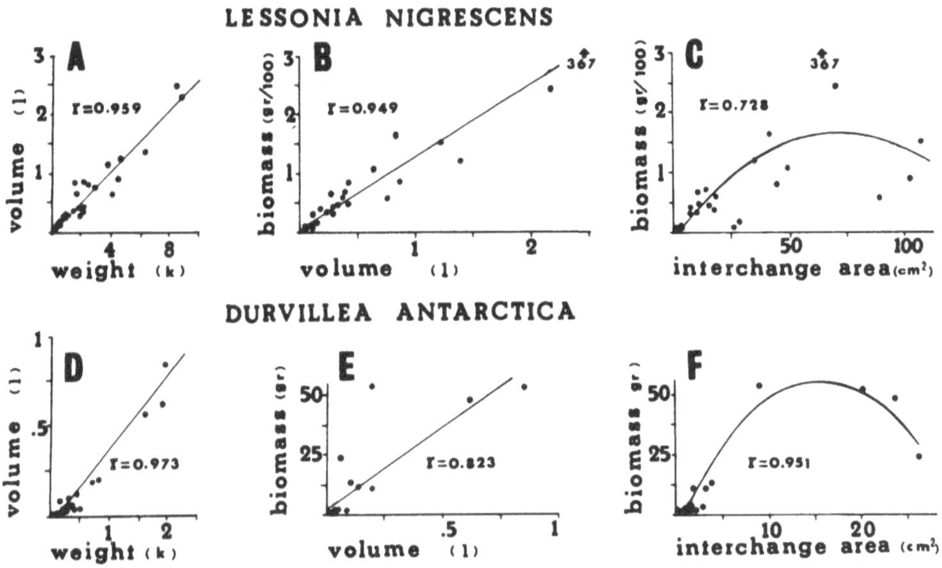

Fig.2. Relationship between parameters of the holdfast and animal biomass. Regression equations: A:$y=251.212x + 27.92$; B:$y=0.12x + 7.019$; C:$y=4.849x - 0.033x^2 - 10.274$; D:$y=392.486x - 53.376$; E:$y=0.069x + 2.523$; F:$y=7.622x - 0.235x^2 - 5.843$. With n=31 for A, n=26 for B,C and n=22 in D,E,F. Volume in cc. All correlation coefficients $p < 0.01$.

in holdfasts with areas of water interchange of intermediate values (Figs.2C,2F). Filter feeders, depending on the volume of water flow through the holdfast, are a high percentage of the total invertebrate biomass found in these algae (58-70%). Their reduction with increasingly larger interchange area seems related to reductions in the shelter offered by the holdfast and to increments in predator densities inside the holdfast.

Cluster analysis shows the existence of three groups of holdfasts in D. antarctica (Fig.3). Two groups of holdfasts, with volume ≤50cc, are dominated each by polychaetes or mussels. A third group of holdfasts, of larger size, is dominated by crabs which apparently replace the two other invertebrate groups as the holdfasts increase in size. This predictable pattern of succession is absent from L. nigrescens.

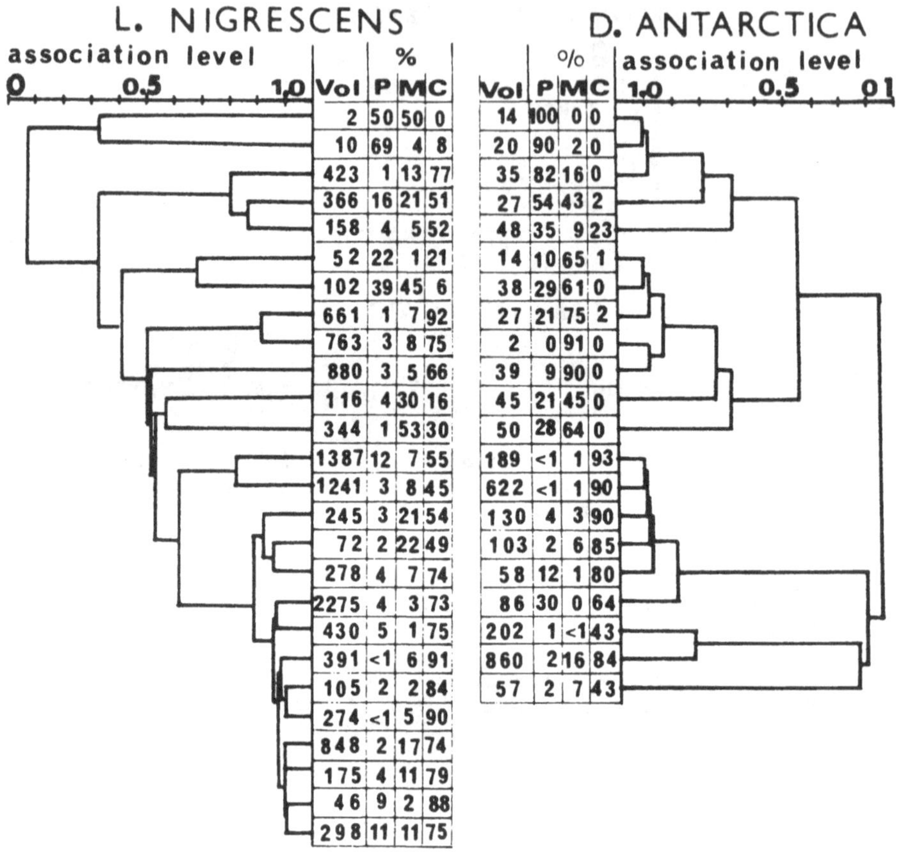

Fig.3. Cluster diagram of holdfasts based on biomass of animals. (Vol = volume in cc., P,M,C are the percentages of the biomass corresponding to polychaetes, mussels and crabs).

The existence of more than one chamber in the holdfast of L. nigrescens probably allows more species to fully utilize the holdfast internal volume without replacement among them. It is not surprising then that the cumulative diversity is higher in L. nigrescens than in D. antarctica (Fig.4). This is so even though both algal species show an essentially similar species assemblage in their holdfast faunal community.

These results also indicate that the general models of colonization

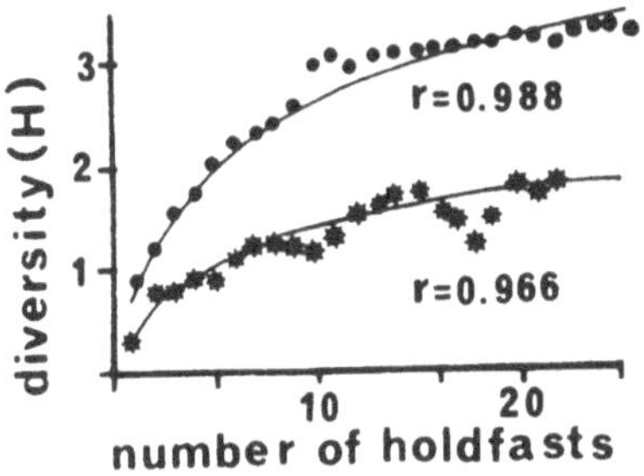

Fig.4. Cumulative diversity in holdfasts of L. nigrescens (●, H=0.857 + 0.812 lnx) and D. antarctica (✱, H= 0.277 + 0.468 lnx).

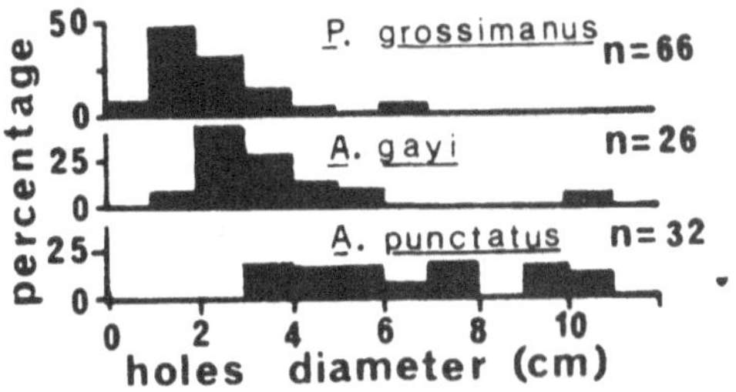

Fig.5. Distribution of 3 species of crabs in holes of different sizes. (Observed in L. nigrescens May and June 1979).

and species succession proposed in the literature (20) do not apply for multichambered algal holdfasts. The existence of several chambers allows the permanence of the first invaders in the community without species replacement at late successional stages. Species replacement, however, might occur in any given holdfast chamber, (providing that the connection with the exterior increases in size with time), as Fig.5 suggests for three species of crabs (Kruskal-Wallis Test (21), $p<0.01$).

In conclusion, this research characterizes invertebrate community development inside brown algal holdfasts to be an atypical process. Invertebrate biomass inside holdfasts increases as the volume increases and as long as this structure provides refuge from wave impact and predation to filter feeders. Colonization processes vary according to the uni or multichambered structure of holdfasts. In uni-chambered holdfasts colonization is a predictable process with species replacement; colonization of multichambered holdfasts is atypical with permanence of first invaders in the community.

Acknowledgements

Financial support from Dirección de Investigación, Pontificia

Universidad Católica. Grant No.78/82 is gratefully acknowledged. Eduardo Nealler, Hernán Castillo, Hernán Bravo, Daniel Moraga, Exequiel González, Juan C. Castilla and Pablo Schmiede helped at different stages of this work. John Sutherland made possible the utilization of a computer Dec-10 for the cluster analysis at Duke University Marine Laboratory during the summer 1978. Victoria Wood checked the final version in English. The British Council provided financial support to one of the authors (J.C.) to attend this Symposium. To all of them we express our gratitude.

References

1. Colman, J.: J. mar. biol. Ass. U.K. 24, 129-183 (1940).
2. Moore, P.G. : J. mar. biol. Ass. U.K. 51, 589-604 (1971).
3. Moore, P.G. : Cah. Biol. mar. 14, 493-518 (1973).
4. Moore, P.G. : J. exp. mar. Biol. Ecol. 32, 53-96 (1978).
5. Jones, D.J. : Helgolander wiss Meeresunters 22, 417-441 (1971).
6. Jones, D.J. : Helgolander wiss Meeresunters 23, 248-260 (1972).
7. Jones, D.J. : Mar. Biol. (berl). 20, 351-365 (1973).
8. Moore, P.G. : J. exp. mar. Biol. Ecol. 13, 97-125 (1973).
9. Moore, P.G. : J. exp. mar. Biol. Ecol. 13, 127-163 (1973).
10. Moore, P.G. : J. exp. mar. Biol. Ecol. 16, 257-300 (1974).
11. Moore, P.G. : J. exp. mar. Biol. Ecol. 10, 59-80 (1972).
12. Bayne, B.L. : Anim. Ecol. 33, 513-523 (1964).
13. Wigham, G.D. : J. mar. biol. Ass. U.K. 55, 45-67 (1975).
14. Canciño, J., Santelices, B. : Biologia Pesquera, Chile (in press)
15. Stephenson, T.A., Stephenson, A. : Life between tidemarks on rocky shores, W.H. Freeman & Co., San Francisco (1972).
16. Brillouin, L. : Science and information theory, Second edition, Academic Press, New York (1962).
17. Pielou, E.C. : Ecological diversity, John Wiley & Sons, New York (1975).
18. Morisita, M. : Mem. Fac. Sci. Kyushu Univ. Ser. E (Biol). 3, 65-80 (1959).
19. Sokal, R.R., Sneath, P.H.A.: Principles of numerical taxonomy. W.H. Freeman & Co., San Francisco (1963).
20. Connell, J.H., Slatyer, R.: Amer. nat. 111, 1119-1144 (1977).
21. Siegel, S. : Nonparametric statistics. McGraw-Hill. Kogakusha Ltd., Tokyo (1956).

PRODUCTIVITY OF *LAMINARIA SOLIDUNGULA* J. AG. IN THE CANADIAN HIGH ARCTIC: A YEAR-ROUND STUDY

A.R.O. Chapman and J.E. Lindley
Department of Biology, Dalhousie University, Halifax, Canada

Introduction

The high productivity of *Laminaria* Lamour. in south-eastern Canada has been well documented (4). Is it possible that kelps in the Arctic are also highly productive in contrast to the phytoplankton (2) and terrestrial vegetation (6)? This paper report measurements made on the rate of production by *L. solidungula* in Foxe Basin in northern Canada. We also report on a comparison of methods used for the estimation of kelp productivity.

Materials and methods

The site selected was situated at 8 m depth on the west side of Turton Bay, Igloolik Island (69°21.9'N81°46.2'W). On the basis of climate and vegetation, this site is classified as High Arctic (1). The site chosen was close to the middle of the depth range of the kelps.

The site was visited in August, September and November of 1978, and in February, April, June and August, 1979. During the 10 month iced-over period, diver-entry holes were cut in the ice.

In August, 1978 all of the kelp specimens were collected in 30 ¼m^2 quadrats placed at random along the 8 m depth contour. Length and weight measurements were made on each plant in each

Xth International Seaweed Symposium
Printed in Germany

quadrat.
Dry weights of tissues were obtained after air drying at 60 C for 3 days when constant weight was obtained. Carbon content was determined using a Hewlett-Packard 185B CHN Analyzer. Linear growth rates were measured on plants for one year beginning August, 1978. The method described by Mann (4) was used. 10 of the 30 plants tagged in 1978 survived one full year. It was thus necessary to add new plants to the group of sporophytes on each site visit. The measured length increments were transformed to biomass increments using the technique of Mann (4). It was found that Production/Biomass (P/B) ratios varied with plant size. Since the frequency of plant sizes in the group used for growth measurements did not correspond to the size frequencies in the population as a whole, it was necessary to weight each P/B ratio by a weighting factor derived from the population size class frequencies (5).

Results

Four kelp species were found in the standing crop measurements made in August, 1978 (Table 1).

Table 1. Standing crop (g fresh weight m^{-2}) of kelp species at 8 m depth, August, 1978. Mean weights $\pm$ standard errors are given for 30 quadrats.

Laminaria solidungula J. Ag.	867.26 $\pm$ 321.4
L. longicruris de la Pyl.	97.0 $\pm$ 56.39
Agarum cribrosum (Mert.) Bory	26.0 $\pm$ 26.0 *
Alaria grandifolia J. Ag.	5.6 $\pm$ 5.6 *

* = single specimen in 30 quadrats

L. solidungula constituted nearly 90% of the kelp biomass and was thus chosen for productivity studies. Insignificant quantities of seaweeds other than kelps occurred at this site. The blades of *L. solidungula* were retained from year to year. We found that plants grew very little between August and November. Therefore, weights of first year blades measured on a unit substrate area basis in August, 1978 give a direct estimate of blade production in the previous year. This was calculated as 17.9 $\pm$ 11.4 gC $m^{-2}yr^{-1}$ (mean $\pm$ 95% confidence limit) given % dry weight as 10.8 $\pm$ 1.4, and % carbon of dry weight as 31.8 $\pm$ 3.1.

Stipes were aged by counting annual growth rings in the discoid holdfast using the method of R. Hooper (pers. comm.). The average stipe production per year per unit surface area could then be estimated. The figure of 1.7 $\pm$ 0.28 gC $m^{-2}yr^{-1}$ was obtained. The combined estimate of blade plus stipe production for 1977-78 was 19.6 $\pm$ 12.2 gC $m^{-2}yr^{-1}$. The relationship between total plant length and total fresh mass is shown in Fig. 1. Using the linear regression shown, the annual length increments of the 10 plants surviving one year were converted to biomass increments.

The relationship between the annual biomass turnover and initial biomass of each plant is shown in Fig. 2. Clearly, the production per initial gram was much higher in the smaller plants. The mean P/B ratio for the 10 plants surviving one year was 0.72 $\pm$ 0.14, giving a production estimate of 21.55 $\pm$ 4.18 gC $m^{-2}yr^{-1}$ for 1978-79. When the data were weighted by size class frequencies in the population, an average P/B ratio of 0.91 $\pm$ 0.129 was obtained giving a production estimate of 26.98 gC $m^{-2}yr^{-1}$.

The daily production estimates for plants surviving between site visits are shown in Table 2. From these data, the annual estimates were computed as 25.74 $\pm$ 6.8 and 31.42 $\pm$ 5.1 gC m^{-2} yr^{-1} unweighted and weighted for population size class frequency respectively.

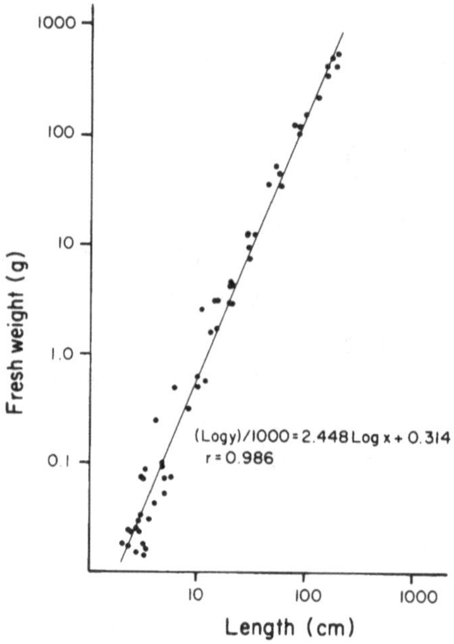

Fig. 1. Relationship between log fresh weight and log length in L. solidungula. Curve fitted by least squares regression.

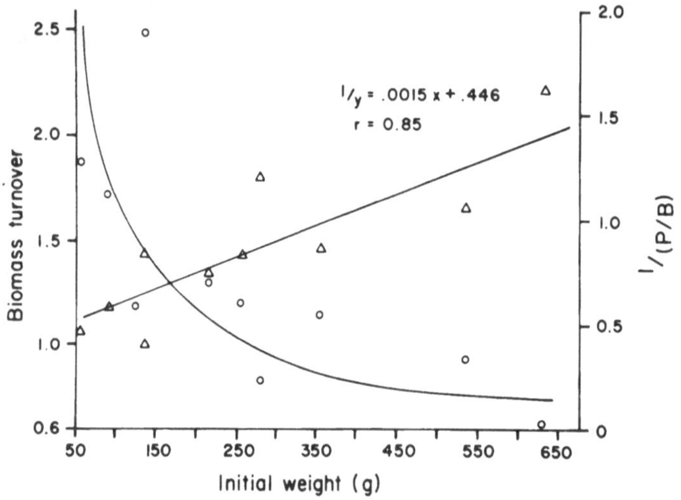

Fig. 2. Relationship between annual biomass turnover and initial biomass (o) for L. solidungula. The curve was fitted by eye. The reciprocal plot (Δ) was fitted by least squares regression.

Table 2. Mean fresh weight production (g g^{-1} day^{-1} $\pm$ 95% confidence limits) for six sampling intervals weighted and unweighted for size class frequencies in the population.

Growth Period	#Days	Production Weighted	Production Unweighted
August-September	41	0.0019±0.0006	0.0016±0.0005
September-November	50	0.0012±0.0005	0.0010±0.0003
November-February	100	0.0019±0.0002	0.0015±0.0004
February-April	47	0.0060±0.0006	0.0042±0.0009
April-June	63	0.0033±0.0006	0.0032±0.0008
June-August	62	0.0039±0.0006	0.0033±0.0009

Discussion

The average standing crop of L. solidungula at 8 m at Igloolik was 867.3 g m^{-2} fresh weight. Production by this species was ca. 20 gC $m^{-2}yr^{-1}$. Both of these figures are much lower than those for St. Margaret's Bay in south-eastern Canada where standing crop was estimated as 16 kg m^{-2} (3) and production as 1.75 kgC m^{-2} yr^{-1} (4). Because of the disparity in the standing crop estimates, the two sites are probably best compared by considering the P/B ratios at the same depth for the dominant species. At 8 m the P/B ratio for L. longicruris plants surviving one year in St. Margaret's Bay was 7.76. In Turton Bay 2500 km to the north the P/B ratio for L. solidungula was 0.72.

In St. Margaret's Bay, the seaweeds are about 10 times as productive as the phytoplankton (4). The productivity of the phytoplankton in Turton Bay is unknown. However, Grainger (2) cites an unpublished MS by Apollonio giving an annual estimate of 15 gC $m^{-2}yr^{-1}$ for Cornwallis Island waters 4° to the north of Igloolik. It seems reasonable to assume then that kelp and phytoplankton production are of the same order of magnitude in Turton Bay, Igloolik Island. Relatively speaking, the kelps are therefore less important as primary producers in northern waters than they are in southern Canada. The methods used in this study of kelp productivity allow an assessment of the accuracy of the technique developed by Mann (4). Direct measurement of L. solidungula production by weighing tissues gave a figure of 19.6 gC $m^{-2}yr^{-1}$. The best estimate based on the conversion of linear growth increments to biomass increments gave a figure of 31.4 gC $m^{-2}yr^{-1}$. However, these measurements refer to two separate years, 1978 and 79. Bearing in mind this qualification, it appears that using the indirect method for L. solidungula at Igloolik gives an overestimate of ca. 60%. When the data are not weighted by the size class frequencies in the population, an overestimate of only 30% is obtained. Whether or not these degrees of overestimation are constant for other localities and species remains to be seen.

Acknowledgements

We would like to thank Dr. Andy Rode and his staff at the Eastern Arctic Scientific Resource Center for their help. Funding was provided by the Natural Science and Engineering Research Council of Canada.

References

1. Bliss, L.C.: Can. J. Bot. 57, 2167-2178 (1979).
2. Grainger, E.H.: In M.J. Dunbar (ed.) Marine Production Mechanisms, pp. 9-30. Cambridge University Press 1979.
3. Mann, K.H.: Mar. Biol. 12, 1-10 (1972).
4. Mann, K.H.: Mar. Biol. 14, 199-209 (1972).
5. Sokal, R.R. and Rohlf, F.J.: Biometry. The principles and practise of statistics in biological research. W.H. Freeman, San Francisco 1969.
6. Webber, P.J.: In L.I. Tieszen (ed.) Vegetation production ecology of the Alaskan Arctic Tundra, pp. 37-112. Springer, New York 1978.

VARIATION IN ANNUAL PRODUCTION BY ASCOPHYLLUM NODOSUM (L.)LE JOLIS WITH DEGREE OF EXPOSURE TO WAVE ACTION.

Roger Cousens
Department of Biology, Dalhousie University, Halifax, Canada.

Introduction

While there are numerous accounts of the productivity of seaweeds over periods of a day or less, there have been relatively few attempts to obtain estimates of their annual production. There are great problems involved in extrapolating instantaneous photosynthetic rates, although Brinkhuis (1) has attempted to do so for saltmarsh ecads of Ascophyllum nodosum. Simpler, more direct methods include finding the difference between the annual maximum and minimum standing crops, and summing the differences in biomass over periods of a number of weeks (1,2).

This paper presents a modification of an existing method by which annual production by A. nodosum can be estimated from a single sample. The method consists of two stages: first the amount of growth made in the year prior to sampling and which is still present at sampling time is estimated, then a correction is made for the quantity which has been lost prior to sampling. The method is then applied to samples taken from a range of exposures to wave action; the results are compared with estimates obtained by the standing crop difference method (yearly maximum minus yearly minimum).

Methods

In Nova Scotia each shoot of A. nodosum generally produces a single vesicle at its apex each spring. The length of shoot

Xth International Seaweed Symposium
Printed in Germany

between successive vesicles - an internode - is thus a record of the past growth of a shoot (3). Let the portion of shoot distal to the most recent vesicle be I_o, the internode adjacent to it I_1, the next internode I_2, and so on. Baardseth (3) assumed that one year before sampling I_1 would have weighed approximately the same as I_o does now, I_2 the same as I_1, etc. Growth in weight over the past year and which is still present at the time of sampling will be given by

$$g = r + n_o\bar{w}_o + \sum_{i=1}^{i_{max}} n_i(\bar{w}_i - \bar{w}_{i-1})$$

where r is the weight of receptacles, $\bar{w}_i$ the mean weight of internode I_i, and n_i the number of internodes.

To divide a sample of reasonable size into all I_i's is very laborious: however, since growth is largely confined to the apical region and the internode below it, a good approximation can be obtained by considering only I_o and I_1. It is also important to distinguish between internodes which have their proximal ends at lateral pits (I_{ip}) and those which join on to further internodes (I_{if}), as the former usually make less growth. Internodes which end at the holdfast are assumed to have contributed a negligable proportion of the total growth and are not considered. The estimate for growth still present then becomes

$$g = r + n_{op}\bar{w}_{op} + n_{of}\bar{w}_{of} + n_{1p}(\bar{w}_{1p}-\bar{w}_{op}) + n_{1f}(\bar{w}_{1f}-\bar{w}_{of})$$

As it stands, g does not account for the appreciable biomass added during the past year but which has been removed prior to sampling. One way to correct for this is to estimate the number of internodes lost during the year (3). An alternative approach can be used if it can be assumed that gains and losses compensate for each other every year, i.e. the population is in steady state. In that case,

$$V + R = r + \Delta V + \Delta R + \Delta B \qquad (1)$$

where V and R are the total vegetative and reproductive growths over the past year respectively, ΔV and ΔR the quantities of those growths lost prior to sampling, r the reproductive biomass at sampling time, and ΔB the biomass lost in the past year but which had been added in previous years. Then

$$V + R - \Delta V - \Delta R - r = \Delta B = g - r \qquad (2)$$

Let the proportion of new growth contained in the biomass lost over the past year be

$$k = (\Delta V + \Delta R)/(\Delta B + \Delta V + \Delta R) \qquad (3)$$

from which $\Delta V + \Delta R = k\Delta B/(1-k)$

Substituting for $\Delta V + \Delta R$, then for ΔB, in (1),

$$V + R = (g - kr)/(1 - k) \qquad (4)$$

Although g and r can be observed directly from a sample, k must be inferred indirectly. Material lost close to one year before sampling contributes much less to k than does material lost just before sampling. Most losses occur during the winter, while growth is mainly in the summer: if a sample is collected in early spring k will therefore be at a maximum. It is unlikely that k will be below the proportion of new growth contained in the spring standing biomass, i.e. probably $k > g/b$ where b is the weight of the sample. In this paper I assume $k=g/b$, bearing in mind that this will likely give an underestimate of production. Finally, then, substituting g/b for k in (4) and adjusting for the size of the sample, net annual production per unit area will be given by

$$P = (V + R)B/b = Bg(b - r)/b(b - g) \qquad (5)$$

where B is the standing crop per unit area at the time of sampling. Equation (5), therefore, summarizes what may be called the Modified Baardseth method.

Samples were collected in mid-April 1979 from a series of locations around an island at Polly Cove, Halifax Co.. The principle difference between sites was in degree of exposure to wave action. An additional sample was taken from an inlet of Dover Soi, 4km from Polly Cove and completely sheltered from wave action. At each site a 10m tape was laid horizontally at a level slightly above mean tide level. The sample from which g was to be estimated was then collected by removing the third closest plant to each of 20-25 random points along the tape.

In this region maximum standing crop occurs in mid- to late-April, and minimum in mid-June immediately after abscission of receptacles. The standing crop at these times was estimated by a non-destructive regression method using the overall length and greatest circumference of bunched plants. The difference between maximum and minimum standing crops was used to give an alternative estimate of g.

Results & Discussion

Fucoid cover was 100% at low tide at all but the two most exposed sites. At the three most exposed sites Fucus vesiculosus and A. nodosum contributed about equally to the standing crop, while A. nodosum occurred in almost pure stands at the more sheltered sites. Standing crop of A. nodosum increased with shelter (Table 1), although there was a slight decrease in extreme shelter where winter ice may be important in limiting algal abundance. Annual dry weight production as measured by the standing crop difference method ranged from 0.05 to 1.05kg. m^{-2}, the values for all sites with full cover not differing

Table 1. Standing crop and production (kg dry wt.m^{-2}) at sites of different exposures to wave action. Sites 1-6 Polly Cove, site 7 Dover Soi. P/B was calculated from the Modified Baardseth production estimates.

	Increasing shelter ⟶						
Site #	1	2	3	4	5	6	7
April Standing Crop (±2 s.e.)	0.67 ±0.34	0.50 ±0.07	1.39 ±0.09	2.04 ±0.13	3.25 ±0.25	3.95 ±0.97	3.52 ±0.61
Modified Baardseth method	0.15	0.35	1.10	1.25	1.37	1.66	1.72
Standing Crop Difference method	0.05	0.26	0.95	0.94	0.84	1.07	1.05
P/B	0.22	0.71	0.79	0.61	0.42	0.42	0.49

significantly from each other ($P>0.05$). The Modified Baardseth method gave consistently higher estimates, rising from 0.15 to 1.75 $kg.m^{-2}$ with increasing shelter.

Data in Soneira & Niell (4) give an estimate of 2.3 $kg.m^{-2}$ by the standing crop difference method for a site in Spain with a peak biomass of 3.59 $kg.m^{-2}$. Westlake (5) adopted a P/B ratio of 0.25-0.33 and estimated production as 2.0-2.6 $kg.m^{-2}$ for a site of 8 $kg.m^{-2}$ standing crop in S.W. Nova Scotia. The highest estimate obtained here, while lower than these estimates, compares favourably with production by temperate terrestrial communities (6). Carbon content was found to be 36% of dry weight, giving a maximum productivity of 630 $gC.m^{-2}.y^{-1}$, which is comparable with a number of marine macro phyte communities (7, 8, but see 9).

Ratios of annual production to April standing crop show peak values at intermediate exposures and range from 0.22 to 0.79. The ratio P/B reflects the efficiency of the population in producing new biomass. Low values at extremely high exposures result from the fronds being very short and soon broken; production is restricted to very short laterals having low growth rates. Low values in sheltered sites result from an increase in self-shading with increasing standing crop, and possibly also from nutrient limitation of growing shoots. In such conditions, although production is high, production efficiency is low.

The two methods compared here both make large assumptions, and their reliability will very much depend on the extent to which these assumptions hold. The standing crop difference method assumes that all growth takes place between June and April, and that all losses occur between April and June. The Modified Baardseth method here proposed assumes that growth is confined to the first two internodes, that linear growth of fronds has been constant over the past two years, and that all new growth on shoots broken has been removed completely. Both methods, therefore, are likely to underestimate g. The calculation of P assumes k=g/b, which will probably underestimate P, and that the standing crop is constant from year to year. Overall, then, it is likely that the estimates given here will be fairly conservative.

The choice of method will depend on the accuracy required and on the time available. The standing crop difference method only takes a few minutes on each of two occasions two months apart and is probably least accurate. The new method takes one person about one week of intensive work but is probably more accurate since it does not entail the unrealistic assumptions that there is no growth from April to June and no losses from June to April.

Acknowledgements

I would like to thank Drs. Chris Pielou and Tony Chapman for their advice and comments, Elaine Wright for assistance with sample analysis, and Marjorie Willison for various help.

References

1. Brinkhuis, B.H.: J.Phycol. 13, 328-335 (1977)
2. Fuji, A. & Kawamura, K.: Bull.Jap.Soc.Fish. 36, 763-775 (1970)
3. Baardseth, E.: Regrowth of *Ascophyllum nodosum* After Harvesting, Inst.Ind.Res.Stand., Dublin 1955.
4. Soneira, A. & Niell, X.: Invest.Pesq. 39, 43-59 (1975)
5. Westlake, D.F.: Biol.Rev. 38, 385-425 (1963)
6. Whittaker, R.H. & Likens, G.E.: *In* H.Leith & R.H.Whittaker (eds.) Primary Productivity of the Biosphere, pp 305-328, Springer-Verlag, New York 1975.
7. Kain, J.M.: Ocean.Mar.Biol.Ann.Rev. 17, 101-161 (1979)
8. Mann, K.H.: Science 182, 975-981 (1973)
9. Chapman, A.R.O. & Lindley, J.E.: This volume.

ECOLOGY DISTRIBUTION AND SEASONALITY OF CRYPTONEMIA HIBERNICA GUIRY ET IRVINE ON THE SOUTH COAST OF IRELAND.

J. P. Cullinane and P. M. Whelan
Botany Department, University College, Cork, Ireland.

The species Cryptonemia hibernica was first described in 1974 based on drift material only, found by Guiry in Cork harbour in 1971. The species was known only from Cork harbour. The morphology, reproduction and seasonal variations have been described from drift material (Guiry and Irvine 1974). The holotype specimen was collected by Guiry at "Camden, at the mouth of Cork harbour" in November, 1971. Following that, an unsuccessful sublittoral search was made by W. F. Farnham in April 1972, at Camden to try to locate attached living material. Since Farnham failed to locate it while diving to a depth of 9m. below Chart Datum (C.D.), Guiry and Irvine suggested that "the species grows at a greater depth."

A general survey of the marine flora and fauna of Cork harbour directed by one of us (J.P.C.) was commenced in 1976 and attached living material of Cryptonemia hibernica was found at numerous locations within the harbour. The authors have concluded that not only is the species widely distributed in the (lower) harbour but that it is the most abundant species at some locations.

Drift material has been observed by the authors at various locations in Cork harbour during the period 1972 to 1980, including all the sites listed by Guiry and Irvine and in addition, Curraghbinny headland, Lough Beg, Fota Island and East Ferry (see Fig. 1). Drift material is undoubtedly most abundant in the Ringaskiddy area.

Xth International Seaweed Symposium
Printed in Germany

Attached living material of the species was first found at Ringabella at the south western extremity of Cork harbour November 20th, 1976, by two scuba divers, John O'Keeffe and Paddy Hyde, amateur phycologists attached to the University Marine Research Group. The material was identified as such by the authors. Since then Cryptonemia has been found growing at a number of locations in the harbour and at all times of the year.

TABLE 1.

Locations (see map)	Dates	Depth in meters											
		½	1	2	3	4	5	6	7	8	9	10	11
1 Ringabella	December 1976	+	+	+									
	April 1977	+	+	+									
2 Poulnacallee Bay	June 1977						+						
(Churchbay)	September 1977				+	+	+						
3 Weaver's Point	June 1977								+				
4 Ram's Head	June 1977			+	+	+	+	+	+	+	+		
(Camden)	August 1977			+	+	+	+	+	+	+	+		
5 Curlane Bank	June 1977		+										
6 North of Spike Is.	June 1977			+	+	+	+	+	+	+			
	September 1977			+	+	+							
7 Spitbank	September 1977	+											
8 Aghada Pier	May 1977		+										
9 Lower Aghada	May 1977		+										
10 Long Point	May 1977						+						
11 Corkbeg	May 1977									+			
	August 1977									+			
12 Fort Davis	May 1977			+	+	+	+						
(Carlisle)	June 1977			+	+	+	+	+	+				
	August 1977			+	+	+	+	+	+	+	+		
13 White Bay	May 1977		+	+	+	+	+	+	+	+	+	+	+
	July 1977		+	+	+	+	+	+	+	+	+	+	+
14 Roche's Point	July 1977			+									

Table 1. lists the locations (names in common usage in parenthesis), dates (1976 and 1977) and depths (below C. D.) for attached Cryptonemia. For numbered locations see map Fig. 1.

At Ringabella where attached plants were first located, there were approximately 50 plants in the entire area (40m x 20m) but Cryptonemia was far more abundant at Carlisle (30 plants per sq. m. at a depth of 3m.) and other locations. At Ringabella it was confined to one rocky outcrop and was absent from similar outcrops only 15m away. At most other locations it was not as restricted. Quantitative studies at Carlisle indicated that the number of plants decreased with depth, e.g. from an average of 30 plants per sq. m. at 3m. to 6 per sq. m. at 9m. However, the latter plants were of a much larger size and profusely branched and measured in situ up to 40 cm. wide.

Figure 1. Map showing the numbered locations of attached plants of Cryptonemia in the (lower) seaward parts of Cork harbour.

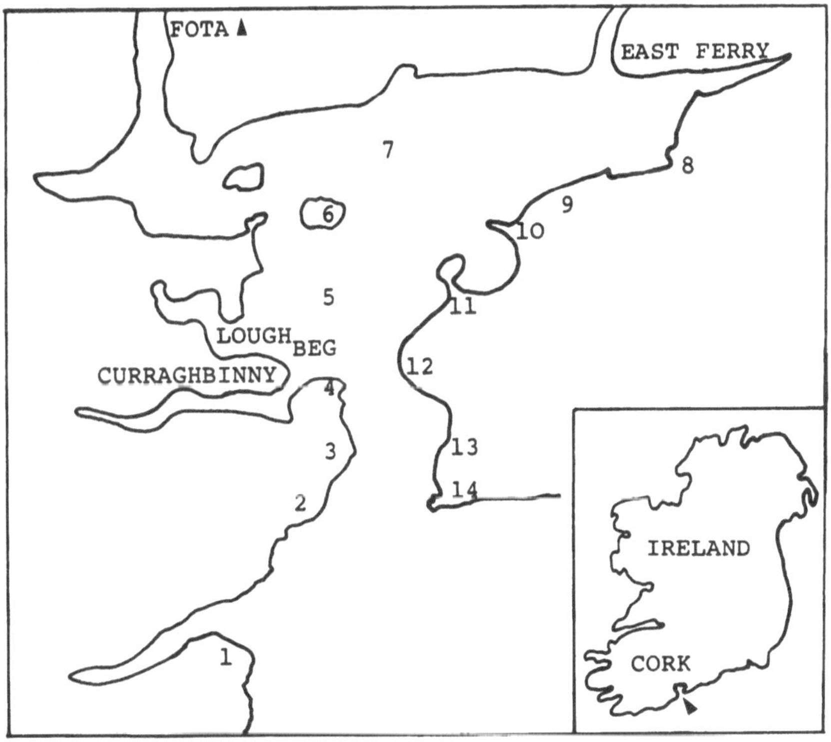

Cryptonemia was found in a variety of habitats, growing on a variety of substrates e.g. on bedrock in Laminaria forests in the immediate subtidal; on Laminaria holdfasts; in stable sand; on empty shells and stones of various sized, some as small as 2½ cm. x 1 cm. and weighing 2.5 grams. Many of these small stones and shells with Cryptonemia attached are mobile and are often abundant in areas of silty sand and strong currents e.g. around Camden headland where the authors have observed such small stones and shells with attached Cryptonemia being carried by the current. The occurrence of Cryptonemia on mobile stones and shells often accounts for it being found at locations such as, Aghada, near Corkbeg and in Zostera beds on the Curlane Bank and at White Bay. However it also grows anchored on non-mobile stones in Zostera beds.

At Ringabella Cryptonemia, unlike other species, is able to tolerate sand abrasion but it is limited to 3 m. depth by the occurrence of coarse mobile sand. It frequently grows on submerged bedrock thinly covered with sand. In such a habitat it shows a preference for the sand-filled depressions between the adjacent rock ridges, rather than the tops of the ridges. This contrasts with other species which show a preference for the tops and sides of the ridges. At locations such as Carlisle at depths of approximately 9 to 10 m., silt and strong currents prevail and in such conditions the animal Antennularia antennina and Cryptonemia were often found to be the two most abundant species along with lesser amounts of Dilsea carnosa (Sch.)O. Kuntze, Laminaria saccharina (L.) Lamour, Brongniartella byssoides (Good. et Woodw.) Schmitz, Dictyota dichotoma (Huds.) Lamour, Polysiphonia nigrescens (Huds.) Grev. and Stenogramme interrupta (C. Ag.) Mont. In this habitat all species including C. hibernica are noticeably pale in colour, probably due to turbidity and consequent reduction in light.

It is a general feature of it's ecology throughout the harbour that *C. hibernica* tends to grow best and consequently is the dominant species in conditions where other species are poorly represented (possibly a feature of an introduced species?). It exhibits a tolerance of sand abrasion, strong currents and reduced light not shown by any other seaweed species in Cork harbour and, here, with *Laminaria saccharina*, it is the deepest growing macro-alga.

The more extensive populations of *Cryptonemia* in the harbour are located as follows (see Fig. 1.); on the western side Ringabella (1), Churchbay (2), Weaver's Point (3) and Camden (4); in the inner harbour, near Spike Island (6) and Spitbank (7); and on the eastern side, Carlisle (12), White Bay (13) and Roche's Point (14).

Cryptonemia has also been located in two other bays west of Cork harbour namely; at Oyster Haven, in March 1979 and Kinsale, in March, 1979. Preliminary investigations of these locations indicate that it is very restricted in the former but in both bays ecological observation agree with those made in Cork harbour.

During the present study, 1976-1980, plants were found growing at depths from ½ to 11 m. However, in years in which lower tidal levels e.g. below C.D. might occur, it is extremely doubtful if *Cryptonemia* would survive in the intertidal. The species occurs in shallow water, 2 m. at Camden, and Farnham's failure to locate it there can, therefore, be attributed more to it's absence at the exact locations where he dived and not to the depth in question. It may be that, at the time Farnham searched for it, *Cryptonemia* may have been far less abundant than it is now.

In the original description of the new species, based on drift material, the surface of the thallus is described as "bullate" (Guiry and Irvine, 1974). Correlations were also suggested between the shape of the blade and seasonality, referring to ovate/obovate blades found from June to August and lanceolate blades found from November to February. Observations made during the present study based on attached plants indicate that the bullate appearance is a feature of drift material. Only two of the several hundred attached plants examined by the authors exhibited this bullate appearance, both occurring in very shallow water (1 m.). The authors are of the opinion that the shape of the blade is related to the type of habitat rather than to seasonal variation. Plants with linear/lanceolate blades tend to occur where they are attached to mobile substrates (e.g. shingle and shells) in areas of strong current, while plants with ovate/obovate and linear/lanceolate blades occur in areas of weak current. Plants are perennial and ovate/obovate and linear/lanceolate blades of various dimensions can be found at all times of the year. There appears to be continuous production of young blades throughout the year.

Nomenclature used was that of Parke and Dixon (1976).

Acknowledgements

We wish to thank John O'Keeffe and Paddy Hyde. We thank Mrs Linda Irvine (British Museum) for her help with the manuscript. The work was financed by the National Board of Science and Technology, Ireland.

References

Guiry M. D. and Irvine L. M. A species of _Cryptonemia_ new to Europe. _Br_. _phycol_. _J_. _9_ 225-237 (1974).
Parke M. and Dixon P. S. Check-list of British marine algae. Third revision _J_. _mar_. _biol_. _Ass_. U.K. _56_ 527-594 (1976).

STUDIES ON THE DEVELOPMENT OF PALMARIA PALMATA (RHODOPHYCEAE) USING IN SITU CONTROLLED CULTURES

Patrick Dion and René Delépine

Biogéographie et Ecologie Benthiques
B.V.M., 7 Quai St-Bernard
75230 PARIS CEDEX 05

Introduction

The red alga *Palmaria palmata* (L.) Stackh. (= Rhodymenia palmata (L.) Greville), is a common species on the rocky shores of the lower part of mediolittoral belt in Brittany. This species is used by the C.N.E.X.O. (Centre National pour l'Exploitation des Océans) in its experimental *Haliotis* nursery (Flassch, pers. communication) (5). Furthermore, its important biomass could play a significant role in the coastal ecosystem. Finally, the life cycle of *P. palmata* was not well understood because only tetrasporophytes and male plants were known in the field. For these reasons, an ecological study was undertaken on this species in order to estimate the influence of the tide level on its development and repartition, using *in situ* controlled cultures which have been successfull in a previous study (1).

Material and methods

- In the following experiment we used a single fertile tetrasporophyte plant, collected on October 26, 1977 at Roscoff (France). It grew at the optimal altitudinal level of development of the *Palmaria* belt.

- The sporeling was obtained from the mature blade placed in a large container (10 liters) of filtered seawater with bubbled air. Spores were released during all the night and settled down homogeneously on corrugated slides. Then the sporecovered artificial substrata were placed in Petri dishes and maintained in the laboratory during twenty days. Germlings were

Xth International Seaweed Symposium
Printed in Germany

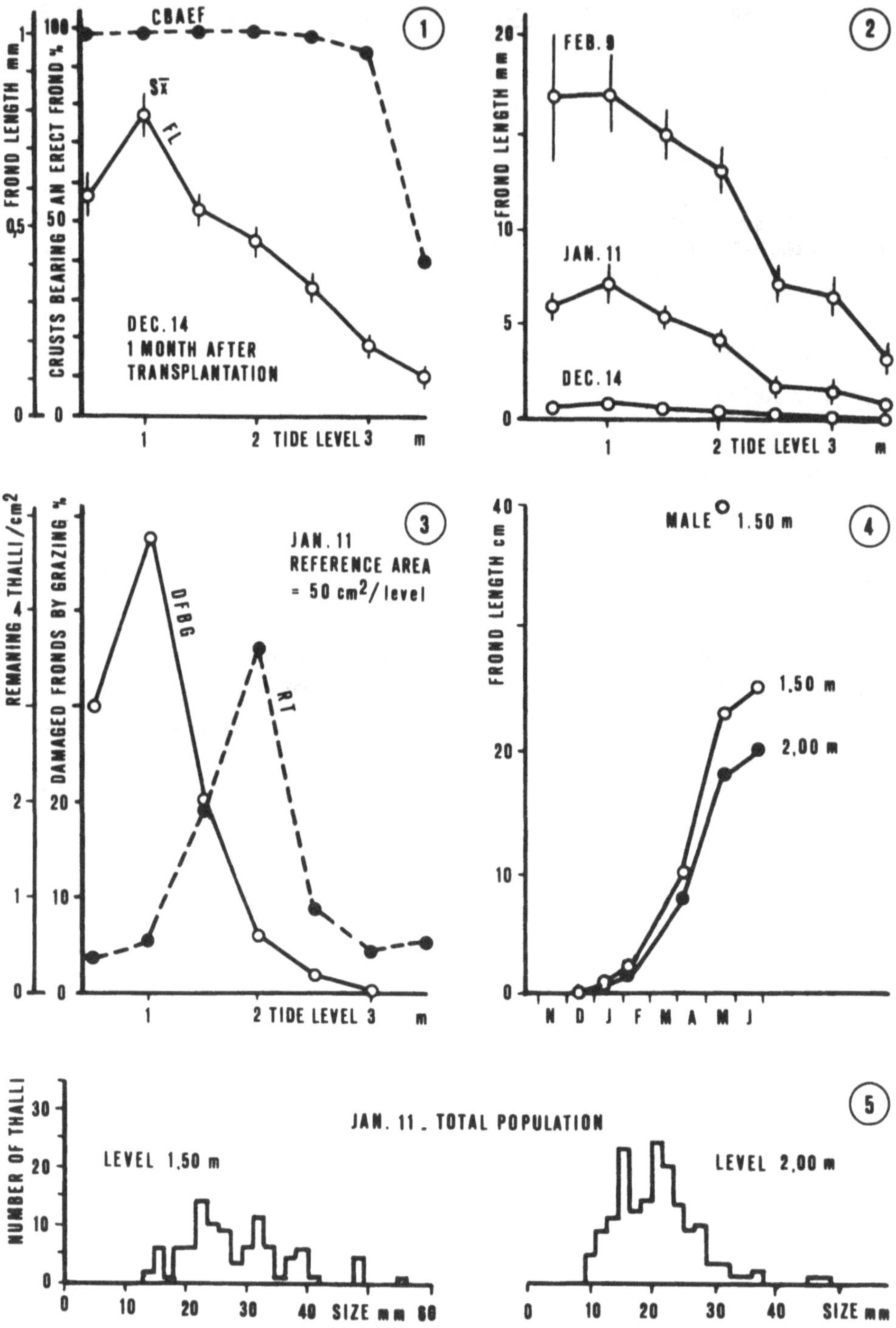
1
FROND LENGTH mm
CRUSTS BEARING AN ERECT FROND %
CBAEF
$S\bar{x}$
FL
DEC. 14
1 MONTH AFTER
TRANSPLANTATION
TIDE LEVEL
m
2
FEB. 9
JAN. 11
DEC. 14
FROND LENGTH mm
TIDE LEVEL
m
3
REMANING THALLI/cm^2
DAMAGED FRONDS BY GRAZING %
JAN. 11
REFERENCE AREA
= 50 cm^2/level
DFBG
RT
TIDE LEVEL
m
4
MALE 1.50 m
FROND LENGTH cm
1,50 m
2,00 m
N D J F M A M J
5
NUMBER OF THALLI
JAN. 11 _ TOTAL POPULATION
LEVEL 1,50 m
LEVEL 2,00 m
SIZE mm

cultivated at 14° C under a Sylvania light (Gro-lux) with a photoperiod of 12/12 L.D.; they received 1200 lux.

- The slides were transplanted in the field on November 14 and 17 using a "PVC slide holder" itself attached on a concrete block as already described (1). They were placed at seven different altitudinal levels of the mediolittoral belt, namely 0,5 m - 1,0 m - 1,5 m - 2,0 m - 2,5 m - 3,0 m - 3,5 m above the hydrographic zero for Roscoff. We used maregraphic records from the "Service Hydrographique et Océanographique de la Marine" to assess the altitude of blocks.

- The length of fronds was measured either directly through a stereomicroscope (for observations on December 14, 1977 and January 11 -February 9, 1978) or from photographic records for following observations (April 7 - May 22 and June 21, 1978).

Results and discussion

1) Production of erect fronds. At the date of their transplantation in the field, the discoïds germlings reach a diameter of 80-100 um (pl. 1/1). Very soon, young erect fronds appear on these discs (pl. 1/2). At the time of the first observations (one month after the transplantation) every crust bears an erect frond in the lower levels, from 0,5 m to 2,0 m; the percentage stays very high for other levels except for 3,0 m where it is particularly low (fig. 1). Thus the growth rate of this species is especially rapid in the lower intertidal zone.

2) The growth of fronds is clearly influenced by altitudinal level. On December 14 the best level is 1,0 m (fig. 1) whiles later both levels - 0,5m and 1,0 m - appear similarly optimal (fig. 2). Thus the ecophysiological optimal levels correspond to infralittoral conditions but we must note that at these levels the wild population is very sparse. The relative importance of emergence/submergence on photosynthesis and respiration has

....

Figures 1 - 5. Data on Palmaria palmata. See the text for discussion.

not been studied as for Iridaea for example (2), but P. palmata could also be better adapted for submergence than emergence. The important drop of the growth between the levels 2,0 m and 2,5 m (fig. 2) could also be due to these physiological characteristics. However we must note that there is a great development of Enteromorpha at these levels. In any way the level 2,5 m appears as a critical one for the ecology of P. palmata.

3) The density of thalli is also affected by tidal level. The density of germlings on slides at the time of transplantation was the same for each level. Two months later (on January 11), the density of remaining thalli is clearly maximal at 2,0 m level (fig. 3). The difference between this optimal density level and the above optimal ecophysiological level could be in relation with the grazing by phytophagous animals. The grazing is quite obvious on fronds (pl.1/3) and the percentage of damaged fronds by herbivorous is maximal at 1,0 m level (fig. 3). As a consequence the best biomass production appears around 2,0 m level in our experiment as it is observed in the field where the maximal density of wild population is also situated.

4) Evolution of populations. At the 1,5 m level, the population (composed of eight remaining thalli in May 1978) shows one male (pl. 1/6), one tetrasporophyte (pl. 1/7) and six sterile plants. Thus, in situ, we obtain a complete cycle (from tetraspore to new mature tetrasporophyte) in seven months. At the 2,0 m level there are no fertile plants and the thalli are less developed (fig. 4, pl. 1/8).
In the populations from these two levels (the only ones where sufficient thalli are staying at the end of the experiment) growth is rapid in spring (fig. 4 - pl. 1/4 and 1/5). After June the growth rate decreased and the thalli lost their pigmentation and became covered with numerous bladlets and hyaline hairs. In July and August, all erect fronds have disappeared and only few basal crusts are still visible. During this period the wild population undergoes the same pigment photodestruction. It is important to note that a drop of protein content is observed in summer (3). Any aquaculture program must take care of the above observations.

5) Life cycle. In our experiment no female plant was observed and on the two fertile plants there were one male and one tetrasporophytic thallus.

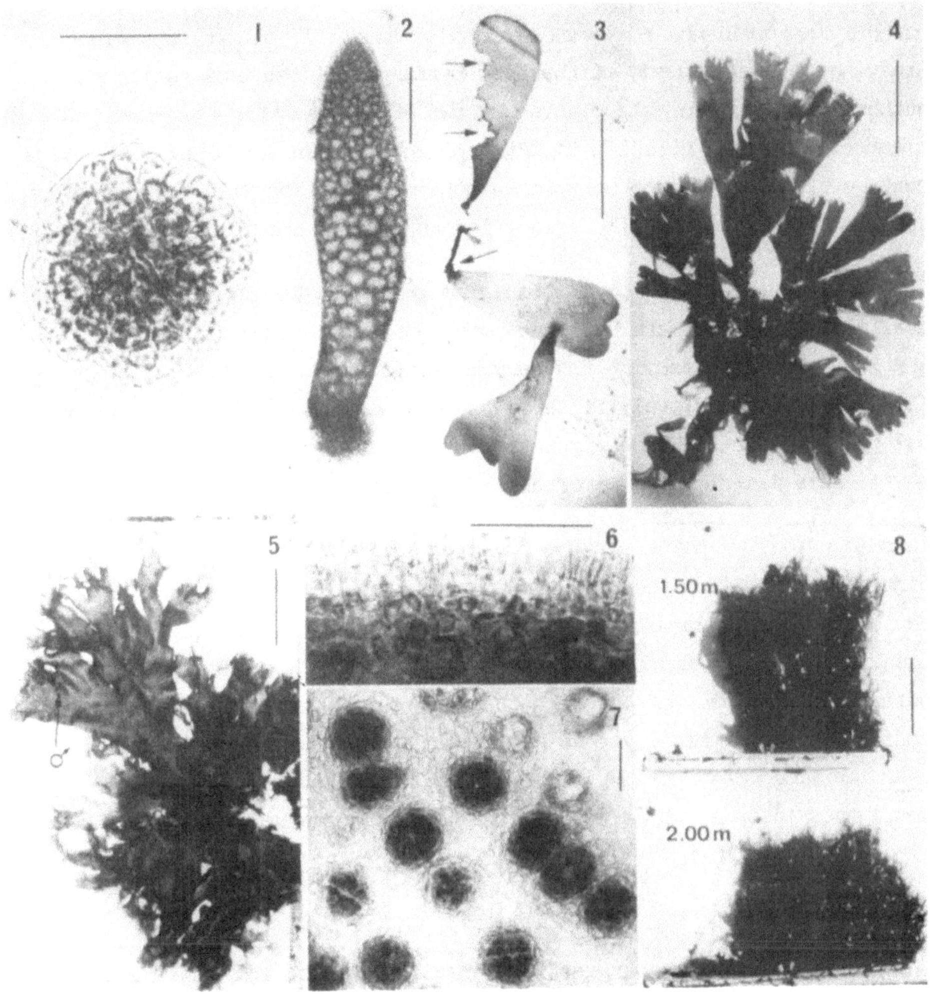

Plate 1. Data on Palmaria palmata : Pl. 1/1. Discoïd germling, twenty days old, at the date of transplantation - November 15, 1977 - Scale = 50 µm ; Pl. 1/2. A young frond crust, one month after transplantation - December 14 - 1,0 m level - Scale = 200 um ; Pl. 1/3. Young fronds damaged by grazing two months after transplantation - January 11, 1978 - 0,5 m level - Scale = 1 cm ; Pl. 1/4. Thalli on their artificial substratum about four months after transplantation - April 7 - 1,5 m level - Scale = 5 cm ; Pl. 1/5. Development of thalli about six months after transplantation - Note the large size of the male thallus - May 22 - 1,5 m level - Scale = 10 cm; Pl. 1/6 and 1/7. Reproductive structures - Transverse section in the male cortex (6) and superficial view of a tetrasporophytic blade (7) - May 22 - Scale = 50 µm ; Pl. 1/8. Comparaison of the vegetation at the 1,5 m and 2,0 m levels, seven months after transplantation - Note the disappearance of the male thallus at 1,5 m level - June 21 - Scale = 10 cm.

Recently the evidence for sexual reproduction in P. palmata was given (4). Studying the development of the four tetraspores from isolated tetrads, the authors obtained two male plants and two other ones which are dwarf female plants giving rise directly to tetrasporophytes. In our experiment the percentage of male/tetrasporophytic is not in opposition with their results, as is also in agreement the larger length of male thallus (fig. 4, pl.1/5).

6) Heterogeneity of the population. The frond length distribution of all thalli (fig. 5), appears very heterogeneous for each population from 1,5 m and 2,0 m levels. Assuming that environmental factors have been similar for all thalli at each level, it seems that the heterogeneity is due to genetical factors, although the populations originated from a single tetrasporophyte. This fact is also important in aquaculture process.

In conclusion, we point out the advantages of such in situ controlled experiment already discussed (1) and the importance of these results for any aquacultural program with Palmaria. We may mention especially either genetical selection to improve the raw material or the relations between specific physiological characteristics and the biological factors as herbivorous for example. Finally this experiment was developing at the same time that was Amoco Cadiz oil spill (March 1978) and all results should be reinvestigated in order to detect any influence of this marine disaster.

Aknowledgments

This work was done in the Roscoff Biological Station and supported by C.N.E.X.O. funds numbers 77/1669 and 78/1844.

References

1. Dion, P., Delépine, R.: Rev. Algol. N.S. 14(4), 327-341 (1979).
2. Quadir, A., Harrison, P.J., DeWreede, R.E.: Phycologia 18(1), 83-88 (1979).
3. Chaumont, J.P.: Botanica Marina 21, 23-29 (1978).
4. van der Meer, J.P., Chen, C.M.: Can. Journ. Bot. 57(21), 2452-2459 (1979).
5. Flassch, J.P., Woitelier, E.: Actes Colloques 4, 287-305 (1977)

EFFECTS OF TEMPERATURE AND IRRADIANCE ON GROWTH AND RESPIRATION OF CHONDRUS CRISPUS STACKH.[1]

C. T. Enright
Marine Colloids, Ltd., Dartmouth, N.S., Canada, B2W 3E0

J. S. Craigie
Atlantic Research Laboratory, National Research Council of Canada, 1411 Oxford St., Halifax, N.S., Canada, B3H 3Z1

Introduction

Interest in the productivity of macroalgae has stimulated reports which include information on growth (1), photosynthetic (2,3) and respiratory rates (2,4,5,6,7) of red algae, but usually data were recorded for plants from nature without adequate knowledge of their preconditioning. Frequently such data are widely variable and thus of marginal value in interpreting physiological events. We report the effects of two environmental variables on photosynthesis, respiration and growth of *C. crispus* raised under defined conditions.

Materials and Methods

Growth rates of *C. crispus* NRCC-T4 were measured in 3-tank systems in a greenhouse (8) at ± 1°C of the set temperatures. Plants grown at 2 & 5, 5 & 8 and 8 & 11°C (experiments 1-7, Fig. 1) were maintained under natural irradiance (Dec. 6-Jan. 24, 1978-9) plus two 300 W flood lamps which were operated between 0600 and 2200 h at 62 cm above each tank. This reduced the time required to obtain easily measurable growth

Xth International Seaweed Symposium
Printed in Germany

at the low temperatures. The other comparisons, experiments 8-20, were conducted with natural light alone from Feb. 27-May 2. The histogram bar heights give average growth rates ± s.d. (n=9); numerals within the bars give average irradiance (PhAR = 400-700 nm) in $MJm^{-2}d^{-1}$. The pH was controlled near 8.0 with CO_2 and 1.75 tank volumes of fresh seawater was added daily with commercial fertilizer (N + P) supplied thrice weekly.

Production vs. irradiance experiments were conducted outdoors in growth troughs (9) partitioned so that multiples of 5 replicates were exposed to 100, 86, 74, 64, 59 and 43% of full sunlight during Jun.-Aug., 1979. Fresh seawater was added at ca. 5 volumes d^{-1} per trough to keep the pH less than 8.5 and temperature at 14 to 17°C. Commercial fertilizer (N + P) was added in increasing amounts during trials 1 to 4 to eliminate colour differences between the exposed and shaded plants.

Oxygen consumption was measured (10) in water samples from six replicate 300 ml dark bottles containing 1-2 g fr. wt. of *C. crispus* incubated with continuous agitation at the designated temperatures. Plants were conditioned in the temperature-growth study for more than 4 weeks before testing. Measurements commenced near the end of the photoperiod (1900 h). The effect of irradiance level on subsequent dark respiration was examined in May and Aug. in parallel plant samples exposed to full sunlight or shaded with layers of fly screen.

Results and Discussion

From 2 to 14°C growth rates improved markedly with the greatest relative change occurring between 2 & 5°C (Fig. 1). The increase slowed between 14 & 20°C and rates became sensitive to irradiance levels. At 23°C growth rates declined.

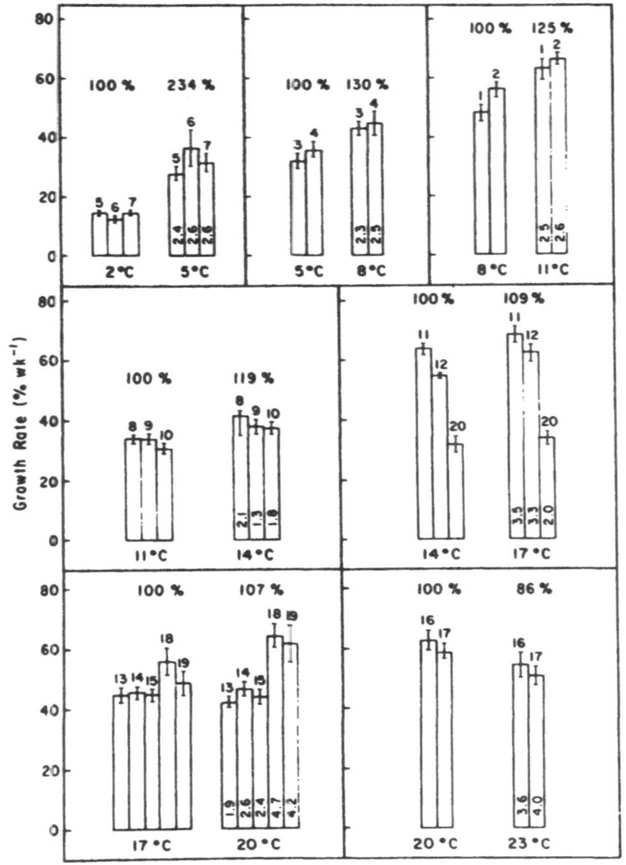

Fig. 1. C. crispus growth rate vs °C.

Table 1. C. crispus production (g fr. wt., $\bar{x}$ ± s.d., n = 5) as a function of a PhAR gradient (%) of natural light, Jun. 1-Aug. 14, 1979.

Days (Mean MJ m^{-2} d^{-1})	100%	86%	74%	64%	59%	43%
9.5 (6.6)	10.2 ±0.2	10.3 ±0.5	10.8 ±0.1	10.5 ±0.8	10.2 ±0.3	9.3 ±0.4
8.5 (6.8)	8.4 ±0.3	8.6 ±0.7	8.8 ±0.2	9.1 ±0.5	8.1 ±0.2	7.1 ±0.4
14.4 (6.1)	20.5 ±1.0	20.4 ±0.5	19.6 ±0.7	22.6 ±1.4	18.5 ±0.7	16.6 ±0.7
10.1 (6.6)	14.0 ±0.3	13.8 ±0.6	14.9 ±1.1	14.4 ±0.9	12.7 ±1.0	12.0 ±0.5
12.4 (5.7)	21.5 ±1.0	18.7 ±1.4	18.3 ±1.8	18.7 ±0.2	18.4 ±1.0	16.0 ±1.0
13.3* (4.0)	8.8 ±0.3	8.4 ±0.5	8.3 ±0.5	7.6 ±1.3	7.5 ±0.4	6.8 ±0.7

*Initial fr. wt. = 4.0 g each; all others = 13.0 g

Light saturated growth was achieved near 4 $MJm^{-2}d^{-1}$ of PhAR (400-700 nm) with virtually no indication of photoinhibition at the highest irradiances (Table 1).

Evidence for variable rates of dark respiration for temperature acclimated plants is shown in Fig. 2a. Regardless of the initial value, however, all rates declined by ca. 27% after 8-9 h of darkness. The dependence of respiration rate on temperature is illustrated for plants grown at three temperatures (Fig. 2b). Plants grown at 1-2°C showed a significantly higher respiration rate for a given temperature than those grown at 9°C, but within each of these groups little day to day variability was encountered. In five experiments plants grown at 17°C were tested simultaneously at two temperatures (dashed lines) to determine a mean $Q_{10} = 2.3 \pm 0.47$ for respiration between 8 & 23°C. The absolute respiration rates of plants maintained at 17°C, however, were unpredictable on different days. An important uncontrolled variable was the prior natural irradiance. Fig. 2c shows the influence of differing levels of irradiance on the subsequent dark respiration rates of otherwise identical plants. On May 6 and Aug. 23 half of a _C. crispus_ culture was screened from bright sunlight prior to 1000 h. Respiration measured at 1900 h (follow dashed line) was greatly diminished for the shaded plants. Data for the 17°C grown plants when replotted as a function of irradiance (Fig. 2c) generally followed the same pattern. The low respiration rates recorded on Apr. 9, 11 and 21 follow respective photoperiods in which the distribution of irradiance was quite irregular due to clouds or partially overcast conditions. The data show that dark respiration rates generally increase with increased irradiance levels in the previous photoperiod, and these rates may be sensitive to the distribution of irradiance within that photoperiod.

In order to estimate the P:R ratio for _C. crispus,_ the maximum growth rates compounded daily were calculated from the data in

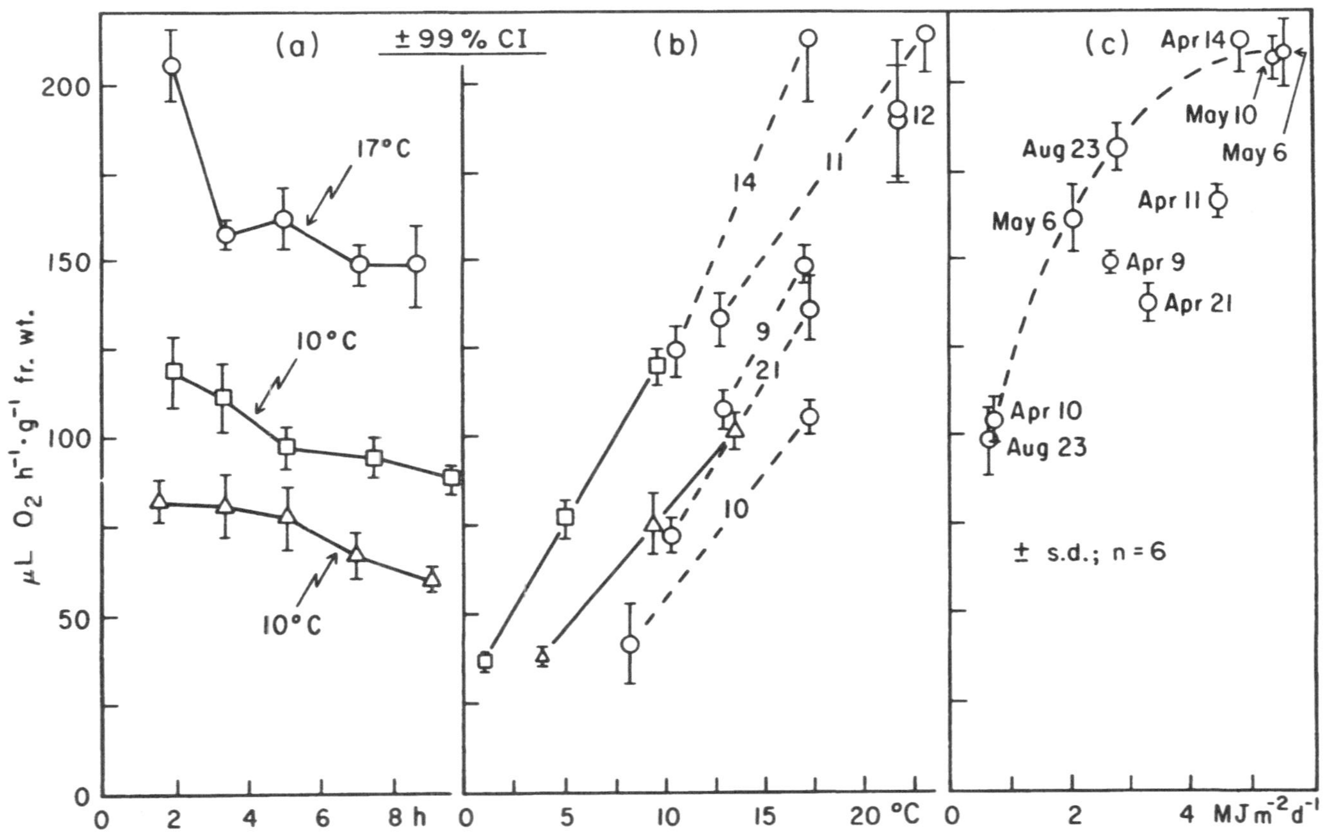

Fig. 2. Dark respiration rates of C. crispus acclimated to 1-2°C (□), 9°C (△), 17°C (○) as affected by: a) Increasing hours of darkness. b) Temperature. Numerals are dates in April 1979; dashes connect parallel plant samples. c) Prior irradiance level (PhAR).

Table 1 as 6.6, 6.4, 7.2, 7.9 and 9.1% d^{-1}, respectively, for each of the six trials. The dry matter was 19% of the fresh weight and contained 28.6% carbon. Thus, 1 g fr. wt. contained 4.53 mmole C, and assuming 8% d^{-1} increase in biomass the net photosynthetic rate would be 362 μmole C $d^{-1} \cdot g^{-1}$ fr. wt. To estimate a gross photosynthetic rate we added the carbon respired (=151 μmole) assuming the dark rate of 150 μL O_2 $h^{-1} \cdot g^{-1}$ fr. wt. (=6.29 μmole O_2 converted at standard temperature and pressure) was constant for 24h. The carbon respired is ca. 29% of the daily total 513 μmole fixed, thus under natural conditions of irradiance at ca. 14 to 17°C the P:R ratio approximates 3.4.

[1]NRCC No. 18255. Supported by a PI/LP contract to Marine Colloids, Ltd., Dartmouth, N.S. We thank T. Moore, P. Shacklock and L. Staples for expert assistance.

References

1. Prince, J. S., Kingsbury, J. M.: Biol. Bull. 145, 580 -588 (1973).
2. Kanwisher, J. W.: pp. 407-420. In Some Contemporary Studies in Marine Science. Ed. H. Barnes. Allen & Unwin, London · U.K. 1966.
3. Ramus, J., Beale, S. I., Mauzerall, D.: Mar. Biol. 37, 231-238 (1976).
4. Buggeln, R. G., Craigie, J. S.: Proc. N.S. Inst. Sci. 27, Suppl: 81-102 (1973).
5. Newton, L., Devonald, V. G., Jones, N. R.: pp. 121-139. In Écologie des Algues Marines. CNRS, Paris · Fr. 1959.
6. Mathieson, A. C., Burns, R. L.: J. Exp. Mar. Biol. Ecol. 7, 197-206 (1971).
7. Ogata, E., Matsui, T.: Jap. J. Bot. 19, 83-98 (1965).
8. Neish, A. C., Shacklock, P. F., Fox, C. H., Simpson, F. J.: Can. J. Bot. 55, 2263-2271 (1977).
9. Enright, C. T.: Proc. Int. Seaweed Symp. 9, 209-218 (1978).
10. Carritt, D. E., Carpenter, J. H.: J. Mar. Res. 24, 286-317 (1966).

DISTRIBUTION AND CONTROL OF THE BROWN ALGA *SARGASSUM MUTICUM*

W. Farnham, C. Murfin*, A. Critchley and S. Morrell
Departments of Biological Sciences and Mechanical Engineering*,
Portsmouth Polytechnic, England.

A. Distribution and Spread

Although *S. muticum* was first reported in European waters from the Isle of Wight (1), it is more likely that the alga was already well established on the Cherbourg peninsula of northern France (2). Its pattern of colonisation along new coastlines has been uneven, but *S. muticum* has spread along the north-west American coast at an average rate of about 60 km/year and along the English south coast at about 30 km/year. To date, only drift plants have been found along the Belgian and Dutch coasts - as far north as Terschelling. We suggest that attached plants may become established within the next few years in the Netherlands and even further northwards. Attached populations have been recently discovered in the Channel Islands.

The reasons for this remarkable spread of *Sargassum* can be partly explained by its distinctive reproductive features. In Europe, the plants undergo maximum vegetative development (formation and extension of primary laterals) from winter to early summer (October to July). Growth is reduced as receptacles develop from April to October. Many of these fertile laterals are cast-off during the summer months, and being buoyant, become planktonic (3). Laboratory observations suggest that *Sargassum* has a potential drift life of up to three months. *S. muticum* receptacles are androgynous (4) and self-fertilization can occur. Viable germlings are released from these floating fronds; thus one fertile fragment could initiate a new population. The movement of detached plants is brought about mainly by wind-generated surface drift and tidal streams. This

Xth International Seaweed Symposium
Printed in Germany

process constitutes an apparently effective method of "marginal dispersal" (5), relying upon the chance events of floating, fertile material releasing viable zygotes at suitable sites for colonization.

Surface drift bottles have been released from Sargassum population centres and have been recaptured at sites of previously recorded drift material and also at recently established populations. The general distributional pattern indicates that Sargassum has spread in Europe mainly by this method, although the movement of vectors such as shipping, pleasure-craft and shellfish may have also been instrumental.

B. Population Studies

The distribution, consolidation and ecological effects of Solent populations have been monitored since 1973 (see 2, 3, 6). Spread of S. muticum within Langstone and Chichester Harbours has been followed since 1975 using infra-red photography (Table I). Mud flats exposed at low water are dissected by a complex system of channels, the bottoms of which consist of flintstones and broken shells, which can be colonized by algae. S. muticum is unable to encroach upon the mud flats because of the lack of suitable substrate and too much desiccation. The cover by the alga within these harbours is less than 1% of their total area. This increase is likely to be maintained for the next few years, as Sargassum continues to spread along the low water channels.

At Bembridge, Isle of Wight, S. muticum has increased in extent but not necessarily at the expense of indigenous algae. Some plants attached to stones are moved by combined drag and buoyancy from the lower eulittoral lagoons into the shallow sublittoral to augment the population which exists down to a depth of 10 m. Over the winter 1979/80, there has been a decline in the Zostera beds in the upper eulittoral and Sargassum sporelings have started to colonise the newly exposed bedrock.

C. Mechanical Removal

The decision to attempt the eradication of *S. muticum* from the Solent has already been discussed (7). Early efforts (1973-1977) involved volunteers

TABLE I. Areas covered by *Sargassum* within Langstone and Chichester Harbours

Site	Cover in hectares (no 1978 data) 1975	1976	1977	1979
Langstone Harbour:-				
Portsea Creek	0.04	0.11	0.15	0.40
Broom Channel	0.06	0.14	0.29	0.92
Great Salterns Quay Channel	0.00	0.06	0.11	0.34
Russell's Lake	0.87	1.23	1.60	5.20
North Hayling Channel	0.00	0.00	0.30	1.01
Chichester Harbour:-				
West Cut	0.09	0.12	0.25	0.63
Swear Deep	0.00	0.00	0.43	1.37
Total area	1.06	1.66	3.13	9.87

removing *Sargassum* by hand (8). This method has been virtually discontinued because it is extremely time-consuming and needs to be repetitive. It can still be used at isolated sites at early stages in colonization where the population is sparse. Herbicides have not been used because of their lack of selectivity.

1) Design concepts

Removal of *Sargassum* from the seabed and transportation ashore for disposal inland were the objectives. Major design restraints included the need for minimal ecological damage combined with low capital and running costs. The approach adopted was to have the clearance devices operated and controlled from the removal vessel (10 m inshore craft), behind which the shoe-mounted operating heads were towed so as to follow the seabed contours.

Removal of attached Sargassum can be achieved by techniques of pulling or trawling, cutting, suction, or any combination of these. Collected material was most easily retained by nets or cages.

2) Removal by trawling

Techniques of inshore fishing could be readily adapted to Sargassum trawling. The weed trawl, developed from inshore fishing trawls, differs in minimizing both contact with the seabed and collection of marine species. Rapid removal of collected seaweed from the trawl is also possible. Trawls having tubular frameworks supported on two 50 mm wide shores with retaining nets were tested. The most effective weed collection heads were up to 3 m long with round bar teeth in two staggered rows mounted from the shoes about 250 mm above the seabed. Some 90% of the Sargassum (1 m long) was removed by this trawl type.

3) Removal by cutting

Mechanical cutting devices used previously on macrophytes (both marine and freshwater) were considered unsuitable for Sargassum in our local sea conditions. The basic trawl design concept was retained, but with cutters replacing the trawling head. Longstroke (75 mm), slow speed (600 r.p.m.) and reciprocating cutters (3 m long) have been tested. Low pressure, open-circuit fluid power transmission systems have been used to drive the cutters from the boat-mounted power source. A water-driven system, although more expensive, is more powerful than an air-driven system and therefore more effective in dense weed. Field trials using 2 m blades have shown that strips could be cut at about 2 knots leaving behind Sargassum 0.5 m long. Collecting the weed in nets was the major efficiency restraint.

4) Removal by suction

Direct removal of the seaweed by suction is likely to cause an unacceptable amount of ecological damage, but this method can be used in marinas and channels which may be dredged. The major design difficulty is in dealing with the weed when tangled and when attached to large stones (150 mm). This approach is unsuitable for large Sargassum (1.5 m). Two principles have been considered: jet-lift, using air or water, and "torque-flow". Our present system employs a vessel-mounted rotodynamic pump, with the suction head mounted on seabed-following shoes. The weed is collected in net cages which are buoyed and released when full, to be towed ashore.

5) Removal by cutting and suction

An air-driven cutter combined with an air-lift pump using exhaust air from the power source and spare capacity from the vessel reservoir can be used for continuous cutting and removal. This combination will help to overcome the problems of local weed storage with cutting and blockage by stones with suction.

D. Monitoring

Use of mechanical methods is being monitored to detect any changes within Sargassum communities, and the regenerative and reproductive abilities of the plants left behind and recolonisation of cleared areas are being followed. The comparative efficiency of different control tests is also being monitored. Material removed during clearance operations is examined (size, weight etc.) and associated species inadvertently collected identified. Monthly samples of plants are being made from previously cleared areas and adjacent uncleared sites; observations of reproductive and vegetative development are assessed.

Over 20 species of macro-fauna were dredged up with Sargassum in Langstone Harbour in August 1979; most abundant were Carcinus maenas (L.) and elvers,

Anguilla anguilla (L.). The dredge mainly removed large plants; smaller plants left behind were not appreciably damaged and later became fertile. During autumn plants decreased in length as fronds were cast. Over winter the lengths of plants in both cleared and uncleared areas became much the same, thus clearance would be necessary the following year. Juvenile plants (1 year) are more numerous in the cleared area and are important in the population next season.

Acknowledgements

We are grateful to the Department of the Environment (CM, SM & WFF) and to the Natural Environment Research Council (AC) for funding this work. Contributions have also been received from local authorities towards *Sargassum* clearance.

References

1. Farnham, W. F., Fletcher, R. L., Irvine, L. M.: Nature, Lond. 243, 231-232 (1973).
2. Farnham, W. F.: In, The Shore Environment: Methods and Ecosystems, Volume 2, (eds. Price, J., Irvine, D., Farnham, W. F.), Academic Press, London. New York (in press).
3. Jephson, N. A., Gray, P. W. G.: In, Biology of Benthic Organisms (eds. Keegan, B. F., Ceidigh, P. O., Boaden, P. J. S.), Pergamon, Oxford (1977).
4. Setchell, W. A. P.: Hong Kong Nat. 2 237-253 (1931).
5. Crisp, D. J.: J. mar. biol. Ass. U.K. 37 483-520 (1958).
6. Lewey, S., Farnham, W. F.: Proc. Intern. Seaweed Symp. 8, (in press).
7. Farnham, W. F., Jones, E. B. G.: Biol. Conserv. 6, 57-58 (1974).
8. Gray, P. W. G., Jones, E. B. G.: Envir. Conserv. 4, 303-308 (1977).

REMARKS ON SUCCESSION IN ROCKY INTERTIDAL SYSTEMS

M. C. Fernández
Department Ecology. University of Oviedo, Spain

F. X. Niell
Inst. Investigaciones Pesqueras. Muelle de Bouzas. Vigo, Spain

J. M. Fuentes
Inst. Español Oceanografía. Muelle Animas. La Coruña, Spain

Introduction

Quantitative structural data on benthic communities growing on cleared surfaces are very scarce (1). In a recent paper (2) the pattern of succession based on quantitative data has been stated. The aim of this work is to describe the behaviour of intertidal communities other than those studied previously (2).

Methods

Two sets of samples have been used in this study, one of them has been taken at Vigo (NW Spain) on a semiexposed coast in a fertile estuary, the second one has been collected in Cape of Peñas (N Spain) on an exposed coast, where the nutrient content in the water is lower than at Vigo. Sampling procedures and the measurements carried out are the same as those made in a previous study (2).

The samples have been collected in the following belts: <u>Fucus spiralis</u> (FC; high on 0 level, h=2,30 m), <u>Fucus</u> <u>vesiculosus</u> (Fv. C.; mean h=2 m) in Vigo; and, at lower levels (h=0,30 to

Xth International Seaweed Symposium
Printed in Germany

Table I. - Dominant species occupying the surfaces.

Time after clearing	2 months	9 months	1 year	2 years
F.C. (a)	Enteromorpha (Not in summer)	Fucus spiralis few Enteromorpha	Regenerated (Not in summer)	All surfaces regenerated
Fv.C. (a)	Enteromorpha	F. vesiculosus, few Enteromorpha	Regenerated	
H.C. (a)	Ulva Enteromorpha scarce	Himanthalia young, Ulva, caespitose reds	Regenerated	
B.C. (b)	Ulva plus Enteromorpha	Corallina, Ulva, Gigartina Halopteris		Regenerated
G.C. (b)	Ulva, Enteromorpha, scarce	Regenerated		
S.C.C. (b)	Ulva dominant	Gelidium latifolium, Ulva, several reds		Regenerated

a) Vigo b) Cape Penas

1,20 m) Himanthalia elongata (H. C.). In Vigo, this species is substituted at Cape Peñas by Bifurcaria bifurcata (B. C.), Gelidium latifolium (G. C.) and Saccorhiza polyschides-Cystoseira sp. (S. C. C.) going from high to low level.

Results

Three phases have been recognized during the colonization of cleared rocks (Table I), the first one is the C. P. consisting of a diatom film with germlings of macrophytes, a second one is the Ulvacean phase in which the algae of this group dominated. During this phase there is a continuous settling of macroalgae (Chaneling phase, C. S. P.) which drives the community to its normal structure (N. S. C.). In general the communities reached the N. S. C. after one year in rich areas, but the process takes two years in Cape Peñas in B. C. and S. C. C.

The number of species increased with time, averaging higher values at low levels in the ria of Vigo, than in other levels (Fig. 1).

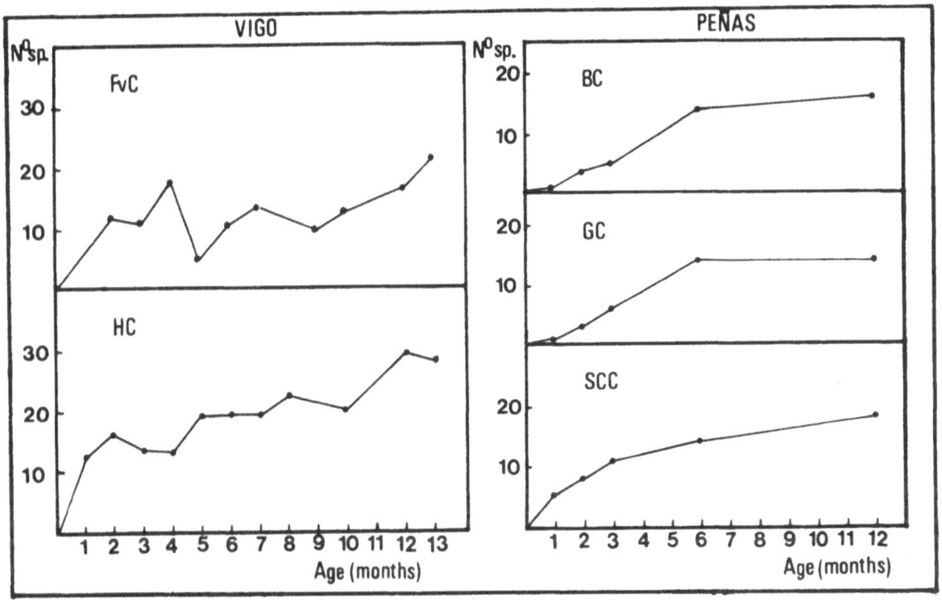

Fig. 1. - Increase of the number of species during the regeneration of the communities.

Changes in biomass are represented in Fig. 2; after one year the biomass in Cape Peñas, the locality in which the quantity of nutrients is lower, does not exceed 250 g m^{-2}, while in Vigo the quantities, even in high level (Fv. C.) exceeded the value of 250 g m^{-2} recorded at Cape Peñas, the level of biomass of the N. S. C. was reached faster in low level than in mid and high ones, and in rich than in poor waters.

Diversity presents several peaks along the succession (Fig. 2). Some of them due to the overlapping of two phases of the colonization, as described by (2). The general trend is an increase of diversity with time.

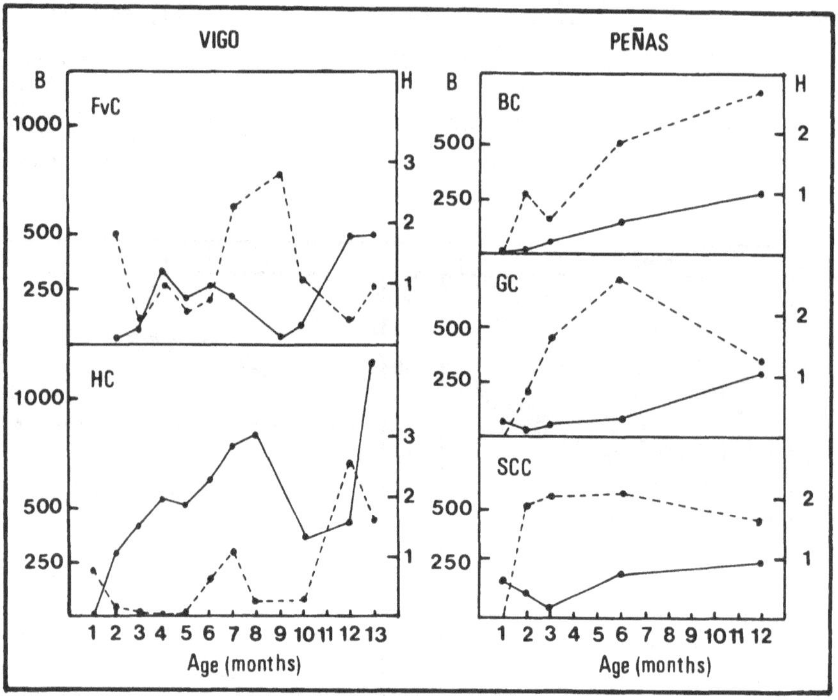

Fig. 2. - Changes of biomass (continuous line) in g m^{-2} and diversity (broken line) along the regeneration of the communities.

Discussion

The structure of the communities first installed after clearing, depends on the season, but the community installed at each level on the coast is predestinated, the community reached is the N. S. C. (3, 4, 9).

The speed in reaching the highest development in a determined level is different at each locality, depending on nutrient concentration in the water. The level influences the success in reaching the N. S. C., depending on the stress caused by the length of emersion.

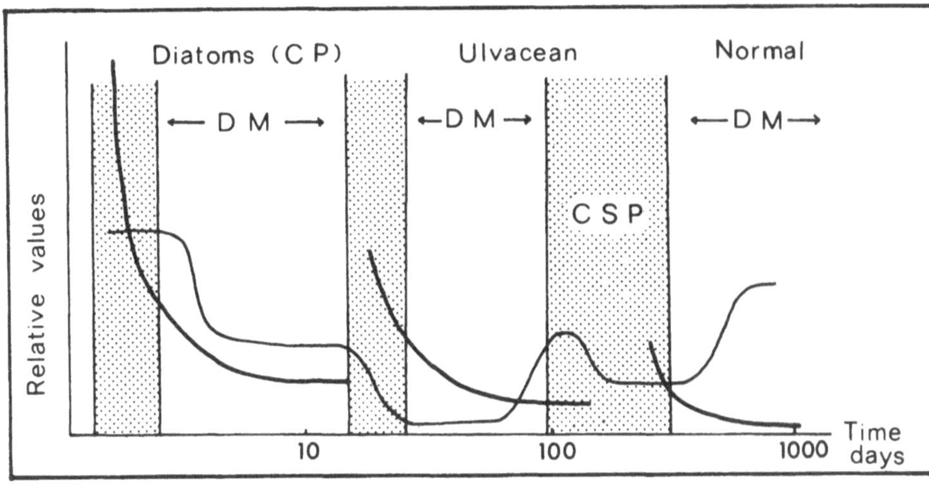

Fig. 3. - Changes of energy flow (B/B coarse lines) and diversity (fine lines) along the recolonization. Shaded areas represent transient moments between development moment (D.M.)

The results obtained here fit well with the model (Fig. 3) presented in a previous paper (2) which suggested some connection with physical theories (7, 8). Two states alternate in succession: these are the named development moments (7, 8), during which the energy flow through the community is minimized, and an improvement of structure could be verified by means several indexes (9), and the transient moments, or overlapping situations mentioned above during which a jump of energy flow is detected. - These are unstable communities quite ephemerous from one state to another.

The jump of energy flow decreases from the earlier transient moments to the older ones, and the energy flow maintained through the communities is greater in earlier than in the more advanced states.

These theories have empirical verification with experimental data exposed in this paper, and in (2, 3, 4, 5) in rocky communities, but also in other systems (11). As stated in (10) these trends seem to be a general behaviour of the communities along the successional processes.

References

1. Foster, M.S.: Mar. Biol., 32: 313-329 (1975).

2. Niell, F.X.: J. Exp. Mar. Biol. Ecol., 36: 185-200 (1979).

3. Fernández, M.C.: Ph. D. Thesis. Univ. of Oviedo (Spain) (1980.

4. Niell, F.X., Varela, M.: unpublished.

5. Fuentes, J.M.: Ph. D. Thesis. Univ. Santiago de Compostela (Spain) unpublished.

6. Huve, P.: Soc. Phycol. France Bull., 19: 101-120 (1974).

7. Prigogine, E.: Allen, P.M., Herman, R. The evolution of complexity and the laws of nature in Laszlo. E. & J. Bierman, 5-63 (1977).

8. Velarde, M.G.: Bull. Fund. J. March, 3-17 (1978).

9. Niell, F.X.: Helgol. Wiss. Meeresunters., 30: 315-332 (1979).

10. Margalef, R.: La Biosfera, entre la termodinámica y el juego Ed. Omega. Barcelona (1980).

11. Wagensberg, J.: Ph. D. Thesis. Univ. of Barcelona (Spain) (1976).

THE GROWTH, REPRODUCTION, HARVESTING AND MANAGEMENT OF PTEROCLADIA PINNATA (RHODOPHYCEAE) IN THE AZORES, PORTUGAL

R.A. Fralick, Plymouth State College, University System of New Hampshire, Plymouth, New Hampshire, U.S.A.
F. Andrade, Faculdade De Ciencias, Universidade De Lisboa, Portugal.

Abstract: The growth, reproduction, harvesting and management of Pterocladia pinnata (Huds.) Papenfuss = P. capillacea were investigated for a period of one year (August, 1979 to July 1980) to determine the maximum sustainable yield of Pterocladia resources for agar-agar production in the Azores. Field and laboratory studies including SCUBA diving and harvest impact analyses were completed. Optimal conditions for the growth of P. pinnata have been defined. Relative paucity of sexual phases for reproduction was observed and confirmed as well as the probability that hydrodynamic fragmentation is a major causal factor in the sucessful vegetative propagation for P. pinnata and a possible causal factor inhibiting the formation of sex organs.

Pterocladia pinnata with an annual harvest of about 1800 M tons is utilized as a resource material for the extraction of 325 tons of high quality agar-agar per year in the Azores. The industry and the Regional Government are concerned over the growth, harvesting extent and management of the Pterocladia resource and its potential limitations for the future. Definition of a maximum sustainable yield (MSY) and knowledge of standing crop stability will allow the expansion of the industry to a production level of 400 M tons of agar-agar in the next few years. Increased production technology is now being considered and implemented by the factories to improve agar extraction.

Xth International Seaweed Symposium
Printed in Germany

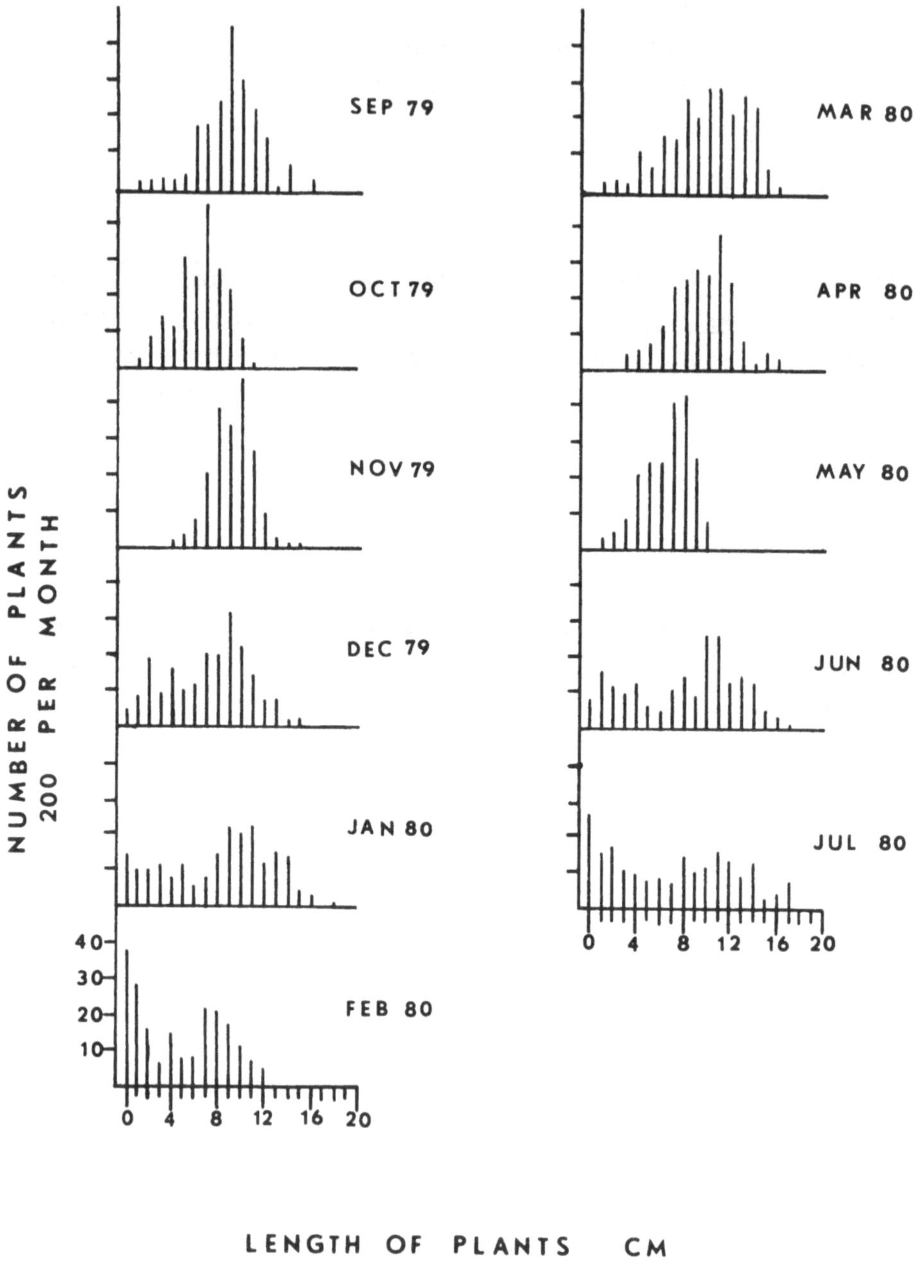
SEP 79
OCT 79
NOV 79
DEC 79
JAN 80
FEB 80
MAR 80
APR 80
MAY 80
JUN 80
JUL 80
NUMBER OF PLANTS
200 PER MONTH
40
30
20
10
0
4
8
12
16
20
LENGTH OF PLANTS CM

Materials and methods: The growth of _Pterocladia pinnata_ was evaluated from monthly collections of 200 plants from homogeneous populations on the island of Terceira. SCUBA diving was used to make the monthly collections and to observe harvesting impacts _in situ_ from depths of -1 to -20 M. The monthly collections were transported to the laboratory and measured within 24 hours. During the measurement each plant was evaluated microscopically for sexual structures using a Wild Dissecting microscope. Examples from the monthly collections were preserved in 4% neutralized Formalin for future study. Management and harvest data was obtained from individual factories and from government statistical records. The harvest statistics were compared with annual agar-agar production to determine their validity.

RESULTS

Growth: _Pterocladia pinnata_ exhibits a pattern of year-round growth in the Azores. Optimal growth and branch proliferation occurs when the water reaches a temperature of between 20-22° C during the summer and early Fall months. This period coincides with maximum solar radiation and day length. Growth is reduced during the winter and early spring. During periods of optimal growth plants of 15-16 cm in length are frequent. The _Pterocladia_ plants grew in abundance at depths of -1 to -5 M.

During the growth studies (Figure 1.) we observed periods of hydrodynamic fragmentation during December, January and June. In each case a major oceanic storm preceded the coll ections for growth studies. The process of fragmentation for _Pterocladia_ apparently had no physiological basis but, consistently occured during storms. Fragmentation may occur during any part of the year depending on meteorlogical conditions. The _Pterocladia_ plants that were longer than 10 cm seemed particularly sensitive to storm induced wave damage. Thus, a critical size of 10 cm or greater becomes important in the evaluation of Pterocladia growth. In addition, hydrodynamic fragmentation may be related to the unsucessful development of sexual structures for Pterocladia. Large Pterocladia plants observed during September, October and November became reduced in size when fragmentation occurred. Large washings of fragmented shore-cast alage can greatly influence the annual harvest of Pterocladia. Following the occurrence of fragmentation a uniform population level was apparent (January). The uniform population level for June could have been associated with a harvesting effort, as well as a storm.

Reproduction: The reproduction of Pterocladia pinnata in the Azores is primarily vegetative. Tetrasporophytic plants were collected at the beginning of October and were conspicuous throughout December. The tetrasporophyte plants occurred in unusually high numbers and accounted for as much as 30% of the Pterocladia plants in the areas we studied. Sexual plants were not observed at any time during the year.

Harvesting: Three methods are used to harvest Pterocladia in the Azores. These include collecting of shore-cast seaweed after local storms; hand and snorkel collecting in shallow water; and diver harvesting from compressed-air equipped boats in deeper water. Each of the methods has its own well defined zones for collecting and do not substantially overlap one another. Some impacts can be anticipated with theincrease in number of compressed-air divers. The impacts are caused by divers filling the collecting sacks quickly and damaging the bases of the plants in the process. Collector instruction is presently being implemented to advise the diver collectors of the proper method of collecting Pterocladia without excessive damage to the regenerative bases of the plants.

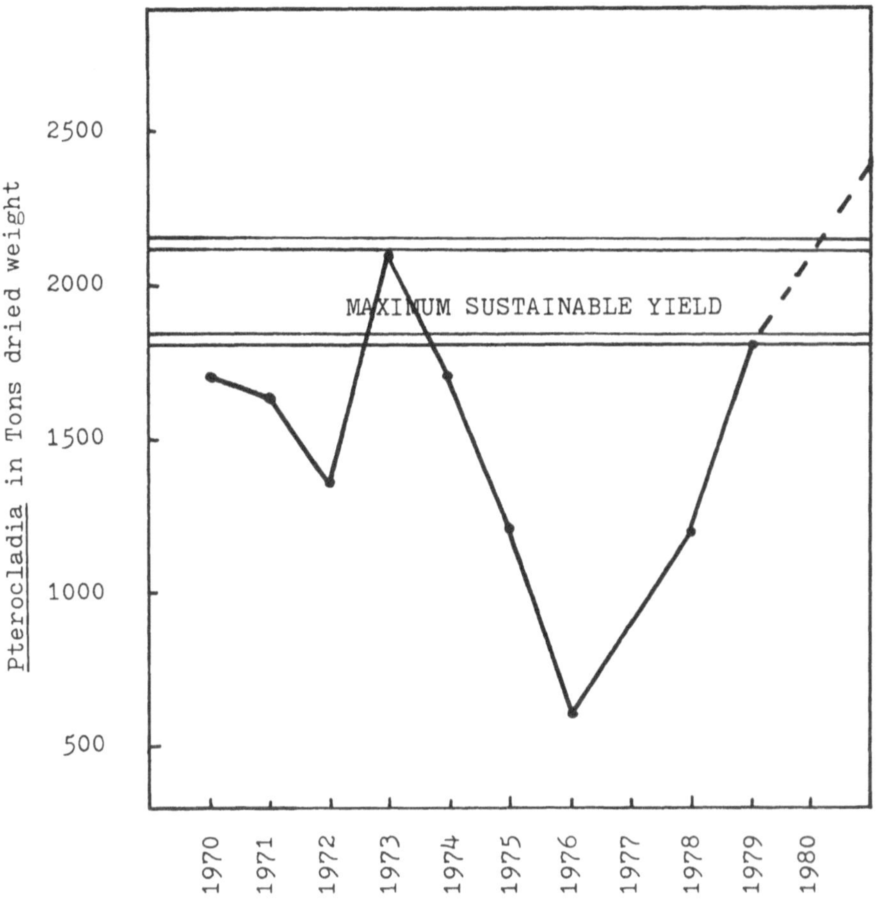

Maximum Sustainable Yield for Pterocladia in the Azores, Portugal

Management: A maximum sustainable yield (Figure 2.) has been proposed for the Pterocladia resources of the Azores. An annual harvest of 2000 M tons of dried Pterocladia can be harvested effectively for many years. However, in light of potential expansion of the agar-agar industry in the future an annual re-calculation was recommended. If the 2000 ton MSY can be maintained for three consecutive years the MSY can be increased by 10% for the fourth year. If the annual harvest cannot be maintained at MSY levels, with an equal collecting effort, for two consecutive years the MSY should be reduced by 10% for the third year. An annual review of the Pterocladia harvest should be conducted to identify any specific changes for any given season. Then, a viable harvest quota can be assigned to specific local areas and not interfere with the overall MSY. This also gives the factories a knowledge of expectations for the forthcoming season. It is important that all known collections of Pterocladia during a given season be recorded by the government, even those collections that are lost to the weather, to insure a critical evaluation of the collecting effort. The harvest data should be reported by:

a.) Individual island and specific location
b.) Quantity and method of the harvest
c.) Quality of the harvest

Conclusions: During the study of growth and reproduction for Pterocladia pinnata in the Azores, our results showed growth to be year-round and reproduction to be primarily vegetative. The growth of Pterocladia varied in rate with maximum thallus elongation during the summer months. Hydrodynamic fragmentation of P. pinnata occurring during oceanic storms accounts for more than 20% of the annual harvest. A critical size of greater than 10 cm in length appeared most sensitive to wave action from oceanic storms. After fragmentation the algae are washed ashore and collected by the harvesters by hand or with various types of nets. The physical impact of fragmentation may enhance the growth of juvenile P. pinnata plants arising from the remaining basal hapteres of the fragmented plants. At the point of fragmentation new branch apices proliferate in a matter of 7 to 10 days. Fragmentation during the spring months may inhibit the formation of sexual organs on the remains of fragmented plants that have not been able to sucessfully grow to maturity.

Several methods of harvesting have been discussed earlier. The method most likely to have an impact on future harvests of Pterocladia is that of the compressed-air divers. Since sexual reproduction for Pterocladia is limited or absent in the Azores, the maintainence of the regenerative basal portions of P. pinnata is critical to future harvests. No impacts were recognized for the collection of shore-cast algae nor were the impacts of snorkel or hand collecting considered significant at this time.

A maximum sustainable yield (MSY) was defined for Pterocladia harvest management in the Azores. Some 2000 dried tons of Pterocladia per year are recommended. Some harvest impacts in localized areas can be anticipated with an increase of diver collectors. But, if the diver collectors are adequately instructed on the proper collecting methods, without damaging the Pterocladia bases the future harvests will be secure. Without adequate instruction the Pterocladia harvest may become uncertain in some local areas.

Acknowledgements: We are indebted to the Regional Government of the Azores for recognizing the need for this study and to the University of the Azores for their co-operation during the study. We particularly thank Correia da Cunha, Luiz Saldanha, Adelaide Lobo, Katy Fralick, Ken Salem, George Leaf and Ray Broderick for their assistance.

This project was funded by the United States Agency for International Development under Contract No. AID/NE-C-1639, Portugal.

COMPARISON OF GROWTH STRATEGY IN *LAMINARIA* POPULATIONS LIVING UNDER DIFFERING SEASONAL PATTERNS OF NUTRIENT AVAILABILITY.

J.A. Gagné and K.H. Mann
Biology Department, Dalhousie University, Halifax, N.S., Canada. B3H 4J1

Introduction

Because of its internal storage capability, *Laminaria longicruris* is able to extend its growth period over several months when light energy or dissolved nitrogen in the water is in short supply (Chapman and Craigie, 1977, 1978; Hatcher et al, 1977). The work reported here was an investigation of the extend to which growth strategy in the species can be adapted to different patterns of nitrogen availability in different habitats.

Materials and Methods

Three sites with contrasting seasonal patterns of nitrogen availability were chosen in Nova Scotia for a year-round study: St. Margaret's Bay, near Halifax, Pubnico, at the south end of the province and Centreville, in the Bay of Fundy.

Each site was visited 11 times between September 1978 and August 1979, and attention was concentrated on *Laminaria longicruris* blades over 100 cm long. On each occasion blade elongation was measured on about 20 tagged plants, as in Parke (1948), and the internal concentrations of soluble nitrogen and of laminaran were analyzed on 5 others, following Chapman and Craigie (1977). Three water samples were frozen in the field and later analyzed for phosphates and ammonia; nitrates and nitrites were measured together, using a hydraxine-copper reduction technique (Strickland and Parsons, 1960; 1972). *In situ* quantum irradiance between 400 and 700 nm and water temperature were continously recorded. Variations in growth

Xth International Seaweed Symposium
Printed in Germany

rate at St. Margarets' Bay and Centreville were tested statistically by a one-way analysis of variance and the Duncan a. posteriori test ($\alpha = 0,05$). The Pubnico data had heterogenous variances and were analyzed by an approximate test of equality of means (Sokal and Rohlf, 1969).

Results

The St. Margaret's Bay plants showed a seasonal growth pattern similar to that described by Mann (1972) and Hatcher et al (1977). Growth was minimal (about 0,4 cm day^{-1}) (Fig. 1a) from September to December, at a time when the supply of dissolved inorganic nitrogen was reduced and internal reserves of nitrogen were almost exhausted (Fig. 2a). The internal reserves of laminaran (Fig. 2a) began to decrease in September, indicating that ambiant light levels (Fig. 1a) were insufficient to meet metabolic requirements. Hatcher et al (1977) had shown that the highest rate of respiration occured at this time, and that the plants were dependent on their reserves of carbohydrates from October to December. In December our plants had minimal reserves of carbon and nitrogen, but with increasing availability of dissolved nitrogen in the water, growth accelarated between December and January, and continued to increase until March, in spite of low light levels and water temperatures. During the winter period of nitrogen abundance, internal reserves of nitrogen were replenished. When ambient nitrogen dropped sharply during the spring bloom of phytoplankton, the plants began to draw on their internal inorganic nitrogen reserves. By April, those inorganic reserves were diminished and the plants began to use their organic nitrogen reserves. In this way they maintained a good growth rate well beyond the time of virtual disappearance of ambient nitrogen. Profiting from increasing light levels, they reached their maximum growth rate in May and June. From July the growth rate declined towards the fall minimum, while the products of summer photosynthesis were used to build up carbohydrate reserves once more.

At Pubnico, both the fall increase and the winter maximum of nitrogen in the water column occured about 2 months earlier than in St. Margaret's

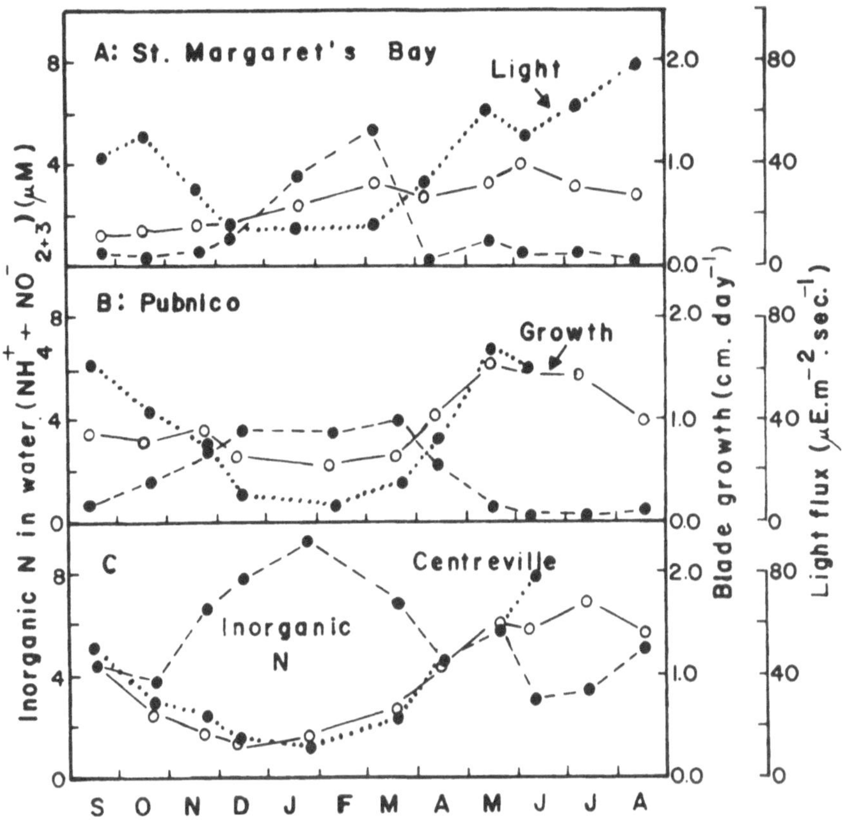

Fig. 1. Light intensity, dissolved inorganic nitrogen in the water and elongation rate of *L. longicruris* blades at 3 contrasting sites in Nova Scotia.

Bay, and a high nitrogen level was maintained for three months, whereas in the Bay there was a peak of short duration. The effect on the kelp was to shift the period of minimum growth from fall to winter (Fig. 1b). During the fall, increases in nitrogen occurred at a time when first solar energy and later carbohydrate reserves could be used to maintain a growth rate close to 1 cm day^{-1}. The carbohydrate reserves were smaller than in the St. Margaret's Bay plants, and were used up faster (Fig. 2b), so that growth was energy-limited from December to March. Stimulated by an increasing supply of solar energy, growth accelerated rapidly in the spring to reach its annual maximum just at the time when nutrients became exhausted in the surrounding waters. Using internal reserves of nitrogen, growth was maintained at a high level for three months, finally declining in August. As in St. Margaret's Bay, reserves of carbohydrates were built up during spring and summer. However, probably on account of the metabolic demands of the prolonged period of rapid growth in early summer, the reserves of carbohydrate did not reach as high a level as in the St. Margaret's Bay plants.

In Centreville, the important feature of the site was an upwelling of nutrient-rich water during summer so that nitrogen was relatively abundant throughout the year (Fig. 1c) and never appeared to limit growth. Indeed, the growth curve followed the light curve almost perfectly, the correlation coefficient being 0,95. The absence of laminaran in the tissues (Fig. 2c) indicated that even during summer growth kept pace with net photosynthesis. Without a carbohydrate reserve, the plants could not, as the two other populations did, maintain a high rate of growth in fall and early winter, when light was low. There was no period of nitrogen limitation, and the plants did not build up internal reserves of inorganic nitrogen.

Discussion

In Nova Scotia waters, the kelp populations are confronted with two potentially limiting factors, light energy and nutrient supply. Nitrogen may be present year-round, as at Centreville, or in short supply for up to 8 months of the year, as in St. Margaret's Bay. It is this pattern of nitrogen availability that determines the seasonality of growth and production of L. longicruris and not the water temperature. Low temperatures may lead to low rates of respiration, thus helping to produce an assimilatory surplus as stated by Hatcher et al (1977), but are not necessary or sufficient to produce high growth rates. Indeed, the Centreville plants had their maximum growth rate during summer, when the water temperature was around 15°C, and growth was minimal at both Centreville and Pubnico during winter. In severely nutrient-limited areas, such as St. Margaret's Bay, the plants have to concentrate their main growth into the period of nutrient availability, i.e. in the winter and up to the time of the spring bloom. By also building up internal nitrogen reserves they are able to prolong their rapid growth for at least 2 months, to take

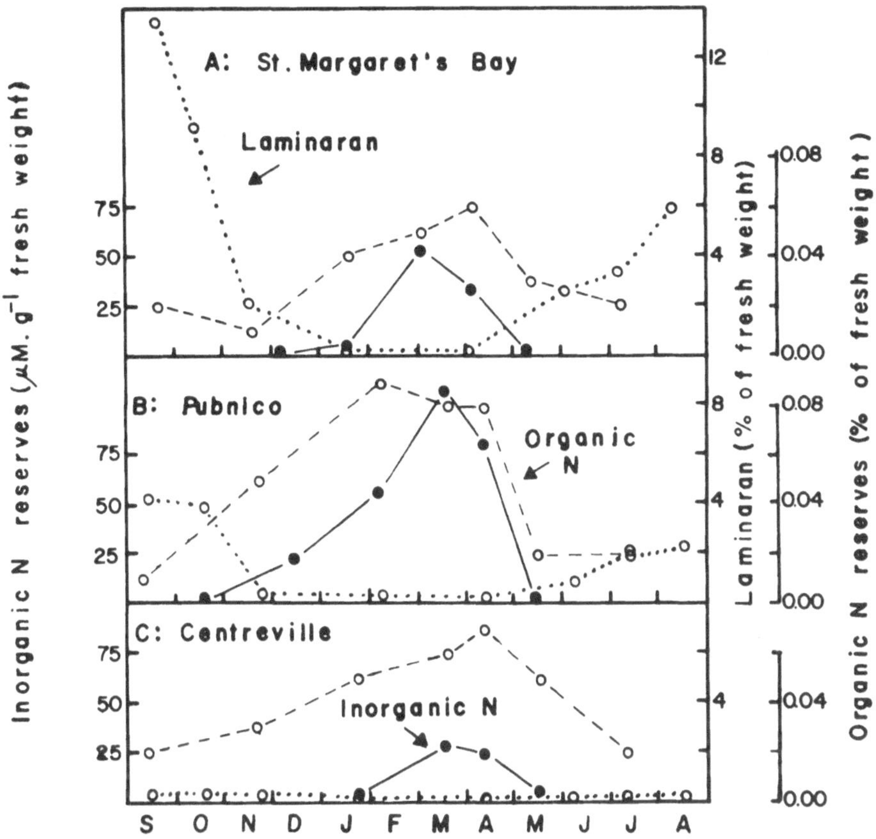

Fig. 2. Internal reserves of nitrogen and laminaran in blades of *L. longicruris* at the 3 sites.

advantage of improving light conditions.

As the period of nitrogen limitation decreases from site to site, the role of light as a limiting factor becomes more important. In St. Margaret's Bay growth is light limited for only a couple of months in late fall. In Pubnico low light conditions limit growth for about 5 months, and in Centreville light is the limiting factor year-round. This shows clearly that under natural conditions in Nova Scotia, growth is not necessarily nitrogen limited in the way suggested by Hatcher et al (1977) for St. Margaret's Bay.

When plants from Centreville were transplanted to Pubnico they grew at a significantly lower rate during the fall. When transplanted to St. Margaret's Bay, they died in October, and in a repeated experiment a second set of transplants died at the same time of year. There is a possibility then that the growth cycle of _L. longicruris_ is not solely imposed by external factors as suggested by Chapman and Craigie (1977). We recognize that if young plants had been transplanted instead of mature plants, they might have responded differently. Nevertheless, the possibility of genetic differences between the populations can not be eliminated. These results and the slow growth rate of the Pubnico plants in early winter indicate that the magnitude of the carbohydrate reserves are important in determining the early winter growth rate, in contradiction with Chapman and Craigie (1978). It thus appears that _L. longicruris_ has a much more flexible growth strategy than previously thought and that genetic controls are not to be ruled out yet.

This paper deals only with rates of blade elongation, and leaves aside questions of production which depend on blade morphology and on population density. It is already apparent that the overall performance of plants at Pubnico in summer is better than the performance of plants in St. Margaret's Bay, although at this time the supply of light and nitrogen at Pubnico is no better than in the Bay. Comparison of production performances can help clarifying such phenomenons and therefore, will be the next step in the analysis.

References

1. Chapman, A.R.O., Craigie, J.S.: Mar. Biol. 40, 197-205 (1977).
2. Chapman, A.R.O., Craigie, J.S.: Mar. Biol. 46, 209-213 (1978).
3. Hatcher, B.G., Chapman, A.R.O., Mann, K.H.: Mar. Biol. 44, 85-96 (1977).
4. Mann, K.H.: Mar. Biol. 14, 199-209 (1972).
5. Parke, M.: J. mar. biol. Ass. U.K. 27, 651-709 (1948).
6. Sokal, R.R., Rohlf, F.J.: Biometry, W.H. Freeman, San Francisco (1969).
7. Strickland, J.D.H., Parsons, T.R.: Bull. Fish. Res. Bd. Can. 125, 61-69 (1960).
8. Strickland, J.D.H., Parsons, T.R.: Bull. Fish. Res. Bd. Can. 167, (1972).

FLUCTUATIONS IN EPIPHYTE BIOMASS FOLLOWING LAMINARIA HYPERBOREA CANOPY REMOVAL

E. Harkin
Department of Marine Biology, University of Liverpool, Port Erin, Isle of Man

Introduction

Manipulative experiments designed to investigate species competition and succession have been a long accepted ecological technique applied frequently to the marine environment (1, 2). Subtidally, these investigations often involved complete clearing or denuding of natural surfaces (3, 4) or the investigation of the effect of invertebrate grazers on the control of algal populations (5, 6). The present investigation involved only the removal of fronds (laminae, blades) of the dominant subtidal alga Laminaria hyperborea (Gunn.) Fosl. in order to determine the effect of the, frond created, canopy upon the relative biomass of the major algal species epiphytic upon the stipe.

Materials and Methods

The Carrick Rock (54°04'17"N, 04°42'36"W), 1.5 km. offshore from Port St. Mary on the S.E. of the Isle of Man, was the site of the investigation. SCUBA techniques were employed to secure 6 mm. courlene rope to L. hyperborea holdfasts using short lengths of ribbon so that a rope network, enclosing a treatment area of 46.8 m^2 with a control of 55 m^2 on each side, was established at a depth of 5.5 m below Lowest Astronomical Tide. Frond removal was carried out in June 1978 and consisted of cutting all L. hyperborea canopy plant (stipe length $\geq$50 cm) fronds such that an arc of approximately 6 cm diameter was left intact. A total of six sets of samples were taken by divers over the 20 month period from the date of clearing. Each set consisted of eight, 1 m^2, quadrats, four from the frond removed area and two from each control. Repeat sampling of quadrats was avoided. The epiflora (named in accordance with Parke and Dixon (7)) on the stipes of the, frond removed, canopy plants from each quadrat was examined as follows. Fresh and dry weights of Palmaria palmata, Phycodrys rubens, Membranoptera alata and either Polysiphonia

Xth International Seaweed Symposium
Printed in Germany

urceolata or Cryptopleura ramosa were measured. All large brown epiphytes were removed intact and only fresh weight measured. All other epiphytes were identified and recorded as present or absent. All significance tests were carried out at $P = < 0.05$.

Results

The epiphytes examined represented, in all cases, at least 95% by weight of the total canopy plant stipe epiflora. When considered collectively (Fig. 1), a significant increase occurred during the summer immediately after frond removal. A further gradual increase during the winter and following spring gave way to a second rapid increase producing a summer peak which declined to almost equal the control plants by the second winter.

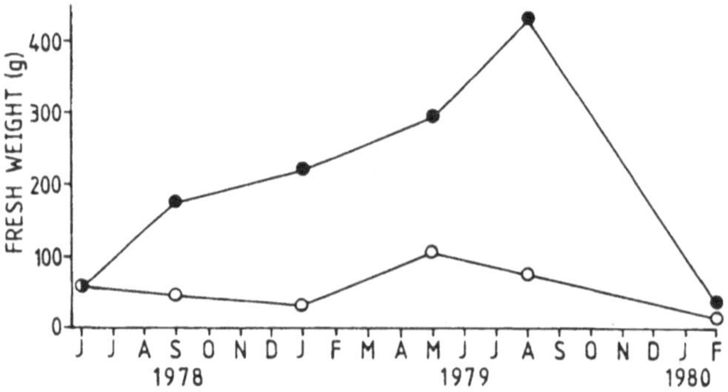

Fig. 1. Mean epiphyte fresh weight (Rhodophyta and Phaeophyta) per canopy plant for Controls (O) and Frond Removal (●).

The initial summer peak in the frond removed plants was caused by the Rhodophyta component (Fig. 2) which was composed of 14 species in June 1978. Phycodrys rubens was mainly responsible for this increase with smaller contributions by Palmaria palmata and Membranoptera alata (Fig. 3). Palmaria, in fact, increased to occupy a stipe length of up to 25 cm with a sweep of up to 30 cm around the stipe while Cryptopleura extended into the positions formerly dominated by Phycodrys and Membranoptera. The Phaeophyta component initially consisted solely of L. hyperborea (Fig. 4) with a mean density of 0.4 plants per canopy plant. This declined on both control and frond removed plants during the first summer but the appearance of Desmarestia (showing a preference for epiphytic L. hyperborea holdfasts) kept the overall level, in the treatment area, constant during this period (Fig. 2).

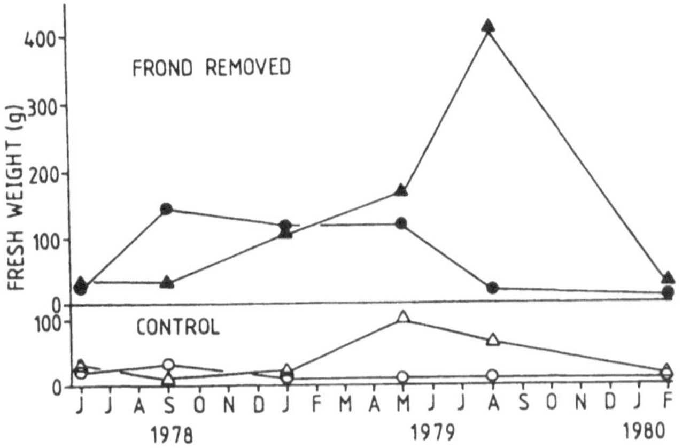

Fig. 2. Mean epiphyte fresh weight per canopy plant for Controls and Frond Removal. Rhodophyta (O, ●) Phaeophyta (Δ,▲).

From September 1978 onwards, the gradual winter increase seen in Fig. 1 for the frond removed plants was caused by the Phaeophyta (Fig. 2) which, by the beginning of the following summer, had just overtaken the Rhodophyta although a comparable level was also attained by the control Phaeophyta. L. hyperborea was mainly responsible for this increase (Fig. 4) but a peak of S. polyschides (showing a preference for epiphytic L. hyperborea holdfasts and stipes) and the continued presence of Desmarestia were also contributory in the case of frond removal while the appearance of L. digitata enchanced the control biomass.

The Rhodophyta and Phaeophyta then took opposite courses during the summer of 1979 (Fig. 2). The former declined from their previously significantly higher levels to become equivalent to the control values while the latter peaked sharply due to the continued increase in L. hyperborea and the appearance of L. digitata and L. saccharina accompanied by a slight increase in Desmerestia. Epilithic S. polyschides also contributed a temporary summer canopy in 1979 to the frond removed area which added to the epilithic L. hyperborea canopy developing from the sub-canopy plants whose fronds had not been removed in June 1978.

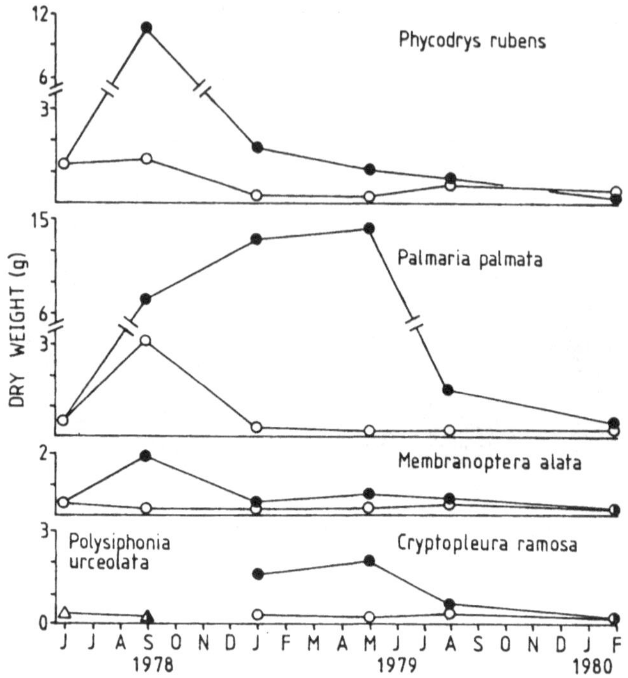

Fig. 3. Mean dry weight per canopy plant for five species of epiphytic Rhodophyta for Controls (O, Δ) and Frond Removal (●, ▲).

Discussion

At the onset of frond removal the stipe epiflora was composed of equal fresh weights of red and brown algae (Fig. 2). However, the Rhodophyta, comprising 14 species, were much more diverse and numerous than the Phaeophyta, represented by a few L. hyperborea plants, so that they were clearly better able to respond fast to the improved conditions, increasing by over 500% to dominate the epiflora within the first season. Palmaria showed a greater percentage increase than Phycodrys over this period but, on account of its greater abundance at the outset, the latter still held the dominant position in September 1978. The winter increase in Palmaria was presumably facilitated by the improved illumination in the absence of frond cover and its confinement to the distal end of the stipes, above all other competitors. The contrasting decline, during the same period, in all other red epiphytes, with the exception of Cryptopleura, both in control and treatment plants reflects the seasonal changes expected in winter and agrees very well with the results of Whittick (8) who noted a similar decline in epiphyte biomass per stipe, commencing also in September, at a range of depths in S.E. Scotland. The winter decline observed here may also have been aided by the fact that by January 1979 Palmaria had grown dramatically, restoring a form of canopy to the distal end of the stipes.

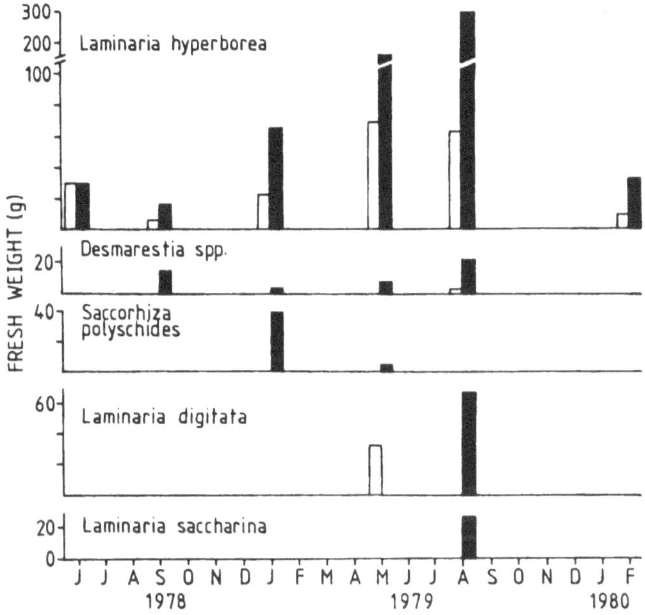

Fig. 4. Mean fresh weight per canopy plant of selected Phaeophyta for Controls (□) and Frond Removal (■).

As pointed out by Kain (9), the perennial nature of the L. hyperborea stipe tends to discourage opportunist epiphytes so that a few species become dominant. This results in a common zonation pattern described by Marshall (10) and Whittick (8) with Palmaria (Rhodymenia), Phycodrys, Membranoptera and Ptilota as the dominant genera. This was also found here with the exception that Cryptopleura was more important than Ptilota which is similar to the situation at a comparable depth in S.W. Ireland (11).

Desmarestia followed by Saccorhiza were the first opportunists to appear after frond removal. These played a similar role in other clearance experiments (3, 4) and their preference here for the holdfasts and stipes of developing epiphytic L. hyperborea, free from the epiphytic Rhodophyta, points to the fact that space more than light had become limiting.

Meanwhile, beneath the red epiphyte bloom epiphytic L. hyperborea sporelings must have been doing very well. They increased in number, significantly, on the treatment plants immediately after frond removal but remained small and grew slowly due to competition from the reds. Their appearance between June and September, outside the sporing season, must infer their presence in the form of gametophytes or early sporophytes at the time of clearing; a conclusion arrived at by Kain (3).

The arrival of summer in 1979 provided the Phaeophyta with suitable conditions for rapid growth and since they were then more numerous and therefore better established than the year before, they were able to outcompete the declining Rhodophyta. However, with their large fronds and distal positions on the stipes, the Phaeophyta were, like S. polyschides, particularly susceptible to the winter storms of 1979-1980 and in some cases the epiphyte burden they created was responsible for the complete disappearance of L. hyperborea plants in the treatment area. This, combined with the canopy produced by the original sub-canopy L. hyperborea plants was responsible for the decline in epiphytic Phaeophyta in the second winter.

Summary

The advantage of being more numerous and diverse allowed the Rhodophyta to benefit immediately from frond removal. Winter conditions and dense Palmaria caused their decline giving the Phaeophyta a chance to establish from early stages already present and from opportunists colonizing free space on other epiphytes. These fast growing Phaeophyta were then favoured during the second summer but winter conditions and a developing epilithic L. hyperborea canopy caused their decline.

The assistance of Dr J.M. Jones, M. Bates, Dept. of Education N.I., and Liverpool University are gratefully acknowledged.

References

1. Dayton, P.K.: Ecol. Monogr. 41, 351-389 (1971).
2. Sousa, W.P.: Ecol. Monogr. 49, 227-254 (1979).
3. Kain, J.M.: J. Ecol. 63, 739-765 (1975).
4. Kain, J.M.: J. mar. biol. Ass. U.K. 56, 267-290 (1976).
5. Jones, N.S., Kain, J.M.: Helgoländer wiss. Meeresunters. 15, 460-466 (1967
6. Dayton, P.K.: Fishery Bull. Nat. Mar. Fish. Serv. U.S. 73, 230-237 (1975).
7. Parke, M., Dixon, P.S.: J. mar. biol. Ass. U.K. 56, 527-594 (1976).
8. Whittick, A.: M.Sc. thesis, Durham University, 139pp. (1969).
9. Kain, J.M.: Oceanogr. Mar. Biol. Ann. Rev. 17, 101-161 (1979).
10. Marshall, W.: Br. phycol. Bull. 2, 18-19 (1960).
11. Norton, T.A., Hiscock, K., Kitching, J.A.: J. Ecol. 65, 919-941 (1977).

BIOMASS AND VERTICAL DISTRIBUTION OF *FURCELLARIA LUMBRICALIS* AND ASSOCIATED ALGAE*

J. E. Holmsgaard

Litex Industri, P. O. Box 7
DK-2600 Glostrup, Denmark

M. Greenwell and J. McLachlan

National Research Council, 1411 Oxford Street
Halifax, Canada B3H 3Z1

Introduction

Furcellaria lumbricalis (Huds.) Lamour. occurs in both the eastern and western North Atlantic and is exploited commercially in northern Europe and Canada. In Europe this species ranges from Spitzbergen and Nova Zemlya (9,15) in the Barents Sea, the Faeröe Islands, round the British Isles (2,5), and south to the Bay of Biscay (7); it has not been recorded from either Iceland or Greenland (3,14). The alga penetrates into the brackish waters of the Baltic Sea to Wasa Archipelago in southern Botnian Bay (6) and to Kotka in the inner Gulf of Finland (8); in both areas the salinities are low (4). In the western Atlantic, *F. lumbricalis* was first recorded in 1933 (1,11) in the lower Gulf of St. Lawrence. The present known distribution of this alga in North America comprises this area and the Atlantic coast of northern Nova Scotia including an isolated, free-living, small population (*in herb.* NRCC).

*NRCC No. 18249

Xth International Seaweed Symposium
Printed in Germany

In Canada *F. lumbricalis* is exclusively sublittoral as is generally the case in Europe although occasionally it occurs there in the lower littoral zone (5,7). Two forms of this species are recognized, the epilithic type which is the main form in Canada, and forma *aegagropila* Reinke which occurs extensively in the Kattegat and the Baltic Sea.

Materials and Methods

Surveys have been conducted, which include recorded vertical distributions and estimations of biomass of epilithic *F. lumbricalis* and associated species, along the following coasts: 1. Swedish west coast, south of Göteborg [1978] (12°24'E, 56°54'N). The substratum is relatively steep, rocky to 11-12 m, and thereafter soft; tidal amplitude is about 0.3 m. 2. Northern shore of Sjaellands Odde (Danish island of Zealand), southern Kattegat [1979] (11°24'E, 55°58'N). The substratum is granite boulders and stone on a sandy bottom; tidal amplitude is about 0.2 m. 3. Northeastern shore of Prince Edward Island, Canada [1978] (62°30'W, 46°30'N). The substratum of alternating friable sandstone and sand slopes gently, about 1 m depth/100 m distance; tidal amplitude is about 1 m.

In these three areas, random samples were obtained by SCUBA divers. Plants were scraped from the substratum by spatula; 0.25 m^2 quadrats were cleared in Canada and Sweden and 0.1 m^2 quadrats in Denmark. The Swedish samples were from 5 specific depths (Fig. 1a) and the algal cover was 100%. The Danish samples were taken from 7 depth zones (Fig. 1b), and the cover was also 100%. The Canadian samples were taken from 6 depth zones (Fig. 1c), and the cover within the quadrat was estimated by the diver. Samples were sorted by major species; those from Sweden were dried at 105°C before weighing whereas samples from the other two areas were blotted and fresh weights measured.

The lower depths for sampling in Scandinavia were limited by an absence of solid substrate, whereas in Canada these limits were set by the deepest occurrence of plants, the substratum remaining solid below these depths.

Results

The results are displayed in Fig. 1. Because of the different method in determining weights and because of differences in percentage cover, the values for absolute biomass are not comparable among the three populations. However, the species constituting the algal communities and their relative dominance and depth distribution can be compared.

In general the same species made up the major portion of the biomass, the notable exception being species of the Delesseriaceae which were absent (Delesseria sanguinea) from or a minor component (Phycodrys rubens) of the Canadian samples. A substantial proportion of the biomass of all 3 areas was made up of F. lumbricalis, and this alga was the dominant species in the Danish area. In the 3 areas the vertical distribution of F. lumbricalis was similar, although the maximum biomass in Scandinavia occurred at about 4 m depth and within the 7.5 - 10 m depth zone in Canada.

Chondrus crispus Stackh., another economically important species, also occurred in the 3 areas. Along Sjaellands Odde, C. crispus was present only as scattered plants from 8 m and deeper, the biomass being insignificant. In Sweden this species was present at all depths, but in small quantities. Along the Canadian coast, C. crispus was, however, one of the major species in all depth zones down to 10 - 15 m. The maximum biomass was at 2.5 - 5 m depth where this species made up more than one-half of the total weight of algae.

Species of Phyllophora were found within a wide range of depths. The most abundant species in the Canadian samples was P. truncata (Pall.) Zinova, as was the case in both the Swedish and Danish samples down to depths of 5 m. In Denmark and in Canada, *P. pseudoceranoides* (Gmel.) Newroth et Taylor occurred only as scattered plants, but as the predominant species of *Phyllophora* in Swedish samples from below 5 m depth. The principal occurrence of *Phyllophora* in both the Canadian and Danish sites was at greater depths than the maximum biomass of either *F. lumbricalis* or *C. crispus*. However, biomass of *Phyllophora* along the Swedish coast was maximal at 4 m, at the same depth as maximum biomass of *F. lumbricalis*.

Discussion

The 3 areas studied generally contained the same major species. Differences in vertical distribution and relative quantities of biomass are undoubtedly the result of several factors, with the most obvious difference being salinity. In the lower Gulf of St. Lawrence the salinity is 27-28°/₀₀, whereas there is considerable fluctuation along the Scandinavian coast. Minimum salinities of Danish and Swedish waters are about 18°/₀₀ and 21°/₀₀ respectively with maxima around 30-32°/₀₀ (12,13). The range of temperatures is about the same in the 3 areas: 17-18°C during summer and ice-covered during winter in Canada and some years in Denmark and Sweden. Turbidity is probably greater in the Scandinavian waters than in the Canadian waters which remain relatively clear along the north shore of Prince Edward Island.

These and other factors, including biotic, influence environmental gradients and affect inter-specific competition. *Chondrus crispus* occurred only in small quantities or was absent at the Scandinavian sites, and low salinities there may be the major limiting factor. In the absence of *C. crispus*, *F. lumbricalis* was found in abundance at shallower depths. However, in the absence of *F. lumbricalis* at greater depths, as in some other areas of Prince Edward Island, *C. crispus* does not occur in abundance. Species of *Phyllophora* may compete with both *F. lumbricalis* and *C. crispus*. In clear waters *Phyllophora* was generally restricted to greater depths (Fig. 1), although in the absence of F. lumbricalis, Phyllophora may directly compete with C. crispus. Moreover, with increasing turbidity, Phyllophora occurs in abundance at much shallower depths.

Acknowledgements

We thank R.-A. Kornfeldt for permitting us to make use of her data from Sweden; Aase Kristiansen for criticism during preparation of the ms; J. Christensen for translating the Russian papers.

References

1. Bell, H. P., MacFarlane, C.: Can. J. Res. 9, 265-293 (1933).
2. Børgesen, F.: The Marine Algae of the Faeröes. Botany of the Faeröes: Part 2. pp. 339-533. Copenhagen 1902.
3. Caram, B., Jónsson, S.: Act. Bot. Isl. 1, 5-31 (1972).
4. Dietrich, G.: Kieler Meeresforschung. 7, 35-69 (1950).
5. Dixon, P. S., Irvine, L. M.: Seaweeds of the British Isles, Vol. 1, part 1. British Museum (Natural History), London 1977.
6. Ehnholm, G.: Mem. Soc. Fauna Flora Fennica 13, 21-25 (1938).
7. Gayral, P.: Les Algues des Côtes Françaises. Deren, Paris 1966.
8. Häyrén, E.: Mem. Soc. Fauna Flora Fennica 4, 50-59 (1928).
9. Kjellman, F. R.: The Algae of the Arctic Sea. Kgl. Svenska Vetensk. Handl. Vol. 20, no. 5. 1883.
10. Kornfeldt, R-A.: Svensk. Bot. Tidskr. 73, 131-138.
11. MacFarlane, C. I.: Bot. Gothoburgensia 3, 151-153 (1965).
12. Monthly Average Values of Hydrological Observations on Swedish Lightships, 1923-1952. Fisheries Board of Sweden, Ser. Hydro. Rep. no. 5.
13. Nautisk Meterologisk Årbog, 1961. Det Danske Meterologiske Institut, Copenhagen 1962.
14. Pedersen, P-M.: Medd. om Grønland, vol. 199, no. 3, pp 1-80.
15. Sinova, E. S.: Algues de la Novaja Zemlia (in Russian). Inst. Hydrologique, Explor. des Mers d'USSR, Fasc. 10. 1929.

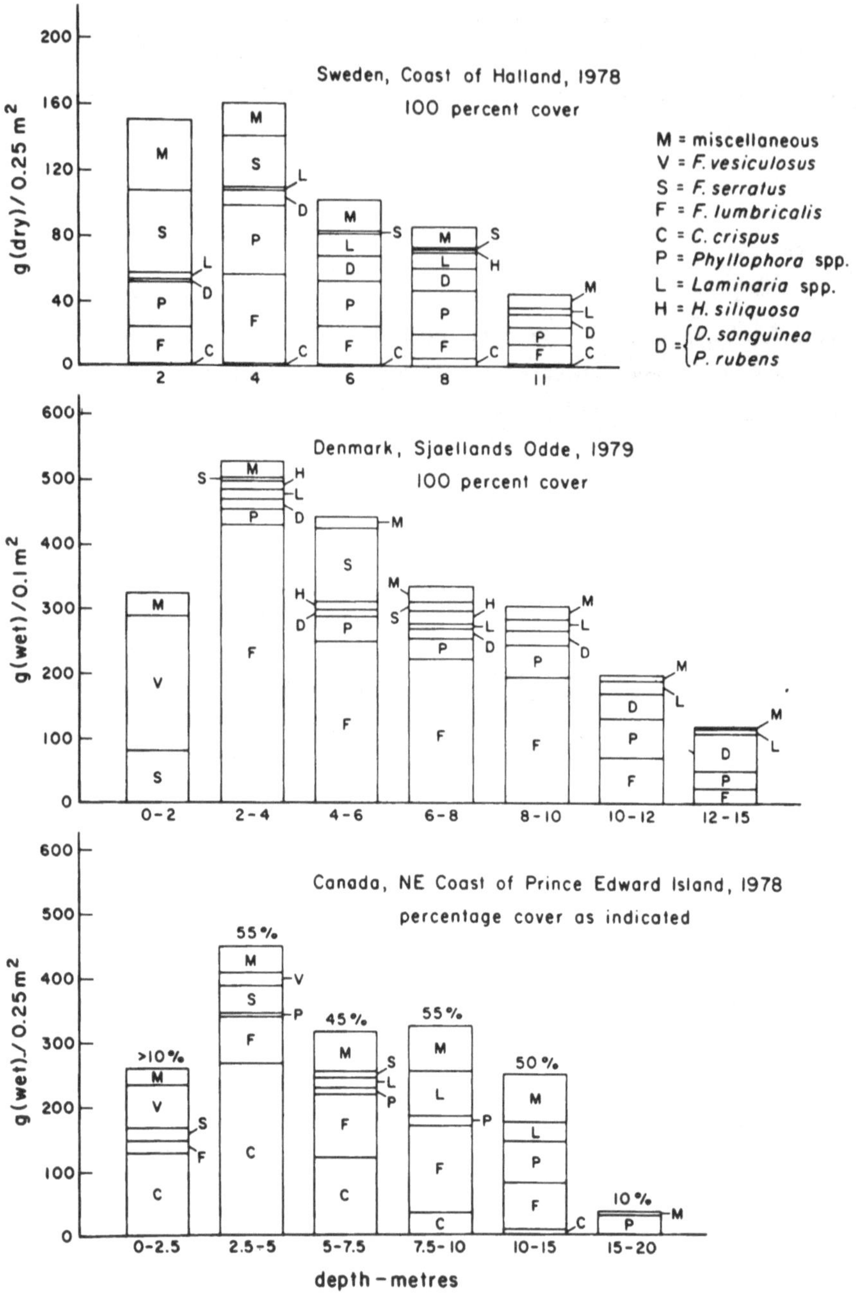

FIG. 1. (A) SWEDEN, DRY WT/0.25 M^2; (B) DENMARK, WET WT/0.1 M^2; (C) CANADA, WET WT/0.25 M^2

THE ALGAL VEGETATION AT A PLATFORM SHORE AT MIKHMORET

B. Lundberg
Department of Botany, The Hebrew University of Jerusalem, Israel

Introduction

Marine ecologists have recently been using ordination techniques to analyze algal vegetation in relation to depth, temperature, salinity and other environmental factors (Boudouresque, 1973; Velimirov et al, 1977; Devinny, 1978; Cinelli et al, 1979). In this paper nodal component analysis (Noy-Meir and Whittaker, 1977) was used to define and distinguish noda (units of vegetation) based on monthly samples of the algal species in each of three previously defined vegetation habitat-units on a platform shore. The habitats are: subhorizontal platforms at about sea level, subvertical walls at depths of 0-2 meter and bottoms at 2-6 meter. The description of the vegetation as obtained by the nodal component analysis will be compared with the summary of vegetation types as found in the predefined habitats.

Methods

At Mikhmoret on the Mediterranean coast of Israel, monthly samples were taken during 1973 and 1974 from a 300 meter long stretch of shore extending about 100 meter into the sea. It included sites exposed to high waves as well as sheltered ones. At least 5 samples were taken from each habitat (platforms, walls and bottoms) with a randomly placed metal rectangle (22x28 cm). The absolute cover of each species was visually estimated in the sea to be determined in the laboratory and transformed to relative cover, i.e. the percentage of the particular algal species of the total algal cover. Monthly relative cover f rom platforms, walls and bottoms were averaged over seasons. Each vegetation type as defined by the combination of habitat and season is characterized (Table 1) by the species

Xth International Seaweed Symposium
Printed in Germany

which are dominant, abundant and common in it (according to cover values). Nodal component analysis as described by Noy-Meir and Whittaker (1977) (principal component analysis of non-centered data followed by varimax rotation of components) was applied to the monthly relative cover data of algal species in habitat units. Each component can be identified with a nodum of vegetation. Normalized projections of species and sites on the component, which usually vary between 0 and 1, are called the affinities of species and sites to the nodum. A nodum is defined by a group of species and sites which have high affinities to it. Noda may reflect gradual trends in the vegetation by overlap as measured by high or moderate affinities of species and sites to more than one nodum. Varimax rotation was repeated with 3 and 7 components. All components were unipolar and meaningful except one at level 7 which was bipolar (difficult to interpret).

Results

The species composition of the algal vegetation fluctuated seasonally. These fluctuations were greatest on the platforms, where Ulva lactuca and Enteromorpha spp. dominated during spring and another group: Jania rubens, Acanthophora nayadiformis and Laurencia papillosa during autumn (Table 1). Species diversity was greatest on platforms during winter and smallest during spring. Seasonality was smaller on walls and bottoms where Corallina mediterranea dominated all seasons except summer and Jania rubens dominated all seasons. Colpomenia sinuosa increased greatly during spring as did Pterocladia capillacea during winter and Peyssonellia squamaria during summer - both only on walls. Some species developed best in one habitat, for instance, Gracilaria bursa pastoris on platforms, Rhodymenia pseudopalmata on walls and Sphacelaria spp. on bottoms. Others were often found in all habitats like Sargassum vulgare and Cladophora spp. or in two habitats: platforms and walls (Gigartina acicularis) or on walls and bottoms (Halimeda tuna and Halopteris scoparia). This reflects gradual changes in species composition between habitats. Within habitats a common element of species was usually found between two subsequent seasons - also a gradual trend. Three major noda accounted for 75% of the total variance (Table 2a). Nodum Jr defined by the group: Jania, Chaetomorpha sp.,

TABLE 1: Dominant, abundant and common algal species during the seasons on platforms, walls and bottoms at Mikhmoret during May 1972 through December 1974. (Mean relative cover of algae per season based on monthly mean values, averaged over seasons.)

Season	Site	MEAN RELATIVE COVER PER SEASON 10% DOMINANT SPECIES	5 - 9 % ABUNDANT SPECIES	3 - 4 % COMMON SPECIES
WINTER	PLATFORM	Ulva lactuca Jania rubens	Enteromorpha spp. Laurencia papillosa Sargassum vulgare Centroceras clavulatum Gracilaria bursa pastoris Gigartina acicularis	Cladophora spp.
WINTER	WALL	Corallina mediterranea Jania rubens Pterocladia capillacea	Gigartina acicularis	Peyssonnelia squamaria
WINTER	BOTTOM	Jania rubens Corallina mediterranea	Bryopsis spp. Polysiphonia spp. Gelidiella tenuissima	Crustose calcereous red algae Halimeda tuna Corallina granifera
SPRING	PLATFORM	Ulva lactuca Enteromorpha spp.		Gigartina acicularis Jania rubens Laurencia papillosa Polysiphonia spp. Cladophora spp.
SPRING	WALL	Corallina mediterranea Jania rubens Colpomenia sinuosa	Pterocladia capillacea Gigartina acicularis Halopteris scoparia	Ulva lactuca Cladophora spp. Rhodymenia pseudopalmata Sargassum vulgare
SPRING	BOTTOM	Jania rubens Corallina mediterranea Colpomenia sinuosa	Halopteris scoparia	Bryopsis spp. Corallina granifera Gelidiella tenuissima Padina spp.
SUMMER	PLATFORM	Ulva lactuca Jania rubens Laurencia papillosa	Acanthophora nayadiformis Cemtroceras clavulatum Sargassum vulgare	Spyridia filamentosa Enteromorpha spp. Hypnea spp. Polysiphonia spp.
SUMMER	WALL	Jania rubens Peyssonnelia squamaria	Corallina mediterranea Sargassum vulgare Pterocladia capilacea	Halopteris scoparia Ulva lactuca Spatoglossum asperum Dilophus spiralis Gigartina acicularis Halimeda tuna
SUMMER	BOTTOM	Jania rubens	Polysiphonia spp. Gelidiella tenuissima Halopteris scoparia Corallina mediterranea Sargassum vulgare	Centroceras clavulatum Sphacelaria spp. Cladophora spp. Padina spp. Spyridia filamentosa
AUTUMN	PLATFORM	Jania rubens Acanthophora nayadiformis Laurencia papillosa	Centroceras clavulatum Sargassum vulgare	Ulva lactuca Gracilaria bursa pastoris
AUTUMN	WALL	Corallina mediterranea Jania rubens	Spatoglossum asperum Halimeda tuna Pterocladia capillacea Peyssonnelia squamaria Sargassum vulgare	Bryopsis spp. Polysiphonia spp. Gigartina acicularis
AUTUMN	BOTTOM	Jania rubens Corallina mediterranea	Polysiphonia spp. Halimeda tuna Sargassum vulgare	Bryopsis spp. Sphacelaria spp. Gelidiella tenuissima

TABLE 2: Noda of vegetation as defined by varimax components. These are based on mean monthly relative cover of algal species at platforms, walls and bottoms, Mikhmoret 1973-1974. (Species listed have affinities greater than 0.5; those with affinities greater than 0.75 are underlined. Asterisk indicates that a high affinity species had low cover in the vegetation.)

TABLE 2A: Three Noda of Vegetation

% of total variance	SPECIES	SITE/SEASON	CODE
25.8	Corallina mediterranea, Pterocladia capillacea, Rhodymenia pseudopalmata, Gigartina acicularis, Halimeda tuna, Amphiroa rigida.	WALL, BOTTOM ALL SEASONS	Cm
32.2	Jania rubens, Chaetomorpha* sp., Acanthophora nayadiformis, Laurencia papillosa, Bryopsis spp., Sargassum vulgare, Cladophora spp., Centroceras clavulatum, Gelidiella tenuissima, Polysiphonia spp.	PLATFORM, AUTUMN, bottom, wall, all seasons	Jr
17.0	Ulva lactuca, Enteromorpha spp., Pterosiphonia pennata*	PLATFORM, winter, SPRING, summer	UlE

TABLE 2B: Six Noda of Vegetation

% of total variance	SPECIES	SITE/SEASON	CODE
26.6	Corallina mediterranea, Pterocladia capillacea, Gigartina acicularis, Rhodymenia pseudopalmata, Halimeda tuna, Amphiroa rigida*.	WALL, BOTTOM ALL SEASONS	Cm
22.3	Acanthophora nayadiformis, Laurencia papillosa, Jania rubens, Centroceras clavulatum, Chaetomorpha* sp., Gracilaria bursa pastoris.	PLATFORM, AUTUMN, wall, bottom, all seasons	AnLp
16.4	Ulva lactuca, Enteromorpha spp., Pterosiphonia pennata*	PLATFORM, winter, SPRING, summer	UlE
7.8	Colpomenia sinuosa, Dilophus spiralis, Halopteris scoparia	wall, bottom spring	Cs
6.6	Sargassum vulgare, Spatoglossum asperum, (Zoobotryon verticillatum), Ceramium* spp.	wall, bottom summer, autumn	Sv
6.1	Polysiphonia spp., Gelidiella tenuissima, Sphacelaria spp.	bottom	P

Laurencia, Acanthophora, Bryopsis spp., Sargassum, Cladophora, Centroceras clavulatum, Gelidiella tenuissima and Polysiphonia spp. was best developed on platforms during autumn but it was also found on walls and bottoms during all seasons. Nodum Cm defined by Corallina, Pterocladia, Rhodymenia, Gigartina, Halimeda and Amphiroa rigida was well developed on walls and bottoms the whole year. The third, slightly smaller, nodum UlE defined by Ulva,Enteromorpha and Pterosiphonia pennata was found on platforms during winter, spring and summer, being very weakly developed in other habitats. Among the six noda defined at a more detailed level Cm and UlE were very close to noda Cm and UlE of level 3 (Table 2b). Nodum AnLp dominated by Acanthophora and Laurencia and with Gracilaria being important was found in the same sites as nodum Jr, hence being related to it by common sites and species. The three noda, Cm, UlE and AnLp account for 65% of the variance while the other three, Cs, Sv and P are small. Nodum Cs is dominated by Colpomenia and nodum Sv by Sargassum. Both developed on walls and bottoms; Cs during spring and Sv during summer and autumn. The Polysiphonia dominated nodum P was found on bottoms. Noda of both levels were closely related, e.g. nodum Jr is related to AnLp, Sv and P. A great deal of overlap was also found within the levels, e.g. Cm was related to Cs by sites on walls and bottoms having high affinities with both. The overlap reflects gradual trends in the vegetation.

Discussion

The vegetation types of the platform differed greatly from those on walls and bottoms. They included less calcareous algae and large populations of Ulva and Enteromorpha and of large fleshy red algae such as Laurencia, Acanthophora, Gracilaria, Hypnea and Gigartina (Table 1). During spring the difference was greatest which was connected with the decrease of Jania on the platforms while Ulva and Enteromorpha increased there, which indicated low species diversity as compared to the higher species diversity at the other habitats. The greater difference in spring between the platform vegetation and that in the walls and bottoms was also reflected by the nodal component analysis. Nodum UlE was found on platforms while other noda defined the wall-bottom vegetation. During summer and autumn, the platform vegetation differed less from those on walls and bottoms because then both Jania and Sargassum had high cover in all three habitats, in which the species diversity as indicated by the number of abundant species was similar. Noda Jr and AnLp reflected this, since they were found in all the three habitats during summer and autumn. The platform vegetation was distinguished by higher affinities and the small seasonal differences on walls and bottoms appeared by the all year occurrence of the noda. By this noda Jr and AnLp seem to grow in the same sites as nodum Cm. The Cm and Jr species groups may be recognized as two separate fractions of the wall-bottom vegetation. Cm algae were found on walls and bottoms exposed to high waves and Jr algae on moderately exposed walls and bottoms (Lundberg, 1980). Noda Cs and Sv reflect the seasonal wall-bottom vegetation of sheltered localities. Nodum P emphasized the filamentous algae found on bottoms usually difficult to recognize. Some trends are revealed by both methods of analysis. For instance, the gradual variation between vegetation units or the greater difference between the platform vegetation and the wall-bottom vegetation during spring as compared to summer and autumn. Each method elucidates or emphasizes certain aspects. Vegetation types show the relative abundance of the important species in each habitat during each season. This information estimates the species diversity for changes in habitats during seasons. Nodal component analysis defines dominant units and emphasizes smaller often overlooked seasonal components.

References

1. Boudouresque, C.F.: Bull. Mus. Hist. Nat. Marseille 33,147-225 (1973).
2. Cinelli, F., Fresi, E., Mazella, L., Ponticelli, M.P.: Giornale Bot. Ital. 113, 3, 173-188 (1979).
3. Devinny, J.S.: Bot. Mar. 21, 357-363 (1978).
4. Lundberg, B.: Ph.D. Thesis, Hebrew University of Jerusalem (1980).
5. Noy-Meir, I., Whittaker, R.H.: Vegetatio 33, 2/3, 79-98 (1977).
6. Velimirov, B., Field, J.G., Griffiths, C.L., Zoutendyk, P.: Helgolander wiss. Meeresunters, 30 495-518 (1977).

Acknowledgments: I want to thank Prof. T. Noy-Meir and Dr. Y. Lipkin, as well as Ms. C. Drobny and Ms. E. Moskowitz for help with the manuscript.

SEASONAL OCCURRENCE OF SCYTOSIPHON LOMENTARIA (SCYTOSIPHONALES, FUCOPHYCEAE) IN RELATION TO ENVIRONMENTAL FACTORS.

Aase Kristiansen

Institut for Sporeplanter, University of Copenhagen,
Ø. Farimagsgade 2D, DK-1353 Copenhagen K., Denmark.

Introduction

Scytosiphon lomentaria (Lyngbye) Link has a world-wide, mainly temperate distribution. It is common in all Danish waters though the salinity decreases from 33°/oo in the North Sea to about 8°/oo around the island of Bornholm in the Baltic. *Scytosiphon* grows mainly in shallow water, i.e. 0-1 m's depth. A combined field and culture study on Danish shores in 1976-77 (2) has given details on its seasonal variation and life history. Observations on the seasonal occurrence in relation to water temperatures were resumed in the unusually cold summer of 1979 and will be discussed here.

Materials and methods

Three localities were chosen for the investigation in 1976-77: Loc. 1, Hvidøre, situated in the Sound (Distr. Su) about 10 km north of Copenhagen. Loc. 2, Gilleleje, a fishing harbour on the north coast of Zealand, in the southern part of Kattegat (Distr. Ke). Loc. 3 is Frederikssund, situated in Roskilde Fiord (Distr. Ks). The three localities represent different types of habitats: Loc. 1 is a littoral-sublittoral stony and sandy area with a mixed *Fucus vesiculosus* - *Zostera marina* vegetation, and a single emergent granite boulder. Loc. 2 is a more exposed jetty, constructed of granite boulders, with all

Xth International Seaweed Symposium
Printed in Germany

zones well developed. Loc. 3 is an extremely sheltered and shallow locality particularly common to the Danish fiords. In 1979 the above mentioned localities were supplemented with Loc. 4: Deget, an islet off Frederikshavn in the northern part of Kattegat (Kn), surrounded by erratic boulders. Loc. 5 is the Swedish bedrock island of Hallands Väderö with a rocky coast (Distr. Ke).

The environmental data used, viz. salinity, surface water temperatures, water level fluctuations, and daylength have been obtained from the Danish Meteorological Institute, Copenhagen.

Results

The investigations 1976-77 showed that erect cylindrical fronds could be found in nature from September to June (Loc. 3) and from March to June-July (Loc. 1 and 2). The growth season always started in March and continued during spring with increasing water temperature and daylength. The population grew in abundance and size, reaching maximum in late April, May, or early June when surface water temperatures ranged from 6° to $14^{\circ}C$ and daylength varied from 14 to 17 h. The best developed *Scytosiphon* vegetation, at its peak consisting of 30 cm high tufts, was found in the fiord locality (Loc. 3).

Water level fluctuations are chiefly caused by the wind as the tidal amplitude is less than 30 cm and often only 10 cm. These fluctuations play an important rôle for the seasonal distribution of the population within a locality (2).

The salinity varies in the localities investigated but apparently without any influence on the seasonal occurrence.

The occurrence period normally ended in June-July in all localities when surface water temperatures (measured at 8 a.m.)

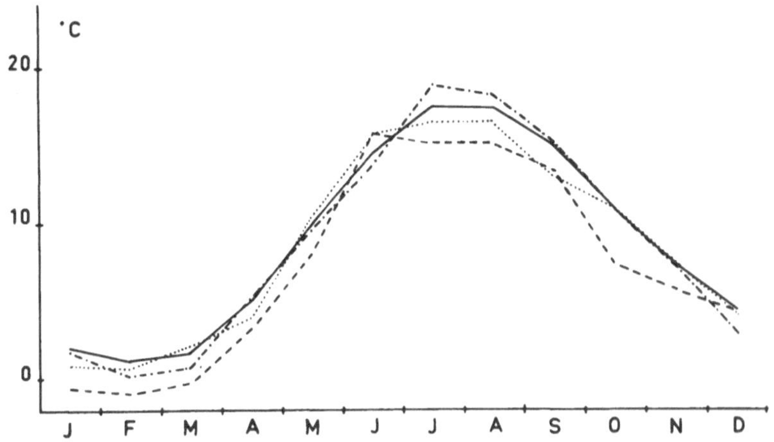

Copenhagen (55°41'N): Surface water temperatures, monthly values based on daily measurements (at 8 a.m.). 1976: -•-•-, 1977: •••••, 1979: -----, Normal: ——— (grand monthly means for the period 1931-60).

were about 17-18°C.

Although the decrease in water temperature followed the same pattern in the three localities during late summer - autumn, *Scytosiphon* reappeared only in the shallow fiord (Loc. 3). Here it was found from late September or October at water temperatures of 12-13°C. The populations persisted during winter, but there was a marked change from old fronds in January - February to many minute, juvenile plants in March (compare 3).

In 1979 nature made an experiment by giving us the coldest July we ever had, with a day and night average temperature of 13.6°C against normal 16.6°C resulting in water temperatures 2-3°C below normal.

In Loc. 4, the islet of Deget, an enormous *Scytosiphon* vegetation was observed on August 18, in the littoral. It consisted of 20-25 cm high tufts, each with few erect fronds. The same locality was visited on June 22, and at that time the fronds were mostly only 10 cm high and with epiphytes. When the islet was visited the previous year, on August 21, no *Scyto-*

siphon vegetation was observed, only scattered specimens, which always persisted particularly on our northern coasts. These observations were followed up in Loc. 5, Hallands Väderö, on September 8. Through the last 8 years we have made university excursions to this island in September or October, and to the very same localities - a small sheltered and shallow bay "Kungshamn" (Loc. 5a) on the south coast bearing a scanty vegetation with *Scytosiphon* corresponding to the Danish locality in the fiord (Loc. 3). Another locality (Loc. 5b) is situated 400 m away east of Kappelhamn where the rocks are covered with a dense *Ascophyllum-Fucus vesiculosus* and *F. serratus* vegetation. Here *Scytosiphon* was never observed in September, except in 1979, when rather high tufts, again consisting of few fronds, grew mixed with long bushy *Cladophora glomerata*, like the late vernal - early summer aspect of Loc. 1.

When this unusual seasonal occurrence of *Scytosiphon* had been noticed, it was natural to re-examine Loc. 1 and 2, which was done on September 20. In Loc. 1 *Scytosiphon* was present as 2-5 cm high plants, judging from their appearance probably remnants of a summer vegetation.

Discussion

These observations make it probable that water temperature is a decisive factor for the seasonal occurrence of *Scytosiphon* in the Danish or South-Scandinavian waters. A lowering of the mean water temperature for July to c. 14.5^{o}C enables *Scytosiphon* to persist in the vegetation through the summer postponing its disappearance from June-July till at least September. As the tufts consisted of relatively few and probably rather old fronds it might be interpreted as if growth is sustained at lower temperatures. Together with these fronds also some small ones were observed, perhaps new fronds?

Observations from adjacent areas shed more light on the occurrence of *Scytosiphon*. On Heligoland, 54°N (1), it is characterized as a spring alga, from March to June. In nearly the same latitude on the east coast of England (3) *Scytosiphon* was found also during summer when August water temperatures were 14.5°C and continued most of the winter. In southernmost Norway, 58°N, it was met with (4) from April, finishing in August when mean water temperatures rose to 17°C from 14°C in July. In Danish waters, 56°N, *Scytosiphon* has mainly a spring - early summer occurrence which in 1979 with the pronounced lowering of the water temperature immediately resulted in a prolonged occurrence period in agreement with the observations from England and Norway.

In agreement with our observations in nature concurrent culture studies (2) showed that erect cylindrical fronds developed as well during LD as SD, independently of salinity and temperature, but sensitive to reduced light intensity, at 15°C. In normal summers *Scytosiphon* is limited by an upper water temperature of 17-18°C in combination with competition on light from the surrounding dense algal vegetation. In the very cold summer the non-increasing water temperature decisively improved the ability of *Scytosiphon* to compete with the still dense vegetation.

The observation that *Scytosiphon* reappears in autumn only in Loc. 3, although water temperature decreases similarly in all the investigated localities, shows that temperature is not the only factor determining the seasonal occurrence. Probably the different locality types favour the surrounding algal vegetation unevenly. The sheltered and shallow fiord or bay has a scanty vegetation consisting mainly of an annual *Petalonia-Scytosiphon-Dumontia* vegetation together with a little *Fucus vesiculosus* (Loc. 3 and 5a) allowing sufficient light, even in September, for the prostrate systems of *Scytosiphon* to produce erect fronds. The other localities (Loc. 1, 2, 4 and 5b) bear a dense vegetation as well in the lower littoral as in the sublittoral often consisting of a *Pilayella* carpet not allowing the prostrate systems to produce erect fronds in autumn.

The prostrate systems (knot filaments or ralfsioid crusts) were only observed in nature (Loc. 1, 2, 3) with certainty from November to March, at least due to identification problems. It seems impossible to distinguish these prostrate systems for example from those of *Petalonia* in the periods when they have no physical connection with erect fronds. But *Scytosiphon* probably persists during unfavourable periods as prostrate systems; these are certainly less sensitive to variations in environmental conditions than erect fronds.

As Danish strains of *Scytosiphon*, as far as we know, are asexual, it is possible that there exist genetically different populations even within limited areas, adapted to different conditions. This question might be answered by forthcoming transplantation experiments.

References

1. Kornmann, P., Sahling, P.-H.: Helgol. wiss. Meeresunters. 29, 1-289 (1977).
2. Kristiansen, Aa., Pedersen, P.M.: Bot. Tidsskr. 74, 31-56 (1979).
3. Roberts, M., Ring, F.M.: Bull. Soc. bot. Fr., Mém. 1972, 117-128 (1972).
4. Åsen, P.A.: Marine benthosalger i Vest-Agder. Thesis, University of Bergen (1978).

TIDE POOL ASSOCIATIONS OF BENTHIC ALGAE IN ICELANDIC WATERS

Ivka M. Munda

Biological Institute, Slovene Academy of Science and Arts,
Ljubljana, Yugoslavia

Introduction

Tide pools on rocky shores represent very particular habitats according to their ecological characteristics (1,3,6). On shores were the tidal ranges are considerable as is the case in Iceland (maximum spring tide of 3.5 m in the South and minimum of 1.5 m in the East) there is a clear gradient in ecological conditions and vegetation patterns from the uppermost towards the lowermost pools and lagoons. In pools situated above the zero line there are wide short-term variations in ecological conditions which are closely correlated with atmospheric factors. Such pools were classified as rock pools (4) as opposed to tide pools which have more stable ecological conditions dependent on local hydrographic factors.

Tide pool associations of benthic algae in Iceland were studied with emphasis on local vertical gradients and on regional differences caused by hydrographic conditions. Comparable data on pool vegetation in the North of Europe are sparse (e.g. 4,16,17). Previous studies in Iceland have proved that tide pool associations are a characteristic feature for the vegetation of different hydrographic districts (7,9,10,11,12,13,14,15). Tide pool associations found in pools above the zero line and in the upper eulittoral were similar all around Iceland, whereas in the mid- and low eulittoral pools regional differences became obvious. In accordance with the hydrographic conditions and vegetation patterns three main regions were distinguished:

Xth International Seaweed Symposium
Printed in Germany

the typical Atlantic water region, including the South, SW and NW up to Hornbjarg; the transitional North Icelandic area and the subarctic area in the East. The Icelandic coast is surrounded by water masses of different origins and characters (2,5,8,18). Between these water masses there are either sharp boundaries or areas of gradual transition which are reflected in more or less pronounced floristic and vegetation limits (11).

Pools at the Level of the Littoral Fringe

Algal associations found in pools above the zero line are alike all around Iceland and characterized by a dominance of green algae, a Cyanophycean undergrowth, a reduced species diversity and low degree of covering. Such pools were lacking in the South when this level coinsides with sandy slopes. Around the Reykjanes peninsula they were situated among belts of dwarf Pelvetia canaliculata, and among populations of Prasiola species, Rhizoclonium riparium and Lichenes in all other areas. Typical associations of this level were those of Enteromorpha intestinalis var. asexualis f. cornucopiae and of E. clathrata the first one floating and the second attached. They could also occur as codominants. Small scale variations were, however, found between the areas, Cladophora species, occurring in aberant growth forms were frequent in such pools in the Southwest. An association of codominant Chaetomorpha capilaris and Rhizoclonium riparium was found at Reykjanes, whereas in the Northwest, North and East Fucus distichus in an extremely reduced growth form was frequently association-forming. Enteromorpha prolifera was a rather rare association found in the NW. In the East, an uplift of Acrosiphonia sonderi and Chaetomorpha melagonium to this level was common. High-lying pools were, however, especially frequent in the East and were on their highest level populated solely by Cyanophyceae.

Pools at the Upper Eulittoral Level

As in pools above the zero line associations found at this level were similar all around the coast. They were still lacking in the South where these levels are covered by sand whereas in other areas they were situated at the levels of either Pelvetia canaliculata or Fucus spiralis, of Ulothrix species or Porphyra umbilicalis. A slight increase in species number was observed here and the Cyanophycean undergrowth was replaced by Hildenbrandia rubra and crustose corallines (e.g. Phymatolithon lenormandii). A common and widespread association was that of codominant Enteromorpha intestinalis and E. compressa. Associations of diverse Cladophora species and of Enteromorpha prolifera were rare. An association with a rather restricted distributional pattern was that of Stictyosiphon tortilis and that of Punctaria plantaginea was rare. Fucus distichus ssp. distichus was a common association but most frequent in the North and East. The physiognomy of the association was different from the one found above the zero level due to a floristic enrichment and increased size of the dominant species. Acrosiphonia spp. and Chaetomorpha melagonium associations were frequent in the East, having a continuous undergrowth of crustose corallines as opposed to their high level associations.

Pools at the Mid- and Lower Eulittoral Levels

Pools situated among Fucacean fields (mid-eulittoral) or diverse low eulittoral algal belts were inhabited by different associations. They were characterized by a high species diversity and degree of covering as well as a three layered stratification. At these levels differences between the coastal districts became obvious. Pronounced differences were found between the associations of the hydrographic districts and small scale variations within them. Some of the tide pool associations exhibited a wide distributional pattern around the coast (e.g. Palmaria palmata, Ulva lactuca, Dictyosiphon foeniculaceus, Petalonia spp., Chordaria flagelliformis, Cladophora oblitterata, Enteromorpha prolifera, Chaetomorpha melagonium and Fucus distichus ssp. distichus), whereas others were restricted.

Along the South Icelandic coast red algae associations prevailed in the pools. Their floristic composition and structure has been described (13). Typical associations of this warmest area were: Chondrus crispus, Ahnfeltia plicata, Dumontia crassata, Cladophora rupestris, Corallina officinalis, Ceramium spp., Ceramium spp.-Cystoclonium purpureum. An Asperococcus fistulosus association was restricted to southern pools and another of Desmarestia ligulata to lagoons. Brown algae associations with a wide distributional range were still rare in the South. Lagoon associations of Laminaria saccharina and L. digitata were on the other hand especially frequent in this area.

In the Southwest (Faxa Bay) some changes were noticeable due to preveilingly soft substrata and lowered salinities in some areas. E.g. in the Borgarfjördur Chondrus crispus was lacking but it appeared again at the Snaefellsnes peninsula. In this fjord the Corallina association was characterized by a rich epiphytic cover of Ulva lactuca and Enteromorpha linza which changed its physionomy. Halosaccion ramentaceum joined the rest of the tide pool associations having a rich epiphytic cover of Ceramium rubrum. A typical association of the SW was that of codominant Cladophora oblitterata and Enteromorpha linza. A Chorda filum association was found in partly muddy lagoons and Polysiphonia nigrescens in sand-covered shallows. While in Borgarfjördur and the Mýrar area Cladophora species dominated in many pools (Cl. oblitterata, Cl. rupestris, Cl. glaucescens) red algae associations were again outstanding along the Snaefellsnes peninsula which has in general a rich vegetation.

In the Northwest, along the west Bardaströnd coast, the tide pool vegetation was again impoverished. The disappearance of the Dumontia incrassata, Chondrus crispus and Ahnfeltia plicata associations is notheworthy along with the presence of Halosaccion ramentaceum and Acrosiphonia spp. association which were subordinate in the vegetation. They occurred side by side with typical Atlantic water associations (Corallina officinalis, Ceramium spp, Ceramium spp-Cystoclonium purpureum, Cladophora rupestris). Associations of diverse filamentous brown algae were well represented in the NW. Here the fjord vegetation was studied (15). A pronounced gradient in the tide pool associations from the head to the mouth of the fjord was found, green algae (diverse Enteromorpha species, Ulva lactuca) as well as Dictyosiphon chordaria, Pylaiella littoralis and Ectocarpus species were restricted to pools of the inner fjord areas.

Along the North Icelandic coast Halosacchion ramentaceum (12), Acrosiphonia spp. Chordaria flagelliformis and Petalonia spp. associations were dominant. Here cold boreal, subarctic and typical Atlantic vegetation features mingle. Atlantic water associations of Corallina officinalis, Ceramium spp. and Ceramium spp.-Cystoclonium purpureum follow the North coast to the hydrographic boundary area in the Northeast. It is further noteworthy that in the Northwest and North Alaria esculenta inhabited deep lagoons beside Laminaria species.

Along the East Icelandic coast, influenced by arctic water, all typical Atlantic tide pool associations were absent. Two new associations appeared here: that of codominant Coilodesme bulligera-Ralfsia fungiformis in pools and the other of Saccorhiza dermatodea in lagoons. Halosaccion ramentaceum, Acrosiphonia spp., Chordaria flagelliformis and Petalonia spp. along with Fucus distichus ssp. distichus were associations typical for the eastern coast. The last named fucoid species increased in size in lower pools.

Observations also revealed that the distributional patterns of some tide pool associations do not always coincide with those of their corresponding low level associations. Corallina officinalis is belt-forming up to the hydrographic boundary area in the NW whereas its tide pool association spreads along the North coast. Halosaccion ramentaceum forms belts in the North and

East while its tide pool association was found also in the NW and SW. Chordaria flagelliformis forms belts in the North and East but its tide pool association is widespread all around the coast.

References

1. Ambler, P., Chapman, V.J.: Trans.Roy.Soc.New Zealand 78,394-409 (1950).
2. Dietrich, G.: Ann.Biol.14, 22 (1959).
3. Feldmann, J.,Davy de Virville, A.: C.R.Ac.Sci.Paris 197, 870-871 (1933).
4. Gustavsson, U.: Bot.Mar. 15, 210-214 (1972).
5. Hermann, F., Thomsen, H.: Medd.Kom.Havundersögelser, Hydrografi 3, 1-87 (1946).
6. Jårnfelt, H.: Verh.Int.Ver.theor.angew.Limnol. 9, 79-101 (1940).
7. Jónsson, H.: Botany of Iceland 1, 1-186 (1912).
8. Krauss, W.:Wiss.Erg.deutsch.Atl.Exp."Meteor" 1926 5,251-490 (1958).
9. Munda, I.:Acta Naturalia Islandica 21, 1-36 (1972).
10. Munda, I.: Bot.Mar. 15, 1-45 (1972).
11. Munda, I.: Bot.Mar. 18, 223-235 (1975).
12. Munda, I.: Bot.Mar. 19, 161-179 (1976).
13. Munda, I.: Bull.Res.Inst.Nedri Ås.Hveragerdi,Iceland 25, 1-69 (1976).
14. Munda, I.: Bull.Res.Inst.Nedri Ås.Hveragerdi,Iceland 28, 1-69 (1977).
15. Munda, I.: Nova Hedwigia XXIX, 281-403 (1978).
16. Round, F.E.: Commentat.Biol. 21, 1-21 (1959).
17. Saugestad, T.: Sarsia 46, 79-96,(1971).
18. Stefánsson, U.:Rit.Fiskideild. Reykjavik 3, 1-269 (1962).

PHOTOSYNTHETIC LIPOSOLUBLE PIGMENTS IN SEAWEEDS, PHYSIOLOGICAL AND ECOLOGICAL MEANING

F. X. Niell
Instituto Investigaciones Pesqueras, Muelle de Bouzas
Vigo, Spain

Introduction

Pigment content in phytoplankton is a usual measure in the study of pelagic communities but it has not been used routinely in studies of benthic communities. The aim of this paper is to state and explain how chlorophyll *a* concentrations change with different variables in benthic studies.

Methods

Materials included algae growing on surfaces of 15 x 15 cm and tissues of *Himanthalia elongata*, *Saccorhiza polyschides* and *Laminaria ochroleuca*. Pigments have been extracted following the classical methods (1, 2) slightly modified (3). The concentration of chl. *a* has been calculated according to Tailling (4), and the ratio between the absorption at 430 and 665 nm has been obtained for each extraction.

Results

I. *Intertidal communities*. There is a high concentration of chl. *a* in intertidal communities which agrees with the high production values (3). Frequently the concentration of chl. *a* m^{-2} was greater than 300 mg m^{-2}, which is considered the

Xth International Seaweed Symposium
Printed in Germany

Table I. Mean values of chl. a m^{-2} and D_{430}/D_{665} ratio in different communities.

Vegetation type (a)		mg chl. a m^{-2}	mg g^{-1}	D_{430}/D_{665}
I (3)	High level Pelvetia	488.33	0.60	2.42
	Low level Himanthalia	1442.93	0.87	2.26
II (7)	High level (b) Pelvetia	144.00	2.35	2.54
	High level (c) Pelvetia	465.50	2.54	2.54
	Low level (b) Himanthalia	364.50	2.36	2.5
	Low level Himanthalia (c)	1027.75	5.39	2.5
III (8)	Colonization phase (d)	10	-	-
	Diatoms (e)	100	-	3.5
IV (9)	Fouling on artificial substrate (b)	100	9	5
	Fouling on artificial substrate (c)	100	2.3	3.25
	Phytoplankton	16(10)	2.47 7.25(11)	2.75 4.25(10)

a) see text. b) measured in the 2nd month. c) measured in the 5th month. d) measured in the 10th day. e) measured in the 20th day. Numbers are bibliographical references.

maximal effective quantity of assimilatory pigment in a mono-layered vegetation (5, 3). Intertidal communities support more than 300 mg m^{-2} of chl. a due to a leaf index averaging a value of 4 (6). The quantity of chl. a per g and the values of D_{430}/D_{665} in different communities are expressed in Table I.

II. Communities on cleared intertidal rocks. The values of chl. a m^{-2} were lower than in communities of type I (Table I). Values of D_{430}/D_{665}, used as a structural index (5), were higher than in communities of type I.

III. Colonization phase. Diatom communities on rocky surfaces. This kind of community appears on the rock, before communities of type II. No values greater than 300 mg chl. a m^{-2} have been measures and the value of D_{430}/D_{665} decreases from very high to low and becomes more and more stable with time (8).

IV. Fouling communities on artificial suspended substrate. This type of community has characteristics intermediate between plankton and benthos. The quantity of chl. a per unit surface area is lower than in communities type I and II. Chlorophyll a concentration per unit biomass is quite similar to those obtained in communities of type II. D_{430}/D_{665}, however, is usually higher.

V. Plant tissue analysis. In Fig. 1 mean values of chl. a per unit surface area and biomass and D_{430}/D_{665} ratio is expressed for different parts of the thallus of three species of seaweed growing in the northwest Spanish intertide. The quantity of chl. a does not exceed 300 mg m^{-2} in any plant tissue, even in those undergoing active growth. Himanthalia elongata tissues show a high concentration of chl. a, as opposed to Saccorhiza polyschides whose values do not reach 200 mg m^{-2}. In Laminaria ochroleuca the tissues considered active averaged higher values than those without meristematic activity.

Discussion

The concentration of chl. a in different communities shows a pattern related to the stability of the substrate and the stage of succession. Young communities show a higher content of chl. a per unit biomass than those at the end of the succession in more stable phases. In tissues of macroalgae,

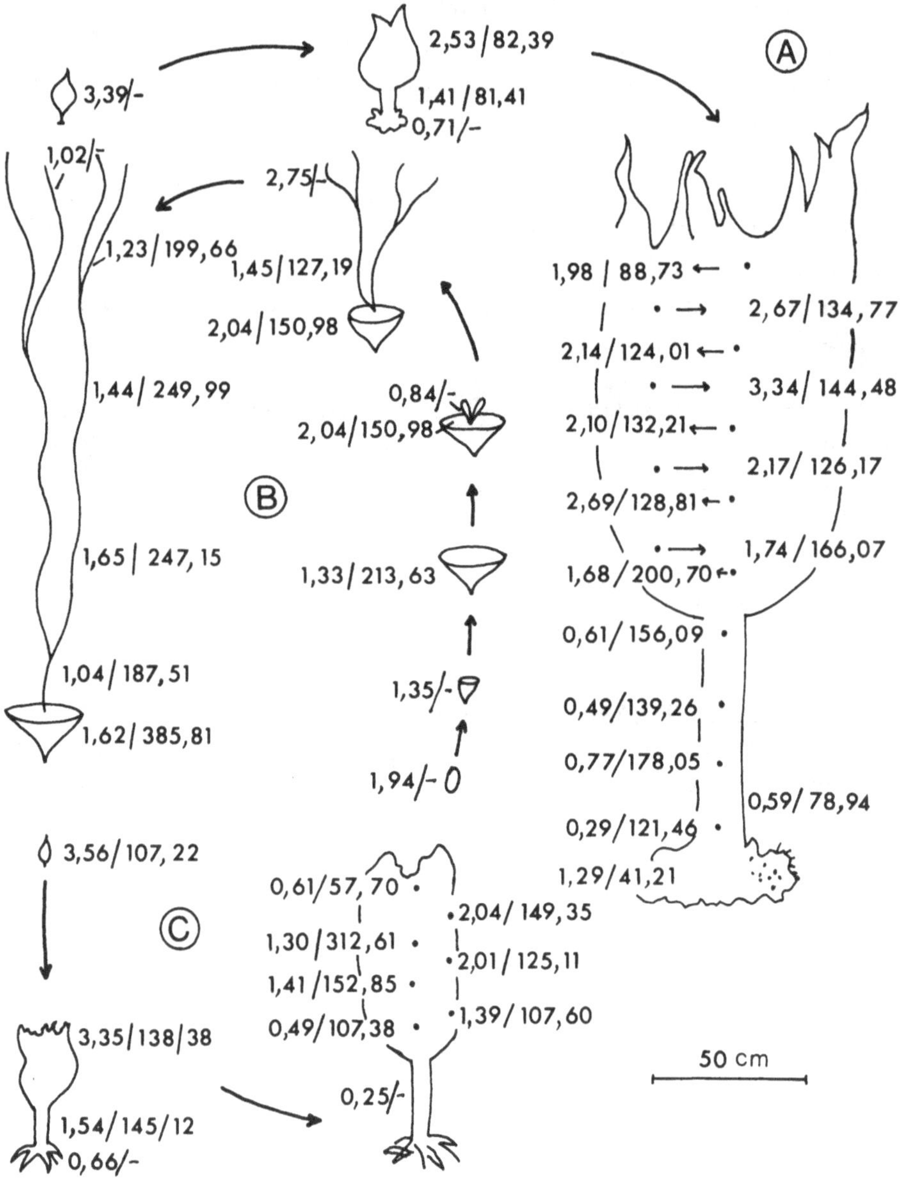

Chl. a per biomass/per surface area in plants of Saccorhiza polyschides A; Himanthalia elongata B; Laminaria ochroleuca C

young individuals have average contents of chl. a greater than 3 mg per g of dry matter. However, the values obtained in tissues of adult plants, were in general lower than 2.5 mg per g even in the meristematic zones. In zones where growth activity is null, chl. a reached very low values as in the stable communities. Values of chl. a per biomass unit are also very high in initial states of the succession, in which the potential of production is very high (7, 8).

D_{430}/D_{665} has been used as a diversity index in several works, inversely related with production (12, 13). In a previous paper it has been stated that in highly productive systems such as the intertidal zone, D_{430}/D_{665} does not reflect the dynamic changes as it does in systems of low productivity. This ratio allows comparison of very different communities, and different stages of changeable communities, such as the plankton community. In fact some authors (12) have used this index in planktonic studies with partial success.

In conclusion there is a parallelism of behaviour of the pigmentary indexes in tissues of algae and in benthic systems: those systems being more productive show higher concentration of chl. a per unit of biomass than those non productive, and the tissues showing a cell division activity have higher values of chlorophyll per biomass than the more inactive parts. Differences between tissues and communities have been recorded when the quantity of chlorophyll is expressed per surface unit, in this case the quantity of chlorophyll in the thallus of the algae is due to a two layered system. It is considered that D_{430}/D_{665} is not sensitive enough to discriminate different stages neither of the communities nor in the thallus of the algae.

References

1. Strickland, J.D.H., Parsons, T.R.: Bull. Fisheries Res. Board of Canada 167, 1-311 (1968).
2. Jeffrey, S.W., Humphrey, G.F.: Biochem. Physiol. Pflanzen 167, 191-194 (1975).
3. Niell, F.X.: Helgol. wiss. Meeresunters. 30, 315-333 (1977).
4. Tailling, J.F., Driver, D.: in Doty, M. (ed.) Conference on Primary productivity measurement. U.S. Atomic Commission Hawaii (1961).
5. Margalef, R.: Ecologia. Omega ed. Barcelona (1974).
6. Lüning, K.: Mar. Biol. 3(3), 282-286 (1969).
7. Niell, F.X.: J. exp. Mar. Biol. Ecol. 36, 185-200 (1979).
8. Niell, F.X., Varela, M.: unpublished.
9. Anadón, R., Niell, F.X., Otero, J., Varela, M: unpublished.
10. Alcaraz, M.: Ph. D. Thesis. University of Barcelona (1977).
11. Strickland, J.D.H.: in Riley, J.P., Skirrow, G. Chemical Oceanography. Acad. Press. London (1968).
12. Margalef, R.: Rapp. Reun. Com. Int. Exp. Mer Medit., 15, 277-281.
13. Vallespinós, F.: Inv. Pesq. 40 (2), 515-532 (1976).
14. Niell, F.X.: Ist Symp. Iberico Bentos. S. Sebastián. In press (1979).

ASPECTS OF THE LIGHT AND NITROGENOUS NUTRIENT REQUIREMENT FOR GROWTH OF CHORDARIA FLAGELLIFORMIS (O.F.MULL) C.AG.

Trevor A Probyn
Biology Department, Dalhousie University, Halifax,
Nova Scotia, Canada B3H 4J1.

Introduction

It has previously been shown that the summer growth of the perennial kelps is limited by ambient nitrate concentrations in Eastern Canada (1,2). For this reason the present study was undertaken to investigate the physiological and ecological strategies adopted by the summer annual seawweed Chordaria flagelliformis that enable it to grow rapidly during periods of limited nutrient availabilty. This paper presents preliminary findings regarding the growth of Chordaria in the field and laboratory in relation to the nitrogenous nutrients and irradiance.

Methods

Growth rates in the field were measured by pairing plants between harvests (7) and are expressed here on an absolute basis. Quantum irradiance was measured as described earlier (3) at a depth of 1-3 m. Ammonium and urea concentrations were determined on unfiltered seawater samples using the methods described in Grasshof (7). Nitrate and nitrite were measured according to Strickland and Parsons (12).

Unialgal cultures of sporophytes were established from young field material (2-5mm) collected in April. It was found that these plants collected during the early stages of growth could

Xth International Seaweed Symposium
Printed in Germany

be isolated in unialgal culture by repeated transfer through sterile seawater. Enriched seawater medium was used in batch cultures (PES McLlachan(11)). In flow through experiments nitrate and ammonium were excluded from the base medium (4).

The irradiance/growth relationship was determined in unialgal batch culture at 8 irradiance levels. Plants were cut basally to a length of 2cm and length increments measured every 2 days for 10 days. The growth/nitrogenous nutrient relationship was investigated in flow through culture (9) using the ammonium and nitrate free base medium, supplemented with either nitrate, urea or ammonium at six concentrations. All experiments were conducted at 10 C and growth saturating irradiance levels. Plants were starved for 2 weeks prior to experimentation to minimize the effect of nutrient reserves on subsequent growth. All plants were trimmed to aproximately equal weight and the biomass increment was measured over a 1 week period. Growth rate is expressed on a relative basis and its relationship to external nutrient concentration is assumed to be described by a Michaelis-Menton equation.

Results

Sporophytes first appeared in April and after a quiescent period displayed a rapid increase in growth rate in late May to peak in mid-July (fig.1). Growth rate decreased towards the latter part of summer and by late Aug early Sept plants began to lose weight and length. This decline in population biomass continued till mid-Nov by which time the plants had all but dissapeared.

The seasonal nutrient fluctuations (fig.1) is similar to earlier data reported for the nearby kelp beds of St Margarets Bay (2). Nutrient levels are low throughout the summer with urea (0.5 μM nitrogen) being quantitatively the most important nutrient. The seasonal availibility of nutrients was inversely

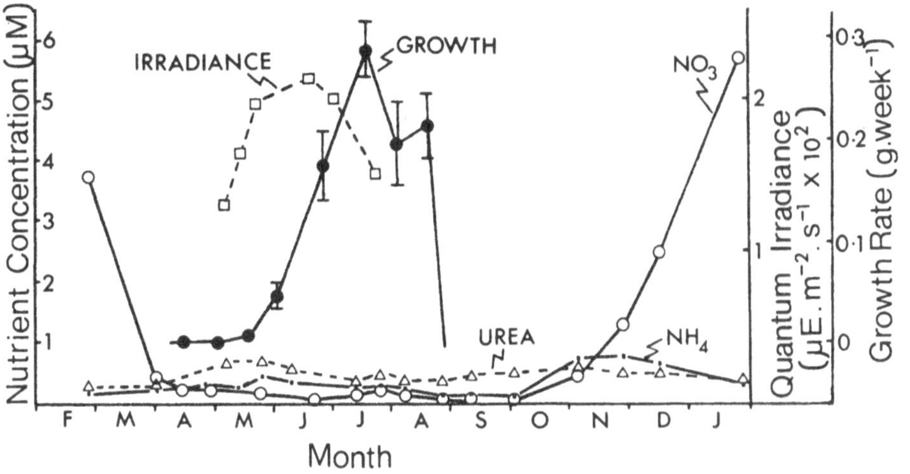

FIG.1 Growth rates of Chordaria flagelliformis (mean ± 95% confidence limits), summer irradiance levels, and the seasonal variation in nitrogenous nutrient concentration at Cranberry Cove, Nova Scotia.

related to the growth pattern of Chordaria.

Continuous flow experiments with either ammonium, urea or nitrate as the limiting nutrient gave estimated half saturation constants for growth (Ks values) of 2.78 for ammonium, 2.06 for urea and 1.32 for nitrate. Growth on ammonium was superior to the other nutrients tested.

Light saturation of growth in the laboratory occurred at an irradiance of 165 $\mu E.m^{-2}.s^{-1}$ (fig.3). In situ irradiance reached a maximum of 215 $\mu E.m^{-2}.s^{-1}$ in mid-July (fig.1).

Discussion

There appears to be a correlation between growth and saturating irradiance. The rapid increase in growth rate in May (fig.1) occurs about the same time in situ irradiance reaches the experimentally determined growth saturating irradiance level

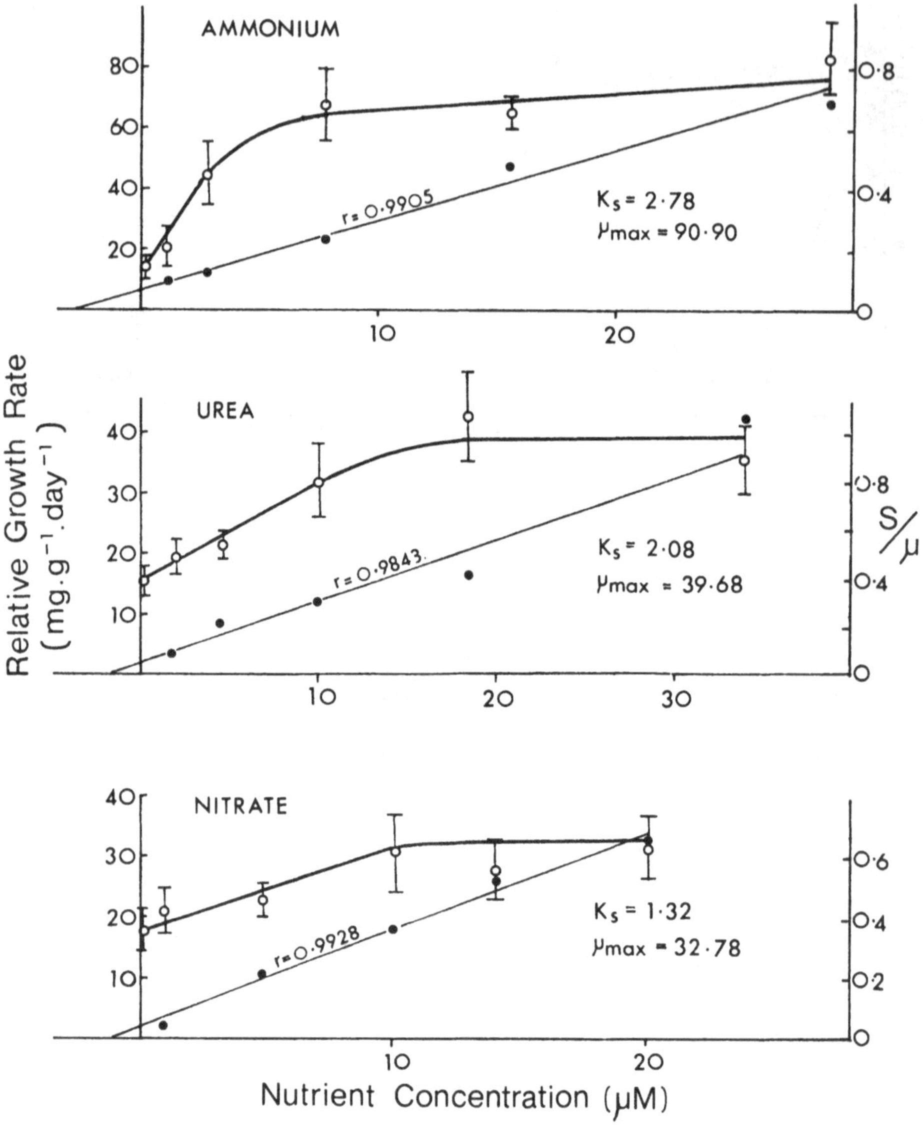

FIG.2 Relative growth rates (μ) of <u>Chordaria</u> sporophytes in continuous flow culture as a function of nitrogen concentration (S): means $\pm$ 95% confidence limits. The linear transformation was performed by regression of S/μ on S.

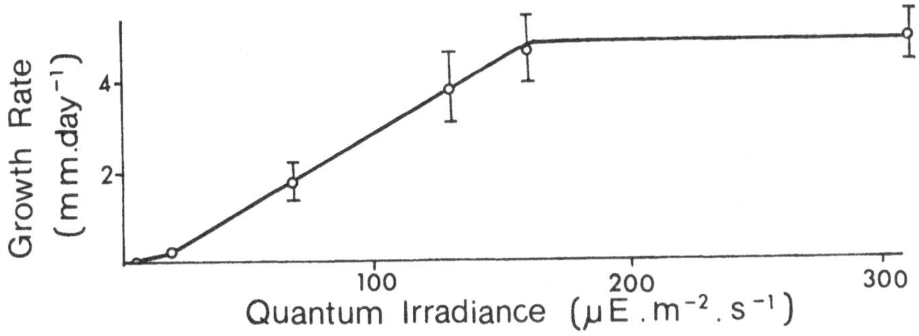

FIG. 3 Growth rates (means ±95% confidence limits) of unialgal batch cultures of Chordaria sporophytes over a range of irradiance levels.

(fig.3). Similarily the decline in growth rate in mid-July corresponds with the fall in in situ irradiance below this level (fig.1). This situation differs from the kelp beds of St Margarets Bay, Nova Scotia, where inorganic nitrogen rather than light limits summer growth (2,6). Kelps are able to alleviate this summer nitrogen limitation by utilizing reserves accumulated during the winter. Such a strategy is not open to the summer annual Chordaria.

The fact that Chordaria is able to grow rapidly under a low ambient nitrogen regime would suggest these plants have impressive capacities to assimilate low nitrogen concentrations Flow through culture experiments however reveal that this is not so. The estimated Ks values for Chordaria are almost an order of magnitude higher than those reported for the red algae Gracilaria and Neoagardhiella (5). Considering the relatively high nitrogen demand in flow through culture it is difficult to see how Chordaria achieves the rapid growth rates observed in the field. Growth saturating nutrient concentrations far exceed ambient levels and even though Chordaria achieves substantial growth at low (including zero) nitrogen concentrations, growth in culture at ambient levels is only 36% of the maximum relative growth rate observed in the field.

The discrepency recorded here may be a result of comparing the explicitly steady state situation in continuous culture with the non-steady situation in the field. Recently McCarthy and Goldman (10) proposed a highly dynamic scheme of nutrient flux whereby the assimilation of small pulses of nutrients (generated from animal excretion or microbial degradation) might allow oceanic phytoplankton to grow at close to maximal rates in nutrient depleted waters.

References

1.Buggeln,R.G.:J.Phycol.10.283-288 (1974).

2.Chapman,A.R.O.,Craigie,J.S.:Mar.Biol.40,197-205 (1977).

3.Chapman,A.R.O.,Lindley,J.E.:Mar.Biol. In press (1980).

4.Chapman,A.R.O.,Markham,J.W.,Luning,K.:J.Phycol.14,195-198 (1978).

5.DeBoer,J.A.,Guigli,H.J.,Israel,T.L.,D'Elia,C.F.:J.Phycol.14, 261-266 (1978).

6.Gagne,J.A.,Mann,K.H.:Proc.Xth Internat.Seaweed Symp.(In press)

7.Grasshof,K.:Methods of Seawater Analysis, Verlag Chemie, Weinheim.. New York.

8.Hunt,R.:Plant Growth Analysis, Edward Arnold, London.(1978).

9.Markham,J.W.,Luning,K.,Sperling,K-R.:Proc.lXth Internat. Seaweed Symp.153-159 (1978).

10.Mc Carthy,J.J.,Goldman,J.C.:Science 203,670-672 (1979).

11.Mc Lachan,J.:Handbook of Phycological Methods. Vol 1, Ed. J.R.Stein. Cambridge University Press (1973).

12.Strickland,J.D.H.,Parsons,T.R.:A Manual of Seawater Analysis. Bull.Fish.Res.Bd.Can.125,61-69 (1960).

BIOLOGICAL DATA ON A SOUTHERN PHAEOPHYCEAE ADENOCYSTIS UTRICULARIS (BORY) SKOTTSBERG, POSSIBLE MATERIAL FOR ALGINATE INDUSTRY.

Bruno de Reviers and René Delépine
Biogéographie et Ecologie Benthiques
B.V.M., 7 Quai St-Bernard
75230 Paris Cedex 05

Introduction

Adenocystis utricularis (Bory) Skottsberg is a characteristic brown alga from the Southern Ocean (1). The thallus is made of one utricle which can be about 10 cm long and several individuals can develop on the same crust forming a cluster. In the Kerguelen Islands, Adenocystis grows abundantly either in the inferior mediolittoral or in the upper part of the infralittoral belts. It occurs also in littoral pools at higher levels. The specimens are usually found on rocks but can also develop as epiphytes on other algae such as Corallina or Chaetangium. Furthermore the species is euryhalin.

Studies, using the same technics as those utilized by Mascré (2) for Durvillea, have been done to determine some constituants of Adenocystis utricularis (fig. 1), from the Kerguelen Islands. For example, the degree in alginic acid, obtained by decarboxylation method (3), is 31 % of the dry weight for large thalli collected in March from infralittoral. These values are interesting when compared to those known for Macrocystis (4) and Durvillea (5). They allow us to consider Adenocystis as an eventual good raw material for the production of alginates. In consequence, the purpose of this work was to investigate the different biological characteristics of the species in order to master its productivity.

Materials and methods

Several biometrical parameters have been used so as to define the morpho-

Xth International Seaweed Symposium
Printed in Germany

POPULATIONS	LENGTH mm	NUMBER THALLI	DRY WEIGHT	ALGINIC ACID	MANNITOL	LAMINAR^N	PROTEIN	ASHES
INFRALITTORAL MARCH 1979	75 _ 150 0 _ 75	42 15	14.45 10.94	30.9 31.7	12.5 7.6	1.5 1.2	9.1 15.9	31.7 34.3
MEDIOLITTORAL APRIL 1979	50 _ 75 25 _ 50 0 _ 25	115 208 73	15.33 17.53 23.95	29.4 26.3 16.3	8.9 7.9 7.8	5.8 1.4 1.2	15.9 17.6 16.6	26.1 34.6 23.9

1

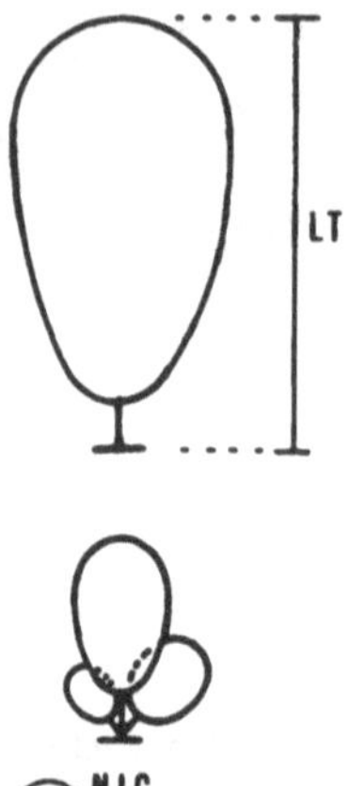

2

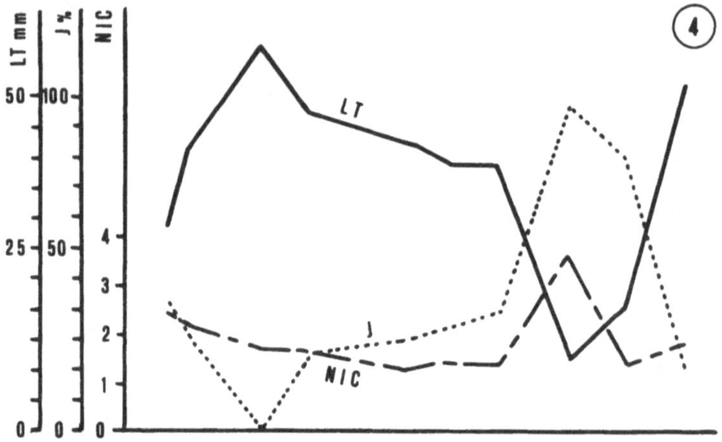

4

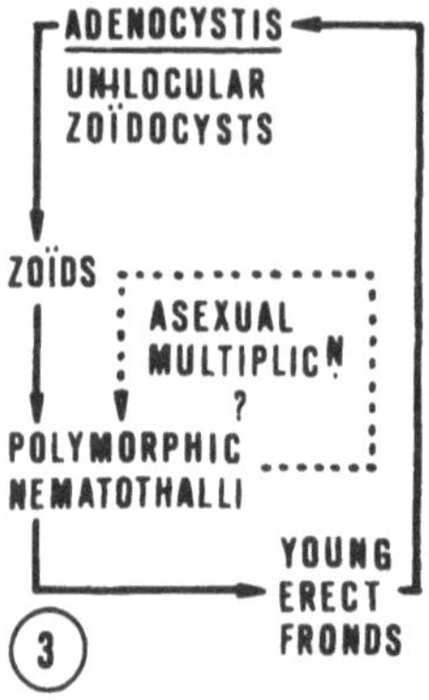

3

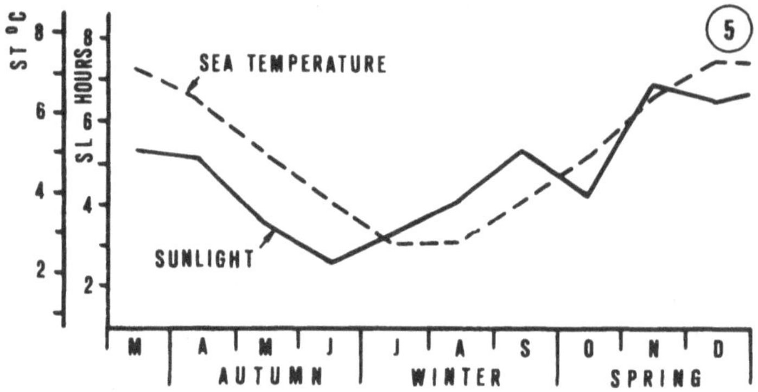

5

Figures 1 - 5 : DATA ON ADENOCYSTIS UTRICULARIS IN KERGUELEN ISLANDS

Fig. 1. Chemical analysis - The dry weight is expressed in percentage of fresh weight, all other constituants in percentage of dry weight; Fig. 2. Two of the biometrical characteristics measured on thalli - LT = length of thallus, NIT = number of thalli in a cluster; Fig. 3. Life cycle; Fig. 4. Periodicity of the mediolittoral population for three parameters between March-December 1977 - J = percentage of juvenile thalli (inferior to 25 mm) Fig. 5. Mean monthly sunlight (SL) expressed in hours per day.

logical characteristics of Adenocystis thalli in the field and in culture (fig. 2). Natural populations have been studied by monthly sampling in defined localities near Port-aux-Français (Kerguelen). Two populations have been distinguished : one in the infralittoral (2 - 3 m below low water level) and the other in the mediolittoral grouping individuals growing either on rocks or littoral pools. The later is more heterogeneous than the former one. Cultures have been established, in von Stosch or Provasoli medium (6), with zoïds from unilocular zoïdocysts. Several clones were obtained from one single zoïd each, these zoïds being shed by the same isolated zoïdocyst. Ecophysiological experimentation was carried on nematothalli generated by zoïds, using enclosures where the temperature and photoperiod were controlled and with different light intensities obtained by neutral filters.

Results

1) Life cycle (fig. 3). In our experiments (7) the zoïds from unilocular zoïdocysts produce nematothalli of various morphology (discoïd, filamentous, intermediary forms) according to the culture conditions. These nematothalli generate erect fronds which show all biometrical characteristics of the parent thalli but remain sterile.Up to now, attemps to count the chromosomes were not successful.

2) Phenology of the natural populations. In the infralittoral population utricles occur only between November and June and the young plants are abundant in November.

In the mediolittoral population utricles occur through the whole year and the young plants are mostly abundant in October. We have no personal acta on these populations for January-February, but in February 1980 young thalli seem to have been numerous (Bazin, personal communication). Figure 4, which corresponds to the mean values of three parameters for each monthly sampling, explicits these observations. The number of individuals in each cluster (NIT) shows a very clear maximal value in October; in the same way the percentage (J) of juvenile plants, inferior in size to 25 mm, accredits that these plants, even though they occur whole the year around, are much more abundant in October. The mean length of the plants (LT) grows rapidly during two periods : October-November and March-April. Thus it seems that, in the mediolittoral population, the young utricles appear in October and develop rapidly during the next spring and summer; furthermore it seems that there is another period of occurence of young plants in February. This would explain the values of LT in March-April-May.

Finally it must be noted that fertile individuals can be found whole the year around in the both populations.

Discussion

1) Life cycle. These studies support previous data (8) about the absence of heteroblasty and the fact that morphological variations of the nematothalli are closely related to environmental conditions.

The adult thalli collected around the Kerguelen Islands shed zoïds which generated nematothalli. These produced directly fronds similar to parent plants. No plurilocular zoïdocysts were observed on nematothalli in opposition to what Naylor (9) and Parsons (10) described. It must be noted that a doubt subsists as an ectocarpoïd plant is always associated with adult *Adenocystis*, and a precise study of this plant is needed. In any case, even if in certain conditions nematothalli can produce plurilocular zoïdocysts, the life cycle of *Adenocystis* is a direct one without alternation of generations in our experiments.

Studies on *Adenocystis* cannot be dissociated from those on the genus *Utriculidium* which is characterized by the presence of plurilocular organs. The eventuality of an alternation between *Adenocystis* as a sporophyte and *Utriculidium* as a gametophyte remained possible (8). Investigations made simultaneously on *Utriculidium durvillei* (Bory) Skottsberg from Kerguelen (7) show that this species has also a direct cycle in our culture conditions : the zoïds shed by plurilocular zoïdocysts give rise to nematothalli; these, without producing sporocysts, generate erect plants which, even though they remain sterile, possess all the biometrical characteristics of the parent plants.

In such conditions, it is advisable to maintain two different taxa for Adenocystis and Utriculidium. However new studies are necessary in the whole area of distribution of the two species in order to clarify the problem of an eventual alternation of generations. For instance, in Kerguelen, there is a large predominance of Adenocystis in comparaison with Utriculidium and there could be a very low percentage of alternating thalli. The influence of geographical distribution on reproductive mecanisms is important in the Pheophyceae as it is well known for Cutleria/Aglaozonia for instance (11).

2) Phenology. The massive occurence of young plants in November and their disappearance in June in the infralittoral population of Kerguelen are to be compared with data obtained in New Zealand (10). There, at Wellington, Adenocystis has been observed between October 1964 and July 1965. In both regions the development of the young utricles during the austral spring is to be correlated with a rise of the temparature and of sunlight (fig. 5; see also Parsons fig. 8 page 62). However, in Kerguelen, the variations of temperature and sunlight are less conspicuous than in New Zealand. Thus the presence of Adenocystis in the field corresponds to a sea temperature of 2-8° C and sunlight of 3-7 hours/day in Kerguelen whiles in New Zealand these values are respectively of 12-18° C and 3-10 hours/day. These lower amplitudes and differences in values could explain the greater stability of the mediolittoral population which occurs the whole year in Kerguelen Islands.

In a general way our data correspond rather well with those in New Zealand. The ecophysiological experiments on the production of erect plants already mentioned (8) have shown that the fronds appeared in great number under strong light and high temperature. The developing utricles during spring and summer in the nature (at time of rise of temperature and sunlight) are in good agreement with the experimental results. The development of utricles in February-March could be due to new nematothalli issued from zoïds shed by plants appeared in October.

In conclusion, the acquired knowledge on chemistry, life cycle, ecophysiological characteristics should motivate mass culture in order to determine rates of production in eventual aquaculture program.

Acknowledgments

This work was financed by "Terres Australes et Antarctiques Françaises". We thank M. Masrcé for chemical analysis and Dr. M. Parsons who kindly sent us his unpubliched thesis on Adenocystis of New Zealand.

References

1. Delépine, R.: C. R. Soc. Biogéogr. 374, 52-68 (1966).

2. Mascré, M.: (Manuscript) "Diplôme Etudes Supérieures" - Université P. et M. Curie - Paris.

3. Larsen, B.: in Handbook of Phycological Methods - Physiological and Biochemical Methods - Ed. J. E. Helleburst and J. S. Craigie. 14, 144-149 - Cambridge University Press 1978.

4. North, W. J.: Nova Hedwigia 32, (1971).

5. Delépine, R., Goubern, M., Hubert, M.: C. R. Acad. Sc. Paris 270, 59-62 (1970).

6. McLachlan, J.: in Handbook of Phycological Methods - Culture Methods and Growth Measurements - Ed. J. R. Stein. 2, 25-51. Cambridge University Press 1973.

7. de Reviers, B.: (Manuscript) Thesis for "Doctorat de Spécialité" in Phycology - Université P. et M. Curie - Paris 1980.

8. Delépine, R., Asensi, A.: Rev. Alg., N. S. 13(1), 43-85 (1978).

9. Naylor, M.: Trans. Roy. Soc. N. S. 83, 295-301 (1955).

10. Parsons, M. J.: Thesis submitted for M. S. in Botany - Victoria University of Wellington, New Zealand 1966.

11. Caram, B.: Rev. Alg. N. S. 12(1/2), 87-99 (1977).

PRODUCTION ECOLOGY OF CHILEAN GELIDIALES

B. Santelices, P. Oliger and S. Montalva

Departamento de Biología Ambiental y de Poblaciones, Instituto de Ciencias Biológicas, Pontificia Universidad Católica de Chile, Casilla 114-D, Santiago, Chile

Introduction

In Chile, three species of *Gelidium* are of primary commercial value. This research provides basic ecological knowledge to improve utilization of these species and evaluates the usefulness of physiological and biochemical criteria for their taxonomic segregation.

The genus *Gelidium* is one of the world most important agar producers. In Chile, three species of *Gelidium* (*G. filicinum* Bory, *G. lingulatum* J. Agardh and *G. spinulosum* Kützing prox.) are exported as raw material for agar production (1)(2). However, little is known about the ecological factors controlling their production ecology. Even their taxonomy is confused (3)(4)(5) because they are morphological variable and occur in close proximity in the intertidal and shallow subtidal. By adequate selection of production ecology studies it was expected to contribute to the taxonomic understanding of the genus *Gelidium* in Chile. Biomass changes, phenology and regeneration capacity were studied in the field, to obtain information for resource management programs (6). The effects of key ecological factors on the growth rate of these species in culture were experimentally compared. Such factors have been useful to define species limits in other Gelidiales (7) and multifactorial compensation fertility sites in the Rhodophyta (8). Since most chilean *Gelidium* are small sized plants with low biomass per unit area, preliminary experiments on mass culture were intended to evaluate specific differences in survival capacity and productivity under free floating conditions. Spe-

Xth International Seaweed Symposium
Printed in Germany

cific differences in quality and quantity of agar extracts were also searched.

Materials and Methods

Field studies were carried out at Pelancura in Central Chile (33° 35'S; 71° 38'W). Here, the Gelidium populations occur 1-2 m above zero level and extend downward up to 1-2 m below MLLW (Mean Lower Low Water). Two 5 m wide permanent transects were established to follow standing stock changes and phenology. In each transect, six 100 cm^2 samples were monthly taken every half a meter of intertidal distance. Gelidiales so gathered were washed with fresh water, sorted under a microscope to species and reproductive phase and weighed dry. Field recolonization capacities were monthly followed after denudation of permanent quadrats.

Experimental thalli were incubated in growth chambers (Forma Scientific, model 220), in glass dishes (Pyrex 3250) filled with 250 ml of enriched seawater (Erdschreiber medium). Temperatures used ranged from 10 to 25°C. Photon flux densities used were 45 and 120 $\mu E\ m^{-2}\ sec^{-1}$. Water movement (0 - 150 revolutions per minute) was provided by rotary shakers (Labline, Orbit Junior). All treatments included 3 replicates and lasted 28 - 30 days. Results (daily growth rates) were subjected to factorial analysis of variance (split-plot) after arcsin transformation.

Mass cultivation was attempted using 5 gallons Pyrex solution bottles (Corning 1595) filled with 16 l of seawater enriched with 5 ppm of $NaNO_3$ and 5 ppm of NaH_2PO_4. Water was kept moving by bubbling filtered air. Growth rates were measured and the culture medium renewed every 7 days. Temperature was maintained at 14°C and radiation at 45 $\mu E\ m^{-2}\ sec^{-1}$ for 12 hours daily. Each experiment lasted 28 days.

Hot water extraction of agar was performed at different cooking times for each species reproductive phase. All extracts were quantitatively filtered hot under pressure through diatomite filter aid. Gel strength was measured using a Marine Colloids Gel Tester with a 10 mm diameter plunger while viscosity measurements of a 1.5% agar solution were determined using a Brookfield Viscometer.

Results and Discussion

Temporal standing stock changes in Gelidium lingulatum and G. filicinum show little seasonality (Fig. 1). They have statistically significant negative correlation with the biomass changes of the other Gelidium species. Field data (Fig. 1) also indicate continuous presence of cystocarpic and tetrasporangial plants. The absence of seasonality in growth and reproduction of the two species was unexpected because of the temperate regime of the area and given the seasonal growth and reproduction of most Gelidiales elsewhere (9)(10). Regular sampling of G. spinulosum was impaired due to the very exposed habitat where it occurs. However, non quantitative sampling indicate presence of tetrasporangial and cystocarpic plants all year round. Recolonization of newly vaccated surfaces by any of the three species is extremely slow. Forty eight months took to G. lingulatum or G. filicinum to recover about 45% of the 100% rocky surface previously occupied. This is consistent with other recolonization studies on the group (11)(12).

The effects of several ecological factors on the growth rates of these species are shown in Fig. 2. Gelidium spinulosum has optimum growth at 15°C. Its growth rate decreases at higher temperatures, high radiation or extended photoperiod (16:$\overline{8}$). Its day-length tolerance can be increased at lower photon flux densities. G. lingulatum also has optimum growth at 15°C and decreasing growth rates at higher temperatures. However it has increasing growth rates at high radiation or under extended photoperiod. G. filicinum shows maximum growth at 20°C, and under higher radiation and extended photoperiod. None of the three species can survive under temperatures above 25°C or under conditions of continuous illumination. Only the growth

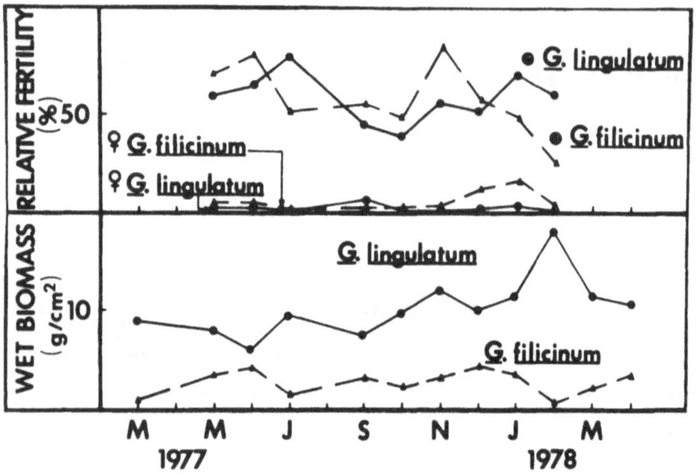

Fig. 1.- Temporal changes in standing stock and fertility of two Gelidium species in Central Chile.

Fig. 2.- Ecophysiology of the Gelidium spp.

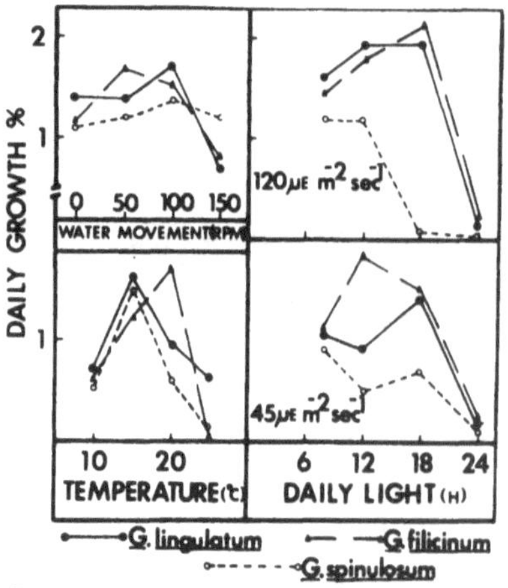

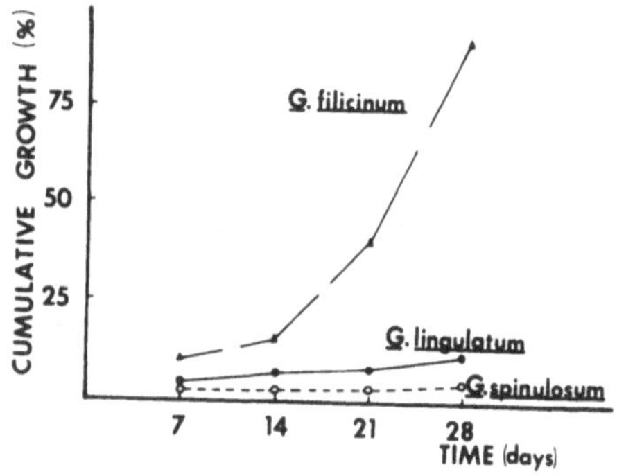

Fig. 3.- Mass culture of three Gelidium species under laboratory conditions.

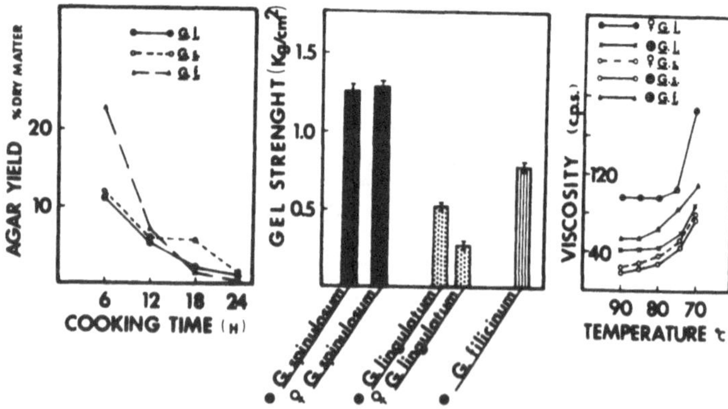

Fig. 4.- Comparative agar properties of the three Gelidium species from Central Chile

(G.l.=G. lingulatum; G.s.=G. spinulosum; G.f.= G. filicinum)

rates of G. lingulatum and G. filicinum are enhanced by water movement variations under the laboratory conditions used, as previously described for other Gelidiales (7).

Mass cultivation in large tanks shows (Fig. 3) Gelidium filicinum to be the species with the fastest growth rate doubling its original weight in about 30 days. The capacity of these algae to grow free floating is related to their ability of adopting a globular habit, devoid of holdfasts and production of profuse, radially-oriented branches. Similar morphological modifications are common in free-floating populations of macro-algae (13). These changes were most evident in G. filicinum in which the thalli become globose after 28 days. Radially branching thalli of G. lingulatum resulted from proliferations of the attachment parts of the thalli as the experiment ended. The thalli of G. spinulosum showed no modifications.

Specific differences in summer agar yields, gel strength and viscosity are shown in Fig. 4. Agar yield ranges from 31 to 20% the species dry weight. Highest agar yield was produced by G. filicinum while the strongest agar was produced by G. spinulosum. G. lingulatum has significant differences in gel quality among karyologically different generations. Such differences have been reported in carrageenan producers (see references in 14), but are yet unreported from agar producers. Although karyological differences in gel strength were reported from Gracilaria confervoides (15), they were smaller than between-habitat differences. In G. lingulatum such differences persist at different habitats and seasons.

In conclusion, this research indicates that the three Gelidium species studied have low biomass per unit area, slow field growth rates and low regeneration capacities. In spite of their continuous reproduction, they do not appear as a promising fishing enterprise in the field. They could survive under free-floating conditions with expected daily growth rates of up to 3%. Adequate handling of key ecological factors for the different species and phases could increase their growth rates in culture, allowing production of raw material for quantitatively different types of agar. Specific differences in the genus become most evident when the growth effects of key ecological factors or the agar characteristics of each species are compared. Thus, taxonomic segregation in the genus can be improved by the use of these physiological and biochemical criteria.

Acknowledgments: It is pleasure to acknowledge the continued support of the International Foundation for Science (IFS), Grant G 112, and of the Dirección de Investigación, Pontificia Universidad Católica (Grant DIUC 82/78) to carry out this research and to attend the Symposium.

References

1. Jaramillo, O.: Biol. Pesquera, Chile 7, 1-112 (1975).
2. Joyce, L., Santelices, B.: Biol. Pesquera, Chile, 10, 3-26 (1978).
3. Acleto, C.: Bol. Soc. Peruana Bot. 6, 1-164 (1973).
4. Howe, M. A.: Mem. Torrey Bot. Club 15, 1-185 (1914).
5. Levring, T.: Lunds Univ. Arsskr N. F. Avd 2, 56, 1-83 (1960).
6. Barilotti, D. C., Silverthorne, W.: Proc. Inst. Seaweed Symp., 6, 255-261 (1965).
7. Santelices, B.: Pac. Sci. 32, 119-147 (1978b).
8. Doty, M. S.: Proc. Int. Seaweed Symp. 9, 35-58 (1979).
9. Santelices, B.: Sea Grant Tech. Rep., Hawaii Univ. 1, 1-111. (1974).
10. Santelices, B.: Phycologia 17, 53-59 (1978a).
11. Lawson, G. W.: Oceanogr. Mar. Biol. Ann. Rev. 4, 405-448 (1966).
12. Northcraft, R. D.: Am. J. Bot. 35, 396-404 (1948).
13. Austin, A. P.: Hidrobiologia 14, 255-277 (1959).
14. Doty, M. S., Santos, G. A.: Aquatic Botany 4, 143-149 (1978).
15. Kim, D. H., Henríquez, P.: Proc. Int. Seaweed Symp. 9, 257-262 (1979).

ECOLOGY, LIFE HISTORY AND NUCLEUS DNA CONTENTS OF *DERBESIA TENUISSIMA* FROM THE CARIBBEAN COAST OF COLOMBIA

R. Schnetter, B. Mohr

Botanisches Institut der Justus-Liebig-Universität Giessen,
D-6300 Giessen, Germany

G. Bula-Meyer

Departamento de Biología, Universidad Tecnológica del Magdalena,
Santa Marta, Colombia

G. Seibold

Strahlenzentrum der Justus-Liebig-Universität Giessen,
D-6300 Giessen, Germany

Introduction

For the identification of the taxonomic position of algae culture experiments are often employed. This method seems especially important in such difficult genera as *Derbesia* that is generally characterized by heteromorphy in its life history (7). Sporophytic plants of a *Derbesia* species collected in Colombia have been cultured under laboratory conditions with the purpose to obtain informations about the variability of thallus form, growth responses on different temperatures and light intensities related to environmental factors of the collection site, and life history.

Materials and methods

Sporophytes of a *Derbesia* species have been collected on *Rhizophora mangle* L. roots in the Ensenada de Chengue near Santa Marta, Colombia, growing intertidal and at a depth of about

Xth International Seaweed Symposium
Printed in Germany

0.5 m below low tide level. In Giessen the algae were cultured in an enriched seawater medium (8, 9) at temperatures of 18°, 20°, 22°, 24°, 26°, 28° and 30° C. As light sources Philips fluorescent tubes (type TL 20 W/29) were used, with a photoperiod of 12 $h \cdot day^{-1}$. Light intensities were 0.5, 3, 10 and 13 $W \cdot m^{-2}$. The medium was changed weekly. Formation of sporangia could be induced by light intensities of 10 and 13 $W \cdot m^{-2}$; in this case the medium was changed every second day. For productivity studies cultures were started with 1 cm long sporophyte filaments and thallus dry weight was determined after 3 weeks. Microspectrophotometric measurements of DNA contents of nuclei were carried out after Feulgen staining (5), for fluorescence measurements nucleus DNA was stained with HO 33258 (Hoechst) (6). Measurements were made using scanning microscopes (Zeiss UMSP 1 and SMP 05, respectively).

Results

Under the employed culture conditions sporophytes show optimal productivity at a temperature of 24° C and a light intensity of 3 $W \cdot m^{-2}$ (table 1). At a temperature of 18° C the growth is insignificant while at 20°, 22° and 30° C thallus growth is considerably low. Within 3 weeks temperatures of both 18° and 30° C cause serious damages in the thalli.

The diameters of filaments of sporophytes cultured under conditions for optimal vegetative growth vary between 45 µm and 89 µm (average of 100 filaments: 61.9 µm, ± 7.3 µm). There is no statistical difference between these values and those from plants collected in the field (average diameter: 57.6 µm). Sporangia are 161 µm ± 19 µm long, 119 µm ± 11 µm wide; their stalk "cells" are 11.5 µm ± 3 µm long. In the sporangia more than 20 zoospores are formed with a diameter of 24 µm ± 2 µm.

In culture zoospores from the above mentioned sporophytes (S I)

Table 1. Thallus biomasses (dry weight in mg; mean values of 5 cultures ± standard error) after 3 weeks as results of different temperatures (° C) and light intensities ($W \cdot m^{-2}$).

	0.5	3	10	13 $W \cdot m^{-2}$
24° C	0.9 ± 0.3	5.2 ± 0.7	3.0 ± 0.5	2.3 ± 0.4 mg
26° C	0.1 ± 0.1	3.8 ± 1.0	2.3 ± 0.9	1.6 ± 0.4 mg

	18°	20°	22°	24°	26°	28°	30° C
3 $W \cdot m^{-2}$	0.2	2.4	2.7	5.2	3.8	3.6	2.5 mg
	±0.2	±0.6	±0.6	±0.7	±1.0	±1.0	±0.2 mg

germinate forming initially a small filamentous stage (rhizoid) that bears later one or more vesicles up to 4 - 10 mm long and 1.5 - 4.8 mm wide. All observed gametophytes are female, the gametangia develop distally and discharge gametes of about 11 µm diameter through an apical pore. From the gametes new sporophytes (S II) are produced by parthenogenesis. Chloroplasts of all plants have pyrenoids. Both types of sporophytes show remarkable differences. The filaments of S II are more slender (35.4 µm ± 7.3 µm) and the sporangia are narrower (145 µm ± 20 µm long, 75 µm ± 12 µm wide) than those of S I. Further, the S II thalli have a reduced vitality and generally decay after formation of sporangia. There are no differences between the gametophytes obtained from the two types of sporophytes. Four successive life cycles could be observed in culture.

Nuclei of both S I and S II have identical average dimensions of 3.4 µm x 2.1 µm. Gametophyte nuclei reach 6.3 µm x 2.4 µm during the migration phase before gametangium formation. Results of microspectrophotometric measurements after Feulgen staining are shown in figure 1. Mean values are 2.7 AU, 2.4 AU and 1.7 AU for S I, the gametophyte and S II, respectively. No statistical differences were found between the main peaks of both types of sporophytes and the gametophyte. Further peaks at 3.0 - 3.7 AU and 4.6 - 4.9 AU are observed in the nuclei of S I and the gametophyte, but not in S II (fig. 1). Relative DNA

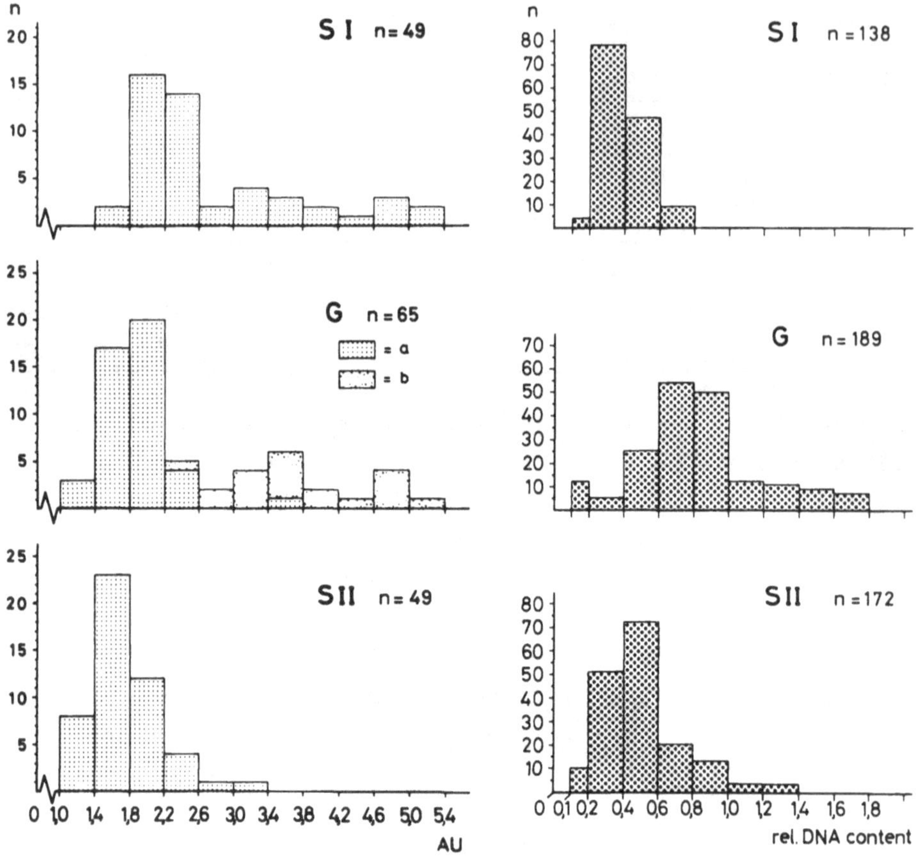

Fig. 1 (left). Absorption units (AU; relative extinction) of nuclei of sporophyte I (S I), gametophyte (G) and sporophyte II (S II); n = number of nuclei; a = nuclei after, b = nuclei during migration phase.

Fig. 2 (right). Relative nucleus DNA contents; abbreviations see Fig. 1.

content measurements realized by means of fluorescence give similar results (fig. 2): no statistical differences exist between the peaks of the two sporophytes and the gametophyte.

Discussion

Filament diameters of the sporophyte of the Colombian *Derbesia* are hardly useful as criterion for taxonomic purpose due to the overlapping ranges of filament sizes of the Atlantic species of this genus (10). Further, the grade of ramification applied for identification of *Derbesia* species (1) was proved to be influenced by environmental factors; in the Colombian plant the grade of ramification increases with light intensity. The length to width ratio of the zoosporangia of the Colombian *Derbesia* is similar to that of *D. tenuissima* (De Notaris) P. L. & H. M. Crouan (see 10), of which parthenogenesis was described (10). Moreover, by the form and size of gametophytes, terminal position of gametangia and presence of pyrenoids, the Colombian plant is referable to *D. tenuissima* hitherto known specially from warm-temperate seas (1, 3, 4). The temperature optimum of thallus growth (24° C) and the decay of the plants at 18° C show that the *Derbesia* that was studied belongs to a thermophile and stenotherm race highly adapted to the temperature regime at the Caribbean coast of Colombia (23.8° - 28.8° C annual oscillation in a bay nearby the collection site; 2).

Life history studies resulted in an alternation of plants with the aspects of sporophytes and gametophytes, but with identical DNA contents of nuclei, and in parthenogenetic sporophyte development from female gametes. Main peaks in the histograms of figure 1 are considered to be corresponding to nuclei in the G_1 phase of the mitotic cycle. Additional peaks in the histograms of sporophyte I and gametophyte result from nuclei in G_2 phase and perhaps polyploid stages. This interpretation is supported by the measurements of migrating gametophyte nuclei that are a short time before their simultaneous division (see 11). Comparison with chicken erythrocytes as standards results in a DNA content of G_1 nuclei of about 0.15 pg.

Because no male gametes could be observed in culture and as the DNA content of nuclei of plants collected in the field is equal to the respective values of gametophytes and S II obtained in culture, it may be assumed that the Colombian race of *Derbesia tenuissima* has lost the male gametophytes. Further, it seems probable that sporophytes as well as gametophytes are haploid. Differences in thallus size, form of sporangia and vitality in both types of sporophytes are not reflected by different nucleus DNA contents. Occasional observations of a more vigorous development of S II thalli growing epiphytic on gametophytes may indicate that the vitality of the sporophyte may be influenced by a still unknown environmental factor.

Acknowledgement

The present study was supported by the Deutsche Forschungsgemeinschaft.

References

1. Børgesen, F.: Kgl. Danske Vidensk. Selsk., Biol. Medd. 5(3), 1-123 (1925).
2. Caycedo, I.E.: An. Inst. Inv. Mar. - Punta Betín 9, 17-44 (1977).
3. Giaccone, G.: WWF Italia, Parco Marino Miramare 6(19), 1-118 (1978).
4. Hamel, G.: Rev. Algol. 5, 383-430 (1931).
5. Jordanov, J.: Acta histochem. 15, 135-152 (1963).
6. Linden, W.A., Baisch, H., Seibold, G.: Pulse Cytophotometry Suppl., 107-111 (1976).
7. Neumann, K.: Botanica Marina 13, 176-185 (1974).
8. Stosch, H.A. von: Proc. 4th Int. Seaweed Symp., 142-150. Pergamon Press, Oxford · London · New York · Paris 1963.
9. Stosch, H.A. von: Laboratory working sheet (1966).
10. Ziegler-Page, J.R.: J. Phycol. 6, 375-380 (1970).
11. Ziegler-Page, J.R., Kingsbury, J.M.: Phycologia 4, 105-116 (1964).

CHEMICAL CONSTITUENTS OF SOME BALTIC MACROALGAE IN RELATION TO ENVIRONMENTAL CONDITIONS

I. Wallentinus

Askö Laboratory, Institute of Marine Ecology, University of Stockholm, P.O. Box 6801, S-113 86 Stockholm, Sweden

The extent to which pigments and nutrients in plants reflect ambient growth conditions has been widely studied in microalgae, but interest has also recently been focussed on macroalgae. This study was aimed at examining seasonal, nutrient-induced changes in some macroalgae, mainly the green alga *Cladophora glomerata*, from different parts of an archipelago area at Trosa, N. Baltic proper. The study area and methods have been described earlier (1-4), and more detailed results are given in (4). The innermost part of this archipelago is moderately polluted, mainly by agricultural run-off, but also by domestic sewage, and has in spring and early summer an excess of nitrates brought from fertilized soils by the river, while phosphates increase in late summer and autumn. In the outer part of the archipelago phytoplankton blooms in spring deplete most of the nutrients, especially nitrogen, and in summer there is a gradual recovery from nitrogen deficiency, but a comparative lack of phosphorus.

Results

1) Nitrogen. *Cladophora* had N contents highly varying with both time and space (Table 1, Fig. 3). By far the highest mean value (41.8 ± 10.2 mg N·g dry wt^{-1}) occurred at the most nutrient-enriched station 2 in the innermost archipelago, and there the values were less dependent on season. In spring and early summer *Cladophora* from the inner area usually had higher N contents than from the outer area, while in summer the values were rather low

Xth International Seaweed Symposium
Printed in Germany

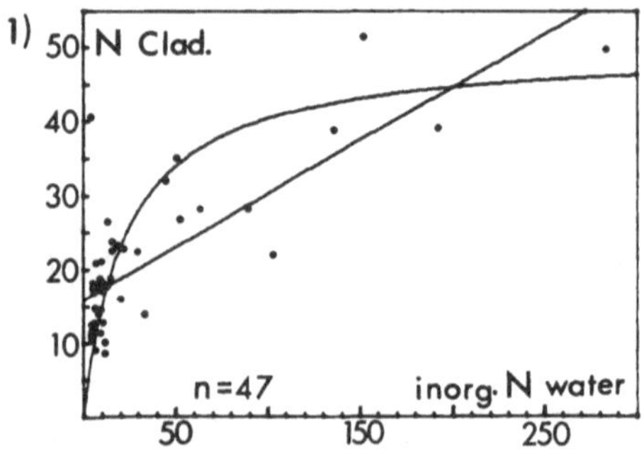

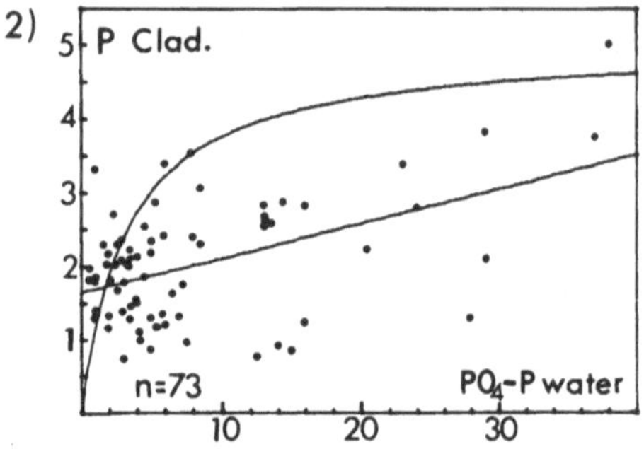

Fig. 1 and 2. Total nitrogen (1) and phosphorus (2) in pooled *Cladophora* samples ($mg \cdot g$ dry wt^{-1}) versus inorganic nutrients ($\mu g \cdot l^{-1}$) in water.

everywhere (except station 2). The increase in autumn often started earlier in the outer area. Even though there was a linear correlation ($r=0.78$, $p<0.001$) with the inorganic nitrogen in water, there is more a trend of hyperbolic relation (Fig. 1). The red alga *Ceramium tenuicorne* from the outer area had higher N contents than *Cladophora*, in spite of growing in the same environment, but the same seasonal trend was found (Fig. 3).

The perennial brown alga *Fucus vesiculosus* (Table 1) had in both areas the highest N contents in apical, active parts during winter and early spring and lowest during summer. The old mid- and basal parts had lower, less seasonal influenced values. *Fucus* transplanted for three weeks in late summer into the innermost area, where they do not grow because of too low salinities in spring (2), increased their N contents by 116-160% at station 2, and by 7-11% at station 3, but no increase occurred at the other stations, despite slightly higher N concentrations in the water.

Table 1. Nutrients and pigments ($mg \cdot g$ dry wt^{-1}) and ratios (by atoms) in the green alga *Cladophora glomerata*, the red *Ceramium tenuicorne* and the brown *Fucus vesiculosus* (apical parts, not transplanted).

Species	area	N	P	C	N:P	C:N	Chl.*a*
Cladoph.	outer	10-37	1.4-3.0	313-399	12-29	13-41	0.2-6.0
"	inner	8-52	0.7-5.4	183-385	8-66	9-49	0.3-11.6
Ceramium	outer	13-59	3.1-5.0	358-365	9-32	7-13	0.2-4.0
Fucus	outer	9-39	1.2-5.7	335-363	14-19	24-43	0.5-3.7
"	inner	9-45	1.0-5.6	350-371	17-38	14-46	-

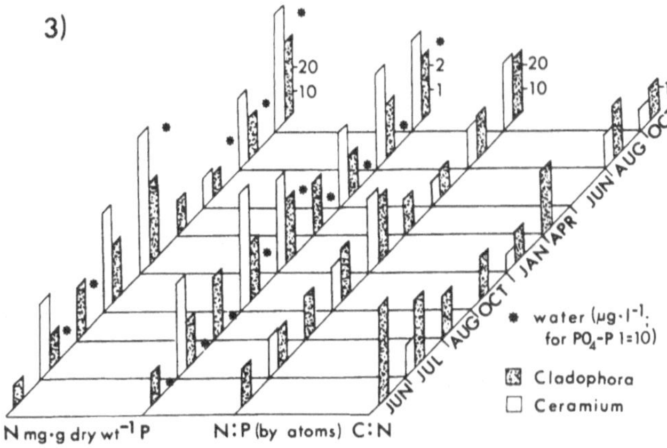

Fig. 3. Total nitrogen and phosphorus, N:P and C:N ratios in *Cladophora* and *Ceramium*, and inorganic nutrients in water; outer archipelago 1972-1973.

2) Phosphorus. P contents also varied greatly in *Cladophora* with space and time (Table 1, Fig. 3), and the mean value was highest at station 2 (3.80 ± 1.49 mg P·g dry wt^{-1}). In spring and early summer *Cladophora* from the outer area usually had higher values than from the inner (except station 2), although P concentrations in the water were lower. In autumn all values increased, especially in the inner area. Linear correlation with inorganic phosphorus in the water was less strong ($r=0.48$, $p<0.001$) than for nitrogen, and no hyperbolic relation was found (Fig. 2). The P contents in *Ceramium* were also higher than in *Cladophora* (Fig. 3), but had the same seasonal trend.

Fucus (Table 1) in both areas had the lowest P content in summer and maxima in late winter to early spring. The P limitation in spring noted for the annual *Cladophora* was not found for the perennial *Fucus*, which can store phosphorus during winter for future usage. The transplanted *Fucus* increased the P content by 47-65% at station 2, and 6-41% at station 3, but almost no increase occurred at the other stations, despite higher P levels in water.

Surplus phosphorus measurements (see 5) in *Cladophora* ranged from 0.31 to 2.75 mg P·g dry wt^{-1} and had a strong correlation ($r=0.85$, $p<0.001$) with total P contents. *Cladophora* from the inner area had values typical of P-limited algae (below 0.6-0.7 mg P) in spring and early summer (Fig. 4), but above P limitation in late summer

4)

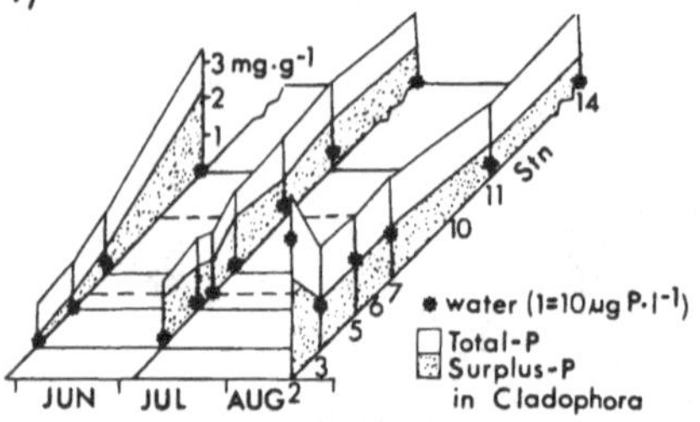

5)

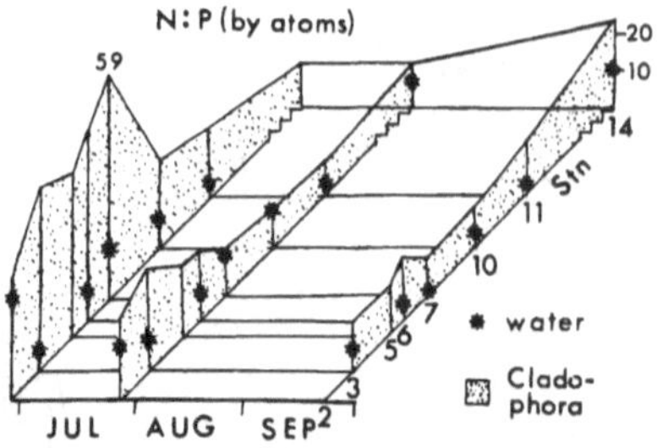

Fig. 4. Surplus and total phosphorus in *Cladophora* and phosphate in water; whole area 1971.

Fig. 5. Ratios of total nitrogen and phosphorus in *Cladophora* and inorganic N:P in water 1970.

and autumn. *Cladophora* from the outer area were only slightly P-limited in late summer. Surplus P in *Fucus* in August ranged from 0.64 to 1.18 mg P and transferred plants showed values lower than *Cladophora* growing there at the time, while in the outer area *Fucus* had more surplus P. In October *Fucus* showed much higher values (3.7-4.6 mg P).

3) Carbon. In the outer area the C content of the algae varied much less seasonally than nitrogen and phosphorus (Table 1). At the innermost station 2 the dark-green *Cladophora* had evenly high C contents, but from most of the inner area the values in *Cladophora* were lowest in spring and early summer.

4) Nitrogen : phosphorus ratio. In *Cladophora* the N:P ratio varied highly with space and time (Table 1, Fig. 5). During spring and early summer samples from the inner area had very high ratios, indicating N excess, especially in years with a vigorous spring flood. In late summer and autumn the ratios dropped to more normal levels. In the outer area *Cladophora* and *Ceramium* had very similar ratios (Fig. 3), indicating N limitation in spring and early summer but rather P limitation in late summer and autumn. There was no linear correlation between N:P ratios in *Cladophora* and those of inorganic N:P in the water ($r=0.22$, $p>0.1$), and although *Cladophora* with high N:P ratios mostly had very little surplus P, the opposite could also be found.

Fucus from the outer area had no drastic changes in N:P ratios with season, and the ratios were similar to those of annual algae in summer, but lower in winter. In the inner area the ratios were higher in spring (but not in winter) than in late summer. However, the spring values were not as extremely high as in *Cladophora* (Table 1), owing to stored phosphorus in the perennial *Fucus*. The plants transferred into the innermost area only increased the ratios at the most enriched station 2.

5) Carbon : nitrogen ratio. The C:N ratio in *Cladophora* (Table 1, Fig. 3) were often much higher than values given for phytoplankton (cf. also 3). They varied with space and time, but since the C contents were more stable than that of N, C:N ratios often gave the reverse picture of the N:P ratios. *Cladophora* from the most enriched station 2 had low C:N ratios, but from the rest of the inner archipelago the ratios were low in spring and early summer, increased during summer and dropped again in autumn. In the outer area the ratios were highest in spring and early summer and decreased towards winter. *Ceramium* had much lower C:N ratios than *Cladophora* from the same date (Fig. 3). Since the N:P ratios were very similar, the C:N ratios are less likely to indicate deficiency, but more differences in storage of carbohydrates versus nitrogen bound into proteins or chlorophyll.

In summer *Fucus* from the outer area had even higher C:N ratios than *Cladophora*, and higher ratios in old parts than in young, indicating more storage of carbohydrates or lipids. Also in the inner area the ratios were high in summer, but lower in spring, and the transplanted *Fucus* had markedly decreased ratios only at the most nutrient-enriched station 2.

Discussion

The ranges of nutrients and pigments given in Table 1 agree well with those recorded from other areas (references in 4; 6-7), although in some polluted areas the P content of *Cladophora* and

Fucus was 2-3 times higher than my maximum values. The high N levels in the red alga *Ceramium* compared to the green and brown algae are found also by others (4, 6). The high N:P ratios in the inner area in spring are higher than reported from naturally growing *Cladophora* and *Fucus* in other areas, but high ratios have been found in cultures of *Cladophora* (7, 8).

The great scattering from linear regressions (Fig. 1, 2) can have several explanations. The low accuracy for water analyses with very low N and P levels, especially for the N:P ratios, might partly contribute. However, excess storage in algae, having already depleted available nutrients in the water are probably more important, as are the turnover times and rates of nutrient supply. The relation between N and P in the water certainly has more influence on the surplus P in algae in the outer area in early summer than the absolute concentrations. The capacity of many algae to use more complex N and P compounds as well (e.g. 9) might also reduce the correlation with inorganic levels. Generally, however, algae growing in the most nutrient-rich media (in time or space) had the highest content of both N and P.

Osmotic stress of low salinities has been found to increase the N contents of algae (e.g. 8, 10), and the low salinities in spring in the innermost archipelago (1, 2) might thus have contributed to the extremely high N:P ratios. Since increases in N and P occurred in *Fucus* transplanted to high nutrient levels also at a time of almost equal salinity regimes, nutrient levels in the media are certainly more responsible.

Literature gives a high variety of deficiency values based on the total nutrient contents in the plants (see 4). If nutrient levels in cultured, nutrient-deficient microalgae (rewiev in 11) were valid also for macroalgae, all my algae would have been extremely P-limited and extreme N deficiency would have occurred even in the samples with very high N:P ratios, as would the chlorophyll *a* levels indicate deficiency in almost all algae. The critical levels (21 mg N and 3.3 mg P) for growth of *Clado-*

phora given in (7) would also mean P limitations for most of my specimens (except from station 2 and some autumn and winter plants), as would N limitations have occurred also in some of the plants with rather high N:P ratios. Contrarily, if the critical levels for phanerogams (12) were applied, my algae would have been N- and P-limited only on a very few occasions, and never P-limited in the outer area. The critical levels given for photosynthesis in *Cladophora* (13), being only slightly higher than the levels given in (12) seem more useful when estimating deficiency here, but their P level in the water, necessary to reach above the critical level, was higher than ever occurred in my outer area. The still lower critical P level in *Cladophora* reviewed in (9) is lower than I ever found.

C:N ratios seem often to be coupled to the degree of nutrient availability (references in 3, 4), but the varying ratios for different algae given above seem more to be species-specific than to describe general nutrient deficiency. Optimal N:P ratios are often considered to be around 15 (by atoms), but although species-specific higher optimal ratios occur in both green (14) and red algae (6), more extreme ratios seem to be indicative of less favourable conditions. For *Cladophora* growth has been found otimal at N:P ratios of 13-29 (7) and 22-24 (8), and photosynthetic efficiency at 14-20 (3), while another green alga, *Enteromorpha compressa*, had optimum for production as low as at uptake ratios of 3-12 (15). In this study the similarity in N:P ratios between annual algae, and perennial *Fucus* during its more intense growth period, implies that here the N:P ratios are a better estimate of ambient conditions than are the mere levels of the seperate elements.

In conclusion, the nutrient conditions in this archipelago, as judged from the results of the algal analyses, show an excess of nitrogen but phosphorus limitation in the inner area in spring and early summer, but rather excess of phosphorus in late summer; the conditions being almost reversed in the outer area. Similar trends have also been found for phytoplankton in other inner archipelago areas of the Baltic (16, 17).

References

1. Wallentinus, I.: Contrib. Askö Lab. Univ. Stockholm 15, 1-138 (1976).
2. Wallentinus, I.: Contrib. Askö Lab. Univ. Stockholm 25, 1-210 (1979).
3. Wallentinus, I.: Proc. 10th Europ. Symp. mar. Biol., Ostend Belgium, Sept. 17-23, 1975. 2, 631-651 (1976).
4. Wallentinus, I.: Part VI of thesis, Univ. Stockholm (1979).
5. Fitzgerald, G.P., Nelson, T.C.: J. Phycol. 2, 32-37 (1966).
6. Kornfeldt, R.A.: Manuscr., Dep. Mar. Bot., Univ. Lund (1980).
7. Gordon, D.M., Birch, P.B., McComb, A.J.: Manuscr., Bot. Dep., Univ. Western Australia, Nedlands (1980).
8. von Wachenfeldt, T.: Rapport till Länsstyrelsen i Hallands län; Dep. Mar. Bot., Univ. Lund (1978).
9. Lin, C.K.: J. Phycol. 13, 46-51 (1977).
10. Munda, I.M., Kremer, B.P.: Mar. Biol. 42, 9-15 (1977).
11. Healey, F.P.: Mitt. Internat. Verein. Limnol. 21, 34-41 (1978).
12. Gerloff, G.C., Krombholz, P.H.: Limnol. Oceanogr. 11, 529-537 (1966).
13. Wong, S.L., Clark, B.: J. Fish. Res. Bd Can. 33, 85-92 (1976).
14. Rhee, G-Y.: Limnol. Oceanogr. 23, 10-25 (1978).
15. Kautsky, L.: Aquat. Bot. (in press).
16. Norin, L-L.: Nat. Swed. Environm. Protect. Bd PM 633, 1-36 (1975).
17. Larsson, U., Hagström, Å.: Manuscr., Askö Lab., Univ. Stockholm (1980).

3. Biochemistry, Physiology

THE DISTRIBUTION AND COMPOSITION OF POLYMERS IN THE MEDULLA OF ASCOPHYLLUM RECEPTACLES

E. Baardseth
Institute of Marine Biology, N-5065 Blomsterdalen, Norway

H. Grasdalen and B. Larsen
Institute of Marine Biochemistry, N-7034 Trondheim-NTH, Norway

1. Introduction

The medulla of Ascophyllum receptacles (spring sample, Leangen, Trondheimsfjord) probably contains two different types of alginate (4). The intercellular liquid, which can be squeezed out directly when the receptacles are sufficiently ripe, consists of inorganic salts (ca. 70 %) and alcohol-precipitable carbohydrates (ca. 30 %) (2). About 90% of the carbohydrates was alginate containing 91% mannuronic acid and 9% guluronic acid. The extremely high mannuronic acid content of the alginate in this liquid has subsequently been confirmed by numerous experiments. After a more or less complete removal of this liquid by squeezing, the remainder of the medulla contained an alginate with a significantly different proportion (64.5%) of mannuronic acid.

The purpose of the present investigation was to follow the extraction process quantitatively by stepwise differential extractions of the medullary pieces, with analysis of the various extracts for mannuronic acid and description of the anatomical changes associated with each extraction. Possible relationships between the numbers of diatoms suspended in the succesive extracts and the dissolved organic matter in the extracts were also investigated.

Xth International Seaweed Symposium
Printed in Germany

2. Material and methods

The choice of *Ascophyllum* receptacle medulla as experimental material was determined by its extreme homogeneity and the ease with which it may be separated from the rest of the receptacles. It consists of a 3-dimensional network of cells with large intercellular spaces. An extraction of the intercellular substance could then be made without further comminution of the medullary pieces. The medulla could be squeezed out directly from the fresh material or after soaking the receptacles in distilled water for a day or two. It was also possible to squeeze out the intercellular substance first and the rest of the medulla afterwards. The receptacles were always fixed in 5% formalin in sea water for an hour before any further treatment.

The fluids used for extraction were determined by the following considerations: The intercellular carbohydrates seem, in the native state, to be closely associated with a fluid resembling sea water (2). Sea water was therefore used as a first extraction medium, as it had a fair chance of extracting the intercellular substance without changing its composition. However, these extractions were rather slow. Even after shaking with sea water four times for 24 hours, some water-soluble material was still left in the medullary pieces. This was easily removed by distilled water and hence the latter was chosen as the next extraction medium. After removal of all water-soluble material, the residue still contained some alginate. This was brought into solution by a treatment with0.2 N HCl for 3 hours, followed by washing in distilled water and, finally, neutralization with NaOH.

The uronic acid composition of the alginates was determined by nuclear magnetic resonance (3).

The intercellular substance of the medulla harbours a dense population of diatoms. The density of these was determined by counting the numbers in samples (3.2 mm^3) of the extracts

Table 1. Stepwise extractions of medullary pieces of Ascophyllum receptacles, giving the dry weight, percentage dry weight in the fresh weight and percentage of ash in the dry matter of the residue after each extraction.

Extraction step nr. and programmes	Dry weight (g)	Percentage dry matter	Percentage ash
1. Starting material	1.0	3.66	23.97
2. Sea water, 1st time	1.4437	5.30	66.85
3. Sea water, 2nd time	1.3230	5.48	67.72
4. Dist.water, 1st time	0.4221	1.61	31.10
5. Dist.water, 2nd time	0.2912	0.83	20.61
6. 0.2 N HCl, dist.water	0.1983	2.65	0.0
7. NaOH neutralization	0.0466	1.56	6.41

(Fuchs/Rosenthal). About 10 samples of this size were examined for each extract.

3. Changes in dry weight, percentage of dry matter and ash with each step of extraction

In Table 1 the results of 7 extraction experiments are given. They are numbered 1-7 and each takes the extraction one step further than the preceding one. They all start with 1 g dry weight of receptacle medulla and the quantitative effect of each extraction step is easily followed. It will be seen that the dry weight of the medulla increases with sea water extractions (step 2-3) in spite of the fact that some organic matter is removed. This is obviously due to an uptake of inorganic salts, resulting in a considerable increase in the percentage of ash. The subsequent extractions with distilled water (step 4-5) bring the percentage of ash down to starting level again, but at the same time a considerable reduction in the percentage of dry matter is noticed. This means that there was an uptake of water which could be observed directly by the swelling of the medullary pieces. The dry weight of the residue was thereby brought down to about 29% of the starting

Table 2. Material and extraction steps as in Table 1. Dry weight and percentage of organic matter after each extraction step. Also giving the percentage of carbohydrate matter determined by the phenol-sulphuric method and the percentage of diatoms brought into suspension after each extraction step.

Extraction step nr.	Organic matter in the residue, g	% organic matter extracted	% carbo-hydrates extracted	% diatoms brought into suspension
1	0.7603	0	0	0
2	0.4786	39.30	40.46	57.59
3	0.4271	46.48	51.83	69.31
4	0.2908	65.50	58.20	79.80
5	0.2312	73.82	65.22	87.26
6	0.1983	78.41	Disregarded	91.65
7	0.0436	100	100	100

weight. With acid and a subsequent washing in distilled water (step 6), the pieces contract again and the ash seems to be completely removed. Very little organic matter is removed at this stage. The final neutralization (step 7) breaks up the medullary pieces, brings what is left of the alginates into solution, and leaves only about 5% dry matter in the residue.

4. Extraction of organic matter

The dry weight of the residue after an extraction minus the weight of its ash will be considered as a measure of its content of organic matter. The values of organic matter defined in this way are given in Table 2 and the decrease from step to step is easily followed, starting initially with 0.7603 g at step 1 and ending up at step 7 with 0.0436 g which is the final, insoluble residue. The organic matter removed from the residue by each extraction step (i.e. the amount brought into solution), calculated as a percentage of the initial content of soluble organic matter (i.e. of 0.7603-0.0436) is also given.

By determining the extracted carbohydrate directly in the solutions with the phenol-sulphuric acid method a similar series of percentages was obtained (Table 2). The two series are not exactly identical, but show parallel changes. Generally speaking, it seems that an extraction with 2 changes of sea water and 2 changes of distilled water brings about 70% of the original organic matter into solution, that 5% is dissolved by the acid treatment, and that the remainder, 25%, could be extracted by neutralization with NaOH. By using distilled water, without a previous sea water extraction, about 80% of the organic matter could be dissolved.

Three analyses of the alginate in the water-soluble fraction gave 95% (march 21), 93% (May 5) and 91% (Aug.26) mannuronic acid whereas two analyses of the alginate in the water-insoluble fraction gave 65% (March 21) and 56% (May 5) mannuronic acid.

5. Diatoms and intercellular substance

The diatom Navicula endophytica Hasle occurs in the intercellular substance of the receptacle medulla of Ascophyllum, but it is not known whether it is associated with the organic fraction, with the inorganic salt solution or with both. Table 2 shows the percentage of diatoms brought into suspension at various stages of extraction. It can be seen that the diatoms are extracted much more easily than the organic matter. This suggests that the diatoms are, at least to some extent, independent of the organic matter.

6. Anatomical changes with steps of extraction

Fig. 1,A shows a section of a part of the medulla of Asco-

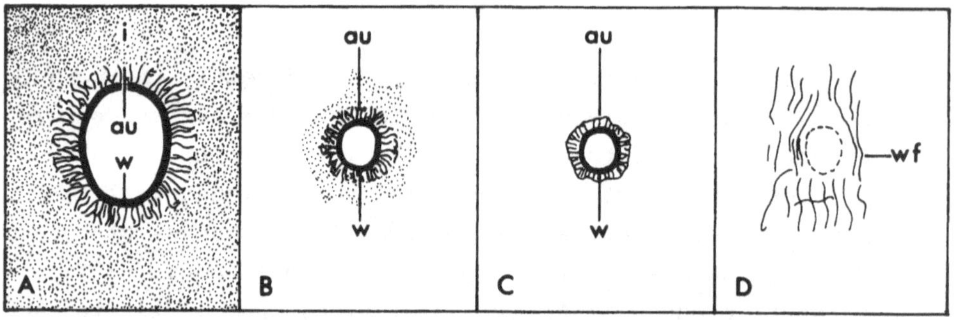

Fig. 1. Schematic drawings of a medullary cell with its surroundings. A native state, before any extractions, B after a water extraction, C after a water and an acid extraction and D after a water and an acid extraction followed by neutralization with NaOH. i intercellular substance, w cell wall, au aureole and wf wall fibres.

phyllum before the extractions, showing the cell wall (w), the fibrillar area immediately surrounding the cell wall (i.e. the aureoles, au) and the intercellular substance (i). All these structures are stainable with toluidinblue and also with hematoxylin after mordanting with e.g. Fe- or Al-salts (1). The presence of alginate in all 3 structures is thus indicated. After a thorough washing with sea and distilled water, the intercellular substance is almost completely removed (Fig.1, B) whereas the walls and the aureoles are left behind, seemingly untouched. With a subsequent acid treatment, a shrinkage of the aureolar area was noticed (Fig.1, C), but without loss of its staining and ion-exchange properties. The final NaOH-neutralization process, however, removed the stainable aureolar and wall material completely at the same time as the wall itself disintegrated into small, non-stainable fibers (Fig.1,D). The intercellular spaces are thus filled with a water-soluble, mannuronic-acid-rich alginate and the walls/aureoles with a water-insoluble, relatively guluronic-acid-rich alginate.

References

1.Baardseth,E.: Proc.Fifth Int.Seaweed Symp., 19-28 (1966)

2.Baardseth,E.: Proc.Sixth Int.Seaweed Symp., 53-60 (1969)

3.Grasdalen,H.,Larsen,B.,Smidsrød,O.:Carboh.Res.68,23-61,(1979)

4.Haug,A.,Larsen,B.,Baardseth,E.: Proc.Sixth Int.Seaweed Symp., 443-451 (1969)

ANALYSIS OF KAPPA-IOTA HYBRID CARRAGEENANS WITH KAPPA-CARRAGEENASE, IOTA-CARRAGEENASE AND ^{13}C N.M.R.

C. Bellion
CECA S.A. France

G.K. Hamer
Xerox Corp. Canada

W. Yaphe
Department of Microbiology and Immunology
McGill University, Montreal, Québec, Canada

Introduction

Kappa (κ) and iota (ι) carrageenans are sulfated polysaccharides extracted from red algae. Both have the same molecular backbone composed of 1-4 linked neocarrabiose units (A-G), 4-0-3,6 anhydro-α-D-galactopyranosyl 1-3-β-D-galactopyranosyl, sulfated in a characteristic manner. The repeat unit in κ-carrageenan, κ-neocarrabiose 4-0-sulfate ($A-G_4$) has a sulfate on galactose C4 while the repeating disaccharide in iota carrageenan, ι-neocarrabiose 2,4-0-sulfate (A_2-G_4), has an extra sulfate on C2 of 3,6 anhydrogalactose. *E.spinosum* and *E.cottonii* have been reported to synthesize almost pure ι- and κ-carrageenans (1). In contrast carrageenans derived from *A.concinna* (2), *Irideae spp* (3,4,5) and other algae in the Rhodophyceae (6) have features of both κ- and ι-carrageenans. The κ/ι hybrid structure of a carrageenan is shown by a peak at 805 cm^{-1} in an infrared spectrum and resonance peaks in a ^{13}C-NMR spectrum for the anomeric carbon atoms (C1) of the 3,6 anhydrogalactosyl residues in κ-neocarrabiose 4-0-sulfate ($A-G_4$, 96.2 ppm) and ι-neocarrabiose 2,4-0-sulfate (A_2-G_4, 93.1 ppm) (2). Another variation in the basic structure of carrageenan is found in *Grateloupia spp* that synthesize methylated, sulfated galactans containing anhydrogalactose (7).

This paper describes at study of the κ/ι hybrid structure of carrageenans derived from species of *Eucheuma*, *Ahnfeltia* and *Irideae*.

Xth International Seaweed Symposium
Printed in Germany

Structural features of the carrageenans were determined by ^{13}C-NMR spectroscopic procedures and hydrolysis with κ- and ι-carrageenases that specifically cleave β1-4 bonds linking disaccharide repeat units in κ- and ι-carrageenans. The cryptic κ/ι hybrid structure of κ- and ι-carrageenans was shown by analysis of enzyme resistant fractions.

Materials and methods

Enzymic hydrolysis. The ι-ase and κ-ase were obtained from marine bacteria and were used to prepare polysaccharide hydrolysates as described elsewhere (8). The polysaccharides were treated twice with the enzymes to ensure complete hydrolysis.
^{13}C-NMR spectroscopy. Proton decoupled ^{13}C-NMR spectra were recorded with a Bruker WP-80/DS FT spectrometer (20.1 MHz) in D_2O solution (30-80 mg/l) at 90°C (nondegraded polymers), 50°C (resistant fractions from enzymatic hydrolysates). A spectral width of 1.2 to 1.5 KHz was used with a pulse angle of 60° to 80° (4 to 5.5 sec) and a repetition time of 0.682 to 0.819 sec. Chemical shifts (ppm) were measured relative to internal methanol and converted to values relative to external trimethylsilane (TMS).
Molecular weight distribution. The enzymic hydrolysates were passed through a Sepharose CL-4B column (Pharmacia K16-40) and 1.15ml fractions were collected. All samples were eluted with 0.5M NaCl (pH 7.0) excepting the non degraded ι-carrageenan from *E. spinosum* which was eluted with 0.1M NaCl. Carbohydrates were determined by the phenol sulfuric method using galactose as a standard (9). Molecular weight values are equivalent to dextran (Pharmacia).
Polysaccharides. The samples analysed were derived from *Eucheuma spinosum* (sample # 31210) and *Ahnfeltia concinna* (from M. Doty and G. Santos), *Ahnfeltia gigartinoides* (from E.L. McCandless), *Irideae ciliata, Irideae laminaroides, Grateloupia lanceola* (from A. Zanlungo), *Eucheuma cottonii* and *E. spinosum* (sample # 712503) from Marine Colloids Inc.

Results and Discussion

1) Enzymatic hydrolysis of carrageenans. The carrageenan samples were treated separately with κ-ase and ι-ase. All of them but *G. lanceola* were hydrolysed by both κ-ase and ι- ase suggesting that the polysaccharides had a hybrid structure. The amount of κ- and ι- type repeat units being related to the degree of hydrolysis with each enzyme. The carrageenan from *G. lanceola* was not degraded by κ-ase or ι-ase suggesting a galactan structure different from that of κ- and ι-carrageenans (7). The κ/ι hybrid structure of carrageenan was analysed more extensively by ^{13}C-NMR spectroscopy and gel permeation chromatography of enzymatic digests.
2) ^{13}C-NMR spectroscopy. The ^{13}C-NMR spectra for ι-carrageenans derived from *E. spinosum* (31210, 712503), fig. 1a, 1b show characteristic signals in the anomeric region for the repeating disaccharide unit (A_2-G_4).

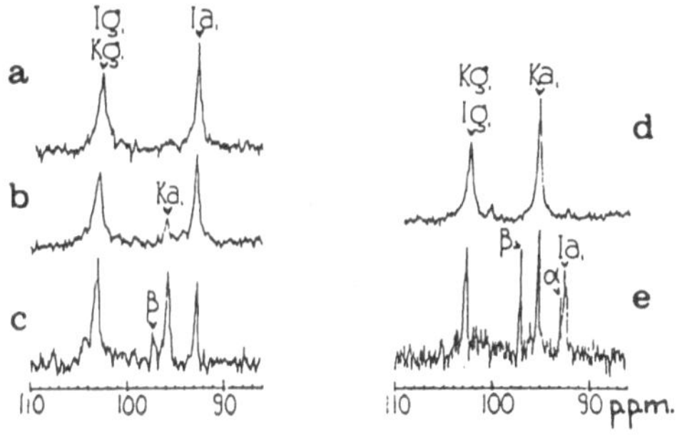

Fig. 1 ^{13}C-NMR spectra in the anomeric region. ι-carrageenan, a. *E. spinosum* (31210), b. *E. spinosum* (712503). κ-carrageenan, d. *E. cottonii*; enzyme resistant fractions: c. *E. spinosum* (712503) after ι-ase treatment, e. *E. cottonii* after κ-ase treatment.

The peaks at 103.0 ppm (Ig_1) have been attributed to D-galactopyranosyl-4-O-sulfate (G_4) and at 92.7 ppm (Ia_1) to 3,6 anhydrogalactopyranosyl-2-O-sulfate (A_2) (2). A minor peak at 95.9 ppm (Ka_1 fig. 1b) is attributed to the 3,6-anhydrogalactopyranosyl (A) residue in the repeating disaccharide (A-G_4) of κ-carrageenan and indicates that ι-carrageenan 712503 has a hybrid κ/ι structure. The intensity of peak Ka_1 increases in the enzyme resistant fraction (RF) of sample 712503 (RF 25-30 %) following extensive treatment with ι-ase (fig. 1c) indicating enrichment of κ-features. The hybrid κ/ι structure of ι-carrageenan 31210 is not seen in the ^{13}C-NMR spectrum (fig. 1a) but after treatment with ι-ase, the RF was enriched with κ-features and had a spectrum similar to sample 712503 (fig. 1c) with a peak of lower intensity for Ka_1 (95.9 ppm). The hybrid κ/ι structure of the enzyme RF (fig. 1c) suggests that the κ- and ι-features are probably linked together in the same molecule. The ^{13}C-NMR spectrum for κ-carrageenan derived from *E.cottonii* shows the characteristic signals for ($A-G_4$) with peaks at 103.0 ppm for G_4 and at 95.9 ppm for A (fig. 1d). Again the hybrid κ/ι structure of κ-carrageenan was revealed only after degradation with κ-ase. The κ-ase RF (RF 20 %, fig. 1e) shows a peak at 92.7 ppm (Ia_1), attributed to the A_2 residue in the repeating unit (A_2-G_4) of ι-carrageenan. The peaks at 93.0 ppm (α) and 97.3 ppm (β) in figs. 1c and 1e, are attributed to C-1 of α- and β-galactose residues at the reducing end of oligosaccharides. The κ-neocarrabiose-4-0-sulfate ($A-G_4$) residues in the RF of κ-carrageenan may be linked with ι-neocarrabiose 2,4 disulfate units ($A_2-G_4-A-G_4$)n being resistant to degradation by κ-ase. The carrageenans derived from *A.concinna* and *A.gigartinoides* were analysed by ^{13}C-NMR (fig. 2).

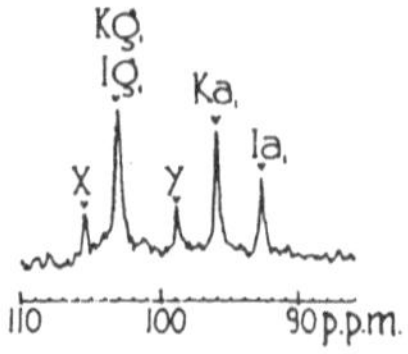

Fig. 2 ^{13}C-NMR spectrum (anomeric region) of non alkali treatment *A.concinna* carrageenan.

Peaks associated with a κ/ι hybrid carrageenan are observed at 103.1 ppm (Kg_1, Ig_1), 95.9 ppm (Ka_1) and 92.7 ppm (Ia_1) (10). Two new peaks at 105.4 ppm (X) and 98.9 ppm (Y) with an integral ratio of peak areas equal to 1 were seen in both samples. The peak intensity ratio for κ:ι:X+Y was 50 %:32 %:18 % for *A.concinna* carrageenan and 51 %:40 %:9 % for *A.gigartinoides*. Following alkali treatment the X and Y peaks were absent and the κ/ι ratio was 50 %: 50 %. The data suggests that X and Y are the characteristic signals for nu-carrageenan $(G_{2,6}-G_4)_n$which is the precursor of ι-carrageenan $(A_2-G_4)n$. Other observations on the KCl solution carrageenan fractions from *Chondrus crispus* and *Irideae* gametophytic plants demonstrated that alkali treatment gave rise to an ι-carrageeenan IR spectrum (peak at 805cm^{-1}). (Personal communications E. McCandless, McMaster Univ. Canada and G. Brigand, CECA, France). Nu-carrageenan may therefore be the common precursor for both κ-and ι-carrageenans since the predicted precursor of κ-carrageenan (mu-carrageenan $(G_6-G_4)n$) was not detected (7).

3) Gel permeation chromatography. Carrageenans derived from *E. cottonii* (κ) and *E.spinosum* (ι 31210) were compared before and after enzymatic hydrolysis by gel permeation chromatography on Sepharose CL-4B (fig. 3).

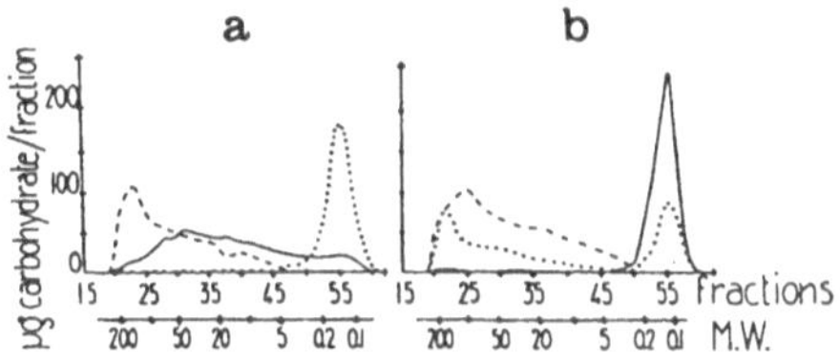

Fig. 3 Molecular weight distribution pattern of enzymic hydrolysates: a.*E.cottonii*,b.*E.spinosum* ; (--) non degraded polymers ; (..) hydrolysates after κ-ase treatment ; (—) hydrolysate after ι-ase treatment: M.W., molecular weight of dextrans (Pharmacia).

The kappa and iota carrageenans showed the same distribution pattern with maximum peaks corresponding to the void volume of the column. The κ-carrageenan was extensively degraded by κ-ase to low molecular weight oligosaccharides and treatment with ι-ase produced oligomers of varying molecular weights (fig. 3a). Following hydrolysis with either κ-ase or ι-ase no high molecular weight fraction was detected, indicating that both κ and ι-features are present in the κ-carrageenan molecules. When ι-carrageenan (31210) is treated with ι-ase, the polymer is completely degraded to low molecular weight material (fig. 3b). In contrast degradation of ι-carrageenan with κ-ase gives rise to a high molecular weight peak corresponding to the void volume and a peak corresponding to low molecular weight oligosaccharides. This suggests that the κ features occur as blocks in the ι-carrageenan molecule. Degradation of these blocks with κ-ase yields high molecular weight fragments of ι-carrageenan. The molecular weight distribution patterns for carrageenans derived from *A.concinna* and *A.gigartinoides* also show that the κ and ι features are located on the same molecule.

In conclusion, the family of carrageenans (κ, ι, ν) are hybrid polymers. The ratio of κ/ι varies in carrageenans derived from different algae. Nu-carrageenan appears to be the common precursor for κ- and ι-carrageenans.

Support by the National Sciences and Engineering Research Council Canada, is acknowledged. We thank M. Doty, G. Santos, U.S.A. ; A. Zanlungo, Chile ; Marine Colloids, Inc. U.S.A. ; and E. McCandless, Canada for samples of carrageenans.

References

1. Anderson, N.S., Dolan, T.C.S. and Rees, D.A.: J. Chem. Soc. Perkin 1, 2173-2176 (1973).
2. Bhattacharjee, S.S. and Yaphe, W.: Mar. Algae Pharm. Sci., ed. by Hoppe, H.A., Levring, T. and Tanaka, Y., Walter de Gruyter Berlin. New York, 645-655 (1979).
3. Yaphe, W.: Can. J. Bot. 37, 751-757 (1959).
4. McCandless, E.L., Craigie, J.S. and Hansen, J.E.: Can. J. Bot. 53, 2315-2318 (1975).
5. Matsuhiro, B. and Zanlungo, A.B.: Bot.Mar. 19, 317-321 (1977).
6. Lawson, C.J. and Rees, D.A.: J. Chem. Soc. Perkin 1, 2177-2182 (1973).
7. McCandless, E.L. and Craigie, J.S.: Ann. Rev. Plant Physiol. 30, 41-53 (1979).
8. Bellion C.: Ph.D. Thesis, McGill Univ. Montreal, Canada.
9. Dubois, M., Gilles, K.A., Hamilton, J.K., Rebers, P.A. and Smith, F.: Anal. Chem. 28, 350-356 (1956).
10. Bhattacharjee, S.S., Yaphe, W. and Hamer, G.K.: Proc. 9th Int. Seaweed Symp. ed. by Jensen, A. and Stein, J.R., Science Press, Princeton, 379-385 (1979).

ISOLATION & PRELIMINARY CHARACTERIZATION OF RECEPTORS INVOLVED IN EGG-SPERM RECOGNITION IN *FUCUS SERRATUS*

J.A. Callow, G.P. Bolwell, L.V. Evans & Maureen E. Callow
Dept. of Plant Sciences, The University, Leeds LS2 9JT, U.K.

Introduction

The initial specific recognition between gametes of a number of organisms is mediated by surface-localised complementary macromolecules. By means of indirect procedures (lectins, polysaccharides, enzymes) we have shown that gamete recognition in *Fucus serratus* is based on an association between fucosyl- and mannosyl-containing ligands on the egg surface and complementary carbohydrate-binding proteins on the sperm surface (Bolwell *et al.*, 1979). In this paper we now report the isolation and partial characterization of factors from egg and sperm with some of the properties expected of fertilization receptors.

Methods and Materials

A. Preparation of the egg receptor for sperm (Bolwell *et al.*, 1980)

The procedure initially requires the preparation of a membrane fraction from eggs enriched in plasma membranes. The membranes were dispersed in 1% SDS in 0.1 M Tris-HCl, pH 7.2 containing 1 mM $CaCl_2$ and 1 mM $MnCl_2$. The SDS was removed from the preparation by salt precipitation and Amberlite XAD 2 resin before applying to a 2 cm^3 affinity column of con A-agarose. The unbound fraction (85% of applied protein) was eluted in tris/Mn^{2+}/Ca^{2+} buffer, the remaining bound fraction, containing all fertilization-inhibiting activity, being specifically eluted by washing with 20 mM α-methyl mannoside.

Xth International Seaweed Symposium
Printed in Germany

B. Preparation of sperm receptor for eggs.

Pelleted sperm were extracted in 2.5 M KCl for 1h. Centrifugation at 100 000 x g for 1h separated a soluble inhibitor of fertilization from a pellet containing protein, carbohydrate and nucleic acid. On dialysis, the soluble inhibitor was precipitated but could be re-dissolved in 1M KCl. On applying to a 3 cm^3 affinity column consisting of desulphated fucoidan covalently linked to Sepharose 6B, all inhibitory activity was bound but was specifically eluted with 0.1M tris-HCl, pH 8.8 containing 1M KCl.

C. Fertilization assays

The presence of putative receptor molecules was assayed by their ability to inhibit fertilization, as described elsewhere (Callow et al., 1978; Bolwell et al., 1980).

Results

Plasma membrane-enriched fractions from Fucus serratus eggs inhibit fertilization in a species-specific manner, by binding sperm (Bolwell et al., 1980). This activity was sensitive to α-fucosidase and α-mannosidase but not to trypsin.

Since fertilization can be inhibited by pre-treating eggs with the mannose and glucose-binding lectin concanavalin A (con A) (Bolwell et al., 1979) attempts were made to isolate the sperm receptor from egg membranes by affinity chromatography on con A-agarose. The majority (85%) of the protein in egg membranes solubilized in SDS does not bind to con A-agarose (Fig. 1). This fraction has no ability to inhibit fertilization. However, the bound fraction, eluted with α-methyl mannoside, exhibited species-specific inhibition of fertilization at low concentration (2 μg protein cm^{-3}). Inhibitory activity was destroyed by α-L-fucosidase (0.02 I.U. cm^{-3}) but was insensitive to trypsin. On SDS-PAGE, the bound, inhibitory fraction gave a number of minor bands and one major band which

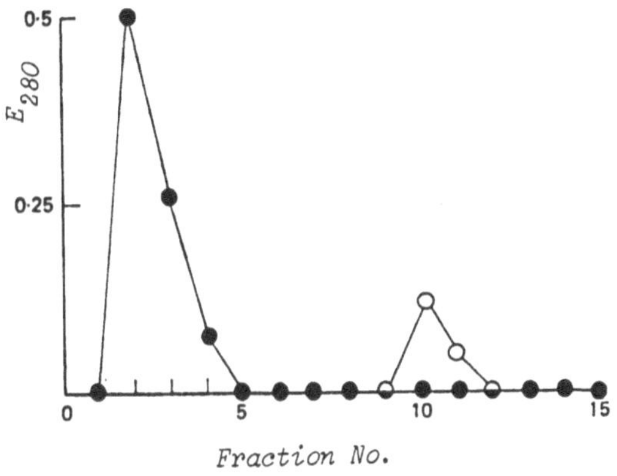

Fig. 1. Fractionation of SDS-treated egg membranes on con A-agarose. • elution profile at 280 nm in tris buffer; o elution profile when 20 mM α-methyl mannoside was added at fraction 8.

was both Coomassie Blue and PAS-positive, with an approximate molecular weight of 30 Kd (Fig. 2). This component was distinct from any of the bands of con A itself (an important distinction since con A can 'bleed' from an affinity matrix, and fertilization is con A-sensitive). Furthermore, the putative receptor inhibits fertilization even in the presence of methylmannoside, which eliminates spurious effects due to any contaminating con A leached from the affinity column.

Sperm treated with 2.5M KCl liberate fertilization-inhibiting activity into a 100 000 x g supernatant, leaving most of the sperm protein, carbohydrate and nucleic acid in a pellet. On dialysis of the supernatant, a white precipitate containing all the inhibitory activity formed. The precipitate was soluble and stable in 1M KCl, inhibitory activity was heat-labile, trypsin-sensitive and species-specific. The activity was removed by affinity chromatography on a fucan-Sepharose matrix or a yeast mannan-Sepharose matrix (Fig. 3). The bound fraction, eluted with 0.1M tris buffer in 1 M KCl at pH 8.8, contained all inhibitory activity in a heat-labile, trypsin-sensitive and species-specific form.

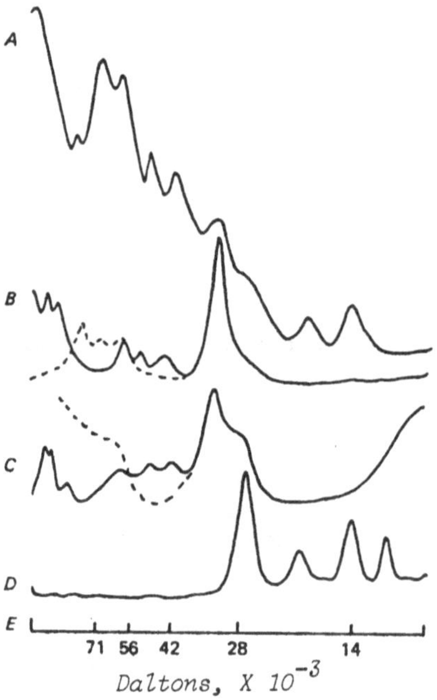

Fig. 2. Densitometer tracings of SDS-polyacrylamide gels, of fractions derived from eggs and con A. A, solubilized membranes stained with Coomassie Blue; B, bound fraction from the con A column, stained Coomassie Blue; C, same as B, PAS-stained; D, 50 μg con A (Sigma) stained Coomassie Blue; E, molecular weight calibration from standard markers.

The inhibitor bound to eggs and its effect could be reversed by washing inhibitor-treated eggs with 100 μg cm^{-3} desulphated fucoidan, before adding sperm. The inhibitor was effective at low concentration, as little as 0.3 μg protein per 2 cm^3 assay causing 50% inhibition of fertilization.

On SDS-PAGE, the inhibitor eluted from the affinity matrix gave one disperse band of mean molecular weight 60 Kd (Fig. 4). The tendency of this fraction to split into two components may reflect proteolysis during purification since only one component was detected in this position before affinity chromatography. The disperse band on the gels did not stain with PAS indicating the probable absence of carbohydrate.

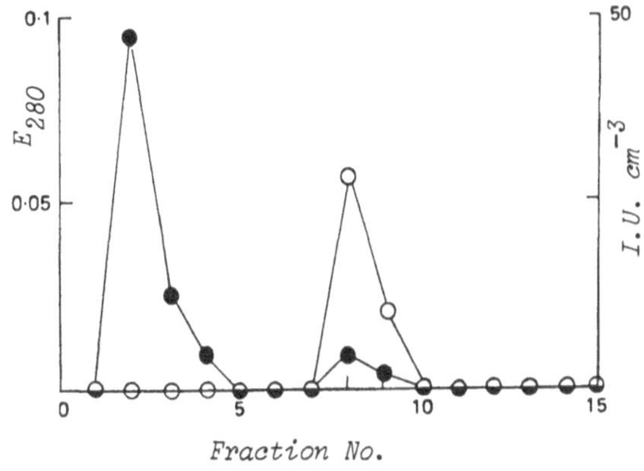

Fig. 3. Elution profile of the affinity chromatography of the 1M KCl-soluble fraction from sperm on desulphated fucoidan coupled to Sepharose 6B. The unbound fraction was eluted with 1M KCl. The bound fraction was eluted with 0.1 M tris-HCl pH 8.8 containing 1M KCl. ●, absorbance at 280 nm; ○, inhibitory activity in fertilization (1 I.U. is the amount of fraction required to give 50% inhibition of fertilization).

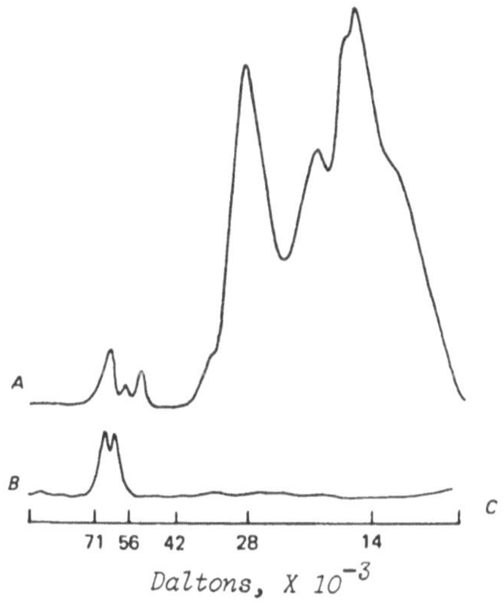

Fig. 4. Densitometric scans of SDS gels stained with Coomassie Blue. A, 1M KCl-soluble fraction of sperm; B, the bound, inhibitory fraction eluted from the fucan-Sepharose affinity matrix; C, molecular weight standards.

Discussion

This paper provides evidence for the isolation of a glycoprotein from egg membranes which exhibits the properties of an egg receptor for sperm, and a lectin-like protein from sperm which exhibits the properties of a sperm receptor for eggs. In both cases, the primary criterion adopted for the presence of receptors has been ability to cause a species-specific inhibition of fertilization at low concentration. In addition, the methods of isolation and purification and their sensitivity to various treatments such as enzymes are in accordance with the expected characteristics of receptors, as determined by previous indirect means (Bolwell *et al.*, 1979). However, before these putative receptors can be convincingly ascribed a role in egg-sperm recognition a number of further criteria must be satisfied. These include:

a) demonstration of their presence on the cell surface,

b) ability to associate *in vitro* in a complementary manner,

c) ability to bind to the appropriate gametic surface.

Acknowledgements

The authors wish to thank the SRC and International Paint Marine Coatings for financial support.

References

Callow, M.E., Evans, L.V., Bolwell, G.P. & Callow, J.A: J. Cell Sci. 32, 45-54 (1978).

Bolwell, G.P., Callow, J.A., Callow, M.E. & Evans, L.V.: J. Cell Sci. 36, 19-30 (1979).

Bolwell, G.P., Callow, J.A. & Evans, L.V.: J. Cell Sci. 43, 209-224 (1980)

CARRAGEENAN ANALYSIS IN APOMICTIC GIGARTINA STELLATA - NEW PUZZLES*

L. C-M. Chen and J. S. Craigie
National Research Council of Canada, Atlantic Research Laboratory, 1411 Oxford St., Halifax, N. S. Canada B3H 3Z1

Introduction

The life history of *Gigartina stellata* (Stackh.) Batt. has been examined by culture studies in vitro and in situ. Germination of carpospores from various populations of *G. stellata* have established three different sequences of morphological development. First is the formation of discoid basal holdfasts from which erect fronds develop directly. The production of carposporophytes in the absence of tetraspores has been reported in populations from the coasts of Nova Scotia, Canada (Chen et al. 1974), Scotland, and Wales, U.K., and Maine, U.S.A. (West & Polanshek 1975), Western Norway (Rueness 1978) and Northern France (Dion & Delepine 1979); Second is sexual reproduction through the formation of discoid *Petrocelis*-like crusts. These produced tetraspores which on germination developed gametophytes, as reported for populations from Maine, U.S.A. (West & Polanshek 1975), and Roscoff, in Northern France (Dion & Delepine 1979); Third involves partial sexual and partial apomictic reproduction, as for populations from Wales (West & Polanshek 1975) and Northern France (Dion & Delepine 1978).

An interesting and useful marker is associated with specific nuclear phases of members of the Gigartinaceae. It was shown

*NRCC 18253

Xth International Seaweed Symposium
Printed in Germany

that erect gametophyte fronds of *Chondrus crispus* and *G. stellata* contained much greater levels of 3,6-anhydrogalactose than did tetrasporophytes (Chen et al. 1973). It is now recognized that gametophytes produce kappa type carrageenan whereas tetrasporophytes form lambda type carrageenan (McCandless et al. 1973). Our previous reports (Chen et al. 1974, Edelstein et al. 1974) on carrageenan analysis in the erect fronds, and the observation of carpogonia in developmental studies confirmed the culture studies which demonstrated that *G. stellata* repeatedly reproduces apomictically to form female gametophytes. Although we have as yet not detected a tetrasporophytic phase in our Nova Scotia *G. stellata*, we cannot exclude this possibility. We have applied culture techniques and carrageenan analyses to further examine this point.

Materials and Methods

Chen et al. (1974) cultured *G. stellata* and reported the apomictic life history. After that, one culture dish with three plants was kept at 5°C with 8:16 h photoperiod and 10 $\mu Em^{-2}s^{-1}$ of cool-white irradiance. Erect fronds with papillae were removed occasionally to prevent new spores from germinating. The medium was changed monthly. In 1980 erect fronds with papillae were collected again from several locations in Nova Scotia including the original sites. The developmental sequence was as originally observed. Plants from other sites are identified in Table 1.

Erect fronds and basal holdfasts from culture were analyzed analyzed periodically for their carrageenan type by infrared spectroscopy (Craigie & Leigh 1978). The anatomy of the basal holdfast was checked frequently by vertical sectioning and comparison with *Petrocelis* from nature.

Table 1. Carrageenan type as determined by infrared analyses for various erect fronds and basal holdfasts.

Dates	Site	Species	Stages	Remarks	Type
17 Apr 74	a	G	gametophyte	fronds with papillae	κ
17 Apr 74	b	G	frond with papillae	viable carpospores	κ
07 Feb 75	b	G	carpospore germinated	holdfast without frond	λ
15 May 79	b	G	carpospore germinated	margin of holdfast	λ
15 May 79	b	G	frond with papillae	viable carpospores	κ
31 Jan 80	c	G	gametophyte	frond with papillae	κ
05 Mar 80	a	G	gametophyte	lower part of fronds near holdfast	κ
15 Mar 80	d	G	gametophyte	fronds with papillae	κ
10 Jul 75	e	P	crustose thallus	with tetraspore	λ
15 Apr 80	f	G	carpospore germinated	holdfasts initiating fronds	κ
15 Apr 80	b	G	carpospore germinated	holdfast near fronds	λ
15 Apr 80	b	G	carpospore germinated	margin of holdfast	λ
15 Apr 80	g	M	gametophyte	fronds	κ
15 Apr 80	g	M	gametophyte	holdfast	κ

a-Finck Cove, Halifax Co.
b-culture No. 701213
c-Peggy's Cove, Halifax Co.
d-Burn's Pt., Yarmouth Co.
e-Gulliver's Cove, Digby Co.
f-culture No. 800208
g-culture No. 701027
G-*Gigartina stellata*
P-*Petrocelis* sp.
M-*Chondrus crispus*

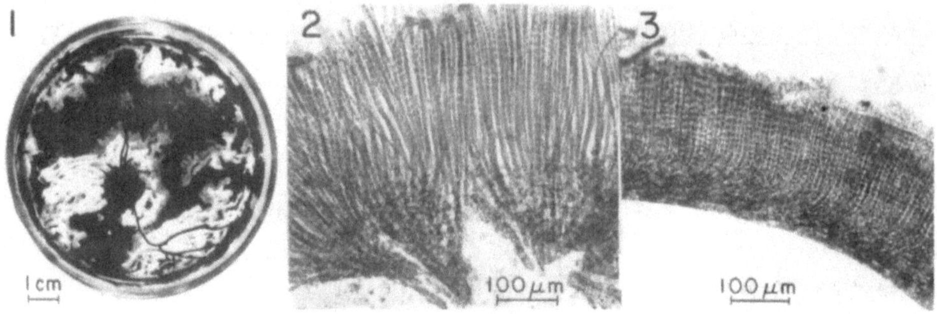

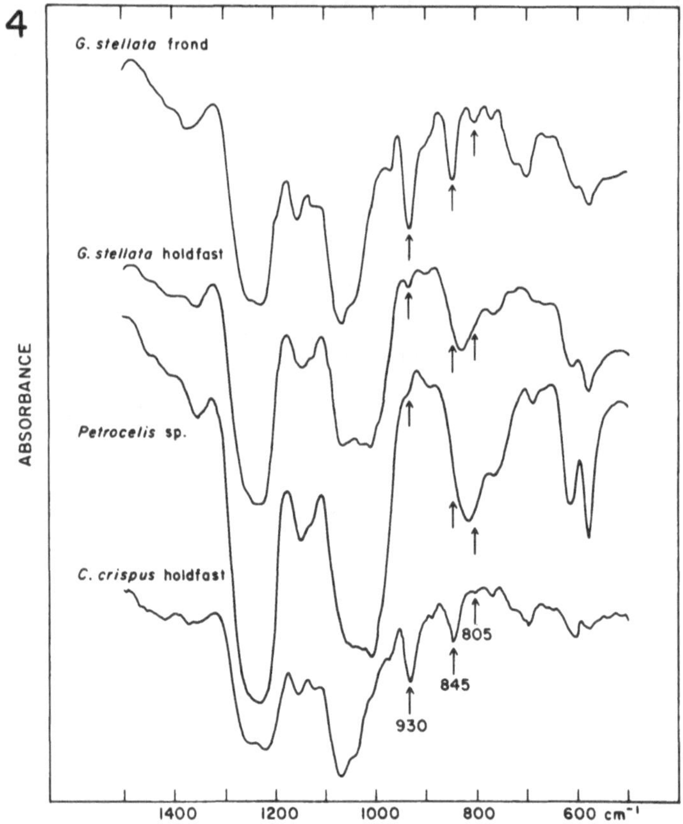

FIG. 1. G. stellata 10 year old culture with fronds and large holdfasts. FIG. 2. Petrocelis sp. squash preparation. FIG. 3. Holdfast from Fig. 1, vertical section. FIG. 4. Infrared spectra of isolated carrageenans.

Results and Discussion

Prolonged culture at 5°C with 10 $\mu Em^{-2}s^{-1}$ and 8:$\overline{16}$h photoperiod suppressed the growth of erect fronds (Fig. 1) while marginal growth of the basal holdfasts continued in characteristic concentric rings. Eventually, the holdfasts coalesced to form a disc approximately 10 cm in diameter. A few erect fronds also developed in a concentric pattern. Regeneration occurred when the old portion of the holdfasts was removed, with new filaments growing from the old crust. Sections of the basal holdfast showed closely compressed filaments of square shaped cells. The filaments adhered firmly to each other, unlike those of *Petrocelis*. The thickness of the holdfast was approximately 120 to 210 μm (Fig. 2), much less than the thickness (*ca.* 1000 μm) of *Petrocelis* (Fig. 3). No tetraspores were ever observed. The carrageenan analyses are shown in Table I. Periodic examination of these basal holdfasts showed that the lambda-type of carrageenan occurred there consistently, while the erect fronds contained the kappa-type of carrageenan. Parallel studies on similarly aged cultures of *C. crispus* (Table I) showed that both the basal holdfast and the erect fronds contain only kappa carrageenan as expected for the same gametophytic phase.

If the type of carrageenan can be interpreted as an expression of ploidy within the Gigartinaceae, then the presence of lambda-type of carrageenan in the basal holdfast of the apomictically cultured plants isolated from *G. stellata* suggests that the basal holdfast is composed of diploid tissue. The analysis of carageenan in *C. crispus* in both holdfasts and erect fronds of the gametophyte phase (grown under the same conditions as *G. stellata*) showed only a kappa-type carrageenan suggesting that in this plant the basal holdfast is composed of haploid tissue as would be expected. Tetraspores isolated from *Petrocelis cruenta* have developed into Gigartina plants (West et al. 1977, Dion & Delepine 1979). Petrocelis which was therefore proposed as the tetrasporophyte contains only a lambda-type carrageenan (Fig. 4). However, the anatomical and morphological features of our G. stellata holdfast are quite distinct from those of Petrocelis crusts (Chen unpublished) both from culture (West et al. 1977; Dion and Delepine 1979) and in nature. In addition we are unable to detect tetraspores in the former, which suggests that they do not comprise the tetrasporophytic phase. Although it would be contrary to the published findings for several other species in Gigartinaceae, it may be argued that the phenotypical expression of the form of the thallus and also the chemistry of the cell wall may be independent of the nuclear phase. No decision on this point can be taken until caryological studies including the determination of chromosome number are completed.

References

Chen, L. C-M., McLachlan, J., Neish, A. C., Shacklock, P. F., J. Mar. Biol. Assoc. U. K., 53: 11-16 (1973).

Chen, L. C-M., Edelstein, T., McLachlan, J., Phycologia 13, 4: 287-294 (1974).

Craigie, J. S., Leigh, C., Handbook Phycological Methods II pp 109 - 131. (Eds. J. Hellebust & J. S. Craigie) Cambridge, New York, N.Y. (1978).

Dion, P., Delepine, R., Rev. Algol., N.S. 14, 4: 327-341 (1979).

Edelstein, T., Chen, L. C-M., McLachlan, J., Phycologia 13, 2: 99-107 (1974).

McCandless, E. L., Craigie, J. S., Walter, J. A., Planta, 112: 201-212 (1973).

Rueness, J., Phycologia 13: 87-90 (1978).

West, J. A., Polanshek, A. R., Phycologia 17: (suppl.) (1975)

West, J. A., Polanshek, A. R., Guiry, M. D., Br. Phycol. J. 12: 45-53 (1977).

CHEMICAL CLEAVAGE OF TANNINS FROM BROWN ALGAE (PHLOROTANNINS)

K.-W. Glombitza, G. Gerstberger, G. Lentz
Institut für Pharmazeutische Biologie der Universität Bonn
D-5300 Bonn, Germany

Introduction

In recent years a large number of polyhydroxy phenols have been isolated from numerous brown algae, which altogether represent dehydropolymerisation products of phloroglucinol. Since at least the higher molecular representatives of this class have the characters of tannins they are known as phlorotannins. On base of the ring linkage and occasionally additional substituents they are classified in 4 groups.

The fucols are oligoaryls, the phlorethols oligoaryl ethers, the fucophlorethols contain oligoaryl and oligoaryl ether elements and the fuhalols are oligoaryl ethers with additional OH groups (1). Beside the lower molecular precursors with 2 to 13 rings mixtures of compounds have been found with 20, 30, 40 or even more rings. Of course due to their large molecular size the separation and structure elucidation has not been achieved until now. The separation and identification of 4 to 7 ring compounds is already very laborious in many cases. The separation of substances with more than 7 rings has been successful only in exceptional cases. On the other hand there are algae in which only high molecular material occurs. In these cases it would be desirable to get an idea of the linkage system and substitution patterns through a more or less controlled degradation to smaller molecules.

Results

A. Cleavage of fuhalols from Halidrys siliquosa

Chemical cleavage was tried earlier by other researchers but with little success. Mostly in analogy to the flavonoids the alkaline cleavage in solution or in the melt was applied (2-6). Phloroglucinol was the only degradation product identified. We could show that fucols, phlorethols and fucophlorethols are not decomposable in alkaline solution even through heating over longer periods. Only the fuhalols of Halidrys siliquosa yield several cleavage products in dependence upon the pH value of the solution, the temperature and the duration of the reaction. Due to their instability the separation and structure elucidation could be made by the acetyl derivates only. Through the combination of low and high pressure column chromatography 7 substances could be isolated from the mixture and investigated by mass spectrometry and ^{1}H NMR-spectroscopy. Substance I is identical with phloroglucinol triacetate, substances II to VII are substituted dibenzofurans (Fig. 1). The here shown structure proposals for VI and VII are not yet positively proved. Due to the uncertainty in the assignement of the aromatic proton signals with the help of increment values in several times substituted compounds and the large number of possible isomers some other substitution patterns are possible in the dibenzofuran ring, which is substituted by a phenoxyl group.

The fact is that from all the phlorotannins only the fuhalols are cleavable, this stands in agreement with the findings of Mayer and colleagues (7). They found that diaryl ethers are only cleavable in alkaline solution if at least one of the aromatic rings carries 3 vicinal OH groups. By the cleavage of 2,3,4-trihydroxydiphenyl ether they could detect phenol only, from 2,3,4-trihydroxydiphenyl ether phenol and pyrogallol. From dehydrodigallic acid only gallic acid could be found. In their paper nothing is written about the detection of dibenzofurans. In the reaction mixture they could show the appearance of quinones which could be isolated by the addition of sulphite. It is also known from different papers (8,9) that quinones in the presence of acids can be condensed with phenols to dibenzofurans. Evidently quinones resulting by the alkaline cleavage of fuhalols have reacted with phenol in a similar way to give dibenzofurans, although the medium is basic. However we have up to now not been successful in correlating the structure of the obtained derivates with the known structures of the fuhalols obtained from Halidrys siliquosa. Nevertheless the alkaline degradation is a suitable fast test reaction for fuhalols, because the mixture produces an UV spectrum with 3 characteristic maxima and the still clearly recognizable two shoulders in the descending part of the spectrum which are all typical for dibenzofurans although it is an addition spectrum of the single components.

B. Cleavage of phenols from Fucus vesiculosus

From the many methods described in the literature for the cleavage of diaryl ethers the most suitable for the treatment of phlorethols and fucophlorethols appeared to be the degradation with sodium in liquid ammonia. Since a large number of free OH groups interferes with this reaction and to differentiate between the OH groups liberated by the reaction and original hydroxy groups the phenols at first must be methylated. The methylation was achieved by phase transfer catalyzed reaction with dimethyl sulphate.

The cleavage was first tried with a few model substances. The difucol hexamethyl ether was recovered unchanged as expected. Diphlorethol pentamethyl ether exclusively produced phloroglucinol trimethyl ether from ring A and phloroglucinol dimethyl ether from ring B. Bifuhalol hexamethyl ether in addition to phloroglucinol trimethyl ether yielded some phloroglucinol dimethyl ether as the demethylation product and the calculated amount of 3,4,5-trimethoxyphenol.

The conditions with trifuhalol octamethyl ether are somewhat

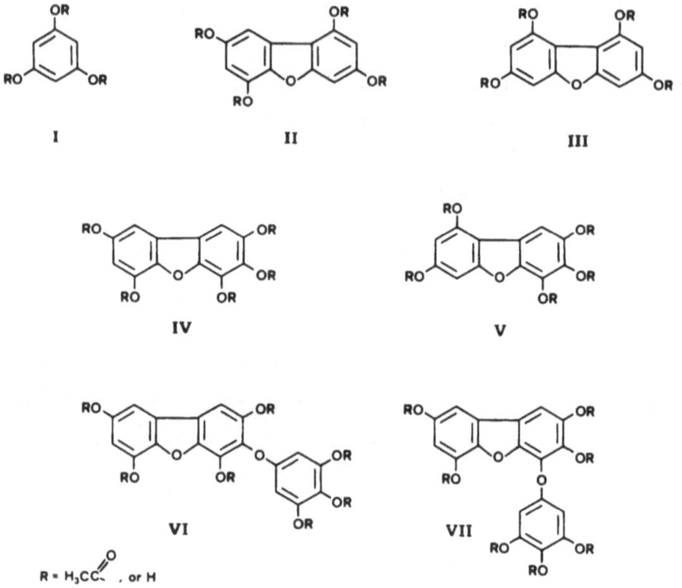

Fig. 1 Products of alkali cleavage of Halidrys phenols

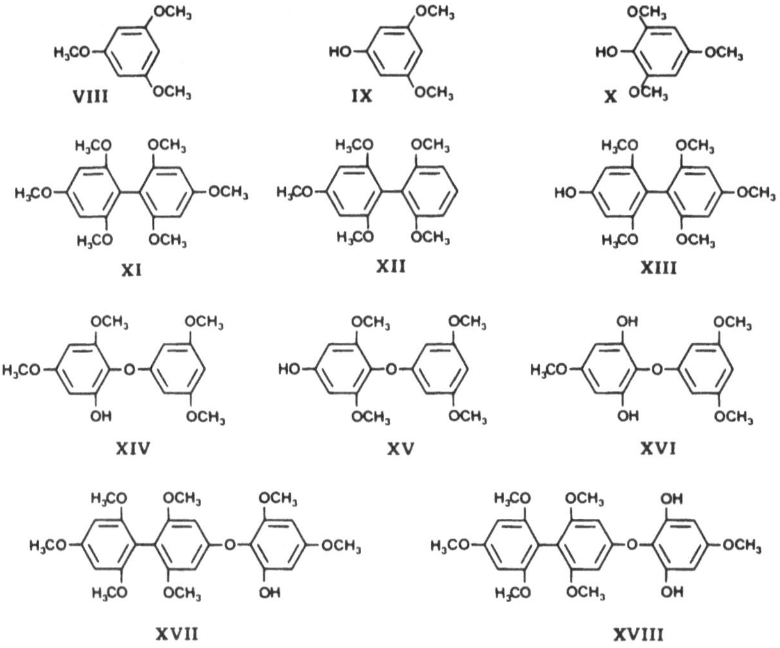

Fig. 2 Products of reductive cleavage of Fucus phenols

more complicated. Beside considerable amounts of unchanged starting material we found the expected one ring fragments of the rings A, B and C as well as one diphenyl ether with a free hydroxy group liberated from the rings B and C and a demethylation compound of the diphenyl ether.

From this experiment we can derive the rule that the diaryl ether compounds prefer to be split up in ortho-position close to two methoxy groups and that the cleavage products with free OH groups cannot be cleaved further.

From the cleavage mixture of phlorotannins from Fucus vesiculosus by the extraction with ethylacetate we can collect a large number of compounds which can be detected on the chromatogram by vanillin sulfuric acid. Some of them could be enriched from the mixture due to their different solubilities in diethyl ether and isolated by low pressure and high pressure chromatography. The structures of 11 substances (VIII-XVIII. Fig. 2) could be elucidated by physical and chemical methods.

Trying to project the elements obtained by the cleavage onto the molecular structures of the phlorotannins known from Fucus vesiculosus we get the following picture.

1. Fuhalols do not appear in Fucus vesiculosus, because 3,4,5-trimethoxyphenol was not found.

2. The diaryl element still carries either all 6 methoxy groups (XI) - in this case the phenoxy group must have been situated meta- to the diaryl bond as in a fucophlorethol (Fig. 3), which has not yet been isolated from Fucus - or a free OH group is found (XIII), then the phenoxy group is arranged in para-position to the diaryl bond as in the known fucodiphlorethol A (10). During the formation of the pentamethoxy biphenyl (XII) the ether bridges must be broken due to unknown reasons between a phenoxyl group and a diaryl unit.

3. The phloroglucinol trimethyl ether (VIII) arises from the

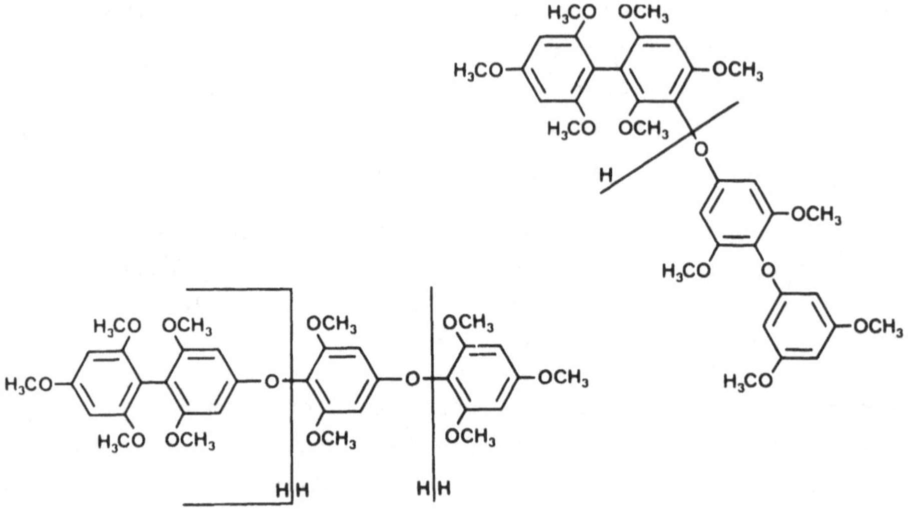

Fig. 3 Reductive cleavage of fucodiphlorethol-A-permethyl ether (left) and an other hypothetical fucodiphlor-etholpermethyl ether (right).

last ring of the fucophlorethols and the phloroglucinol dimethyl ether (IX) arises from the frequently appearing intermediate unit (Fig. 3). If the ether bond of the diphenyl ether remains intact, we get the corresponding fragment with free hydroxy groups in para-position to the ether groups (XV).

4. A fucodiphlorethol as isolated by Craigie et al. (11) from a Canadian <u>Fucus vesiculosus</u> by the cleavage of the terminal phenoxyl group could form substance XVII and by a cleavage in para-position to the diaryl bond it could possibly result in the pentamethoxy biphenyl (XII).

5. A fucodiphlorethol B which till now has only been found in <u>Cystoseira</u> (12) and not in <u>Fucus</u>, could give rise to the diphenyl ether (XIV).

6. The origin of the two cleavage products XVI and XVIII with two free hydroxy groups is still unclear. If they have not arisen through alkyl ether cleavage they must have come from

phlorotannins, that carry 3 vicinal arranged phenoxy substitutions. Such compounds have not as yet been positively identified in Fucus vesiculosus.

It is not possible to state with certainty that other types of linkages which have not been mentioned here, do not exist, since not all the polar substances with two or more OH groups lying in the lower part of the chromatogram have as yet been clarified. Interestingly enough no proof has been found to indicate that the diaryl part appears inside the molecule.

With the described methods the presence of fucophlorethols can be easily proved. The numerous cleavage products can however also be derived from phlorethols. The presence or absence of phlorethols next to fucophlorethols cannot be proved or disproved.

References

1. Glombitza, K.-W. in Marine Algae in Pharmaceutical Science, Edit. Hoppe, H.A., Levring, T., Tanaka, Y., Walter de Gruyter Berlin New York 1979, 303-342
2. Takahashi, T.: Tokyo Kôgyô Shikensho Hôkoku 26, 1-9 (1931)
3. Shirahama, K.: J. Fac. Agric. Hokkaido Imp. Univ. 49, 57-93 (1942)
4. Ogino, C., Taki, Y.: J. Tokyo Univ. Fish. 43, 1-5 (1957)
5. Craigie, J. S., McLachlan, J.: Can. J. Bot. 42, 23-33 (1964)
6. Fretheim, K.: Diss. Univ. Trondheim (1974)
7. Mayer, W., Fikentscher, R., Schmidt, J., Schmidt O.T.: Chem. Ber. 93, 2761-76 (1960)
8. Högberg, H. E., Acta Chem. Scand. 27, 2559-66 (1973)
9. Högberg, H. E., Hjalmarsson, M.: Tetrahed. Lett. 52, 5215-18 (1978)
10. Glombitza, K.-W., Rauwald, H.-W., Eckhardt, G.: Planta Med. 32, 33-45 (1977)
11. Craigie, J. S., McInnes, A. G., Ragan, M. A., Walter, J. A.: Can. J. Chem. 55, 1575-82 (1977)
12. Glombitza, K.-W., Wiedenfeld, G., Eckhardt, G.: Arch. Pharm. 311, 393-99 (1978)

IN SITU INVESTIGATION ON THE ENERGY FLOW IN A BALTIC FUCUS COMMUNITY

Björn Guterstam

Askö Laboratory, University of Stockholm
S-150 13 Trosa, Sweden

Introduction

The Baltic *Fucus vesiculosus* L. communities constitute the bulk of the benthic algal biomass and can be regarded as partial systems of the Baltic ecosystem. They form close colonies with an average 300 g dry wt$\cdot$m^{-2} on rocky and stony bottoms in a depth range from 1 to 4 m as far as the Åland Sea (60°N lat.). The relatively large shallow water areas in the Baltic (9% of the bottom area between 0 and 5.5 m depth) favour colonization by benthic plants. As a result of this they are important contributors to primary production in the Baltic. The aim of this investigation was to follow the energy flow as oxygen metabolism in a typical *F. vesiculosus* community of the north-western Baltic (at Askö 58°50'N, 17°40'E) throughout the year (1).

Methods

An *in situ* method was employed using continuously registering oxygen electrodes (Clark type) in 25-150 l capacity plastic bags (2). The advantages of this method were: 1) The experimental set-up allowed continuous measurements over a time period of days (Fig. 1). 2) The plastic bags constituted a closed system with a transparent plexiglass lid at the top and walls of a flexible plastic material (polyethylene-polyamide) which was essentially impermeable to gases and allowed 95% light transmission. 3) The

Xth International Seaweed Symposium
Printed in Germany

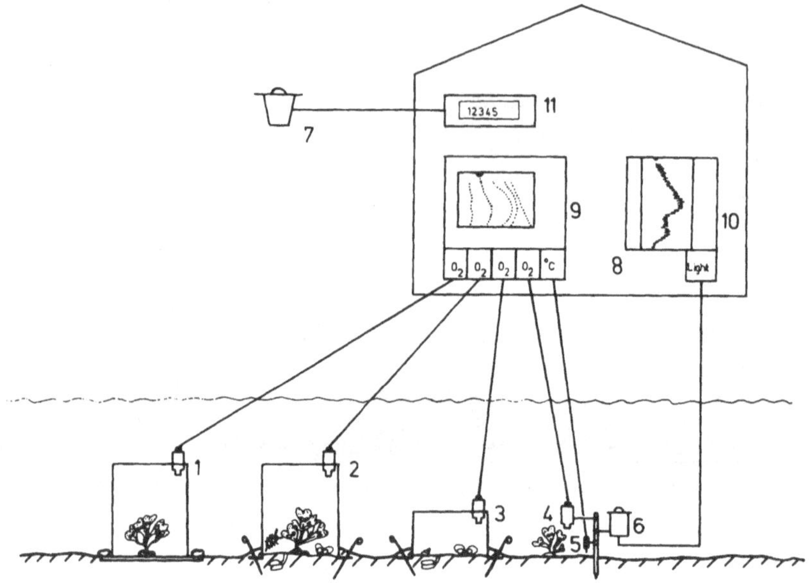

Fig. 1. The experimental set-up for in situ studies of benthic algal metabolism. 1-3 plastic bags with O_2-electrodes 4 O_2 electrode in the water, 5 thermistor, 6 underwater light sensor , 7 solarimeter, 8-11 the electrical supply to the sensors and the recording of data on land.

flexible walls of the enclosures offered the organisms almost natural water movement conditions (3). 4) Temperature and light conditions both above and under water could be registered simultaneously. 5) Nutrient samples i.e. nitrate, nitrite, ammonia and phosphate could be taken with syringes out of the bags during every experiment and analysed in the laboratory.

The experimental set-up was connected by 100 m cables from each electrode to a field laboratory on land, serviced under water by SCUBA diving.

The energy flow in the *Fucus* community was followed by parallel measurements in different enclosures of the community i.e. the *Fucus* community, *Fucus* without its macrofauna and the bottom without *Fucus*. Oxygen metabolism values were converted into energy by using a factor of 3.34 cal mg O_2^{-1} (4).

Results and Conclusions

A picture of the yearly energy flow was obtained from monthly measurements. Pronounced relations to light and temperature could be seen.

The primary production in this system was dominated by *F. vesiculosus* itself. Only in summer could the contribution from other plants be seen, e.g. *Chorda*, *Lithoderma*, *Hildenbrandia* and phytoplankton. The maximum of other plants was found to be about 10% of that of *Fucus*.

The net primary production, given here as the difference between the min. and max. point of the diurnal oxygen curve for *Fucus* (FNP_{LP}), was converted into monthly production values. They were compared in relation to the *in situ* irradiance, the dark respiration of *Fucus* (FR) and the respiration of the consumers (K) (received as the difference between parallel measurements in a *Fucus* community and a *Fucus* without macrofauna enclosure). There was a more than 20-fold increase in FNP_{LP} from December to June (Fig. 2). The low respiration values from February to June indicated an accumulation of energy in the *Fucus* community at that time. In winter and in summer the whole community was balanced between autotrophy and heterotrophy. *F. vesiculosus* demonstrated a much better use of light in winter than in summer. Calculated as % of the *in situ* irradiance the 24 h net production of *F. vesiculosus* (FNP_{24}) changed its light efficiency from 11% in January to 1% in June 1975. The low efficiency in summer was due to light saturation and increased respiration.

Respiration is presented as monthly values in Fig. 2. *Fucus* dark respiration (FR), extrapolated for the whole period, gives an underestimation of its real O_2 consumption. In actuality higher rates of respiration were found as the O_2concentration had increased during the day (1.5). The consumer respiration (K) was better measured with this method since a net for 24 h periods was measured.

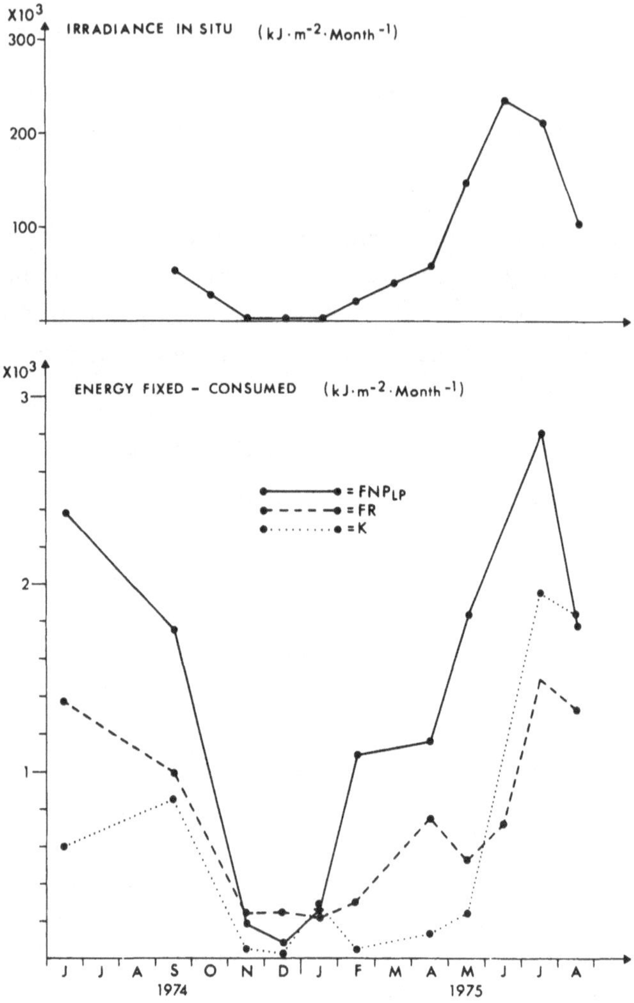

Fig. 2. The yearly change in irradiance, energy fixation and energy consumption in the Fucus community (see text).

The relation between respiration and temperature as they increased from winter to summer is shown in Fig. 3. The Fucus community (FG) showed a close correlation to temperature. Of the total respiration, Fucus (F) itself was the dominating consumer. From December, with 4°C, Fucus was responsible for ca. 80%. At 20°C in summer this had decreased to 40%. The bottom (B) of pebbles and

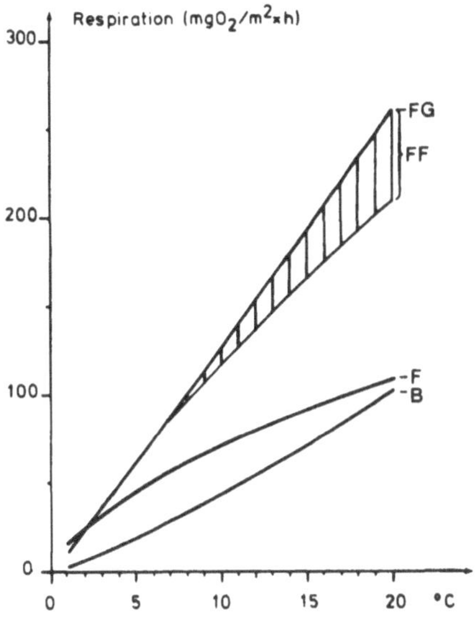

Fig. 3. Respiration as a function of temperature from winter to summer in the Fucus community (see text).

sand with all its growing organisms increased its total respiration from 30 to 40% over the same period. The macrofauna respiration (FF) was calculated by subtraction (FF = FG - F - B shown as dashed in Fig. 3). From April, with temperatures of 6 °C to July with 20 °C this increased from 0 to 18% of the total respiration.

An energy budget for the Fucus community on a yearly basis was calculated for the period 740901 - 750831 (Fig. 4). The energy flow model shows that in 2 m water depth, 53% of the photosynthetic active radiation (PAR) at the surface was available to the plants. 2.2% of this in situ light energy was transformed into chemically bound energy i.e. gross primary production (20.2×10^3 $kJ \cdot m^{-2} \cdot yr^{-1}$). Of this 8.27 $kJ \cdot m^{-2} \cdot yr^{-1}$ was respired by Fucus. The remaining 11.8 $kJ \cdot m^{-2} \cdot yr^{-1}$ was used for the yearly growth of 4.86 $kJ \cdot m^{-2}$ or 300 g dry wt and to feed surrounding organisms. Regarded as an ecosystem, the Fucus community is a producer with the ability to serve other partial systems of the Baltic such as the pelagial, deeper Mytilus banks or soft bottoms.

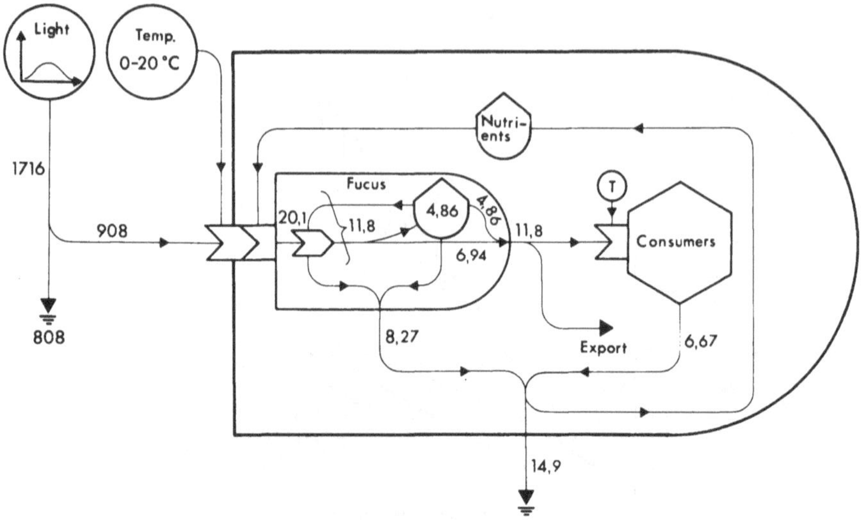

Fig. 4. Energy flow model on a yearly basis for the Fucus community. (Energy flows and biomass in kJ $\cdot m^{-2} x 10^3$.

NP_{LP} was chosen as a comparative value for the net primary production of *F. vesiculosus*. This gave a yearly contribution of 15.6 $kJ \cdot m^{-2}$ (348 $gC \cdot m^{-2}$ with a PQ = 1.2) including all organic material i.e. biomass and dissolved products. Although this is a low productivity compared to other fucoid communities with rates of up to 1000 $gC \cdot m^{-2} \cdot yr^{-1}$ as biomass production (6), it is of the greatest importance for the energy flow in the Baltic.

References

1. Guterstam, B: Reports SFB 95. Kiel Univ. 49, 1-151 (1979).
2. Schramm, W., Martens, V.: Kieler Meeresforsch. 3, 1-6 (1976).
3. Gust, G.: Mar. Biol. 42, 47-53 (1977).
4. Crisp, D.J.: IBP Handbook 16, 197-279, Oxford (1971).
5. Guterstam, B.: 11th European Symposium on Marine Biology, 311-319 (1977).
6. Mann, K.H.: Science 182, 975-981 (1973).

NITRATE AND PHOSPHATE UPTAKE IN SMALL POPULATIONS OF *LAMINARIA DIGITATA* (PHAEOPHYCEAE).

Mentz Indergaard and Arne Jensen
Institute of Marine Biochemistry, N-7034 Trondheim, Norway.

This study was originally initiated to relate daylength, growth and nutrient uptake to some internal parameters of *Laminaria digitata*, the finest alginate-producing kelp of Norwegian waters. At the time the experiments were carried out there was no report available on this basic knowledge. We also considered the experiments as a basis for later studies on cultivation of kelp. In the meantime studies of nutrient uptake by other *Laminaria* sporophytes have been undertaken in the laboratory with discs cut from the thalli (1) and with sporophytes a few cm long (2). The marked difference in physiochemical behaviour between segregated pieces of tissue and whole, intact plants has, however, been pointed out by others (3, 4, 5).

In our experiments young, whole thalli of *Laminaria digitata* (Huds.) Lamour. were carefully detached from their substrate in Trondheimsfjord. Within one hour they were placed in the 13°C cultivation room. During adaptation to low light batch culture the haptera grew and anchored the plants to small rocks within 1-3 weeks. The medium used in the experiments was filtered seawater fortified with nitrate, phosphate and trace metals. The cultivation tank with a volume of 130 l was semi-closed. The input of nutrients was regulated by an adjustable peristaltic pump set according to the daily uptake calculated. Thus it was possible to keep medium concentrations at around 3 μM nitrate and 0.5 μM phosphate. Mixed fluorescent light provided an irradiance of 20 Wm^{-2}. The

Xth International Seaweed Symposium
Printed in Germany

individual plants, around 10 in each experiment, ranged in fresh weight from 8 to 40 g.

Before use the plants were slightly trimmed at the apical end to reduce loss of tissue during the experiment. The culture was continually aerated by compressed air. Sample discs of tissue were taken from the basal and apical parts of the laminas. They were analysed for carbon and nitrogen, and in exp. III also for nitrate and phosphoros. The medium was analysed for nitrate and phosphate according to (6). Nitrate in tissue was similarly analysed in the water extract after drying and grinding, and organic P and polyphosphate were determined according to (7). The fresh weight was calculated by the formula

$$m = m_o e^{\frac{\mu t}{100}} \qquad (1)$$

where m_o = initial fw, m = fresh weight at time t and μ = specific growth rate. The equation assumes steady-state exponential growth. The uptake rates of nitrate and phosphate were calculated as nutrients removed from the medium between each successive analysis.

Some preliminary experiments were run for up to 100 hrs. with starting concentrations in the range 5-28 μM nitrate and 0.5 - 3.6 μM phosphate at L/D 14/10 hrs. No further nutrients were added during the experiments. Results from the light periods were pooled and showed a general increase in uptake rates with increasing medium concentrations. The expected Michaelis-Menten-kinetic relations were not found. The C to N (atomic) ratio in exp. I show large variation (table 1); for apical tissue in the range 23-83 and for basal tissue 19-65. A similar range was found in exp. II and III (table 2).

Nutrient uptake rates for exp. I are shown in Fig. 1 as an example. The medium concentrations varied somewhat during the experiments (see tables 4 and 5), but well within the range occuring in the sea during the growth season.

Table 1. Exp. I. C to N ratio (atomic) in lamina tissue. Values mean of 2-5 samples. Experimental conditions: Light/dark (L/D) regime 14/10 hrs. Total biomass at start 202 g fresh weight.
Ba = Basal tissue. Ap = Apical tissue.

Plant numbers	103 Ba	103 Ap	106 Ba	106 Ap	110 Ba	110 Ap	112 Ba	112 Ap	115 Ba	115 Ap	119 Ba	119 Ap	123 Ba	123 Ap	133 Ba	133 Ap	135 Ba	135 Ap	136 Ba	136 Ap	138 Ba	138 Ap
Start	57	40	44	35	28	49	66	69	57	83	26	31	33	49	28	25	20	23	36	37	23	35
3 days	65	30	34	34	34	46	45	59	45	57	30	31	40	40	30	26	20	27	34	31	28	30
7 days	32	32	33	35	31	40	37	47	33	43	27	31	28	34	29	23	19	30	30	26	26	32

Table 2. Exps. I, II and III. C/N (atomic) in lamina tissue, ranges and mean values. Ba= basal and Ap = apical tissue.
Exp. II: L/D 6/18 hrs., total fresh weight at start 171 g.
Exp. III: L/D 0/24 hrs., total fresh weight at start 231 g.

Exp. no.		Sample range	Mean values Start	3 days	7 days
I	Ba	19-65	37±5	37±3½	30±1½
	Ap	23-83	43½±5½	37±3½	34±2
			Start	4 days	16 days
II	Ba	21-56	33±2	29±2	35+7
	Ap	24-74	51±4	44±3	36±3
			Start	30 days	
III	Ba	16-30	23±1½	19±1	
	Ap	20-52	35½±5	32±3½	

The growth rate μ, based on increase in fresh weight calculated from equation (1) is shown in table 3.

Table 3. Specific fresh weight increase μ.

Exp.	$\mu \pm$ SE	L/D (hrs.	Range of μ
I	2.3 ± 0.2	14/10	1.3 - 3.1
II	0.8 ± 0.2	6/18	0.4 - 2.0
III	0.2 ± 0.05	0/24	0 - 0.5

The dependence of the nitrate and phosphate uptake rates on the longevity of the day is given in tables 4 and 5.

Table 4. Nitrate uptake rates for *L. digitata* at different L/D regimes. Temp. 13°C. Irradiance approx. 20 W/m².

Exp.	Substrate concentration (μM)	Uptake rate (μg N/g fw/h)	L/D (hrs.)	Remarks
I	3	6.9 ± 0.2	14/10	Light period
	"	6.3 ± 0.4	"	Dark period
	"	1.3 ± 0.1	0/24	After 10 days cont. darkness
II	4 ± 1	6.5 ± 0.8	6/18	Light per.
	"	4.6 ± 0.5	"	Dark per.
	"	1.6 ± 0.5	0/24	After 6 days cont. darkness
III	4 ± 1	0.5 ± 0.2	0/24	30 days cont. darkness

Table 5. Phosphate uptake rates for *L. digitata* at different L/D regimes. Temp. 13°C. Irradiance approx. 20 W/m².

Exp.	Substrate conc. (μM)	Uptake rate (μg P/g fw/h)	L/D (hrs.)	Remarks
I	0.5	0.7 ± 0.1	14/10	Light per.
	"	0.7 ± 0.1	"	Dark per.
	0.4	0.1 ± 0.02	0/24	After 10 days cont. darkness
II	0.4	0.6 ± 0.1	6/18	Light per.
	0.4	0.7 ± 0.1	"	Dark per.
	0.4	0.1 ± 0.04	0/24	Cont. darkness
III	0.1	0.2 ± 0.04	0/24	Cont. darkness

Because of technical restraints we chose to run the experiments as semibatch cultures. It is almost impossible to obtain axenic cultures when transplanting macroalgae. A treatment of the plants with antibiotics was found undesirable. After 7-10 days of growth in L/D 14/10 hrs., visible epiphytes were observed. At one occasion all epiphytes on plants and walls of cultivation tank were scraped off after 15 days and found to represent 2 % of the *L.digitata* biomass. To estimate unwanted phytoplankton and bacterial growth the turbidity was controlled during the experiments. The medium was changed completely when the optical density compared to distilled water at 885 nm reached 0.01 in a 10 cm cell. If a population of phytoplankton should influence the nitrate concentration of the medium the density would be in the order of 10^7 cells/ml. Another source of possible influence on nutrient concentration in the medium is the deterioration of apical tissue. All plants were, as mentioned earlier, slightly trimmed to reduce losses of apical tissue during the tests. In exp. III apical tissue contained on the average 430 μg nitrate-N/g dryweight. To raise the nitrate concentration in the tank by 0.5 μM would require lysis of about 10 g fresh weight, i.e. 5-10 % of the whole kelp biomass. This "self-fertilizing" process might have some effect in natural, dense kelp beds in summer time when the decaying of apical tissue increases with raised sea temperatures (8).

Fig. 1. Exp. I. Uptake rates for nitrate and phosphate calculated as N and P. At each point rates are calculated over the period from the preceding point. Shaded area indicate dark periods.

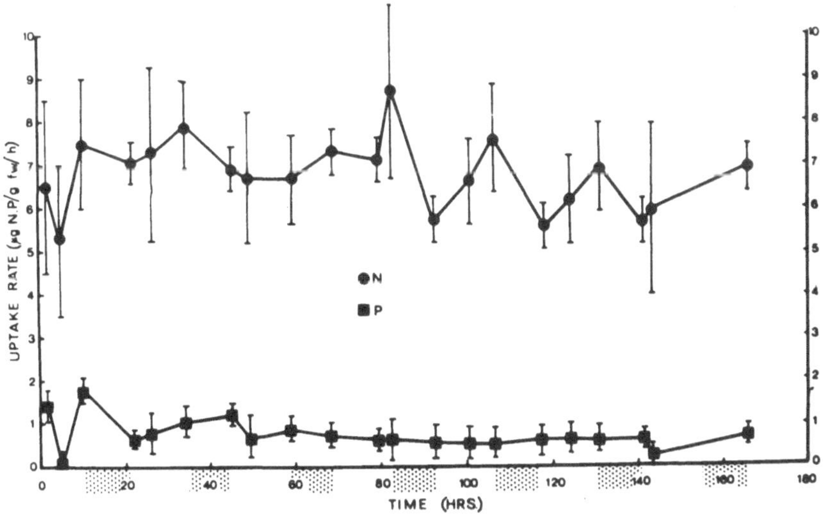

Since the plants were relatively small and were trimmed at the apical end, a large percentage of their biomass consisted of the growth zone. This might explain the initial observations of the growth that made it reasonable to choose the growth equation given by formula (1). The specific fresh weight rate μ increased with the daylength, but even after 30 days in continuous darkness the plants were still gaining weight and removing nutrients. The data in table 2 are in accordance with the general impression of large individual growth variation in even small populations (5-20 plants) which is evident from the work of other groups (9,10,11,12). Taking exp. I as an example the specific growth rate varied from 1 to 3. If this difference was effective during a growth season *in situ* it would result in tremendous differences in production between the plants. Similar individual variability also occurred in the C to N ratios. The C to N ratio decreased within each experiment both for apical and basal tissue as nutrients were absorbed. The lowest value observed here was 14, which might be near the limit considering the high content of polysaccharides in *Laminaria*. C to N ratio of phytoplankton with adequate nutrient supply is around 6-7 and increases to 35 in N-deficiency (13).

Uptake rates were similar in light and dark periods of the 14/10 hrs. L/D regime. The uptake in the dark was however reduced when the light periods were made shorter. In the dark the uptake rates also seemed more sensitive to variations of the substrate concentrations. The nitrate uptake rates corresponded well with those found for *L.longicruris* (1). Phosphate uptake rate varied only within the range 0.2-0.6 μg P/g f w/h from continuous darkness to the 14 hrs. day. The difference between nitrate and phosphate uptake rates became obvious in the atomic ratio N to P of nutrient removed from the medium, ranging from 4 in continuous darkness to 24 in light periods.

These experiments have pointed to a very pronounced individual variation in growth rate and C to N ratios between seemingly identical plants. With applied aquaculture in mind these variations should be clarified and used to promote yields. We are attempting to do this first by observing only the growth of some tens to hundreds of plants. Selection of the fastest growing specimens will be made for further study of chemical content and for breeding experiments. We hope this approach, instead of starting a full-scale analysis program of several hundred plants randomly selected, will be more rewarding both scientifically and commercially.

References

1. Harlin, M.M., Craigie, J.S.: J.Phycol. 14(4), 464-467 (1978).
2. Chapman, A.R.O., Markham, J.W., Lüning, K.: J. Phycol. 14(2), 195-198 (1978).
3. King, R.J., Schramm, W.: Mar.Biol. 37(3), 209-213 (1976).
4. Hatcher, B.G.: Mar.Biol. 43, 381-385 (1977).
5. Wheeler, P.A.: J.Phycol. 15(1), 12-17 (1979).
6. Strickland, J.D., Parsons, T.R.: Fish.Res.Board Can. Bull. 167 (1972).
7. Solorzano, L., Strickland, J.D.H.: Limnol. & Oceanogr. 13, 515-518 (1968).
8. Mann, K.H.: British Ecological Society, Symposia, 17,247-269, Blackwell Sc. Publ., Oxford (1976).
9. Sundene, O.: Nytt.Mag.Bot. 11, 83-107 (1964).
10. Mann, K.H.: Mar.Biol. 14, 199-209 (1972).
11. Chapman, A.R.O., Craigie, J.S.: Mar.Biol. 46(3), 209-215 (1978).
12. Lüning, K.: Mar.Ecol.Progress Ser. 1(3), 195-207 (1979).
13. Sakshaug, E., Holm-Hansen, O.: J.exp.mar.Biol.Ecol. 29, 1-34 (1977).

SOME ENZYMATIC PROPERTIES OF POLYMANNURONIC ACID 5-EPIMERASE IN BROWN ALGAE

M. Ishikawa* and K. Nisizawa

Department of Fisheries, College of Agriculture and Veterinary Medicine, Nihon University, Shimouma-3, Setagaya-ku, Tokyo, 154 Japan

*Present address: Hayashibara Company Ltd. 2-3, Shimoishii, Okayama, 700 Japan

Key-words: Polymannuronic acid 5-epimerase, Alginate lyase, Optimum pH, Optimum temperature, Molecular weight

Introduction

It was found first by Larsen et al. (1,2) and Haug et al. (3) that a kind of epimerase is existent in a culture medium of Azotobacter vinelandii. The epimerase was thought to be capable of epimerizing the mannuronic acid residue (M) of alginate produced by this bacterium to guluronic acid residue (G) on the polymer level. Subsequently, these authors detected similar enzyme activity, i.e. mannuronic acid 5-epimerase (abbreviated as PM5ase) activity in an extract from a brown alga, Pelvetia canaliculata (4). The fact that much higher radioactivity was found in G than M after the incubation of reaction mixture containing tritiated water was considered to be a strong evidence of the existence of PM5ase (2,4).

On the other hand, we found in a previous work (5) that absorbance by carbazole-H_2SO_4 reaction of a reaction mixture containing algal alginate lyase (abbreviated as ALase) and an alginate of higher content of M (abbreviated as M-rich alginate) increased gradually during incubation while absorbance by thiobarbituric acid reaction steeply increased. Moreover, we

Xth International Seaweed Symposium
Printed in Germany

also found in other previous works (6,7) that the M/G ratios of alginate in reaction mixture of bacterial ALase decrease in parallel with incubation upon measurement by ion-exchange column chromatography according to the method of Larsen et al. (8). Thus, the result showed the occurrence of an apparent conversion M to G of alginate used as substrate during incubation. At this time, we, however, explained this fact that the bacterial ALase attacks preferentially the glycosyl bond connecting to M of alginate molecule to convert it into an unsaturated uronic acid residue which was destroyed by acid hydrolysis on the measurement of M/G ratio of the substrate.
In view of these results, we tried to obtain further information about the enzymatic properties of algal PM5ase under the reaction condition which is not affected by coexisting ALase. However, advanced purification of PM5ase was unsuccessful this time, we investigated the properties using enzyme solutions obtained after salting out with ammonium sulfate but under checking whether ALase activity in them is negligibly low or not. The results obtained are presented in this paper.

Results

1). PM5ase activity measured by ion-exchange column chromatography

An aliquot of 50 ml was removed before and after incubation from an incubation mixture containing 5.68 mM of a polymannuronate fragment (abbreviated as SM), an enzyme extract from frozen fronds of Ishige okamurai stored at -20°C for several days after harvest and the standard buffer (consisting of 0.1 M Tris-HCl buffer, pH 7.7, 3.4 mM $CaCl_2$ and 0.1 mM DTT), and their acid hydrolysates were prepared by the method of Fischer and Dörfel (9). After adjustment to pH 8.0, they were individually fractionated by Dowex 1x8 column chromatography according to Larsen et al. (8), and the eluates corresponding to M and G peaks were determined by carbazole-H_2SO_4 method to obtain the

M/G ratios. The M/G ratio of 3.5 at the start of incubation decreased to 1.5 after 24-hr incubation . Such a low M/G value as 3.5 at the start must be due to the G residues preexistent in alginate fragments contaminating the enzyme extract.

2). PM5ase activity measured by GLC

Each 100 ml was removed at appropriate incubation periods from reaction mixture containing M-rich alginate, an enzyme extract from fresh fronds of Ishige sinicola and the standard buffer. The substrate in each reaction mixture removed was recovered by gel filtration on a Sephadex G-75 column (30x350 mm) and hydrolyzed with H_2SO_4 to prepare the samples for GLC according to the method of Perry et al. (10).

From the areas constructing M and G peaks, the M/G ratios of 19.5 for the alginate before incubation decreased to 16.0 and 12.8 with the alginates recovered from the reaction mixture at 24- and 48-hr incubations, respectively.

3). Optimum pH

Using an enzyme extract from frozen fronds of Hizikia fusiformis, pH optimum was estimated by the routine procedure. The enzyme solution contained 0.66 mg of Lowry-protein (11) per ml. An optimal PM5ase activity measured by carbazole-H_2SO_4 method was found at pH 7.7.

4). pH stability

An enzyme extract from frozen fronds of Ishige okamurai containing 2.08 mg of Lowry-protein per ml was prepared. Parts of it was adjusted to various pH's and stored at 4°C for 24 hr. Then, PM5ase activities in the standard reaction mixture were measured by carbazole-H_2SO_4 method. The PM5ase activity was stable in the pH range of 4 to 10, and most stable around the optimum pH.

5). Optimum temperature

An enzyme solution was obtained from frozen fronds of Ishige sinicola and it contained 0.43 mg of Lowry-protein per ml. Each standard reaction mixture containing this enzyme solution was incubated at each of every 10°C-raising temperature from 10°C to 40°C for 24 hr and then PM5ase activities of them were measured by carbazole-H_2SO_4 method. The PM5ase activity was highest at 30°C.

6). Thermal stability

Each given amount of the enzyme solution used in the experiment 5 was individually heated at each of every 10°C-raising temparature from 30°C to 60°C for 10 min, and PM5ase activities of them were measured under the standard condition by carbazole-H_2SO_4 method. The PM5ase activity was considerably sensitive against heat and it was almost completely lost by heating at 60°C for 10 min. In another series of experiment in which two parts of enzyme solution used above were separately stored at 4°C and at -45°C and their PM5ase activities were measured at intervals. The former part of enzyme solution lost almost half of its activity during 15-day storage while the latter one lost only 20 % of its activity for the same period.

7). Substrate specificity

Using an enzyme which was prepared from frozen fronds of Hizikia fusiformis and contained 0.62 mg of Lowry-protein per ml, PM5ase activities for SM, M-rich alginate and SG (a polyguluronate fragment) were measured by the standard procedure after 24-hr incubation. The activities measured by carbazole-H_2SO_4 method for M-rich alginate was approximately 1.5 times higher than that for SM and no measurable activity was found for SG.

8). Effect of metal ions.

Effect of various metal ions on the PM5ase activity was investigated using an enzyme solution from frozen fronds of Ishige sinicola. The enzyme solution contained 2.27 mg Lowry-protein per ml. The PM5ase activities measured after 24-hr incubation by carbazole-H_2SO_4 method in the presence of 10 mM metal ions were in the order of $Mg^{2+} > Mn^{2+} = Ca^{2+} >$ no-metal $> Fe^{3+} > Zn^{2+}$. When the activity of no-metal mixture is taken as 100 %, that with Mg^{2+} was about 130 %, those with Mn^{2+} and Ca^{2+} about 120 % and those with Fe^{3+} and Zn^{2+} about 38 % and 10 %, respectively. Thus, Mg^{2+} and Mn^{2+} ions behaved as activators while

Fe^{3+} and Zn^{2+} as potent inhibitors, although this was slightly different pattern from that obtained by Madgwick et al.(12). p-Mercuribenzoate inhibited so remarkably that the PM5ase may be suggested to be a SH-enzyme.

9). Michaelis constant

Using the same enzyme solution as used in the experiment 8, the Michaelis constant of PM5ase for SM was measured according to the method of Lineweaver-Burk plots. Km value estimated was 7.3×10^{-4} and substrate inhibition was observed at higher concentrations of SM.

10). Molecular weight

An enzyme solution containing 2.34 mg of Lowry-protein per ml was obtained from frozen fronds of Ishige okamurai. The standard curve was made by using several proteins such as horse cytochrome (mol. wt., 12,500), chymotrypsinogen A of bovine pancreas (mol. wt., 25,000), egg albumine (mol. wt., 45,000), bovine serum albumine (mol. wt., 68,000) and rabbit muscle aldolase (mol. wt., 158,000) as reference proteins and using a Sephadex G-100 (Super fine) column (20x800 mm) as gel filtration apparatus. The molecular weight of PM5ase was thus estimated to be 62,000.

11). Localization of PM5ase activity in algal fronds

Growing area of a single Eisenia bicyclis grown more than 2 years was separated from other part of the fronds, and 2 enzyme solutions were prepared from them. They contained 2.64 mg of Lowry-protein per ml and 7.28 mg per ml, respectively. The PM5ase activity of the growing area was 17 times as high as that of the other part when they were expressed by the activity per mg protein in the enzyme extracts(Fig. 1).

References

1. Larsen, B., Haug, A.: Carbohyd. Res. 17, 287-296 (1971).
2. Larsen, B., Haug, A.: Carbohyd. Res. 20, 225-232 (1971).
3. Haug, A., Larsen, B.: Carbohyd. Res. 17, 297-308 (1971).

4. Madgwick, J., Haug, A., Larsen, B.: Acta Chem. Scand. 27, 3592-3594 (1973).

5. Shiraiwa, Y., Abe, K., Sasaki, S. F., Ikawa, I., Nisizawa, K.: Botanica Marina 18, 97-104 (1975).

6. Nisizawa, K., Sasaki, S. F., Saigo, M.: Proceedings of the 7th Internat. Seaweed symp. (ed. by Nisizawa et al.), 485-490, Univ. of Tokyo Press, Tokyo (1972).

7. Min, K. H., Sasaki, S. F., Kashiwabara, Y., Umekawa, M., Nisizawa, K.: J. Biochem. 81, 555-562 (1977).

8. Larsen, B., Haug, A.: Acta Chem. Scand. 15, 1397-1398 (1961).

9. Fischer, F. G., Dörfel, H.: Zeit. physiol. Chem. 301, 224-234 (1955).

10. Perry, M., Hulyalker, P. K.: Canad. J. Biochem. 43, 573-584 (1965).

11. Lowry, O. H., Rosebrough, N. J., Farr, A. L., Randall, R. J.: J. Biol. Chem. 193, 265-273 (1951).

12. Madgwick, J., Haug, A., Larsen, B.: Botanica marina. 21, 1-3 (1978).

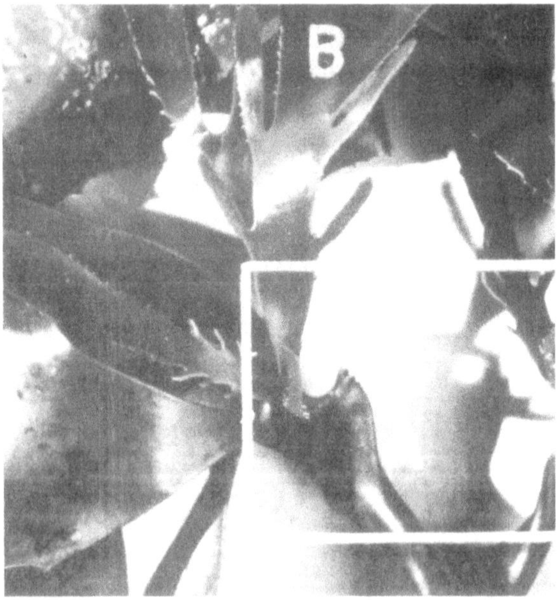

Fig. 1.Photograph of a part of Eisenia bicyclis. A; growing area, B; residual parts.

DETERMINATION OF URONIC ACID COMPONENTS IN ALGINIC ACID

M. H. Ji, W. D. Cao and L. J. Han
Institute of Oceanology, Academia Sinica, Qingdao, China

Introduction

Since Fischer (2) in 1955 had ascertained the presence of L-guluronic acid in addition to the D-mannuronic acid in alginic acid, many contributions were presented on the influence of M/G ratio in alginate on the ion exchange properties of alginates (10) and the absorption of radioactive Sr and other heavy metals in animals (7,8,11), and on the fractionation of M and G by partial hydrolysis to elucidate the heterogeneity of alginic acid (9), etc.

From the fifties in China alginates from Sargassum and then from Laminaria were produced industrially based on the experimental results (4,5,11). With a view to studying the chemical and physical properties and some applications of alginates, it is much important to understand the contents of mannuronic acid (M) and guluronic acid (G) and their ratios in alginates from different species of Chinese brown seaweeds. At first we studied the method of preparation of standard M and G, and then the colorimetric conditions for uronic acids with phenol-H_2SO_4 method, the destroying ratios of M/G during acid hydrolysis and the M/G ratios in alginates of some Chinese brown seaweeds were investigated.

Methods

1) Sodium alginates. They were extracted from seaweeds using

Xth International Seaweed Symposium
Printed in Germany

Na_2CO_3 solution at 70-75°C, and dehydrated with alcohol after bleaching, acidifying and dissolving in alkali (11).

2) Determination of M/G ratios. The M/G ratios in sodium alginate samples were determined essentially according to Haug's method using Dowex 1 x 2 anion exchange resin column and the colorimetric conditions as illustrated below. The M/G ratios obtained were calibrated by multiplying 0.68 (see following sections).

Results and discussion

1) Preparation of M and G solution and their standard curves for determination. The above-mentioned method for M/G ratio determination was taken as the method for preparation of M and G solution using enlarged amounts of sample and reagents, without using the Fischer's cellulose column and pyridinium eluent (2). Thus, 4 g of Na alginate were hydrolyzed with 80% H_2SO_4 at room temp. for 18 hrs, and then the hydrolyzate was diluted to 2 N H_2SO_4 and further hydrolyzed in a sealed glass tube for 5 hrs in boiling water bath. After neutralization with $CaCO_3$ and filtration, the solution was passed through Dowex 1 x 2 anion exchange resin (200-400 mesh) column (50 x 4 cm) in acetate form, and fractionated by a gradient elution with 10 l 0.5-2 N acetic acid. Each fraction collecting 120 ml of eluates was ascertained by paper chromatography and combined in two portions, the former being guluronic acid and the latter mannuronic acid. The combined M or G assayed was evaporated in vacuum to a small volume, and allowed to pass through a cation resin column to eliminate cations. The effluent was again evaporated to syrup, and 30 ml of EtOH were added to dissolve the syrup. Such treatment was repeated 2-3 times in order to expel the acetic acid. Upon adding 30 ml of water , a small amount of anion resin(OH form) was added to discharge the color and residual anions.

After filtration the filtrate was treated with NaOH to pH 8, sufficiently transforming the lactones to sodium uronate. The solution was allowed to pass through a cation exchange resin column, and the acid liberated was titrated with 0.1 N NaOH to confirm the exact concentration of each uronic acid. An aliquot of uronic acid solution was taken to prepare the standard curves of mannuronic (M) and guluronic(G) acids as

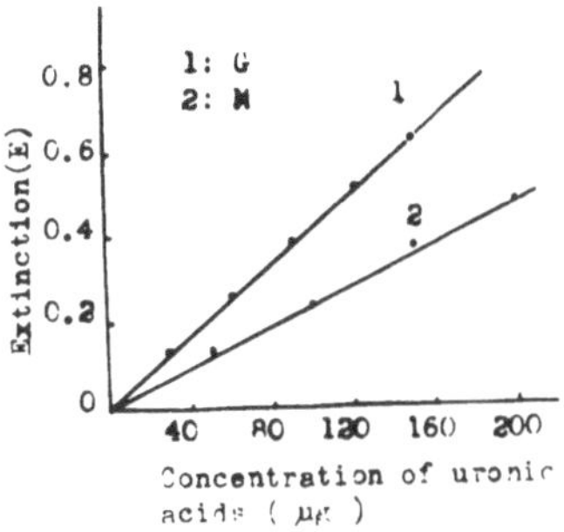

Fig. 1. Standard curves of mannuronic acid (M) and guluronic acid (G)

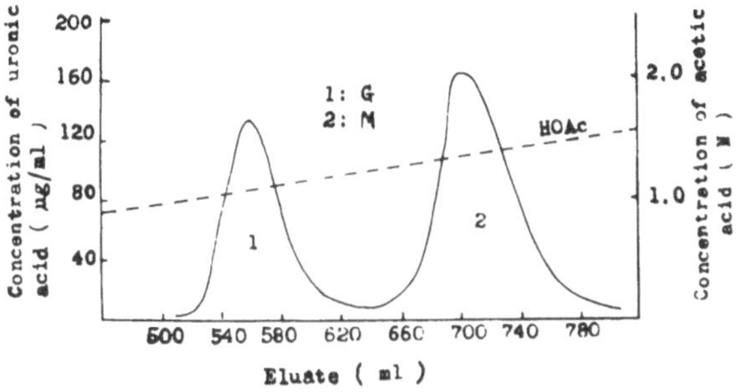

Fig. 2. Distribution of M and G in the eluates of acetic acid

shown in Fig. 1. Thus obtained M and G solution exhibited the single spot on paper chromatogram with Rf value of 0.42 and 0.38, respectively. The mannuronic acid obtained either by this method in solution or by the other method in crystals showed the same extinction values as the same concentration was measured in spectrophotometer. From the pattern of the distribution of M and G of Na alginate in HOAc eluates on colorimetry as seen in Fig.2, the peaks of two uronic acids were distinct, and similar to that of Larsen (6).

2) Determination of the destroying ratio of M and G. The standard solution containing M or G was evoporated to syrup, and submitted to hydrolysis by above-mentioned procedure. The recovery ratio of two uronic acids was 1.46 on an average. All M/G ratios obtained were calibrated by multiplied 1/1.46 or 0.68, close to 0.66 by Haug (3).

3) Colorimetric conditions. Up to now, we haven't found the re-

ference concerning the colorimetric conditions for determination of M and G, except that for free sugars (1), therefore we studied and confirmed the optimum conditions as follows. 2.0 ml of fraction solution containing either M or G were transferred to a ground stoppered glass tube (Ø 2 cm), and 0.04 ml of 80% phenol solution was added. The tube after shaking was maintained at 50°C of water bath, and 5 ml of conc. H_2SO_4 were rapidly poured into it. After setting for 20 min, vigorous shaking and setting again for 20 min, the colorimetric measurement was carried out at 485 mμ.

4) M/G ratios in alginates from some Chinese brown seaweeds. Using the ascertained method the M/G ratios in Na alginates from some Chinese brown seaweeds were determined. It seems that the Na alginates from *Sargassum* spp., the perennial algae, usually have the lower M/G ratios, i.e. with more richer content of guluronic acid than that from *Laminaria*, an annual alga (Table 1).

Table 1. M/G ratios in sodium alginates from Chinese brown seaweeds

Species	Locality and date of collection	M/G ratios in Na alginate
Sargassum pallidum	Qingdao 1978, 5	1.46
Sargassum kjellmanianum	Qingdao 1978, 4	1.25
Sargassum hemiphyllum	Guangdong province 1978 spring	1.06
Sargassum tenerrinium	Guangxi province 1978	1.53
Sargassum henslowianum	Guangdong province 1978	0.82
Sargassum patens	" 1978	1.59
Sargassum siliquastrum	" 1978	1.13
Laminaria japonica	Commercial product from Qingdao 1978	2.01

References

1. Dubois, M., Gilles, K.A., Hamilton, J.K., Ribers, P.A., Smith, F.: Anal. Chem. 28, 350-356 (1956)
2. Fischer, F.G., Dörfel, H.Z.: Hoppe-seyler's physiol. Chem. 302, 186-203 (1955)
3. Haug, A., Larsen, B.: Acta Chem. Scand. 16, 1908-1918(1962)
4. Ji, M.H., Shi, S.Y.: Studia Marina Sinica 1, 182-189 (1962)
5. Ji, M.H., Shi, S.Y., Pu, S.Z., Zhang, Y.S.: Ibid. 3, 77-101 (1963) (in Chinese with English abstract)
6. Larsen.B., Haug, A.: Acta Chem. Scand. 15, 1397-1398 (1961)
7. Patrick, G., Car, T.E.F., Humphreys, E.R.: Inst. J. Radiat. Biol. 12, 427-434 (1967)
8. Silva, A.J., Fleshman, D.G., Shore, B.: Health Physics 9, 245-251 (1970)
9. Simienescu, Cr.I., Popa, V.I., Rusan, V., Liga, A.: Cellul. Chem. Technol. 2, 213-225 (1975)
10. Smidsrød, O., Haug, A.: Acta Chem. Scand 22, 1989-1997 (1968)
11. Tanaka, T., Waldron-Edward, D., Skoryna, S.C.: Can. Med. Ass. J. 99, 169-175 (1968)
12. Tseng, C.K., Ji, M.G.: Studia Marina Sinica 1, 133-151 (1962)

THE VARIATION IN CONTENTS OF VARIOUS STATES OF AMINO ACIDS IN *PORPHYRA YEZOENSIS*

M. H. Ji, S. Z. Pu and Z. Q. Niu
Institute of Oceanology, Academia Sinica, Qingdao, China

Introduction

Porphyra from the ancient time has been appeciated by the Chinese people as a favourable subsidiary food. Now the *Porphyra* was cultivated from the North to the South of China on a large scale. By the practical experiences it was well known that the quality of *Porphyra* varies closely with the culture region, growing season and the fertility of the seawater and usually is assessed on the basis of the apparent color and brilliancy, whereas the chemical assessment on its quality is still not initiated in this country. This work was done with the aim of providing with our own results about the change features of nitrogeneous compounds including various states of amino acids in *Porphyra yezoensis*, which is the main cultivating species in Qingdao, to meet the demand of the cultivating field and ecological, physiological and nutritional studies on it.

Materials and methods

Porphyra yezoensis cultivated on the floating raft was collected monthly from Jan to Apr from three culture regions in Qingdao, typically differing in the fertility of the seawater (2), H region refers to water with high nutrient content(200-300 μg NH_4-N/l),M region to water with middle nutrient content (about 100 μg NH_4-N/l) and L region to water with low nutrient

Xth International Seaweed Symposium
Printed in Germany

content (30-50 μg NH_4-N/l). The collected specimens were dried at 50°C, pulverized and passed through 100 mesh.

For the determination of free amino acids, the algal powder were extracted with 75% EtOH at roomtemp., and for that of total amino acids the algal powder were hydrolyzed with 6 N HCl at 105°C. Then the extractive and hydrolyzate were evaporated and decolorized, the filtrates being submitted to the amino acid analyzer (Hitachi 835-30)

Results and discussion

The content of nitrogen in Porphyra essentially depends upon the amount of NH_4-N in seawater of the culture region. As shown in Fig. 1, the order of N contents in Porphyra in these culture regions is: H region > M region > L region. Seasonally the contents of total N in Porphyra in any region were gradually dropped after Jan. But in H region it slightly increased from 6.25% in Jan to 6.63% in Feb, and then dropped too, while Noda (5) gave the total N content in the same algae reaching 7.6% in Jan/Feb and decreased afterwards. The Porphyra from these culture regions appeared dark-violet in color and brightly brilliant for the sample from H region, and yellow in color and pale for that of L region, being coincident with both the N content in Porphyra and the NH_4-N content in seawater of the field.

The seasonal variation of the contents of sixteen free amino acids including Ala, Glu, Asp, Pro, Thr, Ser, Gly, Val, Ile, Leu, Phe, Met, Lys, Arg, Cys and $CySO_3H$ as well as NH_3 and Tau were determined. Among them Ala was always the predominant one, followed by Glu and Asp. As can be seen from Fig.2, the contents of these amino acids successively decreased with the growing period, the order of decrease being akin to the pattern of the change of total N. Ala varied from 3% in Jan to 0.73% in Apr. Japanese reports also showed the order of decrese was Ala>Glu>

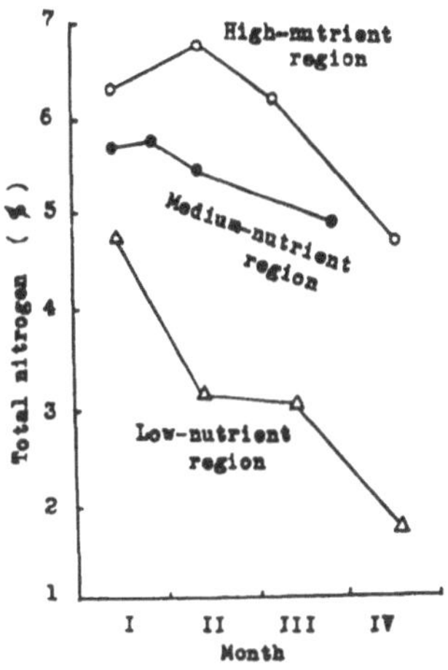

Fig. 1. Total nitrogen in Porphyra yezoensis

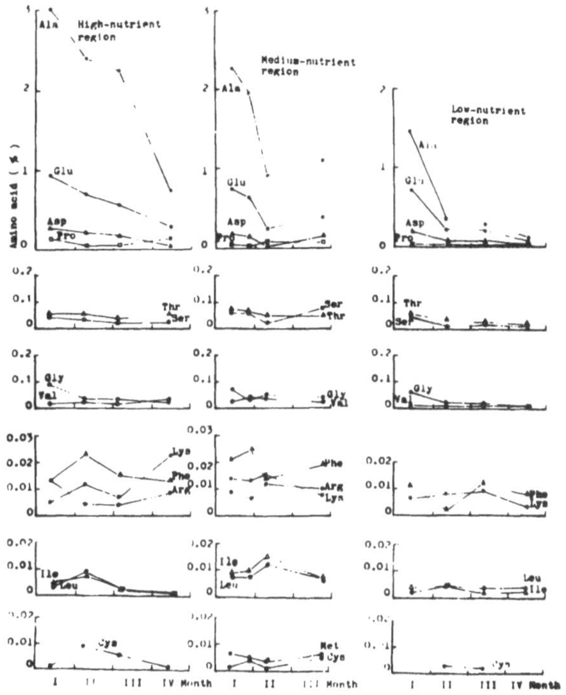

Fig. 2. Free amino acids in Porphyra yezoensis

Asp both in Porphyra tenera (7) and P. yezoensis (6). It seemed that these three free amino acids would be very active at initial stage of metabolism in Porphyra. The contents of Pro and other amino acids fluctuated not conspicuously. The mechanism of amino acid metabolism in Porphyra seemed to be somewhat different from that in Laminaria (4), in the latter the order of the contents of free amino acids being Asp>Glu>Ala.
The percentages of free amino acid N/total N in Porphyra varied from 10.4% in Jan to 4.1% in Apr in H region, from 8.4% in Jan to 5.6% in Mar in M region, and from 7% in Jan to 1.9% in Apr in L region.

Among the total amino acids including free amino acids and proteinous amino acids, eighteen amino acids were quantitatively determined. As shown in Fig.3 in all three culture regions the contents of Ala, Glu, and Asp were more richer than those of

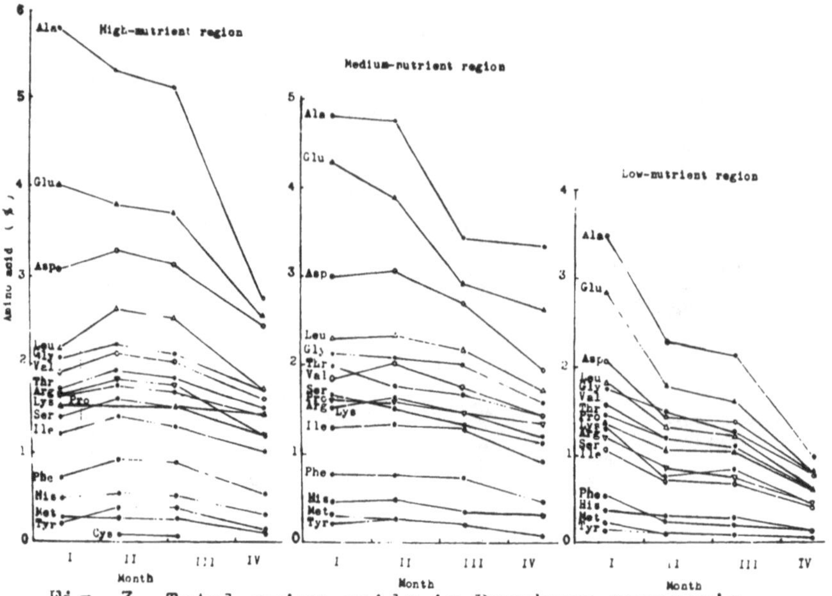

Fig. 3. Total amino acids in Porphyra yezoensis

others with the same order as in free state,i.e. Ala>Glu>Asp, and fluctuated essentially dependent on their contents in free state and also on the change of the content of chromoprotein, which is mainly composed of amino acids (1) in Porphyra. According to Yamakawa (8), the contents of phycoerythrin and phycocyanin in Porphyra tenera are high in Dec/Jan, reaching 1.4% and 0.7%, respectively, and then drop with the growing season. The content of Ala in phycoerythrin amounts to about 12% of the weight of protein (1). The amount of these three amino acids exhibited about 40% of the total amino acids, the contents of Ala and Glu reaching 5.8% and about 4% respectively in Jan at H regions. The tendency of their seasonal variation, i.e. successive decreasing with growing period, was comparable with that of free amino acids, but some amino acids such as Asp, Leu, Ile, Val, Ser, Thr, Phe and Tyr increased slightly from Jan to Feb and Mar, being parallel to the change of the total N. The contents of total amino acids amount to about 30% of the weight of Porphyra (as seen from H region), whereas the calculated crude protein from them exhibited about 40%, being slightly lower than 48% as reported by Noda (5) in the same species. In a word, Porphyra may be recognized as the sole species with the highest contents of amino acids and protein in marine algae.

The content of taurine in hydrolyzate was in general lower than that in alcoholic extractive, probably due to its hydrolysis in some degree. Its content appeared to decrease successively with the growing period. In contrast with this, the amount of cysteic acid in hydrolyzate was usually higher than that in alcoholic extractive. It was probable that the cystine both in protein state and in fee state during hydrolysis might be oxydized to cysteic acid, resulting in the increase of the amount of the latter.

References

1. Fujiwara, T.: J. Biochem. (Japan) 43, 195-203 (1956)
2. Institute of Oceanology, Academia Sinica (ed.): Artificial Cultivation of *Porphyra yezoensis*. Kue Xue Chu Ban She 186 pp. (1978) (in Chinese)
3. Ito, K.: Bull. Jap.Soc. Sci. Fish. 29, 771-775 (1963)
4. Ji, M.H., Pu, S.Z., Cao, W.D., Zhang, J.Z.: Studia Marina Sinica 11, 7-23 (in Chinese with English abstract)
5. Noda, H.: Bull. Jap. Soc. Sci. Fish. 37, 391-396 (1971)
6. Noda, H., Horiguchi,Y., Araki,S.: Ibid. 41, 1299-1303 (1975)
7. Takagi, M.,Kuriyama, M.: Bull. Fac. Fish. Hokkaido Univ. 10, 72-76 (1959)
8. Yamakawa, K.T.: Bull. Jap. Soc. Sci. Fish. 18, 478-482 (1953)

METABOLIC IMPLICATIONS OF ß-CARBOXYLATION IN BROWN SEAWEED

B.P. Kremer
Universität zu Köln, Seminar für Didaktik der Biologie
D-5000 Köln 41, Germany

Introduction

In macrophytic brown algae photosynthetic CO_2 fixation via ribulose-1,5-bisphosphate carboxylase (RubPC; EC 4.1.1.39) is paralleled by a high potential for dark carbon assimilation. This CO_2 fixation is exclusively catalyzed by phosphoenolpyruvate carboxykinase (PEPCK; EC 4.1.1.32)(1) and usually exceeds 5% of photosynthesis as has been substantiated by simultaneous measurements of in vivo and in vitro carboxylation (2-4). In members of the Laminariales the respective capacities for photosynthetic and dark CO_2 assimilation follow a distinct longitudinal profile along the frond axes with maximum activity of PEPCK (dark fixation) in the ontogenetically young tissue of the growing frond region (5). Transversal profiles of carbon fixation have been derived from analyses of different tissues of Laminaria blades (6). This peculiar spatial pattern of CO_2 assimilation which also implies certain seasonal fluctuations (7) required further elucidation with respect to its metabolic significance.

Results

Phosphoenolpyruvate carboxykinase (performing ß-carboxylation of phosphoenolpyruvate (PEP) to oxaloacetate) along with malic dehydrogenase and aspartate 2-oxoglutarate aminotransferase account for relatively strong ^{14}C-labelling of

Xth International Seaweed Symposium
Printed in Germany

aspartate and malate not only after dark incubation of frond samples, but even during photosynthesis. In order to determine the origin of the enzyme substrate, PEP, aspartate ^{14}C-labelled during photosynthetic and dark CO_2 fixation has been chromatographically isolated from Laminaria hyperborea and then oxidized to ß-alanine by eliminating the -carboxyl using the ninhydrin procedure. More than 70% of ^{14}C originally confined to photosynthetically ^{14}C-labelled aspartate was removed indicating that the bulk of radiocarbon is located in the C_1-atom. Considerably less ^{14}C is released from C_1 of aspartate after dark CO_2 fixation (Fig.1). During photosynthesis the ^{14}C-labelled α-carboxyl is therefore preferentially provided by ^{14}C-labelled phosphoglycerate yielding the substrate of PEPCK by a simple metabolic conversion. PEP precursing CO_2 fixation in the dark must be derived from other metabolic sources.

Special attention has been attributed to the correlation of mannitol metabolism and dark CO_2 fixation particularly in the young, growing fronds of Laminaria spp. Fig. 2 compares some physiological features of young and old frond areas. In growing tissue photosynthetic ^{14}C-labelling of mannitol is considerably less effective. However, high activities of mannitol-1-phosphate dehydrogenase are found in that frond region. Furthermore, it should be noted that daily dark respiratory loss of mannitol in young frond areas exceeds that of old frond regions. These features fit well together if it is assumed that in young fronds mannitol-1-phosphate dehydrogenase is operating in the oxidative direction. This would establish a more intense mannitol consumption in the younger rather than in the older blades.

Oxymetric determinations of O_2 uptake confirm this assumption (Fig.2). Moreover, it was shown that ^{14}C-mannitol exogenously supplied to Laminaria and Fucus tissue slices is converted to mannitol-1-phosphate and to further hexose and

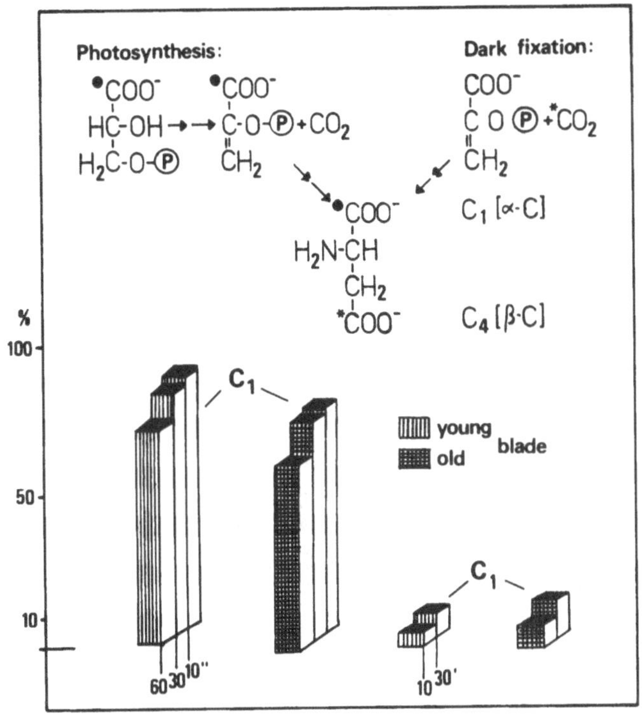

Fig. 1. Percentage of ^{14}C confined to the C_1-atom of aspartate after photosynthesis (10-60 s) and dark fixation (10-30 min).

triose phosphates. This suggests that dark respiratory (glycolytic) degradation of mannitol (representing the primary low-molecular weight carbohydrate reserve of the Phaeophyceae so far examined, cf. (8)) yields PEP and thus the substrate for the considerable ß-carboxylation observed in brown seaweed (2-5).

PEP is not only carboxylated by PEPCK, but is mainly converted by pyruvate kinase (PK) to enter the Krebs cycle for further oxidation. Hence it is likely that both enzymes, PK and PEPCK, compete for PEP as a substrate. In order to prove this the Krebs cycle was blocked by monofluoracetate (10^{-5} to 10^{-3} M). It was observed that the metabolic inhibition was accompanied by a significant enhancement of dark CO_2 fixa-

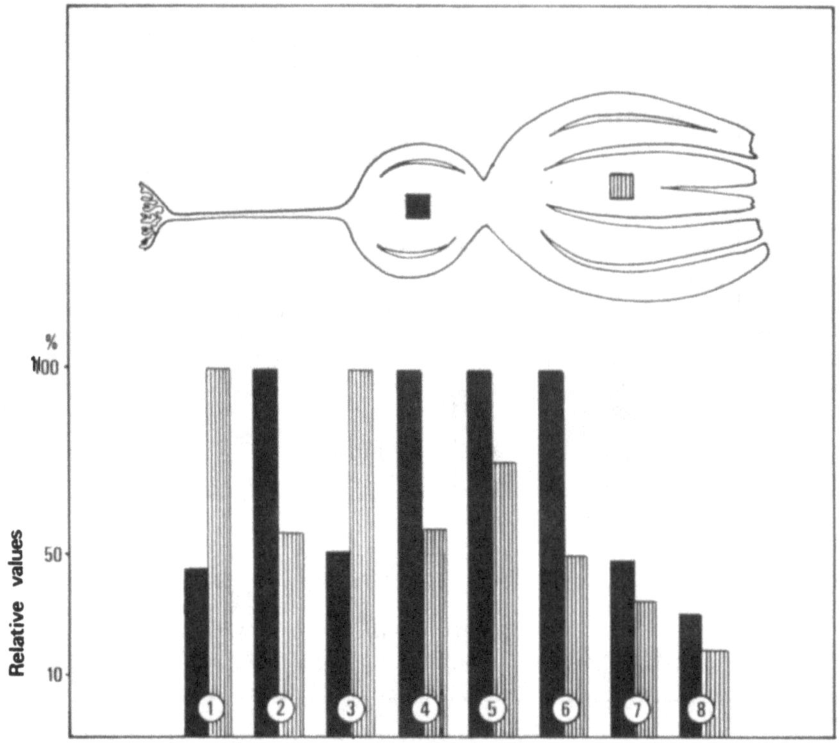

Fig. 2. Laminaria hyperborea. Comparison of some physiological features of young (■) and old (▥) frond areas. 1 Photosynthesis, 2 dark fixation, 3 photosynthetic labelling of mannitol, 4 activity of mannitol-1-phosphate dehydrogenase, 5 daily decrease of mannitol reserve, 6 dark respiration, 7 dark fixation in % of dark respiration, 8 enhancement of ß-carboxylation by 10^{-3} M monofluoracetate. For absolute terms cf. (2-5).

tion: Obviously, a considerable proportion of PEP which did not flow through the Krebs cycle was converted into intermediary C_4-compounds. This result confirms that PEP required for dark CO_2 fixation is derived from mannitol catabolism and suggests that the competition of PK and PEPCK for PEP probably involves a regulatory site depending on whether intermediary compounds or metabolic energy is required in the appropriate frond tissue.

Brown algae exhibiting appreciable rates of ß-carboxylation usually show low extent of photorespiration as determined by the Warburg O_2 effect. In this particular regard, members of the Phaeophyceae seem to ressemble vascular plants showing C_4-photosynthesis. In such plants the high affinity of phosphoenolpyruvate carboxylase (PEPC) for CO_2 is thought to be responsible for the low extent of photorespiration. Since PEPCK basically provides the same metabolic function as PEPC it might be assumed that in brown macroalgae the appreciable ß-carboxylation accounts for the minor extent of the Warburg O_2 effect (9).

Since aspartate and malate occur among the early ^{14}C-labelled photosynthates of brown seaweed (receiving their ^{14}C from ^{14}C-phosphoglycerate) and the same dicarboxylic acids accumulate ^{14}C in the dark, the question arises as to what extent ß-carboxylation participates in photosynthetic CO_2 fixation. If ^{14}C recovered from aspartate and malate after photosynthesis is taken as a measure of ß-carboxylation, then the proportions of PEPCK activity in the light and in the dark can indirectly be compared. It may be seen from Table 1 that

Table 1. _Laminaria digitata_. Comparison of photosynthetic ^{14}C-labelling of aspartate + malate and rates of dark carbon fixation. Data expressed on a dry weight (dr wt) basis.

Incubation	nCi ^{14}C in asp + mal	Dark fixation (nCi ^{14}C g^{-1} dr wt)
10 s	40.7	27.3
30 s	113.5	78.4
60 s	283.8	242.6

ß-carboxylation proceeds at a slightly higher rate in the light than in the dark, thus suggesting that the supply of PEP via phosphoglycerate is more effective than its derivation from mannitol catabolism in the dark. Dark carboxylation via PEPCK is thus fully integrated into light dependent carbon assimilation and does not interfere with carboxylation by RubPC.

Dark carbon fixation partly compensates for the dark respiratory loss of organic carbon in members of the Phaeophyceae (see Fig.2). It is therefore justified to include terms of dark fixation into calculations of gross productivity of brown seaweeds (10).

References

1. Akagawa,H., Ikawa,T., Nisizawa,K.: Plant Cell Physiol. 13, 999-1016 (1972).
2. Kremer,B.P., Küppers,U.: Planta (Berlin) 133, 191-196 (1977).
3. Kremer,B.P.: J.Phycol. 15, 244-247 (1979).
4. Kremer,B.P., Markham,J.W.: Planta (Berlin) 144, 497-501 (1979)
5. Küppers,U., Kremer,B.P.: Plant Physiol. 62, 49-53 (1978).
6. Kremer, B.P.: Mar.Biol. (in press)(1980).
7. Küppers,U., Weidner,M.: Planta (Berlin) 148, 222-230 (1980).
8. Kremer,B.P.: Proc.Intl.Seaweed Symp. 9, 421-428 (1979).
9. Kremer,B.P.: Planta (Berlin) 150 (in press) (1980).
10. Johnston,C.S., Jones,R.G., Hunt,R.T.: Helgoländer wiss. Meeresunters. 30, 527-545 (1977).

EXTRACTS FROM SOME RED AND BROWN SEAWEEDS OF THE PHILIPPINES

E.C. Laserna, R.L. Veroy, A.H. Luistro and G.J.B. Cajipe
Marine Sciences Center, University of the Philippines
Diliman, Quezon City D-3004 Philippines

Introduction

Although studies on the nature and properties of seaweed extracts date back to several decades ago, the subject continues to receive much attention (1-10). The present study on the spectral and physico-chemical properties of polysaccharides from several species of red and brown seaweeds forms part of a bigger program on the evaluation of extracts from Philippine seaweeds.

Results

Agar. Agar was extracted from the following species: *Gelidiella acerosa* (Forsskal) Feldmann and Hamel, *Gracilaria verrucosa* (Hudson) Papenfuss, *Gracilaria coronopifolia* (J. Agardh) and *Corallopsis sp.* after the procedure of Hong et al (11). Extraction was also carried out under alkaline conditions.

Overall, the IR spectra of the agar obtained from these seaweeds compare quite well with the spectrum of commercial agar from Sigma Chemical Company (Fig. 1A). Agar is known to be a lightly sulfated polysaccharide mixture consisting primarily of galacto and 3,6 - anhydrogalactopyranosides. Thus, it is interesting to note that although the band at 1250 cm^{-1} which is characteristic of carbohydrate sulfate esters (12) appears in each of the spectra, there appears to be a variation in the intensity of this band with algal source. The band is weaker in the spectra of commercial agar (a) and of *G. acerosa* (b) and is much stronger in the spectra of *G. verrucosa* (c),

Xth International Seaweed Symposium
Printed in Germany

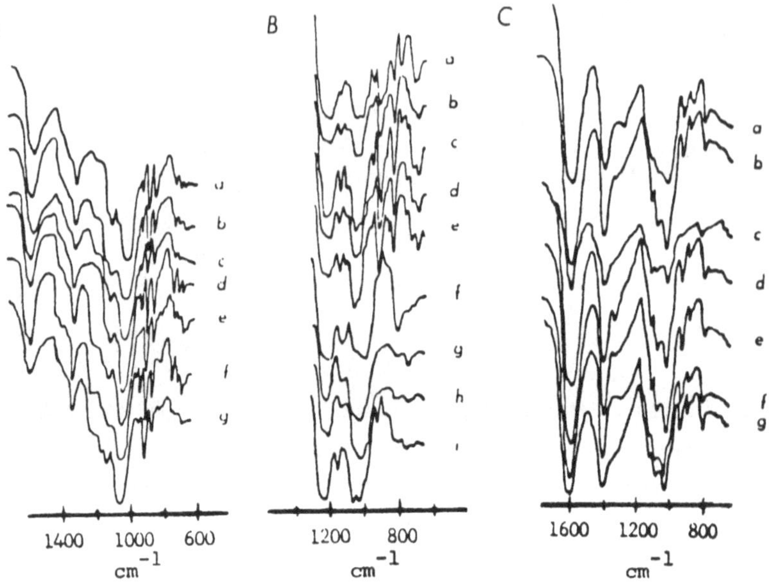

Fig. 1. IR spectra of A) agar from a. Sigma, b. G. acerosa (neutral), c. G. verrucosa (neutral), d. G. verrucosa (basic), e. G. coronopifolia (neutral), f. G. coronopifolia (basic), g. Corallopsis sp. (neutral); B) carrageenan from a. E. spinosum (basic), b. E. striatum (basic), c. Hypnea sp. (basic), d. H. cervicornis (neutral), e. H. cervicornis (basic), f. G. pistillata (Sigma), g. A. muscoides (neutral), h. A. muscoides (basic), i. A. spicifera (basic); C) alginate from a. Sigma, b. H. triquetra, c. H. clathratus, d. T. conoides, e. T. ornata, f. S. polycystum, g. S. duplicatum.

G. coronopifolia (e) and Corallopsis sp. (g), suggesting that the agar from these latter species contain a greater amount of sulfate ester which apparently decreases when agar extraction is carried out under alkaline conditions as indicated by the reduced band intensity in spectra d and f. The relatively low sulfate content (10%) of agar is further indicated in all the spectra by the very weak and poorly defined band in the region between 800 and 850 cm^{-1} where primary and secondary equatorial and axial sulfate groups absorb. The presence of the 3,6 - anhydrogalactose moiety is confirmed by the presence of a peak at 930 cm^{-1} in all the spectra. An increase in the intensity of this peak appears to accompany the above-mentioned decrease in the intensity of the peak at 1250 cm^{-1}. This is to be expected in view of the observation that in carbohydrate sulfate esters, removal of a sulfate group results in the formation of a 3,6 - anhydro-

Table 1. Physico-chemical properties of agar extracts[a)]

Seaweed	% Yield	Gel strength[b)] (g/cm^2)	Viscosity[c)] (cps)	% Sulfate	% 3,6 - AG[d)]
G. acerosa	12.6	520	108.9	2.22	32.58
G. verrucosa	20.8	428	18.5	4.90	32.58
G. coronopifolia	19.0	215	20.5	4.23	26.41
Corallopsis sp.	40.9	145	64	7.26	27.91

[a)] Values reported are for agar extracted under neutral conditions; [b)] Measured with a 0.431 in. plunger at room temperature; [c)] Of a 1.5% solution at $75^{o}C$; [d)] AG = anhydrogalactose

galactose residue (13). Table 1 lists some of the physico-chemical properties of the agar extracted from these seaweed species. Agar appears to form a stronger gel the lower its sulfate content. All gel strength values reported in this paper refer to an aqueous solution that is 2% in the hydrocolloid and 0.2% in KCl. It also appears that a high 3,6 - anhydrogalactose content as determined by the method of Yaphe and Arsenault (14) can be correlated with high gel strength. The viscosities obtained however do not appear to follow a readily decipherable pattern.

Carrageenan. Extracts from the following seaweeds: Hypnea sp., Hypnea cervicornis (J. Agardh), Acanthophora muscoides (Bory) and Acanthophora spicifera (Vahl) Boergesen were investigated and compared with kappa-carrageenan from Eucheuma-striatum (Schmitz), iota-carrageenan from Eucheuma spinosum (L.) J. Agardh and commercial lambda-carrageenan from Gigartina pistillata. The IR spectra of the carrageenans extracted from these species by methods reported earlier (10, 15) are given in Fig. 1B. H. cervicornis has been shown by other workers (8b) to contain kappa-carrageenan, a fact borne out by the strong similarity of spectra d and e to that of an authentic sample (b). The spectrum of the extract from a species also belonging to the genus Hypnea (c) is rather interesting in that it exhibits peaks characteristics of both kappa and iota-carrageenan. The peak at 805 cm^{-1} attributable to the "extra" sulfate (12) at C-2 of the anhydrogalactose unit in iota-carrageenan appears in spectrum C although at a reduced intensity along with the peaks at 850 and 930 cm^{-1}.

Table 2. Physico-chemical properties of some carrageenans

Seaweed	% Yield	Gel strength (g/cm^2)	Viscosity[e)] (cps)	% Sulfate	% 3,6 - AG[f)]
E. spinosum[a)]	27.50	36[c)]	85	31.86	19.37
E. striatum[a)]	31.96	130[d)]	230	21.32	18.49
Hypnea sp.[b)]	20.00	120[d)]	35	21.33	23.00
H. cervicornis[b)]	21.00	640[d)]	19.6	17.72	26.96
A. muscoides[b)]	14.65	non-gelling	10	21.49	5.39
A. spicifera[a)]	11.85	non-gelling	15.65	36.63	8.98

[a)] Extracted under basic conditions; [b)] Extracted under neutral conditions; [c)] Measured with a 0.845 in. plunger at room temperature; [d)] Measured with a 0.431 in. plunger at room temperature; [e)] Of a 1.5% solution at 75°C; [f)] AG = anhydrogalactose

Thus it is possible that this Hypnea species contains either a mixture of both kappa- and iota-carrageenan or a variant of either carrageenan. Indeed the sulfate content of the carrageenan extracted from this species is lower than that of the conventional iota form (Table 2). The markedly higher intensity of the peak at 930 cm^{-1} in the spectra of the Hypnea extracts (c - e) implies that the hydrocolloids from these seaweeds contain a relatively high amount of 3,6 - anhydrogalactose. An independent determination of 3,6 - anhydrogalactose bears this out (Table 2).

The extracts from both A. muscoides and A. spicifera are very similar and appear to be a type of lambda-carrageenan on the basis of their IR spectral characteristics (f - i of Fig. 1B). Spectrum i has been discussed in a previous report (10), the main points being the fairly resolved peaks at 1050 and 1075 cm^{-1} and the anhydrogalactose peak at 930 cm^{-1} which hardly shows in the spectrum of the commercial sample (f). The fairly strong band at around 800 cm^{-1} in the spectrum of commercial lambda-carrageenan (f), which is attributable to the presence of different types of sulfate ester groups, weakens and broadens in the spectra of Acanthophora extracts. Whether these spectral properties can be correlated with the data presented in Table 2 remains to be seen. The relatively high sulfate content (21-37%) of the extracts are to be expected. However, the anhydrogalactose values are atypical. Lambda-carrageenan normally contains about 35% sulfate and only trace amounts of anhydrogalactose (16). The aqueous

Table 3. Yield and viscosity of alginates

Seaweed	% Yield	Viscosity* (cps)
H. triquetra	22	77.0
H. clathratus	2.3	27
T. conoides	10.6	36
T. ornata	19.3	74.5
S. duplicatum	18.7	41.5
S. polycystum	23.1	14.5
Commercial alginate	-	6100.0

* Of a 1.5% solution at room temperature.

solutions of both Acanthophora extracts are typically non-gelling. Their viscosities (Table 2) however are lower than the 325-384 cps of a similar solution of the commercial sample.

Alginate. Sodium alginate was extracted by the method of Haug (17) from the following brown seaweeds: Hormophysa triquetra (Kutzing), Hydroclathrus clathratus (Ag.) Howe, Turbinaria conoides (J. Agardh) Kutzing, Turbinaria ornata (J. Agardh), Sargassum duplicatum and Sargassum polycystum (C. Agardh). The IR spectra of the extracts were compared with the spectrum of a commercial alginate sample (Fig. 1C). All the spectra appear to be nearly identical, characterized by the exceptionally strong absorption of the carboxylate group at 1600 cm^{-1}. The bands between 1100 and 1000 cm^{-1} appear to be better resolved in the spectra of the extracts from the aforementioned seaweeds.

The alginates extracted from the above seaweeds form non-gelling aqueous solutions which exhibit drastically reduced viscosities relative to a similar solution of the commercial alginate from Sigma (Table 3). These alginates are thus limited to low viscosity applications. The extraction conditions used are now being carefully studied to determine how they affect the physical properties of the end product and how the reported degradative effects of phenolic substances present in most brown seaweeds can be minimized (17). Preliminary chromatographic results show that these alginates contain a high proportion of guluronic acid residues.

Acknowledgment

The authors thank the Ministry of Natural Resources for their generous support of this project.

References

1. Anderson, N.S., Dolan, T.C.S., Rees, D.A.: J. Chem. Soc. Perkins I, 2173 - 2176 (1973).
2. Lawson, C.J., Rees, D.A., Stancioff, D.J., Stanley, N.F.: J. Chem. Soc. Perkins I, 2177 - 2182 (1973).
3. Mshigeni, K.E., Semesi, A.K.: Bot. Mar. 20, 233 - 237 (1977).
4. Semesi, A.K., Mshigeni, K.E.: Bot. Mar. 20, 239 - 242 (1977).
5. Mshigeni, K.E., Nzalalila, E.V.: Bot. Mar. 20, 443- 447 (1977).
6. Kim, D.H., Hendriquez, N.P.: Proc. Intl. Seaweed Symp. 9, 257 - 262 (1978).
7. Grasdalen, H., Larsen, B., Smidsrod, O.: Proc. Intl. Seaweed Symp. 9, 309 - 317 (1978).
8. Santos, G.A., Doty, M.S.: a) J. Am. Pharm. Sci. 64, 1704-1706 (1975); b) Proc. Intl. Seaweed Symp. 9, 361 - 367 (1978).
9. Whyte, J.N.C., Englar, J.R.: Proc. Intl. Seaweed Symp. 9, 437 - 443 (1978).
10. Cajipe, G.J.B., Laserna, E.C., Veroy, R.L., Luistro, A.H.: Bot. Mar. 23, 69 - 70 (1980).
11. Hong, K.C., Goldstein, M.E., Yaphe, W.: Proc. Intl. Seaweed Symp. 6, 473 - 482 (1969).
12. Anderson, N.S., Dolan, T.C.S., Penman, A., Rees, D.A., Mueller, G.P., Stancioff, D.J., Stanley, N.F.: J. Chem. Soc. (C) 602 - 606 (1968).
13. Rees, D.A.: J. Chem. Soc. 1821 - 1832 (1963).
14. Yaphe, W., Arsenault, G.P.: Anal. Biochem. 13, 143-148 (1965).
15. Laserna, E.C., Cajipe, G.J.B., Veroy, R.L., Luistro, A.H.: Kalikasan, Phil. J. Biol. 7, 110 - 116 (1978).
16. Smith, D.B., O'Neill, A.N., Perlin, A.S.: Canad. J. Chemistry 33, 1352 - 1360 (1955).
17. Haug, A. Report No. 30, Norwegian Institute of Seaweed Research, Oslo 1964.

THE TURN-OVER OF SULPHATES ON THE LAMBDA-CARRAGEENAN OF THE CELL-WALLS OF THE RED SEAWEED GIGARTINALE : CATENELLA OPUNTIA (GREV.).

G. de Lestang-Brémond & M. Quillet.
Laboratoire de Physiologie Végétale, Université Catholique de l'Ouest, B.P. 808 - 49005 Angers Cédex (France).

Introduction

The sulphated fucoidan of *Pelvetia canaliculata* (Brown Seaweed, Fucaceae) is submitted to the fast turn-over of a part of its sulphates, while immerged (Lestang and Quillet 1974, 1976). This mechanism contributes to preserving the seaweed from the intrusion of an excess of sodium coming from the ambient seawater.

In order to improve the extension of this process among marine algae, we experimented the Red Seaweed *Catenella opuntia* (Grev.), which is quite different from the Brown Algae, as well as regard to its cell-walls' composition and to its physiological behaviour.

The mucilage secreted by *Catenella opuntia* is constitued by the association of an iota and a lambda-carrageenan, *i.e.* by polygalactans bearing semi-ester sulphate groups. The cell-walls of *Catenella* are nevertheless the seat of a turn-over of its sulphates. This one slightly differs from that of the cell-walls of *Pelvetia* : by its rapidity and by the nature of the carrying nucleotide which is here methylated.

We proved too that only one of the two polysaccharides is concerned by the process : the lambda carrageenan.

A. Material and methods

<u>Material</u> : The choice of the Red Seaweed Catenella opuntia, which is small (2 cm) has been conditioned by the necessity to operate within a small vo-

Xth International Seaweed Symposium
Printed in Germany

lume of radioactive seawater, and nevertheless on a population of entire living seaweeds, numerous enough to counterbalance their individual differences.

There are not many wide natural patches of Catenella opuntia. However we were able to use over several years those living in a small island near St Jacut (Cotes du Nord, France). These seaweeds which generally settle on over hanging shadowing rocks were carefully detached by hand, and immediately thoroughly washed until free from slime and numerous organisms which live among and against them. An immersion of a few seconds within a highly diluted solution of a phosphate free detergent ("Decon", Merck) and a violent agitation during one hour in frequently renewed seawater are indispensable to obtain a clean harvest. They were carried in an ice box to the laboratory.

Those harvested samples were kept alive several weeks by a succession, similar to the moving of the sea, of seawater baths (at 15° C) and emersions on a vertical bearing in a cool and shadowed room. In these conditions, the seaweeds slowly burgeoned or emitted spores as in their natural state. Each sample was examined with a binocular magnifying glass to test its cleanliness an physiological state ; most of our experiments were carried out on tetrasporophytes.

The method used to demonstrate the reality of the turn-over of the sulfates in Catenella opuntia is the classical method of incorporation of radioactive $^{35}SO_4$ by living immersed seaweeds.

<u>$^{35}SO_4$ supply and incorporation (pulse)</u> : Identical lots of numerous seaweeds (including roughtly one hundred sprigs), as similar as possible, kept immerged until the experiment, were quickly blotted on a filter paper and weighed (750 mg). They were immerged in filtered seawater (20 ml) containing 200 to 250 uCie of $Na_2{}^{35}SO_4$, and placed on a shaking tray during increasing durations (from I to 20 min.), in the light (20 000 ergs.cm^{-2} s^{-I}) at a temperature of about 20° C.

At the end of this labeling, the seaweeds were quickly collected on a fine meshed trellis of polypropylene, squeezed out and washed within 10 seconds by 100 ml of filtered seawater, drained and again squeezed out through this tissue and projected in boiling ethanol (400 ml). There were no more than 20 seconds between the exit of the radioactive bath and the fixation.

Chase experiments : Identical lots of seaweeds (750 mg) were labeled at their maximum during 20 min., squeezed out and washed in conditions identical to above, then immerged into 400 ml of filtered seawater, agitated and renewed every 90 seconds for the first baths and after several minutes for the last ones, when the chase was reaching its end.

Chase was effected during I min. 30 to 10 minutes.

At the end of the suitable duration, the seaweeds were collected, quickly washed and fixed as above.

Extraction of the labeled polysaccharides : In order to be sure that cell-walls' polysaccharides are truly the subject of the turn-over, measurements of radioactivity were carried out on those purified polysaccharides.

To ward off the plausible objection of a labelling due to the absorption of a small radioactive molecule on the carrageenans, we effected 3 types of verifications :

- several successive extractions in water were carried out : we observed that the last polysaccharides extracted are as radioactive as the first ones.
- chromatography on a column of Sephadex G 50 (elution NaCl 2M) and electrophoresis on paper (Whatman 3 MM ; buffer pH 5,5 acetic acid, pyridine M/20 ; 2 h 30 ; 6 $V.cm^{-I}$; elution by boiling water) show that the zones containing the polysaccharides remain radioactive.

The best extractions of labeled polysaccharides were obtained in hot water at pH 7,5 (80 ml ; $KHCO_3$) in which we dipped the grounded seaweeds boiled in ethanol, contained in centrifugations pots placed in a boiling water bath, with agitation, during 45 min. After centrifugation, the clear supernatant was concentrated to half its volume, and poured into 3 volumes of ethanol containing small amounts of $CaCl_2$.

Purification and counting : The purification of the above floculats was obtained by diffusions in successive baths of 70 and 100° GL ethanol, in order to discard the free nucleotides and residual lipids which could have subsisted.

After drying under vacuum and grinding, each lot provided 40 to 50 mg of a white powder.

The radioactivity was measured by redissoving 12 mg of this powder in 3 ml of water, the counting being effected on 0,400 ml of this solution dissolved in 7,5 ml of ACS scintillating mixture (Amersham), with a Beckman LS 7000 scintillator (I).

B. Results and discussion

As well for the incorporation of sulphate and for the excretion of it, the process is fast.

A lot of experiments were carried out. One of them is represented on figure I. One can see that the maximum of radioactivity is nearly reached within 5 minutes. But nearly all of it is lost within the next 5 min., when the labelled seaweed is immersed in non radioactive seawater :

Activity for 100 mg
of polysaccharide
extracted with some
Na hexametaphosphate

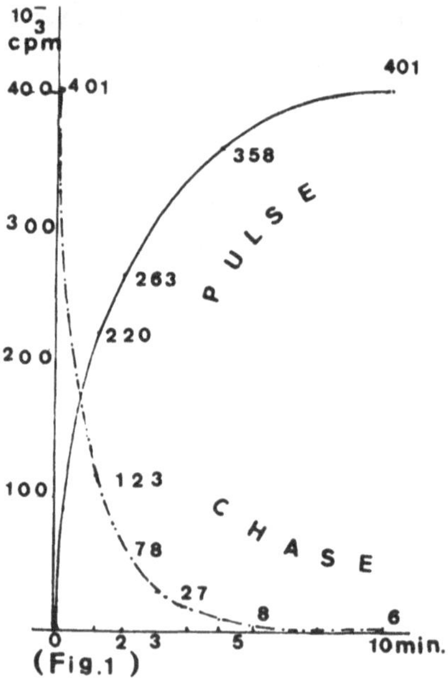

(Fig.1)

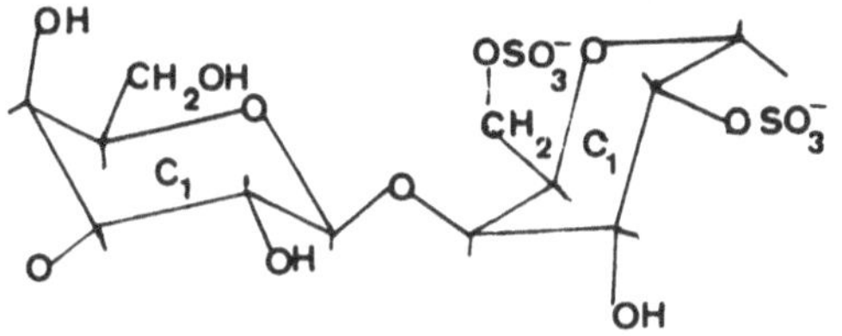

λ Carrageenan

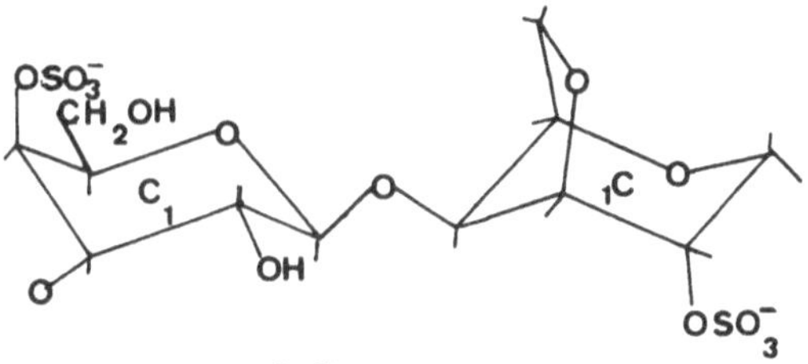

(Fig.2) i Carrageenan

Barring the experimental errors, the graphs of pulse and chase are nearly symetrical.

One can be struck, however, by the weakness of the labeling compared to the weight of carrageenan : the observation is constante in the different experimentations we tried.

One of the first explanations is the lability of the semi-ester bonds concerned. We observed that a number of them were cutt off during the purifications with ethanol we had to do.

But a second reason, more interesting may be, is that only one of the polysaccharides of the cell-wall is involved in the turn-over. It is the lambda carrageenan.

In fact, when the radioactive powders extracted (which are mixtures of iota and lambda carrageenans) are leached by KCl 0,5 M, which for the most part solubilises the lambda (Stancioff, 1965), it appears that the soluble part bears the 9/10 of the incorporated radioactivity, although it represents only I/8 of the mixture : for 100 mg of the dialysed lambda carrageenan 240.10^3 cpm ; 26.10^3 for the iota. Carrageenans' separation being never complete (obvious from I.R. spectrography) we assume that the lambda carrageenan is the main, and even exclusive, subject of the turn-over. Stereochemistry gives an explanation : the lambda carrageenan only bears equatorial $-O-SO_3$ more favorable than the axial ones to enzymic activity.

C. Conclusion

The fixation of a SO_4 ion, by a semi-ester bond, with an -OH of the lambda carrageenan, evidently requires a carrier of sulphate rich in chemical energy.

This one : the MeCDPS has been isolated (cf next paper).

From the fact that analogous turn-over of sulphates do occur in the different Classes of Algae : Red and Brown, it can be expected that it is a rather general process. Present researches in our laboratory on a Green marine Alga will allow to verify this assertion.

This turn-over seems to by, as in *Pelvetia*, a pre-regulation mechanism which reduces the amount of Na^+ moving towards the cytoplasm : the expulsed SO_4^{--} carrying with them away Na^+ ions.

This turn-over is certainly not opposed to the existence of others regulation systems (v.g. the Na/K pump located in the plasmalemma) no more to the far slower sulphation process of native polysaccharides in Golgi vesicles.

References.

de Lestang G. et Quillet M. : Physiol. vég. 12 (2), 199-227 (1974).
de Lestang-Brémond G. et Quillet M. : Physiol. Vég. 14 (2), 259-269 (1976).
Stancioff D.J. : U.S. Pat. 3, 176, 003 (1965).

Acknowledgments : to M. Le Guillanton (U.C.O.) and M. Davers (CHU) for the kind loan of their I.R. spectrograph and Beckman scintillator.

SEASONAL VARIATION OF THE CHLOROPHYLL CONTENTS OF *GRACILARIA* CULTIVATED IN TAIWAN

Chuen-Ying Liu*, Chen-Yen Wang** and Shang-Shyng Yang**

Departments of Chemistry* and Agricultural Chemistry**, National Taiwan University, Taipei, Taiwan, R.O.C.

Abstract

All *Gracilaria* samples tested contained chlorophyll a and R-phycoerythrin as the major chlorophyll and biliprotein, respectively, while the chlorophyll content was similar among the various samples, quantitative differences did exist in different seasons and locations. The chlorophyll content of *Gracilaria* had a positive statistical correlation with sunshine period, air temperature and magnesium content while agar yield, jelly strength and 3,6-anhydrogalactose content exhibited a negative correlation. Therefore, we could predict the agar quality from the chlorophyll content of *Gracilaria* cultivated in Taiwan.

Introduction

The seaweed *Gracilaria* is used commercially to supplement manure in agricultural fertilizer, as a food, and as a source of agar-agar (5). The past 22 years has been a steady growth in the cultivation of this seaweed in southern Taiwan. The use to which a particular crop of *Gracilaria* is put depends primarily on the quality of agar-agar that can be extracted. In order to develop a more quantitative measurement of agar-agar quality, undertook this study to correlate chlorophyll content with agar-agar quality. In a previous study, we were able to show that the quality of agar-agar, as measured by jelly point, jelly strength, and 3,6-anhydrogalactose content depends on the season and location of *Gracilaria* cultivation (9,10). Presently, we have extended these findings to show that chlorophyll content

Xth International Seaweed Symposium
Printed in Germany

also depends on the environmental conditions and that agar-agar quality is inversely related to chlorophyll content.

Materials and Methods

1. Seaweed: *Gracilaria verrucosa* (*G. confervoides*) were collected from the Chiayi, Tainan, Tungkong and Chiatung areas, and magnesium content, extraction of agar-agar, jelly strength and agar yield were described in previous (9).

2. Chlorophyll determination: Total chlorophyll and chlorophyll a contents were determined essentially by Arnon's methods, and the following equations were used (1): Total Chlorophyll (mg/l) $= 20.0A_{645nm} + 8.02A_{663nm}$
Chlorophyll a (mg/l) $= 12.7A_{663nm} - 2.69A_{645nm}$

3. 3,6-Anhydrogalactose determination: Two ml of 1% agar solution was mixed with 10 ml of resorcinol reagent (11) and incubated 4 minutes at room temperature, 10 minutes at $80^{o}C$ in a water bath, and finally, an additional 5 minutes at room temperature, 3,6-anhydrogalactose content was determined from the absorption at 550nm.

Results and Discussion

1. Absorption spectra of *Gracilaria* extracts: The absorption spectra of the pigments extracted from various seaweeds with 90% acetone had similar patterns which, by and large, differed only in the intensity of individual peaks (Fig.1). All spectra had peaks at 664 nm, 608 nm, 568 nm, 534 nm, 503 nm, 408 nm, and 336 nm with shoulders at 484 nm, 456 nm, and 386 nm. The peak at 664 nm, 608 nm, 568 nm, 534 nm, 503 nm and 408 nm indicates that chlorophyll a is the major chlorophyll in seaweed cultivated in Taiwan. Chlorophylls b,c and d appear to absent or in low quantity since peaks at their absorption maxima--- respectively, 642 nm, 628 nm and 696 nm --- were not observed in the spectra. We have assigned the peak at 495 nm to R-phycoerythrin. This appears to

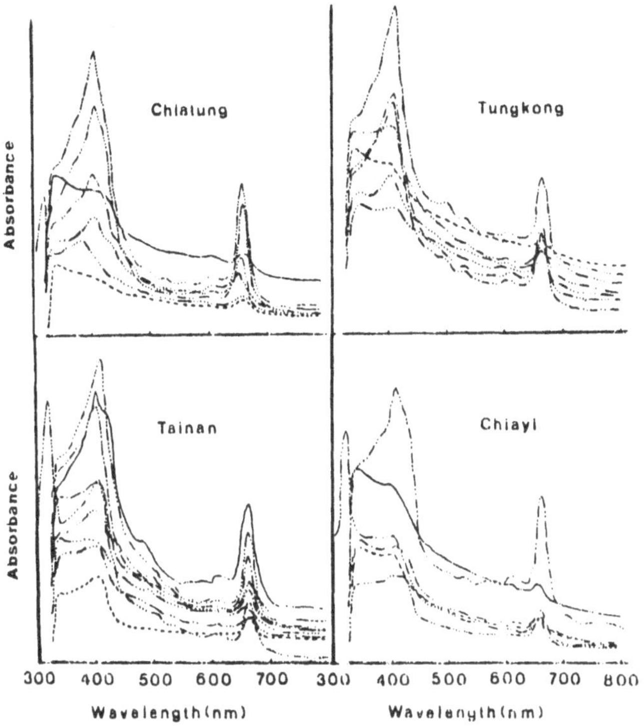

Fig. 1. Absorption spectra of chlorophyll extracted from *Gracilaria*.

———— : August 1977; ········ : September 1977;
—·····—····· : October 1977; —·—·— : November 1977;
—··—··—·· : December 1977; —···—··— : February 1978;
—····—···· : March 1978; —······—······ : May 1978;
—·······— : July 1978.

be the major biliprotein since we were unable to detect significant quantities of the other phycoerythrins and phycocyanins as indicated by the low absorption at 540-560 nm, 450 nm and 600 nm. We conclude, therefore, that in *Gracilaria* grown in Taiwan, chlorophyll a is the major chlorophyll and R-phycoerythrin is the major biliprotein.

2. Chlorophyll and magnesium content: The seasonal variation of the total chorophyll content found in *Gracilaria* grown in the Chiayi, Tainan, Tungkong and Chiatung areas is presented in Fig. 2a. Values varied from 29 to 200 mg/l with maximal values found in samples harvested October 1977 in the Tainan, Chiatung and Tungkong areas and March 1978 in the Chiayi area. In *Gracilaria* harvested in Chiatung and Tungkong

areas, the chlorophyll content fluctuated rapidly, showing peaks during October 1977, December 1977, March 1978, and July 1978. The chlorophyll content of samples harvested in the Tainan area increased from July 1977 to a peak in October 1977, decreased to February 1978, then increased through to the end of the observation period in August 1978. Those samples harvested in the Chiayi area exhibited peaks of chlorophyll content during October 1977 and March 1978 with a minimum during January 1978.

The seasonal variation of chlorophyll a content (Fig.2b) was similar to that of the total chlorophyll content. Values ranged from 15 to 70 mg/l. The discrepancy in the values for total chlorophyll and those for chlorophyll a are due mainly to the calculation method. Inference from biliproteins introduces an over estimation of total chlorophyll. Since chlorophyll a is the major chlorophyll (see above) and the ratio of chlorophyll a content to total chlorophyll content remains fairly constant, ranging from 40% to 60% (Fig. 2c). We used total chlorophyll content for the correlation studies. Although this value is perhaps less accurate than the chlorophyll a content, total chlorophyll is the more simply derived value and the one expected to be used in the marketplace. So in subsequent correlations this value will be emphasized.

The seasonal variation of the magnesium content (Fig. 2d) was, in general, similar to that of the chlorophyll content with one importent difference: The February peak of chlorophyll content was not reflected in magnesium content. This phenomenon is most probably due to the reduced magnesium concentration in the *Gracilaria* ponds caused by extensive rains during this period.

3. Correlation among agar-agar quality, chlorophyll content, and environmental conditions: Agar-agar quality was assessed by measuring the jelly strength (Fig. 2f) and 3,6-anhydrogalactose content (Fig. 2g) of agar extracted from *Gracilaria* obtained from the various areas during the study period. The chlorophyll content of *Gracilaria* had a positive statistical correlation with sunshine period, air temperature and magnesium content, while agar yield, jelly strength and 3,6-anhydroga-

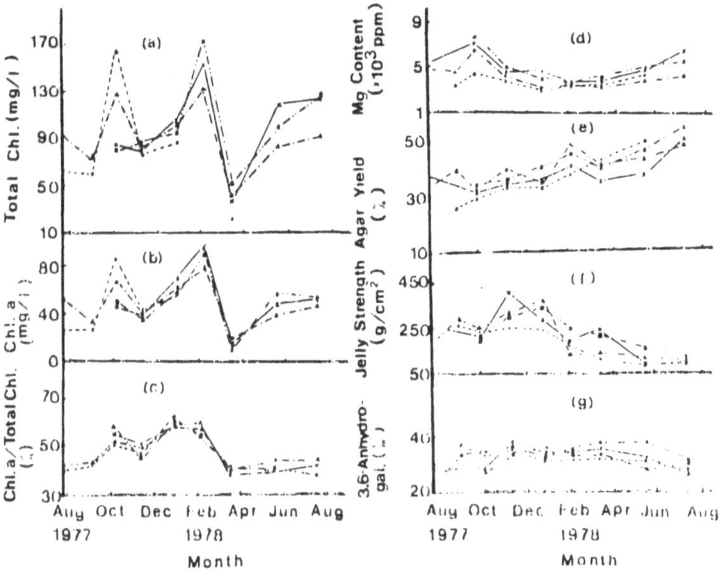

Fig. 2. Magnesium, total chlorophyll and chlorophyll a contents of *Gracilaria* and quality of agar-agar produced.

(a) Magnesium content.
(b) Total chlorophyll content.
(c) Chlorophyll a content.
(d) Ratio of chlorophyll a content to total chlorophyll content.
(e) Agar yield.
(f) Jelly strength.
(g) 3,6-Anhydrogalactose content.

•---------• : Chiayi; ▲—·—·—▲ : Tainan;
x———x : Chiatung; ■—··—··—■ : Tungkong.

lactose content exhibited a negative correlation.

Chlorophyll content of *Gracilaria* depend on environmental conditions (2-4,6-8). In this study, it indicated that sunshine period, growth temperature, growth rate, agar yield, quality of agar-agar and chlorophyll content had a very close relationship among them. Whenever the chlorophyll content was known, the quality and yield of agar-agar could be predicted.

Acknowledgment

The authors express their gratitude to Drs. H.H. Wang and Y.M. Chiang for helpful discussion, and National Science Council of Republic of China for financial support.

References

1. Arnon, D.I.: Pl. Physiol. Lancaster, 24, 1-15 (1949).
2. Beale, S.I., Appleman, D.: Pl. Physiol. Lancaster, 47, 230-235 (1971).
3. Bogorad, L.: Physiology and Biochemistry of Algae, 385-408, Ed. by R.A. Lewin, Academic Press, New York (1962).
4. Brown, T.E., Richardson, F.L.: J. Phycol., 4, 38-54 (1968).
5. Chen, G.C.: Joint Commission on Rural Reconstruction Fisheries Series, 25B, 56-67 (1977).
6. Halldal, P., French, C.S.: Pl. Physiol. Lancaster, 33 , 249-252 (1968).
7. Meek, J.C.: Algal Physiology and Biochemistry, 161-175, Ed. by W.D.P. Stewart, Blackwell Scientific Publications, Oxford London Edienburgh Melbourne (1974).
8. O'Kelley, J.C.: Algal Physiology and Biochemistry, 610-635, Ed. by W. D.P. Stewart, Blackwell Scientific Publications, Oxford London Edienburgh Melbourne (1974).
9. Wang, C.Y., Yang, S.S.: Proc. Natl. Sci. Counc. ROC., 4, 78-86 (1980).
10. Yang, S.S.: ROC-US Symposium on Economic Seaweed, June 5-7, 1978, Guam, 16-17 (1978).
11. Yaphe, W.: Analyt. Chem., 32, 1327-1330 (1960).

A RAPID INFRA-RED CARBON DIOXIDE ANALYSIS SCREENING TECHNIQUE FOR PREDICTING GROWTH AND PRODUCTIVITY OF MARINE ALGAE.*

N. D. H. Lloyd
Genu Products Canada Limited, Halifax, Canada B3K 5M1

J. L. McLachlan
Atlantic Research Laboratory, National Research Council Canada
Halifax, Canada B3H 3Z1

R. G. S. Bidwell
RR #1, Wallace, Canada B0K 1Y0

Introduction

The selection of fast-growing strains of marine algae for cultivation has in the past necessitated time and space consuming productivity measurements of strains under cultivation conditions. These growth measurements may take at least several months and require relatively large quantities of material (several kg), which have to be grown up from a single plant prior to the experiments (1). Such productivity measurements under cultivation conditions are essential for final testing of strains before putting them into large-scale production, but there is a real need for a rapid technique for the initial screening of large numbers of plants to assess their potential.

We have developed a screening technique that takes only a few minutes per plant, and requires only about one gram of material. This technique is predictive of the annual productivity of the strain. It also has several other useful applications for the cultivation of marine plants.

*NRCC #18250

Xth International Seaweed Symposium
Printed in Germany

Methods

The technique involves the measurement of the rate of photosynthesis (PS) of the material using an Infrared Gas Analysis system, but with the volume of water around the plants reduced to a minimum by placing them in a stream of water-saturated air rather than in water (2,3,4). This eliminates the two main problems associated with measurements of PS in water:

1. The long equilibrium times involved, which are usually several hours per measurement.

2. The need to maintain pH levels absolutely constant in solutions. Any change in pH will result in CO_2 influx or efflux rates that may be larger than the PS rates being measured. In our system, the very small quantity of water present will not emit or absorb CO_2 at a rate large enough to affect the measurement of the PS rate.

The gas analysis system is shown in Fig. 1. The plant material is suspended on nylon mesh in a water-saturated gas stream of known CO_2 concentration, and the PS rate is calculated from the difference in CO_2 concentration before and after passing over the plants, measured differentially by an Analytical Development Company Series 225 Infrared Gas Analyser. An equilibrium PS rate is obtained in about five minutes, and all the algae tested so far will maintain this rate for 20-30 minutes before showing signs of desiccation.

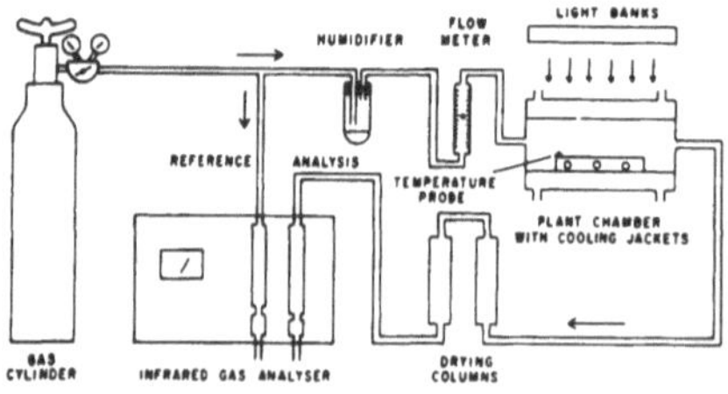

Fig. 1 The gas analysis system.

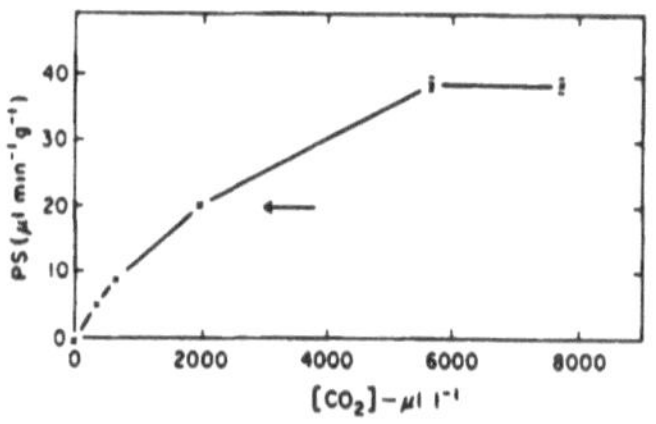

Fig. 2 CO_2 response of light - saturated PS in *Chondrus crispus*.

The selection of the correct CO_2 concentration is of key importance to our technique. The PS rate depends on the CO_2 and HCO_3^- concentration in solution. These concentrations, and the pH, must be maintained at normal levels (pH 8, 2mM HCO_3^-) in order to maintain natural and steady PS rates. This is achieved by adding CO_2 as rapidly as it is removed by PS. The rate of solution of CO_2 from air concentrations of CO_2 is not fast enough to maintain normal CO_2 and HCO_3^- concentrations in solution. For this reason elevated CO_2 concentrations in the gas stream are required. Fig. 2 shows the CO_2 response of light-saturated PS in Chondrus crispus. A CO_2 concentration of about 2000 $\mu\ell\ \ell^{-1}$ will maintain PS rates of C. crispus the same as found in seawater solutions having a pH of 8 and bicarbonate concentration of 2mM, but the exact concentration will vary with experimental conditions and algal material.

Results and Discussion

We routinely make two kinds of measurements on our cultivation material, PS capacity (PS_{CAP}) and PS rates at low light levels (PS_{LL}).

PS_{CAP} measurements are made at light saturation and represent the maximum PS rate attainable under optimum conditions. They are used to monitor the health of the plants, and to study the effects of different treatments such as herbicides, nutrients, or growth conditions. Two examples are given here:

1. Seasonal Variation in PS_{CAP}. Fig. 3 shows weekly-averaged PS_{CAP} measurements made at 15° on samples of C. crispus during tank cultivation. There is a very obvious summer decline which recovers during the winter. We now know that this seasonal fluctuation is not apparent in measurements made at 25°, indicating that some temperature-dependent step in the photosynthetic cycle declines in activity during the summer.

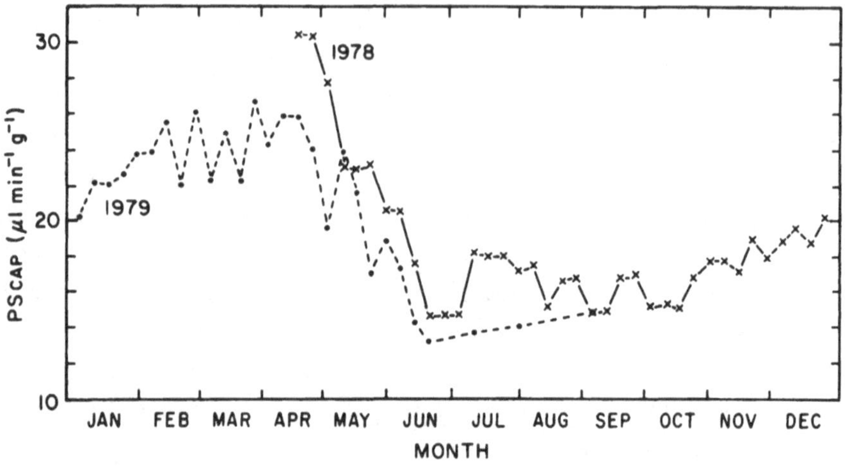

Fig. 3 Seasonal variation in PS_{CAP} at 15° in C. crispus from cultivation tanks during 1978 and 1979.

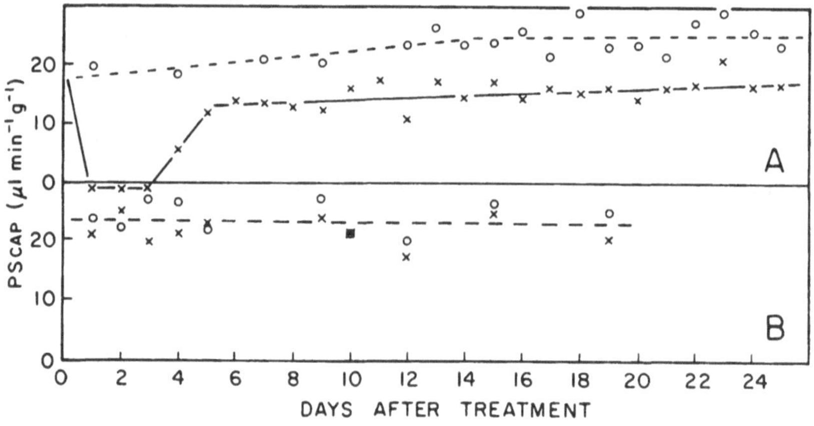

Fig. 4 PS_{CAP} following treatment of C. crispus with herbicides A and B. X = Treated O = Control.

2. Effect of Herbicides on PS_{CAP}. Fig. 4A shows the PS_{CAP} of C. crispus following the application of a herbicide (A) to attempt to control epiphytes. The data show that the PS_{CAP} is depressed to negative values for the first three days, and remains below control levels for a month. This herbicide was not considered suitable for routine use. Fig. 4B shows the PS_{CAP} following the application of

another herbicide (B). This herbicide had no detectable effect on the PS_{CAP} of C. crispus, and would clearly be the more suitable herbicide to use, assuming equal herbicidal activity of A and B.

PS_{LL} measurements are used to predict the annual productivity under cultivation conditions of a strain. The light level at which the measurements are made is similar to that at which individual plant pieces in the growth cabinet will show the same percentage growth rate as plants in outdoor tanks. At this light level the PS rate is directly related to the growth rate.

In order for this technique to be useful, two criteria must be satisfied:

1. The measurements must predict the annual productivity under cultivation conditions.

2. The measurements must be able to be made year-round, or at least during a substantial part of the year.

A year-long study of 6 strains of C. crispus showed that both these criteria are met:

Fig. 5 shows the correlation between the PS_{LL} measurements (mean of all months) and the production of total solids over the period December to September (10 months).

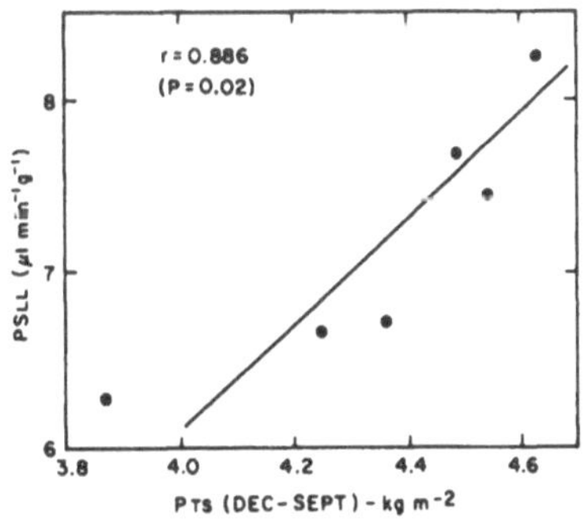

Fig. 5 The correlation between PS_{LL} values and the production of total solids in 6 strains of C. crispus.

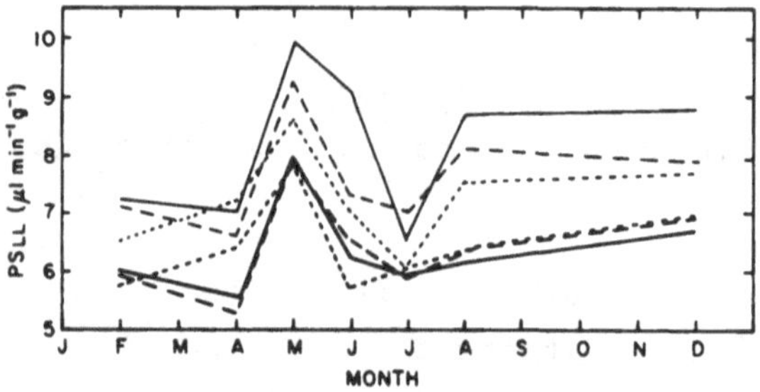

Fig. 6 Annual variation in PS_{LL} in 6 strains of *C. crispus*.

The correlation is highly significant ($p = 0.02$) and indicates that the PS_{LL} measurements do predict annual productivity with useful accuracy.

Fig. 6 shows the PS_{LL} values for these strains over most of the year. There is some variation in rate, but the order of ranking of the strains varies extremely little. This indicates that the measurements can be made at any time of year and still predict the annual productivity.

Conclusions

We have developed a rapid technique with many useful applications for the cultivation of marine algae. It can be used to monitor the health of the plants, to study the effects of various environmental or culture conditions, and to screen for fast-growing strains.

References

1. Waaland, J.R.: Proc. Int. Seaweed Symp. 9, 241-247 (1978).
2. Dodd, W.A., Bidwell, R.G.S.: Nature 234, 45-47 (1971).
3. Bidwell, R.G.S.: Can. J. Bot. 55, 809-818 (1977).
4. Lloyd, N.D.H., Canvin, D.T., Bristow, J.M.: Can. J. Bot. 55, 3001-3005 (1977).

MICRONUTRIENT UPTAKE BY JUVENILE MACROCYSTIS PYRIFERA (L.) C. Ag.

S. Manley and W. North

California Institute of Technology
Kerckhoff Marine Laboratory
Corona del Mar, California, USA 92625

Introduction

This laboratory has been engaged in research to delineate the physical and chemical parameters needed for the maximization of growth by *Macrocystis pyrifera*. Micronutrient requirements of *M. pyrifera* gametophytes (1) and the effect of trace nutrient additions to cultures of sporophytes have been reported (2). In this paper, we report some uptake parameters for iodide, iron, zinc, cobalt, and manganese by blade tissue of juvenile *Macrocystis* sporophytes in an artificially defined seawater.

Materials and Methods

Culturing. Juvenile sporophytes of *Macrocystis pyrifera* (L.) C. Ag. (5-8 cm in length) cultured from spores in flowing seawater were placed in offshore surface seawater renewed daily and supplemented with 20 uM NH_4Cl and 2 uM K_2HPO_4. The water was vigorously recirculated at 15°C in a 40 l Plexiglas aquarium. Plants were cultured up to their first secondary division after which they were discarded. Light intensity was maintained at 3.0×10^4 ergs $cm^{-2}sec^{-1}$ (cool white fluorescent lamps). Discs, diameter of 1.01 cm and total tissue area of 1.602 cm^2, were cut from blades and placed in batch culture for 16 hrs at 15°C. Light intensity was 9×10^3 ergs $cm^{-2}sec^{-1}$. Discs were then used for uptake experiments.

Xth International Seaweed Symposium
Printed in Germany

Artificial seawater. The artificial seawater used for preconditioning tissue discs and for the uptake experiments was Morel's (3) Aquil, modified by Kuwabara, as an optimal medium for *M. pyrifera* gametophytes (1).

Radionuclides. The radionuclides obtained from ICN (Irvine, CA) were: 1) I-125 as NaI in 0.05 N NaOH, carrier free (cf) 90-100% isotopic enrichment (IE), 2) Zn-65 as $ZnCl_2$ in 0.5 N HCl, cf 70% IE, 3) Ni-63 as $NiCl_2$ in 0.5 N HCl, specific activity (sp act) 11.2 mCi/mg Ni, 4) Fe-59 as $FeCl_3$ in 0.5 N HCl, sp act 13.6 mCi/mg Fe, 5) Co-57 as $CoCl_2$ in 0.5 N HCl, cf 70% IE, 6) Mn-54 as $MnCl_2$ in 1.0 N HCl, cf 70% IE.

Uptake experiments. All containers (linear polyethylene) used to store Aquil solutions were thoroughly cleaned to prevent contamination from leaching. Handling of cf solutions was done as to prevent loss from adsorption. All handling of tissue was with Teflon forceps. Polystyrene beakers were used for uptake experiments. A tissue disc, from the bubbling culture, was placed for 5 min in 4 ml Aquil devoid of the micronutrient being studied, then dipped into 40 ml identical Aquil, thence into 4 ml of stirring Aquil containing the radioisotope (pH adjusted to 8.1) at 10-12°C. Surface illumination was 1.3 x 10^4 $ergs \cdot cm^{-2} \cdot sec^{-1}$. After the appropriate uptake interval, the tissue was dipped in 4 ml wash solution, placed in a separate 4 ml stirred wash solution (4-5°C) for a predetermined time, and then counted. The following wash solutions and times were used to remove >90% free-space micronutrients: iodide, $[I] = 1 \times 10^{-6}M$ in Aquil, 10 mins; iron, $1 \times 10^{-5}M = [Fe^{+3}$-EDDHA] (ethylenediamine-di-o-hydroxyphenylacetic acid) in Aquil, 30 mins; nickel, zinc, cobalt, manganese, $1 \times 10^{-5}M = [Pb^{+2}]$ in Aquil, 20 mins. Precautions were taken to minimize trace metal contamination. Analysis by inductively coupled plasma (Vetter Research, Costa Mesa, CA) of the uptake solution after a mock experiment (minus tissue, micronutrient) revealed no contamination. Adsorption of micronutrients on the surface of the beaker during uptake was negligible. Microscopic examination revealed no algal or fungal contamination of the blades. The stirring rates were well above water motion saturation as determined for iodide. Net uptake (after free-space wash) as a function of time was determined; 10 mins was well within the linear range for all micronutrients studied. The effect of darkness, DNP (2,4-dinitrophenol), and DCMU (3-(3,4-dichlorophenyl)-1, 1-dimethylurea) on the uptake of the micronutrients was determined. Tissue discs were preconditioned in DNP and DCMU bubbling Aquil culture for 30 mins and/or 1 hr before uptake experiments. Discs were cut 3, 6, and 12 cm from the base of the blade to determine if uptake was uniform throughout the blade. Effect of micronutrient concentration on uptake was also determined ([trace metal] to [EDTA] ratio kept constant). All experiments were conducted with Aquil containing both labeled and/or unlabeled micronutrient.

Tissue preparation/counting procedure. The tissue discs were held for 12-15 hrs in a capped borosilicate glass vial with 3 ml of NCS tissue solubilizer (Amersham, Arlington Heights, IL) in a heated water bath at 50°C. A thin, transparent, disc-shaped cell wall debris remained thereafter. Samples were bleached by adding 1 ml of 10% w/w benzoyl peroxide in toluene and by exposure to direct sunlight for 5-7 hrs and were heated in a water bath (35-40°C) and dark adapted for 24 hrs to eliminate chemiluminescence and phosphorescence. Ten ml of OCS scintillation cocktail (Amersham) was then added and the vial counted. Quench correction for tissue samples was determined by production of quenched standards. A known amount of radionuclide solution was placed on top of a series of 10 discs in vials, to which 3 ml of NCS was added after 30 mins. The tissue was then treated as described above. Average maximum counting efficiencies were: I-125, 80±3%; Zn-65, 15.6±0.9%; Ni-63, 81±7.5%; Fe-59, 85±2%; Co-57, 70±3%; Mn-54, 26.6±1.8%. Solutions were then quenched and counting efficiency determined by the external standard-channels ratio method (ESCR). This procedure is an approximation since ESCR measures the solution efficiency and not the degree of quenching occurring on the undissolved remains (solid phase) if radioactive ions were indeed at all bound. Counting efficiencies for a given radionuclide varied over a very narrow range, no larger than 5%. Radioactive solutions were counted by liquid scintillation in Aquasol-2 cocktail (New England Nuclear, Boston, MA). Liquid samples quench correction was determined for each radionuclide by ESCR. The maximum counting efficiencies obtained were I-125, 92%, Zn-65, 40.2%; Ni-63, 90.0%; Fe-59, 88.8%; Co-57, 84.6%; Mn-54, 26.0%.

Results and Discussion

Free-space uptake (FSU), primarily due to the cation exchange properties of the cell wall (largely alginic acid and sulfated polysaccharides) (4), represents a significant proportion of the cationic gross uptake after 10 mins (Table 1). Differences in percent FSU amongst the cations primarily arose from net uptake rates since binding affinities of Co, Zn, and Mn were very similar (5). Uptake of the nonmicronutrient Ni was overwhelmingly from cell wall binding; 95% at 100 nM (net uptake = 50 $f\text{-moles}\cdot cm^{-2}\cdot min^{-1}$). Ascophyllum nodosum similarly takes up Sr^{+2} (4). Iron was less readily exchangeable ($[Ca^{+2}]$ at 25 uM did not remove all Fe^{+3} in <40 mins). The slow exchange rate may reflect a greater affinity for Fe^{+3} by the ion exchange sites or be related to cellular uptake mechanism. Iodide exhibited significant FSU which was exchangeable and not considered water-free space. FSU for iodide was due to anion exchange sites, either released from cells ruptured during cutting or from glycoproteins associated with the cell wall. The uptake velocity of Co, Zn, Mn, and I was a function of distance from the base of the blade which either may reflect a physiological difference from cell age (rapidly dividing near base) or perhaps a slight increase in blade thickness. An increase in I^- uptake rate at the blade tip was not explained. Iodide uptake by M. pyrifera was a light independent, active

Table 1. Uptake Parameters for Micronutrients
a - after 10 min uptake period @ 10 nM for Zn, Co & Mn.
b - numerically indicated distance (cm) along the blades from the stipe. c - @ 50 uM, see text. L = Light D = Dark.

Micronutrient		I	Fe	Zn	Co	Mn
Conc in Aquil (nM)		100	350	170	70	30
Major species (%)		I^- (100)	FeEDTA (100)	ZnEDTA (100)	CoEDTA (99)	MnEDTA (65)
[a]FSU, % gross		8	19	20	79	76
Net uptake velocity (f-moles·cm-2·min-1)	[b]3L	1100±69	1001±106	66.8±6.2	21.5±8.7	1.72±0.32
	6L	722±24 (100%)	1120±62 (100%)	50.9±4.2 (100%)	2.8±0.2 (100%)	1.29±0.22 (100%)
	12L	1009±59	1090±95	51.1±4.5	4.7±0.6	1.37±0.39
	6D	722±13 (100%)	1170±7 (104%)	41.9±3.9 (82%)	1.8±0.2 (64%)	1.06±0.07 (82%)
	[c]DNP	378±9 (52%)	280±34 (24.9%)	40.2±2.1 (79%)	1.8±0.3 (64%)	1.08±0.02 (84%)
	[c]DCMU	-	701±13 (62.2%)	47.7±6.2 (94%)	1.9±0.2 (68%)	1.32±0.25 (102%)

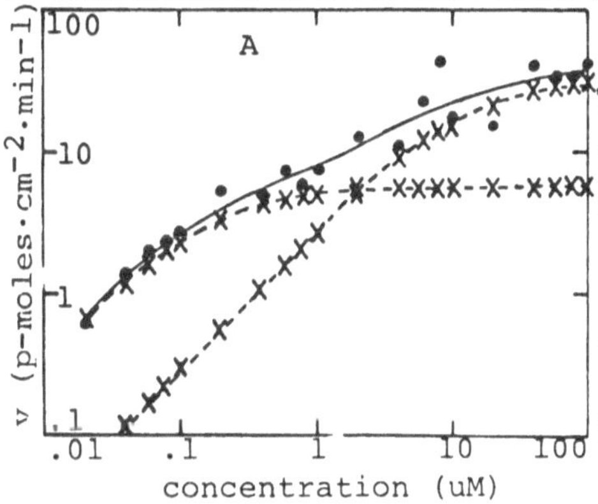

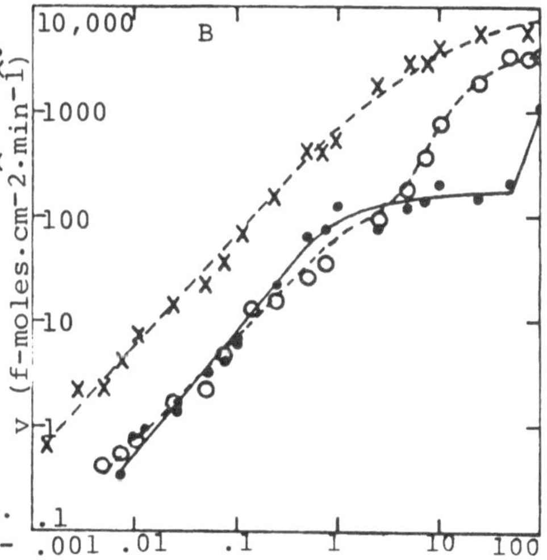

Fig. 1. Uptake velocity vs concentration. a) iodide, ·—· data; x---x, derived by successive approximation, (8) K_{mII} = 0.16 uM, K_{mI} = 11.3 uM, V_{maxI} = 33.3 p-moles·cm^{-2}·min^{-1}, V_{maxII} = 5.65 p-moles·cm^{-2}·min^{-1}. b) Zn^{+2}, x---x, K_m = 6.5 uM, V_{max} = 4.44 p-moles·cm^{-2} min^{-1}; Co^{+2}, ·—·, K_m = 1.05uM, V_{max} = 0.13 p-moles·cm^{-2}·min^{-1}; Mn^{+2}, o---o, K_m = 39.6 uM, V_{max} = 4.95 p-moles·cm^{-2}·min^{-1}.

process, not involving iodate and was similar to other algae in this regard (5,6). Analysis of v vs [I^-] (Fig. 1a) revealed a clear deviation from Michaelis-Menten kinetics, displaying biphasic uptake that may, but not necessarily, reflect the presence of two separate carriers (Fig. 1a). A K_m = 4.3 uM was reported for Fucus ceraniodes although the study was incapable of detecting biphasic uptake (5). Iron uptake was active and light independent. A 1 hr preconditioning in DNP and DCMU decreased the rate significantly but may have caused general plant weakening and thus lower uptake. The decrease was, however, dramatic and did not parallel lack of light. Preliminary experiments supported the mechanism of a ferric to ferrous reduction prior to uptake (Manley, unpublished). It is questionable whether Co, Mn, and Zn uptake was active since effects of DNP and DCMU were not dramatic and may represent general plant weakening. Uptake was light-stimulated and carrier-mediated, with a biphasic appearance for Co and Mn (Fig. 1b). A. nodosum displayed a K_m for zinc of ~3 uM (4) while Laurencia corallopsis yielded a K_m for cobalt of 0.17 uM (7).

References

1. Kuwabara, J.S.: J. Phycol. (in press - 1980).
2. North, W.J., Wheeler, P.A.: in Proc. of the IXth Inter. Seaweed Symp. A. Jensen, J.R. Stein eds., 67-78. Science Press, Princeton (1979).
3. Morel, F.M.M., Rueter, J.G., Anderson, D.M., Guillard, R.R.L.: J. Phycol., 135-141 (1979).
4. Skipnes, O., Roald, T., Haug, A.: Physiol. Plant. 34, 314-320 (1975).
5. Klemperer, H.G.: Biochem. J., 67, 381-390 (1957).
6. Shaw, T.E.: in Physiology and Biochemistry of Algae. R.A. Lewin ed., 247-251. Academic Press, N.Y. (1962).
7. Bunt, J.S.: J. Phycol. 6, 339-343 (1970).
8. Spears, G., Sneyd, J.G.T., Loten, E.G.: Biochem. J., 125, 1149-1154 (1971).

CARRAGEENANS OF SPECIES IN THE GENUS PHYLLOPHORA

Esther L. McCandless, John A. West and Catherine M. Vollmer.

Department of Biology, McMaster University, Hamilton, Ont., Canada and Department of Botany, University of California, Berkeley, Ca, U.S.A.

Introduction

A number of red algae have been identified as potential sources of the gelling galactans, κ-or-ι-carrageenans. Analysis of carrageenans found in 16 spp. in the Gigartinaceae and 19 spp. in the Phyllophoraceae indicates that the usual pattern in these families is biochemical alternation of generations with the gelling carrageenans restricted to gametophytes as reported first in *Chondrus crispus* (1). The present paper describes the carrageenans in 3 spp. of the genus *Phyllophora* including *P. crispa* (formerly *P. nervosa*), the source of so-called "Russian agar".

Materials and Methods

Specimens of *Phyllophora crispa* were supplied by M. Guiry and J. A. West, *P. pseudoceranoides* by J. Fiore, S. Schatz, J. Sears, and C. Bird, and *P. truncata* by J. Fiore and S. Schatz. With one exception of a presumptive tetrasporic culture derived from carpospores of *P. crispa* (JAW culture 1465), these were field specimens. (The culture has not reproduced, hence its phase is not proven.) Carrageenans were extracted, isolated, and characterized by chemical analysis, infrared spectroscopy, and immunochemistry as described previously (1,2). Antibodies were raised in goats by injecting *Chondrus crispus* κ-or λ- carrageenans or λ- carrageenan from *Petrocelis midden-*

Xth International Seaweed Symposium
Printed in Germany

dorfii (3). The anti-κ-antibody reacted also with ι- carrageenan (3). Fresh specimens of P. truncata were also fixed with Karnovsky's fixative followed by OsO_4 and were imbedded in Spurr's resin, as described by Gordon-Mills and McCandless (4). Thin sections were treated with specific antibody followed by FITC labelled fluorescent antibody to goat serum and examined under epiillumination in a Leitz fluorescence microscope.

Results and Discussion

Chemical analysis and infrared spectroscopy. Fifteen samples of Phyllophora spp. were extracted, 2 of P. crispa (1 ♂ and 1 presumed tetrasporic), 9 of P. pseudoceranoides (2 ♀, 2 tetrasporic and 5 vegetative), and 6 of P. truncata (3 ♀ and 3 trimmed nemathecia). Carrageenan content of these samples was 37.1± 8.6% (standard deviation) of dry defatted algal weight. Anhydrogalactose (AG) levels were slightly lower than in Chondrus crispus (1), 19.2± 5.2% of carrageenan weights (12 samples). One exceptionally low value, 5%, was obtained from nemathecia of P. truncata; tetrasporic P. pseudoceranoides and the other P. truncata nemathecial samples did not give values significantly different from those of known gametophytes. The mean sulphate value for all (13) samples analysed was 18.9± 8.3% with slightly higher values in tetrasporic material. The differences were certainly not so great as those between the two phases in C. crispus or other members of the Gigartinaceae. There were insufficient samples of known sporophytes and gametophytes to justify statistical analysis.

Infrared spectra of all carrageenans from P. pseudoceranoides and P. truncata showed absorption peaks at 805, 845, and 930 cm^{-1}, attributed to 3,6-AG-2-SO_4, Gal-4-SO_4, and 3,6-AG, respectively (Fig. 1). The spectra strongly resembled those of ι- carrageenan from Eucheuma spinosum.

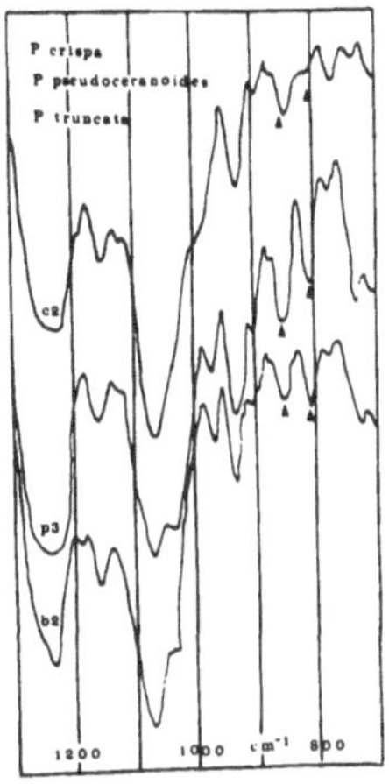

Fig. 1

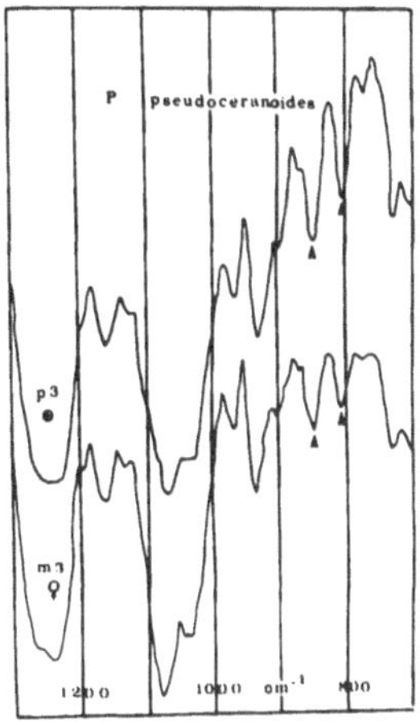

Fig. 2

Fig. 1, IR spectra of Phyllophora carrageenans, top to bottom, P. crispa ♂, P. pseudoceranoides ⊕, P. truncata ♀. Fig. 2, IR spectra of Phyllophora pseudoceranoides carrageenans, ⊕ and ♀ . Arrows indicate peaks at 805, 845, and 930 cm^{-1}.

On the basis of IR spectroscopy, no distinct differences between sporophyte- and gametophyte-derived carrageenans could be claimed (Fig. 2). The spectra of P. crispa were distinctive, entirely lacking the 805 cm^{-1} peak (Fig. 1). This is not unexpected in view of the demonstration by Yarotsky et al. (5) that this carrageenan contains 6-SO_4 on the 3-linked Gal, and less 4-SO_4 than ι-or κ- carrageenan.

Immunological evidence. Agarose immunodiffusion plates were used to test the precipitability of carrageenans from different phases of the three Phyllophora spp. Two different anti-λ-carrageenan preparations reacted with P. truncata nemathecia but not with P. truncata stipe (♀) from which nemathecia had been removed. Carrageenans from P. pseudoceranoides sporophyte reacted, as did a vegetative specimen. Precipitin lines of those Phyllophora carrageenans which reacted and of C. crispus λ-carrageenan merged, suggesting identical reactive groups within the different preparations (Fig. 3, left). All samples tested, from ♀ and sporophytic P. pseudoceranoides, nemathecia and stipe tissue (♀) of P. truncata, and male P. crispa reacted with anti-κ- carrageenan (Fig. 3, right).

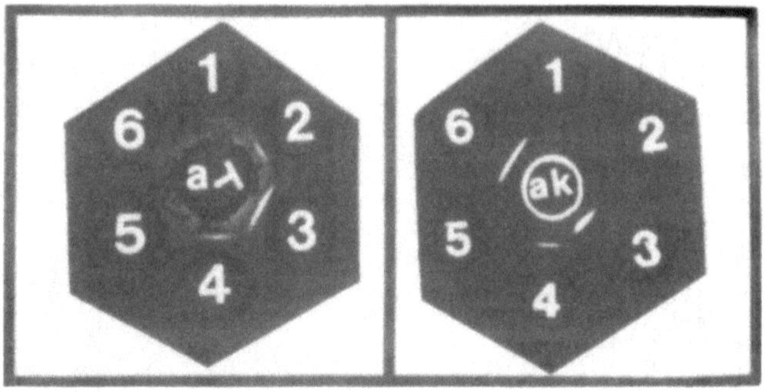

Fig. 3. Immunodiffusion plates. Left anti-λ-carrageenan in centre well, (1) P.p. vegetative, (2) C. crispus λ-carrageenan, (3) P.t. nemathecium, (4) P.p. tetrasporic, (5) P.t. stipe, (6) C. crispus κ-carrageenan. Right, anti-κ-(ι)-carrageenan in center well, (1) P.t. nemathecium (2) C. crispus λ-carrageenan, (3) P.p. ♀ , (4) P.c. ♂ , (5) P.t. stipe, (6) C. crispus κ-carrageenan. (Some lines are faint and may not reproduce well).

However, the precipitin line from C. crispus κ-carrageenan spurred rather than merged with those from P. pseudoceranoides and P. truncata, presumably because these spp. make ι-rather than κ-carrageenan. Similarly, the line from P. crispa carrageenan spurred rather than merged with adjacent lines, thus was shown to differ immunochemically from the other two spp., confirming other evidence.

Fluorescent antibody studies carried out on P. truncata nemathacia demonstrated the presence of both ι-type and λ-carrageenan in cell walls and matrix. Fig. 4 shows the distribution of these two polymers, in similar locations in a nemathecium. The fluorescence associated with ι-carrageenan (Fig. 4, a and b) was distinctly brighter than that associated with λ-carrageenan (Fig. 4, c and d); the latter photos were prepared with longer development time to enhance the image. These results agree with evidence from immunodiffusion and infrared spectroscopy, all showing a predominance of ι-carrageenan. Thus far we have been unable to get a good preparation of stipe tissue of P. truncata, so the question

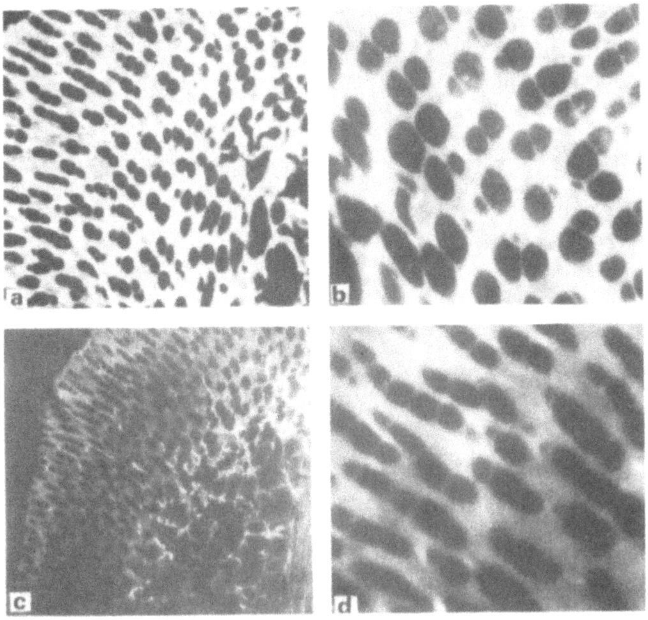

Fig. 4. Phyllophora truncata nemathecium (a) and (b), stained with FITC-anti-ι-carrageenan; (c) and (d), stained with FITC-anti-λ-carrageenan. Note staining of walls dividing cells, in (b) and (d).

whether both polymers are present in gametophytic tissue is unanswered.

Solubility studies. As additional evidence that Phyllophora carrageenans are ι-type, we dialyzed 6 samples prepared in H_2O against 0.3 M Ca Cl_2. Firm gels were formed with carrageenans from vegetative P. pseudoceranoides, ♂ of P. crispa, and stipe tissue of P. truncata. Less solid gels were formed in 0.3 M KCl. Little or no gelation occurred with tetrasporic P. truncata or P. pseudoceranoides specimens. The IR spectra of the non-gelling carrageenans contained the peaks characteristic of ι-carrageenan, and in the one case looked identical to known ι-carrageenan; in the other case, peaks were present but broader. These results are hard to reconcile with the other data. One wonders whether the presence of small amounts of λ-carrageenan in these preparations inhibited gelation.

Summary

In both phases of life cycle of both *P. pseudoceranoides* and *P. truncata* the chief polysaccharide is iota carrageenan. A small amount of λ-carrageenan is present in sporophytic *P. pseudoceranoides* and nemathecia of *P. truncata* but if present in gametophytes was undetectable. These spp. differ from other members of their family by failing to show a clear biochemical alternation of generations, although evidence of some difference was observed. *P. crispa* carrageenan differed from that of the other two species.

Grants from NSERC (Canada), N.S.F. to M.B.L., Woods Hole, MA, and U. of CA Sea Grant Program are acknowledged.

References

1. McCandless, E.L., Craigie, J.S., Walter, J.A. Planta 112, 201-212 (1973)
2. Hosford, S.P.C., McCandless, E.L. Can. J. Bot. 53, 2835-2841 (1975)
3. DiNinno, V., McCandless, E.L. Immunochem. 15, 273-274 (1978)
4. Gordon-Mills, E., McCandless, E.L. Phycologia 14, 275-281 (1975)
5. Yarotsky, S.V., Shashkov, A.S., Usov, A.I. Bioorgan. Chem. 4, 745-751 (1978)

ENZYMES FROM *PSEUDOMONAS CARRAGEENOVORA*. APPLICATION TO STUDIES OF CARRAGEENAN STRUCTURE

M. W. McLean and F. B. Williamson

Department of Biochemistry, University of Aberdeen, Marischal College, Aberdeen, Great Britain, AB9 1AS

Introduction

Carrageenans are a family of sulphated D-galactans from various species of red algae. One type, κ-carrageenan is defined as the repeat unit: 3,6-anhydro-D-galactopyranosyl-α(1-3)-4-O-sulphato-D-galactopyranosyl-β(1-4).

In early enzymological studies of carrageenan degradation, Weigl and Yaphe [1] isolated a κ-carrageenase from *Pseudomonas carrageenovora*. They showed that this enzyme hydrolysed β-linkages in the polymer to produce a series of oligosaccharides homologous to neocarrabiose 4-O-sulphate. In further studies with the same organism, they discovered a glycosulphatase which desulphated the neocarrabiose 4-O-sulphate to neocarrabiose [2].

Recently we purified the glycosulphatase and characterised its action on neocarratetraose 4-O-disulphate [3]. This oligosaccharide was semi-desulphated to a monosulphated neocarratetraose, which was shown to be sulphated on the reducing terminal neocarrabiose unit.

A third enzyme was encountered when we challenged the cell-free extract of *Pseudomonas carrageenovora* with [^{14}C]neocarratetraose 4-O-monosulphate [4]. This substrate was hydrolysed to neocarrabiose and neocarrabiose 4-O-sulphate. The route of degradation of neocarratetraose 4-O-disulphate by *Pseudomonas carrageenovora* is shown in Fig. 1.

Xth International Seaweed Symposium
Printed in Germany

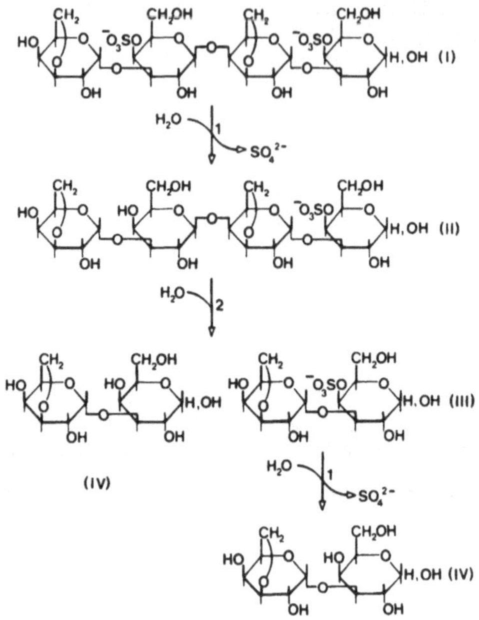

Fig. 1 Pathway of degradation of neocarratetraose 4-O-disulphate (I). Neocarratetraose 4-O-monosulphate (II), neocarrabiose 4-O-sulphate (III), neocarrabiose (IV). Enzymes: 1 glycosulphatase, 2 neocarratetraose 4-O-monosulphate β-hydrolase.

We will describe the use of the three enzymes in the degradation of a commercial carrageenan for the preparation of gram quantities of crystalline neocarrabiose. Similarly, an assay for the estimation of kappa content of carrageenan samples will be discussed.

Although the purification of the three enzymes has been described [1,3,4,5], the partial, co-purification of the glycosulphatase and tetrasaccharidase will be detailed here. This preparation is suitable for the degradation of κ-carrageenan oligosaccharides to neocarrabiose, and estimation of κ-carrageenan content.

Results

Co-purification of Glycosulphatase and Tetrasaccharidase. *Pseudomonas carrageenovora* (NCMB 302) was grown, harvested and disrupted as previously described [5]. The cell-free supernatant from the bacteria of 3x1 dm^3 cultures was chromatographed on Sephacryl S-200. Fractions were assayed for glycosulphatase and tetrasaccharidase [3,4]. The positions of elution of the two enzymes are always practically coincident, the glycosulphatase and tetrasaccharidase having K_{av} values of 0.23 and 0.26 respectively (Fig. 2). Thus each enzyme is purified by a factor of approximately 12, is free of small

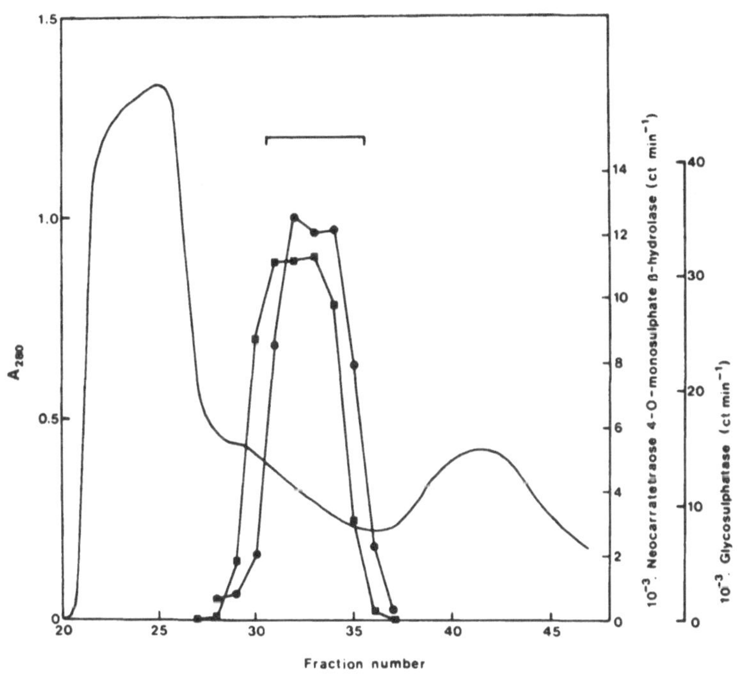

Fig. 2 Chromatography of cell-free supernatant from disrupted *Pseudomonas carrageenovora* on Sephacryl S-200 (102 x 3 cm), buffer: 0.50 mol dm^{-3} NaCl, 0.020 mol dm^{-3} Tris, 5 mmol dm^{-3} $MgCl_2$, pH 7.5. A_{280} (——), tetrasaccharidase (●), glycosulphatase (■). Bar indicates fractions pooled.

molecular weight material, and also free of κ-carrageenase [4]. 62 mg of protein were recovered in the pooled fractions (72 cm^3) which were stored at -20°C after adding BSA (2 mg cm^{-3}).

β-Neocarrabiose. Commercial carrageenan (10 g, Danagel CX) was digested with κ-carrageenase [5] and the oligosaccharides fractionated on Sephadex G-25 (Fig. 3A). The disaccharide and tetrasaccharide were pooled, NaCl/imidazole added [4] and then mixed enzyme (9 mg). Neocarrabiose production was followed by an ion-exchange assay [4], the eluent being assayed with anthrone. At 36 h, with a yield of 98%, the reaction solution was lyophilized, and then dialysed against water. The dialysate was freed of SO_4^{2-} with $BaCl_2$ solution and then chromatographed on Sephadex G-25 (Fig. 3B). The neocarrabiose was crystallized from ethanol/water. Fine needles, yield 2.00 g. Melting point 169-170°C, $[\alpha]_D^{20}$ +85° initially; +96°, 3 h (c=1.0, H_2O). ^{13}C-NMR (D_2O, ppm external TMS) 93.3 (J_{CH} 169.8 Hz, C-1α), 95.1 (J_{CH} 165.4 Hz, C-1'), 97.5 (J_{CH} 156.6 Hz, C-1β). Neocarrabiitol, prepared by $NaBH_4$ reduction of neocarrabiose, was isolated as a syrup after desalting on Sephadex G-25. ^{13}C-NMR, 98 0 (J_{CH} 164.2 Hz, C-1').

Assay for κ-carrageenan. A sample of carrageenan (approx. 1mg) was dissolved in imidazole buffer (0.2 cm^3, 0.025 mol dm^{-3} imidazole, 0.10 mol dm^{-3} NaCl, 2 mmol dm^{-3} $CaCl_2$, pH 7.5), the solution kept at 40°C and κ-carrageenase (1 μg) added in imidazole buffer (10 mm^3) containing BSA (1 mg cm^{-3}). After mixing, the solution was maintained at 25°C. At 10 min an aliquot (50 mm^3) was taken for total carbohydrate. A second aliquot was added to mixed enzyme (20 mm^3), and a third to buffer (20 mm^3) without enzyme as a control. After incubation (3 h), the digest and control were diluted with water (0.3 cm^3) and then ethanol (0.7 cm^3). Precipitated material was pelleted

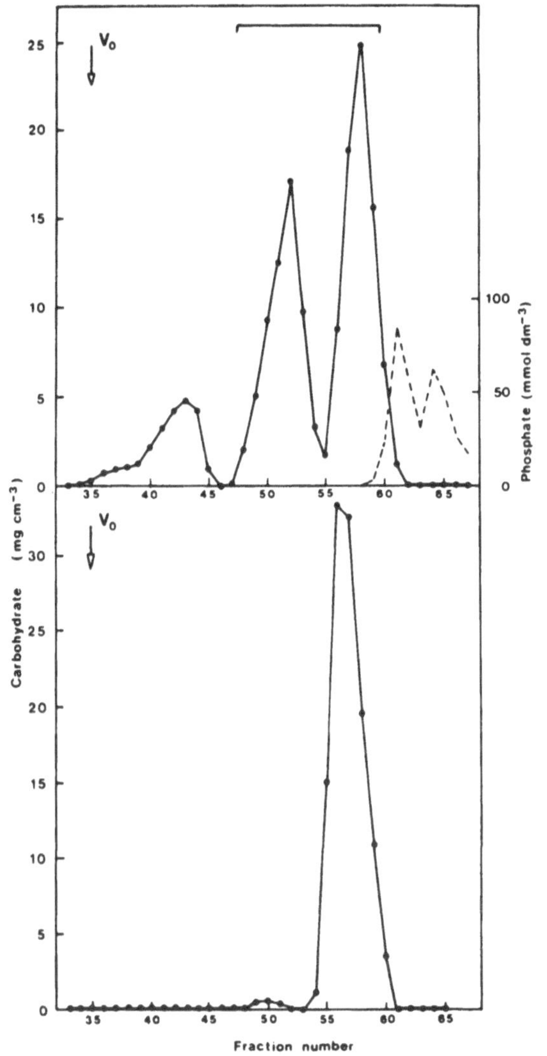

Fig. 3 A (Top) Chromatography of κ-carrageenan oligosaccharides on Sephadex G-25 superfine (139 x 3.5 cm) equilibrated with 0.10 mol dm^{-3} NaCl, 0.025 mol dm^{-3} sodium acetate, pH 5.0. Bar indicates fractions pooled.

B (Bottom) Chromatography of reaction products of degraded tetrasaccharide/disaccharide pool from A. Sephadex G-25 superfine (139 x 3.5 cm) equilibrated with water.

Carbohydrate (●), phosphate (---).

and the supernatant decanted. The precipitated material was extracted with aqueous ethanol (70%, 3x1 cm^3) and the pooled supernatants evaporated. The residues were dissolved in water (4x1 cm^3) and applied to beds of AG 1X8 Cl^- (0.25 cm^3). The eluents were evaporated, the residues taken up in water (1.5cm^3) and anthrone reagent (3.0 cm^3) added. The kappa content was expressed as the anthrone value of the mixed enzyme digest (less control) as a percentage of the total carbohydrate value (Table 1). A more sophisticated assay will include a resorcinol assay [6] on the total carbohydrate to correct the 100% anthrone value for galactose 6-O-sulphate content.

Table 1.

Product	Percent kappa
Danagel CX	62
REX 5401	40
Sigma C-1013	76
Auby Gel X52	7
REX 7701	1

References.

1. Weigl, J., Yaphe, W.: Can. J. Microbiol. 12, 939-947 (1966).
2. Weigl, J., Yaphe, W.: Can. J. Microbiol. 12, 874-876 (1966).
3. McLean, M. W., Williamson, F. B.: Eur. J. Biochem. 101, 497-505 (1979).
4. McLean, M. W., Williamson, F. B.: submitted for publication.
5. McLean, M, W., Williamson, F. B.: Eur. J. Biochem. 93, 553-558 (1979).
6. Yaphe, W., Arsenault, G. P.: Anal. Biochem. 13, 143-148, (1965).

THE EFFECT OF CALCIUM-ION CONCENTRATION ON POLY-D-MANNURONATE C-5-EPIMERASE.

R. Ofstad and B. Larsen
Institute of Marine Biochemistry, The University of Trondheim, N-7034 TRONDHEIM-NTH, Norway.

Introduction

Sodium alginate is a binary copolymer of D-mannuronate (M) and L-guluronate (G) residues arranged in a blockwise pattern along the linear chain (1-3). An enzyme preparation isolated from the growth medium of the soil bacterium *Azotobacter vinelandii* by precipitation with ammonium sulphate (4), when incubated with alginate from brown algae in the presence of calcium ions, epimerized D-mannuronic-acid residues to L-guluronic-acid residues in the polymer chain (5). The enzyme-modified alginate was of the same general type as previously described for marine alginates and for *Azotobacter* alginates (6).

The physical properties of alginate depend not only upon the uronate composition, *i.e.*, the M/G-ratio, but also upon the monomer-sequence distribution in the copolymer (7). Thus, it is important to identify factors affecting both the epimerase activity and the monomer-sequence distribution. Haug and Larsen observed that the M/G-ratio in the polyuronide produced by *Azotobacter vinelandii* was strongly dependent upon the calcium-ion concentration in the growth medium (6). There was also a strong correlation between D-mannuronic acid C-5-epimerase activity and the calcium-ion concentration in the incubation mixture (4).

Xth International Seaweed Symposium
Printed in Germany

Materials and methods

Enzyme assay I: Alginate solution (90 % mannuronic acid; 7.5 mg in 2 ml H_2O) was mixed with collidine buffer (pH 6.8; 6 ml), 34 mM $CaCl_2 \cdot 2\ H_2O$ (2 ml), enzyme preparation (0.125 ml), and water (to 10 ml). The alginate concentration in the solution was then 3.75 mM. The incubation temperature was 37°C. *Enzyme assay II*: Alginate solution (0.25 % w/v; 98 % mannuronic acid; 0.5 ml) was mixed with collidine buffer (2 ml), 34 mM $CaCl_2 \cdot 2\ H_2O$ (0.35 ml), enzyme preparation (0.2 ml), and water (to 3.5 ml). The alginate concentration in the solution was then 1.8 mM. The incubation temperature was 20°C. The incubation time, calcium-ion concentration, and the amount of enzyme preparation were varied as described in the text. The solution was analysed by the Dische carbazole method (8). If P is the ratio between the colorimetric readings at time t and at the start of the reaction respectively, then $P \sim G/M$, because guluronic acid gives a higher intensity of colour than mannuronic acid. (P-1) is thus proportional to the degree of conversion, and this value was chosen as a measure of enzyme activity.

Further characterization of the epimerized alginate, was achieved by ^1H-n.m.r.-spectroscopy (9). With this method the four possible doublets in the binary copolymer, *i.e.* an M-unit followed by an M (MM), an M-unit followed by a G (MG), a G-unit followed by an M (GM), and a G-unit followed by a G (GG), were determined.

Results

The activities of enzyme-1 and enzyme-3, isolated by ammonium-sulphate precipitation from two different batches of bacteria were tested. The correlation between enzyme activity and calcium-ion concentration is shown in Fig. 1. In these experiments, enzyme assay I was followed with 0.125 ml enzyme solution and an incubation time of ½ h.

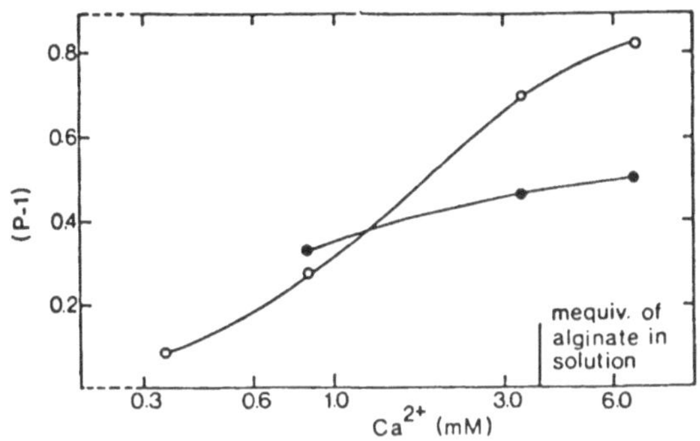

Fig. 1. Enzyme activity (P-1) as a function of calcium-ion concentration. Enzyme-1: ●—●. Enzyme-3: O—O.

As seen from the Figure, the enzymic activity increased over the range of calcium-ion concentrations. However, the relative increase in activity was markedly different for the two batches of enzyme. Even though enzyme-1 had the highest activity at the lower calcium-ion concentrations, enzyme-3 had the highest activity at the higher concentrations.

Alginate epimerized in 0.85, 3.4, and 6.8 mM Ca^{2+} was analysed by ^{1}H-n.m.r.-spectroscopy. The incubation mixture was as in enzyme assay I. Enzyme preparations from three different batches of bacteria were used. The guluronic acid content (F_G) (*i.e.* the degree of conversion) and the guluronic acid containing doublets ($F_{GM} \simeq F_{MG}$ and F_{GG}) are given as mole fractions in Table 1. A marked correlation was observed between the F_{GM}/F_{GG}-ratio in alginate containing approximately 25 % G-units and the calcium-ion concentration in the incubation mixture. Of the total amount of G introduced, the proportion occurring in GM-doublets increased markedly with calcium-ion concentration, indicating that guluronic acid was introduced as a neighbour to guluronic acid when the calcium-ion concentration was low and as a neighbour to mannuronic acid when the calcium-ion concentration was high. Table 1 also shows that the distribution of the GM- and GG-doublets varied with the batches of enzyme used in the epimerizing reaction. At approximately the same degree of transformation, enzyme-2 gave a larger proportion of G as neighbour to G than the other two enzymes. Enzyme-3 introduced the largest part of G as a neighbour to M.

Table 1. Doublet frequencies (F_{GM} and F_{GG}) in alginate epimerized to a guluronic-acid content of approximately 25 % with different calcium-ion concentrations in the incubation mixture. F_G, F_{GG} and F_{GM} are given as mole fractions.

Enzyme batch	Enzyme amount (ml)	Incubation, (h)	$[Ca^{2+}]$ mM	F_G	F_{GG}	F_{GM}	$\frac{F_{GM}}{F_{GG}}$
1	0.125	3	0.85	0.24	0.14	0.10	0.7
"	"	2	3.40	0.24	0.11	0.13	1.2
"	"	2	6.80	0.26	0.10	0.16	1.6
2	0.125	2	0.85	0.34	0.25	0.09	0.4
"	"	1½	3.40	0.24	0.16	0.08	0.5
"	"	2	6.80	0.30	0.18	0.12	0.7
3	0.250	6	0.85	0.38	0.27	0.11	0.4
"	"	2	3.40	0.31	0.08	0.23	2.9
"	"	1½	6.80	0.22	0.05	0.17	3.4

The alginate solution used as substrate contained about 10 % of guluronic acid residues. For an enzyme of this type, the ideal substrate would be a polymer consisting solely of mannuronic-acid-residues. The alginate solution was therefore, after a very mild acid-degradation, fractionated by precipitation with calcium chloride. The major part of the guluronic-acid units were precipitated under these conditions, leaving in solution a polyuronate containing about 98 % mannuronic acid. In the following experiments this solution was used as a substrate, to minimize the influence of pre-existing G-units on the initial rate of epimerization. At the same time we changed the experimental procedure to the previously described enzyme assay II. It should be noted that the ratio between the alginate and the calcium-ion concentration was the same as in the previous experiments.

The activities of three batches of enzyme (4, 5, and 6) were tested, and Fig. 2 shows the degree of conversion, (P-1), as a function of time for enzymes 4 and 5 in 0.34 mM and 3.4 mM calcium chloride. In Fig. 3 (P-1) is plotted as a function of enzyme concentration for enzyme-6, using incubation times of 1 h and 3 h. Below the value of 0.5 (corresponding to a degree of transformation of 25 %), (P-1) shows a satisfactory linearity both with time and enzyme concentration. This agrees with previous results (4).

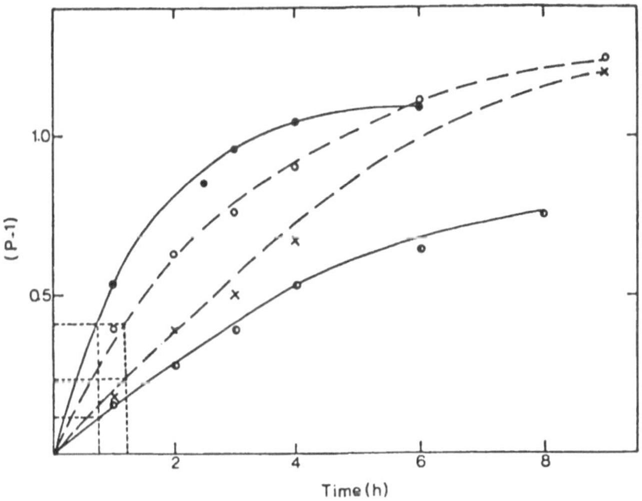

Fig. 2. Enzyme activity (P-1) as a function of time.
Enzyme-4 in the presence of 0.34 mM Ca^{2+} (x---x) or 3.4 mM Ca^{2+} (O---O).
Enzyme-5 in the presence of 0.34 mM Ca^{2+} (◐——◐) or 3.4 mM Ca^{2+} (●——●).

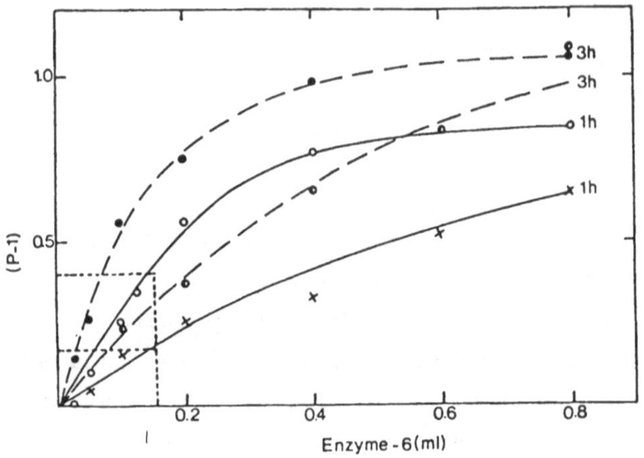

Fig. 3. Enzyme activity (P-1) as a function of enzyme concentration.
Incubation time of 1 h in the presence of 0.34 mM Ca^{2+} (x——x) or 3.4 mM Ca^{2+} (O——O).
Incubation time of 3 h in the presence of 0.34 mM Ca^{2+} (◐---◐) or 3.4 mM Ca^{2+} (●---●).

The Figures show that the activities of the enzymes increase when the concentration of calcium ions increases from 0.34 mM to 3.4 mM. As observed earlier, the relative increase in activity was markedly different for different batches of enzyme. Experiments were carried out to establish whether this observed difference between enzymes was caused by different specific activities of the enzyme preparations. According to the data presented in Figs. 2 and 3, we decided to incubate the alginate substrate with 0.20 ml, 0.17 ml, and 0.15 ml of enzymes 4, 5 and 6, respectively, to give a degree of conversion of ca. 0.4 at 3.4 mM Ca^{2+} for all three enzymes. The incubation time was 2 h and 1 h, at 0.34 mM Ca^{2+} and 3.4 mM Ca^{2+}, respectively. Even though the three enzymes gave almost the same (P-1) values at the higher calcium-ion concentration, they gave significantly different degrees of conversion at the lower calcium-ion concentration. In Table 2 this is given as the ratio between the values of (P-1) obtained at 3.4 mM and 0.34 mM Ca^{2+}.

Table 2. The ratio, (P-1) between the enzyme activities at 3.4 mM and 0.34 mM Ca^{2+} for three different batches of enzyme.

Enzyme batch	Enzyme amount (ml)	Incubation time (h)	$[Ca^{2+}]$ mM	(P-1)	$\frac{(P-1)_{3.4}/h}{(P-1)_{0.34}/h}$	$\frac{0.4}{(P-1)_{0.34}}$ [a]
4	0.20	1	3.40	0.37	2.0	1.9
"	"	2	0.34	0.37		
5	0.17	1	3.40	0.41	3.9	3.6
"	"	2	0.34	0.21		
6	0.15	1	3.40	0.35	2.3	2.2
"	"	2	0.34	0.30		

[a] Calculated by dividing the (P-1) values (0.4) at 3.4 mM Ca^{2+} by the (P-1) value at 0.34 mM Ca^{2+} after the same incubation period (Fig. 2) or using the same amount of enzyme (Fig. 3).

A corresponding ratio may be calculated from the data presented in Figs. 2 and 3. Selecting a (P-1) value of 0.4 at high calcium-ion concentration, the corresponding ratio may be calculated by dividing this value (0.4) by the (P-1) values at low calcium-ion concentration, obtained after the same incubation period (Fig. 2). The same was done with data from Fig. 3, but dividing by the (P-1) values obtained with the same amount of enzyme. These values are listed in the last column in Table 2 and are seen to be in very good agreement with the experimental values discussed above.

Discussion

From the experimental results obtained so far, it seems as if the concentration of calcium ions in the incubation mixture is a factor affecting both the epimerase activity (*i.e.* the uronate composition), and the monomer-sequence distribution in the copolymer. Low calcium-ion concentrations seem to favour homopolymeric sequences of guluronic acid. Alginate epimerized at high concentrations of calcium ions contained, on the other hand, a considerably higher proportion of G-units in GM- sequences.

The most surprising result was that the sequence distribution in the epimerized alginate varied with the enzyme preparation used in the reaction. Even after taking into account the marked effect of calcium ions on the total activity as well as on the sequence-distribution, it still seems very difficult to explain the difference between different batches of enzyme in terms of a single, enzymatic reaction. On the other hand, a simple explanation of this phenomena may be that the epimerizing reaction is carried out by two different enzymes. One, the GG-enzyme, introduces guluronic acid as a neighbour to guluronic acid and another, the GM-enzyme, introduces guluronic acid as a neighbour to mannuronic acid.

Table 1 demonstrates that, at low calcium-ion concentrations, the GG-doublets are favoured, while the GM-doublets are favoured at high calcium-ion concentrations. This may indicate that the concentration of the calcium ions affects the activity of the GM-enzyme more than the activity of the GG-enzyme. If we assume that calcium ions affect the activity of the GM-enzyme very strongly and that the GG-enzyme is very little affected, the behaviour of the system as a whole may be easily understood. Batches of enzyme that contain most GG-enzyme will have nearly the same activity both at high and low concentrations of calcium ions. This can explain the behaviour of enzymes 1, 2 and 4. On the other hand, enzyme preparations consisting mostly of the GM-enzyme will show marked differences in their activities at high and low calcium-ion concentrations (enzymes 3, 5, and 6). These will, consequently, also give a marked difference in sequence-distribution (enzyme-3).

Whether the polymannuronic acid is epimerized by one or two or even more enzymes, cannot be answered definitely until the enzyme preparation is purified.

References

1. Haug, A., Larsen, B., and Smidsrød, O.: Acta Chem. Scand., 20, 183-190 (1966).
2. Larsen, B., Smidsrød, O., Painter, T., and Haug, A.: Acta Chem. Scand., 24, 726-728 (1970).
3. Haug, A., Larsen, B., and Smidsrød, O.: Carbohydr. Res., 32, 217-225 (1974).
4. Haug, A., and Larsen, B.: Carbohydr. Res., 17, 297-308 (1971).
5. Larsen, B., and Haug, A.: Carbohydr. Res., 20, 225-232 (1971).
6. Larsen, B., and Haug, A.: Carbohydr. Res., 17, 287-296 (1971).
7. Haug, A., Myklestad, S., Larsen, B., and Smidsrød, O.: Acta Chem. Scand., 21, 768-778 (1967).
8. Knutson, C. A., and Jeanes, A.: Anal. Biochem., 24, 470-481, (1968).
9. Grasdalen, H., Larsen, B., and Smidsrød, O.: Carbohydr. Res., 68, 23-31 (1979).

THE INFRARED SPECTRA OF CARRAGEENANS EXTRACTED FROM VARIOUS ALGAE

S. Peats

Marine Colloids Division, FMC Corporation
P. O. Box 308, Rockland, Maine 04841 USA

Introduction

Smith and Cook (1) used the term _lambda_ fraction and _kappa_ fraction to designate the soluble and insoluble fractions, respectively; obtained by precipitation of carrageenan at a definite KCl concentration, 0.125 _M_. Rees and coworkers (2-4) used the terms _kappa_ and _lambda_ as well as introducing other terms to denote carrageenans possessing a definite chemical composition. Carrageenans approaching the idealized structures outlined in Fig. 1 occur naturally in certain algae and provide reference points for understanding the more commonly occurring heterogeneous types.

In this paper the term _kappa_ family or gelling fraction will be used to define that fraction that is insoluble in 0.33 _M_ KCl according to the method of Stancioff (5). The term _lambda_ family or nongelling fraction will be used to define that fraction that is soluble in 0.33 _M_ KCl.

Research by various workers has done much to demonstrate that chemical differences exist among carrageenans from the different stages of the life history of some red algae. For the genera _Iridaea_, _Chondrus_ and _Gigartina_ chemical and infrared analyses have shown that the haploid gametangial stages produce carrageenans of the _kappa_ family while the diploid tetrasporangial stages produce carrageenans of the _lambda_ family.

Xth International Seaweed Symposium
Printed in Germany

Figure 1. CARRAGEENAN STRUCTURES

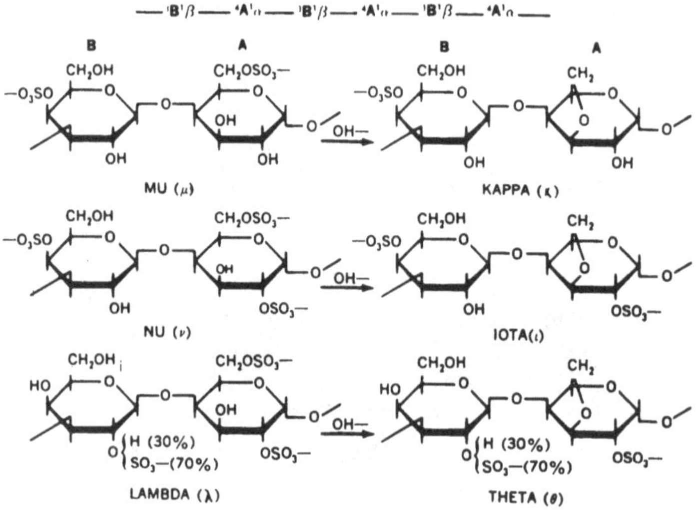

Experimental

On receipt of the algae (Table 1) the thalli were sorted to separate the gametophytes from the tetrasporophytes. The gametophytes were extracted using a $Ca(OH)_2$ procedure (6) and the tetrasphorophytes were extracted using a low NaOH procedure (7). Where separation was not possible both extractions were performed on the mixed fronds. The KCl fractionation was that used by Stancioff (5).

The extractives were washed with an alcoholic Na_4 EDTA solution to exchange them principally to the Na^+ form. The exchanged extractives were then analyzed for ester sulphate (7) and 3,6-anhydrogalactose (3,6-AG) content (8). Galactose was estimated from apparent 3,6-AG and ester sulphate values using an iterative procedure, and the molar ratios expressed on a total hexose basis.

The films for infrared analysis were produced by drying down 3.0 mL of an aqueous solution containing 2 mg extractive/ mL in Teflon wells 0.63 cm deep and 2.52 cm diameter (7).

Table 1.

ALGAE USED IN STUDY

I) Furcellariaceae

Furcellaria fastigiata (Hudson) Lamouroux

II) Hypneaceae

Hypnea musciformis (Wulfen) Lamouroux

III) Phyllophoraceae

Gymnogongrus furcellatus (Ag.) (J. Ag.)
Ahnfeltia concinna J. Ag.

IV) Soleriaceae

Agardhiella tenera (J. Ag.) Schmitz
Eucheuma muricatum (Gmel.) Web. V. B.
Eucheuma uncinatum S. & G.
Eucheuma isiforme (C. Ag.) J. Ag.
Eucheuma striatum Schmitz
Eucheuma gelatinae J. Ag.

V) Gigartinaceae

Chondrus crispus Stackhouse
Chondrus ocellatus Holmes
Iridaea flaccida (S&G) Silva
Iridaea cordata (var. cordata) (Turner) Bory
Iridaea cordata (var. splendens) (S&G) Abb.
Iridaea crispata Bory
Iridaea laminarioides Bory *(Iridophycus boryanum)* (S&G)
Gigartina bracteata (Gmel.) S&G
Gigartina volans (C. Ag.) J. Ag.
Gigartina pistillata (Gmel.) Stackhouse
Gigartina leptorhynchos J. Ag.
Gigartina stellata (Stackh.) Batters
Gigartina johnstonii Dawson
Gigartina harveyana (Kütz) S&G
Gigartina exasperata Harv. and Bail.
Gigartina chamissoi (Mert.) J. Ag.
Gigartina canaliculata Harv.
Gigartina acicularis (Wulf.) Lamouroux
Gigartina skottsbergii S&G
Rhodoglossum affine (Harv.) Kyl.
Rhodoglossum californicum (J. Ag.) Abb.

INFRARED AND CHEMICAL DATA FOR THE GELLING CARRAGEENANS

Algae	Peak Ratio 805 cm^{-1}/845 cm^{-1}	Galactose/ Hexose	3,6 AG/ Hexose	R-O-SO_3^-/ Hexose
Agardhiella tenera	0.81	0.60	0.40	0.83
Chondrus crispus	0.16	0.58	0.42	0.58
Eucheuma striatum	0.02	0.54	0.46	0.54
Eucheuma gelatinae	0.00	0.60	0.40	0.24
Eucheuma muricatum	1.01	0.58	0.42	0.92
Eucheuma uncinatum	0.73	0.61	0.39	0.86
Furcellaria fastigiata	0.00	0.63	0.47	0.26
Gigartina bracteata	0.36	0.54	0.46	0.66
Gigartina canaliculata	0.23	0.55	0.45	0.59
Gigartina stellata	0.26	0.60	0.40	0.63
Gigartina volans	0.15	0.55	0.45	0.54
Gymnogongrus furcellatus	0.60	0.60	0.40	0.72
Hypnea musciformis	0.02	0.60	0.40	0.51
Iridaea laminarioides	0.36	0.54	0.46	0.73
Rhodoglossum californicum	0.22	0.56	0.44	0.56

Results and Discussion

Stancioff and Stanley (9) have made several peak assignments for both the gelling and nongelling fractions.

1) Gelling fractions. The major difference among the spectra is the relative size of the peak at 805 cm^{-1}. Therefore, a ratio of the peak height at 805 cm^{-1} (3,6-AG-2-sulphate) to that at 845 cm^{-1} (galactose-4-sulphate) was calculated using methods outlined by Wright (10). These ratios were then compared with the chemical data (Table 2) and a graphical representation of these data is presented in Fig. 2.

Fig. 2. COMPARISON OF INFRARED AND CHEMICAL DATA

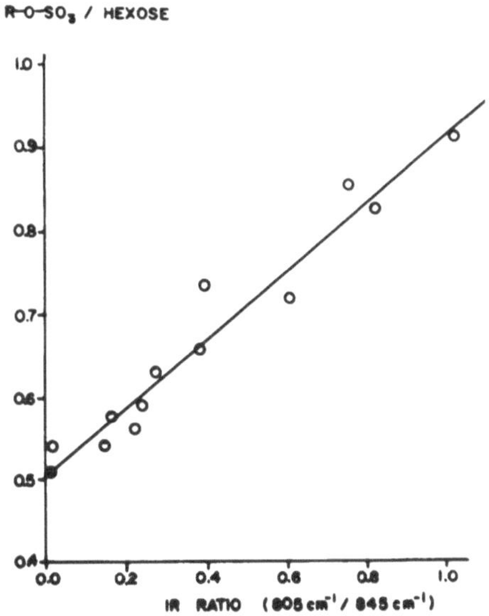

Inherent in this correlation are a number of assumptions. Firstly, that the extraction procedure used modifies all the 1,4 linked galactose-6-sulphate moeities to 3,6-anhydrogalactose moieties. Secondly, that the 1,3 linked galactose units are completely sulphated at Carbon 4. This assumption is certainly not true for the extractives of Eucheuma gelatinae and Furcellaria fastigiata (11) and consequently these values were omitted from the correlation.

Despite these potential inaccuracies the correlation appears to be sufficiently significant to allow the prediction of chemical composition from the infrared spectra. There appears to be no relationship between taxonomic classification of the algae and the chemical structure of their extractives.

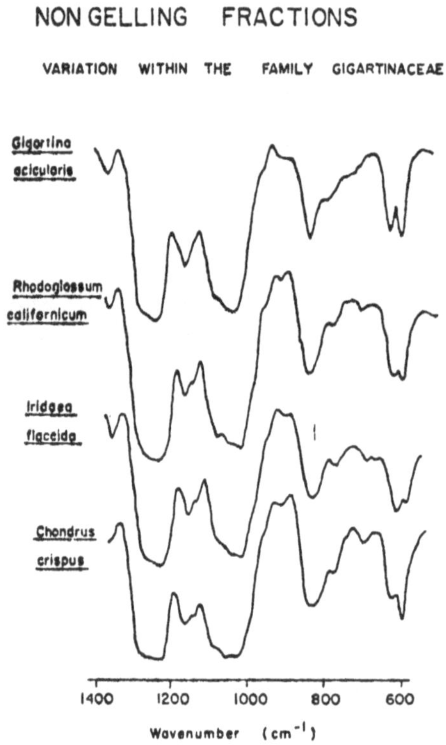

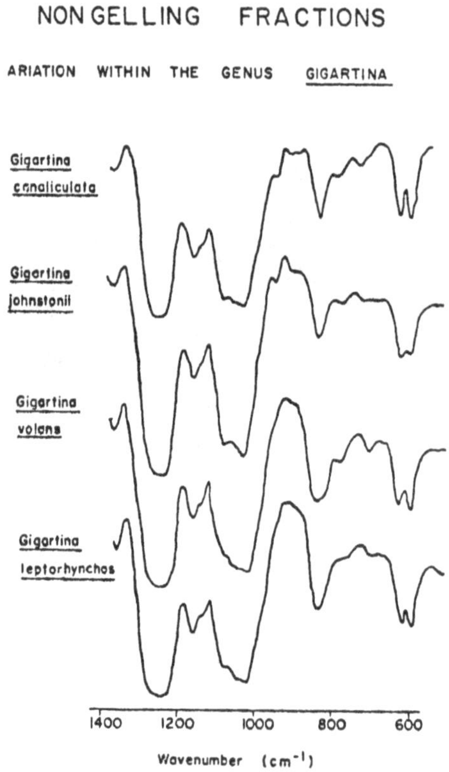

2) Nongelling fractions. The only family that gives an appreciable amount of a true nongelling fraction is Gigartinaceae. The peak of interest in the nongelling fractions is that centered around 825 cm^{-1}. Penman and Rees (11) suggested the term xi carrageenan, the structure of which is similar to that of lambda carrageenan but has no 1,4 linked galactose-6-sulphate. Lambda carrageenan will give rise to a broad peak at 825 cm^{-1} due to the presence of a peak at 830 cm^{-1} (galactose-2-sulphate) and 820 cm^{-1} (galactose-6-sulphate) whereas xi carrageenan will give rise to a much sharper peak.

McCandless (12, 13) used a combination of infrared spectroscopy and immunological techniques to characterize the structural differences in extractives from several species. All the *Gigartina* species studied possessed a narrow peak at 825 cm^{-1} whereas the other species studied possessed a much broader peak at 825 cm^{-1}.

This study indicated that the extractives of *Rhodoglossum*, *Iridaea* and *Chondrus* species studied, all possessed a broad peak at 825 cm^{-1} indicative of a *lambda*-like carrageenan (Fig. 3). Most of the *Gigartina* species analyzed have a sharp peak at 825 cm^{-1} indicative of a *xi*-like carrageenan although extracts from *Gigartina bracteata*, *Gigartina leptorhynchos* and *Gigartina volans* have much broader peaks (Fig. 4).

Kim (14) and West (15) have both questioned the classification of the genera within the family Gigartinaceae using solely differences in gross morphology. Kim has attempted to reclassify this family with respect to similarities in the morphology of the reproductive structure of the individual algae.

The inability to use infrared spectra for taxonomic classification still needs further study. This inability to differentiate could, however, be a result of the general difficulty that taxonomists have encountered in classifying the genera within the family Gigartinaceae and not necessarily a variation of the chemical structure of the nongelling fraction within an individual genus.

References

1. Smith, D.B. and Cook, W.H. (1953) Arch. Biochem. Biophys. 45, 232.
2. Anderson, N.S., Dolan, T.C.S., and Rees, D.A. (1968) J. Chem. Soc. (C), 596.
3. Anderson, N.S., Dolan, T.C.S., Penman, A., Rees, D.A., Mueller, G.P., Stancioff, D.J. and Stanley, N.F. (1968) J. Chem. Soc. (C), 602.
4. Lawson, C.J., Rees, D.A., Stancioff, D.J., Stanley, N.F. (1973) J. Chem. Soc. Perkin I, 2177.
5. Stancioff, D.J. (1965), U.S. Patent 3,176,003 (to Marine Colloids, Inc.).
6. Stanley, N.F. (1963), U.S. Patent 3,094,517 (to Marine Colloids, Inc.).
7. Dawes, C.J., Stanley, N.F., and Stancioff, D.J. (1977), Botanica Marina XX, 137.
8. Yaphe, W. and Arsenault, G.P. (1965), Anal. Biochem. 13, 143.
9. Stancioff, D.J. and Stanley, N.F. (1969), 6th International Seaweed Symposium, Santiago de Compostela, Madrid, 595.
10. Wright, N. (1941), Industr. Eng. Chem. 13, 1.
11. Penman, A., and Rees, D.A. (1973), J. Chem. Soc., Perkin I, 2182.
12. Evelegh, M.T., Vollmer, C.M. and McCandless, E.L. (1978), J. Phycol. 14, 89.
13. Hosford, S.P.C. and McCandless, E.L. (1975), Can. J. Bot. 53, 2835.
14. Kim, D.H. (1976), Nova Hedwigia XXVII, 1.
15. Polanshek, A.R., and West, J.A. (1975), J. Phycol. 11, 4, 434.

THE MeCDPS, A CARRYING SULPHATE'S NUCLEOTIDE OF THE RED SEAWEED CATENELLA OPUNTIA (GREV.).

M. Quillet & G. de Lestang-Brémond
Laboratoire de Physiologie Végétale, Université Catholique de l'Ouest,
B.P. 808 - 49005 ANGERS Cédex (France).

Introduction

This paper completes the observations we publish in this Symposium, about the ester sulphates' mobility of the cell walls' lambda carrageenan in *Catenella Opuntia* (Grev.). The cleavage of the semi-ester bond supposes a little energy and only a hydrolase. On the contrary, the sulphation on the carrageenan free -OH requires a lot of chemical energy, probably ATP, and sulphate's carriers together with enzymic effectors.
This sulphate's carrier has just been isolated and analyzed. It is slightly different from the sulphate's carrier we found in the Brown Seaweed *Pelvetia canaliculata* (the CDPS). Like it, it is a cytidilic nucleotide, but here the pyrimidic base is methylated..

The sulphation's polysaccharide mechanism is not simple. It involves steps of sulphation of the nucleotide, joined to a supply of chemical energy, that we are studying. The nucleotide we are describing here is the active form of the carrier.

A. Methods of extraction and purification

Principle : In order to obtain some tens of mg of the sulphated nucleotide, we cumulated the product of several lots of fresh *Catenella* extracted in boiling salted water.

Those samples had been first fixed in boiling ethanol (I min.), ground, and then their salts, carbohydrates and non sulphated nucleotide removed by 5

Xth International Seaweed Symposium
Printed in Germany

successive diffusions in strongly agitated cold water (I min. each). The non sulphated nucleotide is thus almost completely discarded.

The carrageenan solubilized together with the sulphated nucleotide is floculated in ethanol. The nucleotide remains in the alcoholic phase. This last one, joined to alcoholic washings of the carrageenan, needs a first purification. Alcoholic phases are evaporated to dryness ; the nucleotide is recovered by a several hours extraction, in a BBS micro-extractor, in boiling 85° GL ethanol. The content of this alcoholic extract is at last purified by passing through an active charcoal column.

Elutions by water, then by aqueous ethanol (up to 15° GL) eliminate various impurities which still remain. It is only in 25° GL ethanol containing 5 % NH_4OH that pure nucleotide is liberated.

After elimination of the alcohol and NH_4OH the concentrate crystallizes in small refringent needles.

Experimental :

1°) Fixation : for 70 g Catenella, 250 ml boiling ethanol (I minute).

2°) Diffusion : 5 baths of 200 ml of cold water for 10 g of dry fixed thallouses.

3°) Extraction : in 100 ml of 0,4 M NaCl ; 15 min. at 100° C. Concentration to I/4 and floculation in 2 volumes of ethanol.

4°) Preliminary purification : draining with 85° GL boiling ethanol, in a BBS micro-extractor (4 h), of the dried residue of the alcoholic solutions.

5°) Purification on charcoal (Norit ; 10 g) mixed with Celite (Merck 8II9 ; 10 g) activated according to Ivor Smith' method (1969) successive elutions by water and diluted ethanol (5 to 15° GL) (1000 ml each). Progress of the elution followed by UV spectrographical and chemical procedures.

6°) Elution of the nucleotide with 1500 ml of 25° GL ethanol containing 5 % NH_4OH or with 30 % aqueous pyridine (50 ml). Those solvents were eliminated by cautious evaporation under diminished pressure, with H_2SO_4 in the receptor flask.

B. Analysis of the nucleotide

The constituants of the active sulphated nucleotide have been progressively analysed.

The cytosine: The characteristics of the UV nucleotide spectra (max 275 nm ; min 250 nm) exclude the hypothesis of a puric nucleotide ; they indicate a pyrimidic, probably cytidilic, component.

The cytidilic base has effectively been prepared by drastic hydrolysis, in concentrated perchloric acid (70 %, I h at 100° C).
By chromatographies on paper (isopropanol, water, HCl ; 65/35/0,2 M) ; elution of the fluorescent band) or on charcoal (elution pyridine 30 %) we were able to isolate a few mg of a pure product, whose characteristics are typical of a 5-Methyl-Cytosine, perchlorate form (Shugar and Fox, 1952) :

λ max = 272 at pH 7,7

λ max = 287 at pH 13 (hyperchrome effect)

very small effect of acidity. (cf.fig.1)

The identification of this methylcytosine has been confirmed by different methods:

a) the measure of its melting point: 270°C (as expected).

b) the I.R. spectrography (with 4 mg in a KBr disc), which enabled us to recognize the pyrimidic cycle (strong absorption at 800 and I650 cm^{-1}; broad one between 950 and I000 cm^{-1});
to confirm the presence of the $-CH_3$ group (intense sharp peak at 2860 and bands around I380 and I450 cm^{-1};
and to exclude the hypothesis of a phenolic substance.

c) the chromatography on paper (solvant isopropanol, water, HCl; 65/35/0,2 M) together with a sample of cytosine prepared by hydrolysis of an authentic Merck cytidin monophosphate, showed the faster migration of our cytidilic base.

d) the obtention of characteristic picrate, cristallized in yellow orthorhombic prismatic needles; melting point: 286°C.

The methylation of cytosine accounts for the solubility of the nucleotide in chloroform (which can be used for its extraction) and for its high affinity for charcoal, since elution requires either large volumes of 25° ethanol containing 5 % NH_4OH or aqueous pyridine (30 %).

The ribose : The very strong bond between ribose and methylcytosine is more easily cleaved after sulfonation of this one (I h at 20° C in concentrated H_2SO_4).

The identity of the ribose has been recognized by co-chromatography with an authentical ribose (thin-layer polycarbonate Kodak ; sovant butanol, acetic acid 10/0,I). Revelation of chromatograms and evaluation of the amount of ribose have been effected with sulphuric anthron. We found I ribose for I sulphate.

The phosphates : Phosphates have been characterized and evaluated by the method of Fiske modified by Horecker et al, 1940), in a sulphuric hydrolysate of the nucleotide (H_2SO_4 25 %, 175° C, 2 h). The amount of PO_4 we found indicates that there are 2 moles of phosphates for I of sulphate. The site of those phosphates will be determined with phosphatases and RMN studies.

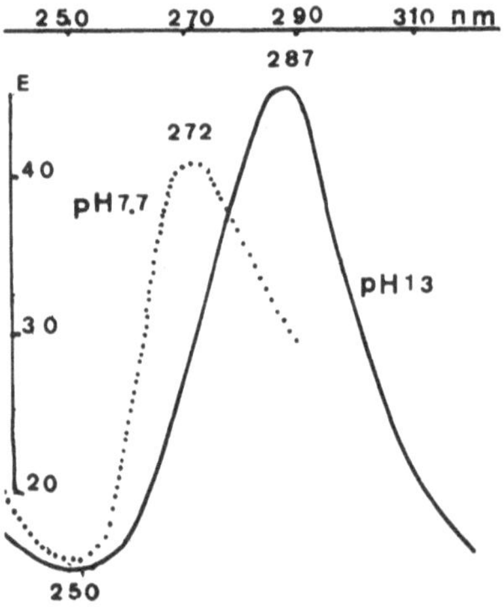

Figure I.

U.V.spectra of the Methylcytosine from the nucleotide of Catenella.

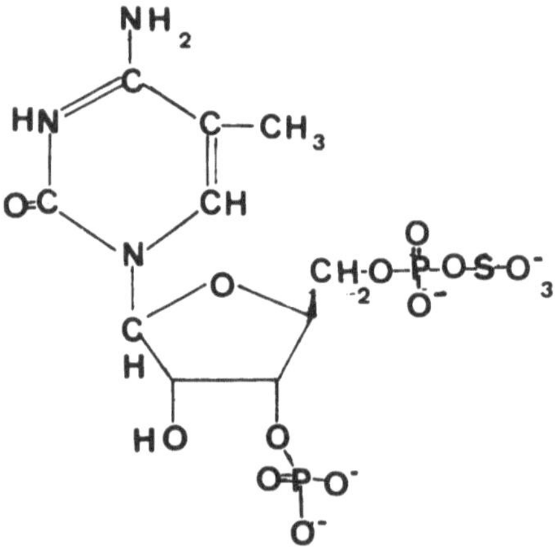

Figure 2.

Possible structure of the MeCDPS.

The sulphate : It is characterised, in the hydrolysates, by its precipitation with Ba, in strongly acidic conditions (HCl pH 2).

It has been dosed after mineralisation, in a sealed tube, with HNO_3 containing a small amount of $HClO_4$ and NaCl (18 h at 110° C), by the chloranilate method (Schafer, 1967).

All these results enable us to conclude at the final analysis :
Methylcytosine, ribose, phosphates, sulphate, in the molar ratio I/I/2/1.
Shortly : MeCDPS (figure 2).

Conclusion

Methylcytosine is an infrequent pyrimidic base.
To our knowledge, it has been signaled only inside the cells (transfer RNA, *Euglena*' nucleus, *Ochromonas*' cytoplasm). It is remarkable to find it here, in the cell walls.

This methyl CDPS, sulphating nucleotide rich in energy, could also be the sulphating agent in the Golgi vesicles, of neosynthetized carrageenans. But the relative abundance of this nucleotide, out of comparison with the very slow production of the polysaccharide in living *Catenella* (which increases only of some I/10 mm monthly) seems to indicate that it has another function : that to permit the turn-over described in the paper we join here, which obviously requires a very intense consummation of energy rich bonds.

It is remarkable, too, to see that in an emerging species of Red Algae, systematically very far from the Fucaceae, we find again an analogous mechanism of regulation of the moving of Na^+ ions towards the plasmic membrane. Researches we have started on *Cladophora rupestris* will show if this mechanism is applicable even to Green Alga.

Ref. : Horecker B.L., Ma T.S. and Haas E. : J. Biol. Chem. 136, 775 (1940),
de Lestang G. et Quillet M. : Physiol. Vég. 12 (2), 199-227 (1974),
Schafer H.N.S. : Anal. Chem. 39 (4) 1719-1926 (1967),
Shugar D. and Fox J.J. : Biochim. Biophys. Acta 9, 199 (1952),
Smith I. : Chromato. and electroph. techn. (I) 306 Wil. Heinemann. Med. Books.LTD (1969).

PHYSIOLOGICAL RESPONSES OF *PORPHYRA* UNDER STEADY STATE AND FLUCTUATING SALINITY REGIMES

Robert H. Reed
Department of Biological Sciences, University of Dundee, Dundee DD1 4HN, Scotland, U.K.

Introduction

Porphyra purpurea (Roth) C. Ag. (Rhodophyta, Bangiales) occurs characteristically in the intertidal zone of marine rocky shores, being subjected to extreme variations in salinity, according to the state of the tide and prevailing climatic conditions.

The present study outlines some of the more important features of recent investigations concerning osmotic adaptation in *P. purpurea* under steady state hyposaline conditions (see 1-5). Basic data are also presented for physiological responses under *quasi*-estuarine salinity regimes, using laboratory apparatus to model salinity fluctuations similar to those found in fully-mixed estuaries (3).

Materials

P. purpurea was collected from the shore at Aberffraw, Anglesey, U.K. and maintained in the laboratory as previously described (1,2). All experiments were performed using discs of plant material excised from healthy, non-reproductive areas of thalli of similar size and pigmentation. Experiments were conducted within seven days of collection, since longer term laboratory studies indicated that thalli remained healthy under laboratory conditions for a maximum of fourteen days (2).

Xth International Seaweed Symposium
Printed in Germany

Methods

Cellular volumes were estimated photomicroscopically (2); photosynthetic activity was assayed polarographically (3); intracellular inorganic solutes (primarily K^+ and Cl^-, see 6) were determined by chemical means, using flame photometry for K^+ and coulometric titration for Cl^- (5) and also using radioisotopic techniques ($^{42}K^+$ and $^{36}Cl^-$, as described in 6); galactosyl-glycerols were estimated using gas-liquid chromatography (4); intracellular amino acids were determined by colorimetric means, using a standard ninhydrin technique (5).

Quasi-estuarine salinity regimes were modelled in the laboratory by means of an LKB 11300 Ultragrad Gradient Mixer, in conjunction with a solenoid valve and mixing chamber (3). The salinity regime used in the present study was a sinusoidal cycle, complete in twelve hours. The amplitude of the salinity change was from x 1 to x $^1/16$ seawater. All experiments were performed using discs of plant material which had been maintained under *quasi*-estuarine conditions for two cycles (24 h).

Results

Changes in intracellular concentration of the major solutes of *P. purpurea* are shown in *Figure* 1. Galactosyl-glycerols, amino acids and KCl comprise over 90% of the total internal osmotic potential across the salinity range x $^1/16$ to x 1 seawater. Inorganic KCl predominates in x 1 seawater, while organic solutes increase in relative importance in hyposaline media.

The total intracellular solute concentration in x 1 seawater is sufficient to balance the external osmotic potential, with little or no cell turgor. However, as the external salt concentration is decreased, the total internal solute concentration is not reduced to a comparable extent (*Figure* 1). This indicates that the primary response to dilution stress is an increase in cell turgor, as previously suggested (2,4). Cell turgor is not returned to its original value, even after two days in hyposaline media, showing that *P. purpurea* does not possess a mechanism for turgor regulation.

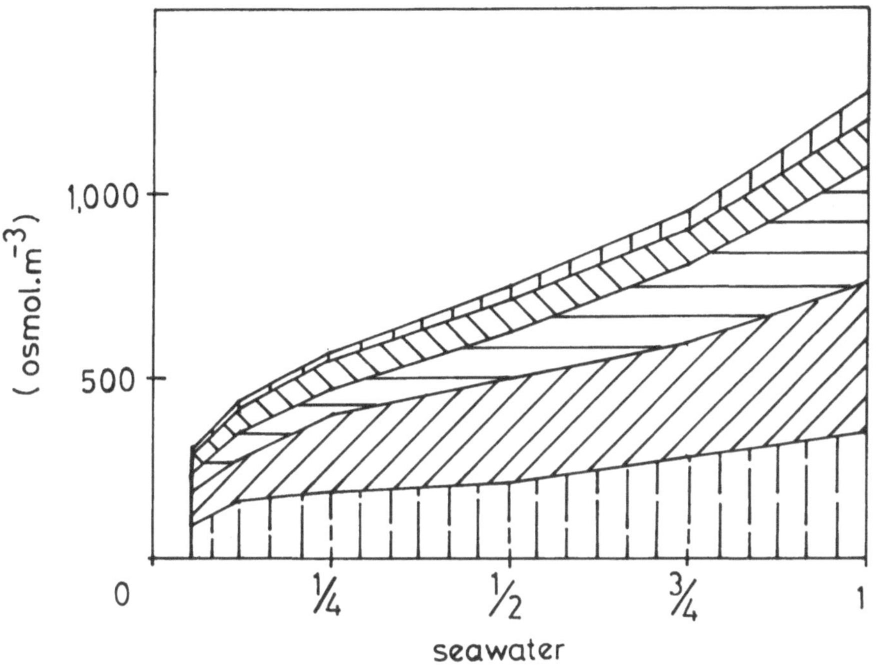

Figure 1. Concentrations of the major intracellular solutes of *P. purpurea.* [pattern] galactosyl glycerols, [pattern] K^+, [pattern] Cl^-, [pattern] amino acids, [pattern] other solutes (see 1,5).

Physiological responses of *P. purpurea* under *quasi*-estuarine conditions are contained in *Figure* 2. Cell volume is shown to increase and decrease during a single sinusoidal cycle of reduced salinity (*Figure* 2*a*). Photosynthetic activity shows the reverse trend, decreasing and then increasing over a similar twelve hour period (*Figure* 2*d*).

The absolute levels of all intracellular solutes (K^+, Cl^-, galactosylglycerols and amino acids; *Figure* 2*b*, *c*, *e*, *f* respectively) diminish during the first half of the cycle and increase during the second half, exhibiting qualitatively similar trends to photosynthetic activity (*Figure* 2*d*). Changes in intracellular KCl content during a *quasi*-estuarine cycle are more pronounced than those exhibited by organic osmotica (galactosylglycerols and amino acids).

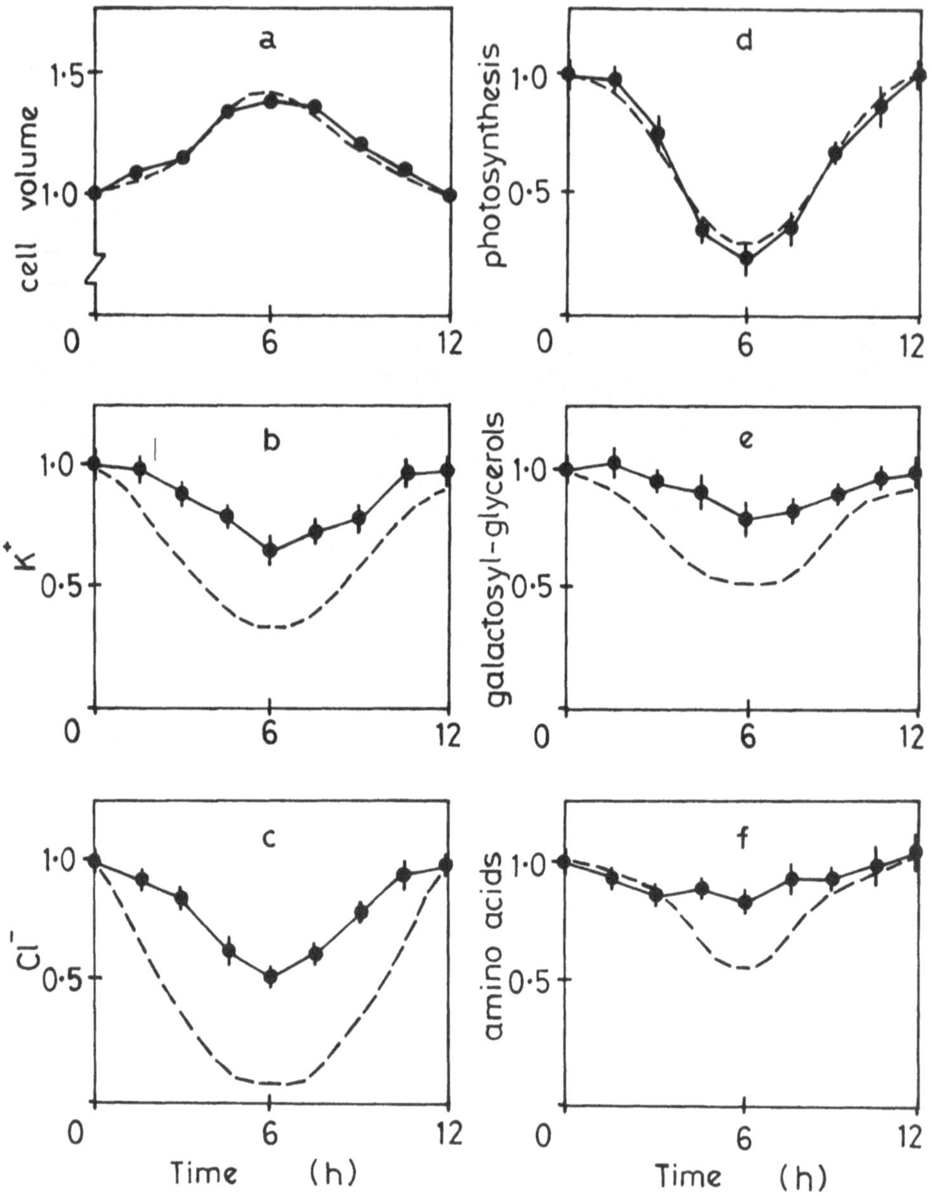

Figure 2. Physiological responses of *P. purpurea* during a single sinusoidal cycle of reduced salinity. Data are shown for (*a*) cell volume, (*b*) K^+, (*c*) Cl^-, (*d*) photosynthetic activity, (*e*) galactosyl-glycerols and (*f*) amino acids. All values are expressed relative to a control value of 1.0. Dotted lines indicate response predicted from steady state data.

Net photosynthesis will serve as a major source of energy, which is required to power ion transport mechanisms (both accumulative and exclusive, see 6) as well as providing the fixed carbon which is necessary for synthesis of organic osmotica. The reduction in photosynthetic activity observed during *quasi*-estuarine cycles of reduced salinity will be accompanied by an increase in the minimal energy required to maintain the observed ion distribution across the cell membrane (5), Cells of *P. purpurea* must commit a larger fraction of their energy budget to processes involving ion transport under steady state or fluctuating hyposaline conditions. The decrease in cellular galactosyl-glycerol content observed in dilute seawater media may therefore reflect an increase in energy demand coupled to a decrease in overall energy production. Such changes will form an integral part of the strategy of osmotic adaptation in *P. purpurea*, and may not necessarily be involved in "osmoregulation" *sensu stricto*.

The extent of change under *quasi*-estuarine conditions varies according to the individual parameter studied (*Figure* 2). Cell volume and photosynthetic activity follow courses which might have been predicted from steady state experiments (*Figure 2a, d*). However, inorganic and organic solutes show a damped response, *i.e.* changes during a sinusoidal cycle of reduced salinity are less than predicted from steady state hyposaline data (*Figure 2b, c, e, f*). This may indicate that the mechanisms responsible for initiating changes in internal solute concentration respond differently to steady and fluctuating signals.

An understanding of the divergent responses under steady state and *quasi*-estuarine conditions will require further data concerning the exact nature of the detector and effector systems in hyposaline media.

Acknowledgement

Financial support was provided by the Science Research Council, the Royal Society, the University of Liverpool and the University of Dundee.

References

(1) Reed, R.H.: The osmotic responses of *Porphyra purpurea* (Roth) C.Ag., *Ph.D. Thesis, University of Liverpool* (1979).

(2) Reed, R.H., Collins, J.C., and Russell, G.: The effects of salinity upon cellular volume of the marine red alga *Porphyra purpurea* (Roth) C.Ag., *J. Exp. Bot., 31* (*in press;* 1980).

(4) Reed, R.H., Collins, J.C., and Russell, G.: The effects of salinity upon galactosyl-glycerol content and concentration of the marine red alga *Porphyra purpurea* (Roth) C.Ag., *J. Exp. Bot., 31* (*in press;* 1980).

(3) Reed, R.H., Collins, J.C., and Russell, G.: The influence of variations in salinity upon photosynthesis in the marine red alga *Porphyra purpurea* (Roth) C.Ag. (Rhodophyta, Bangiales), *Z. Pflanzenphysiol., 98,* 183-7.

(5) Reed, R.H., Collins, J.C., and Russell, G.: The effects of salinity upon ion content and ion transport of the marine red alga *Porphyra purpurea* (Roth) C.Ag., *J. Exp. Bot., 32* (*in press;* 1981).

(6) Reed, R.H., and Collins, J.C.: The ionic relations of *Porphyra purpurea* (Roth) C.Ag. (Rhodophyta, Bangiales), *Plant, Cell, Envt., 3* (*in press,* 1980).

COMPARATIVE PHOTOSYNTHETIC STUDIES ON BENTHIC SEAWEEDS

III. Effect of Light Intensity on Photosynthesis of Intertidal Benthic Red Algae

C. K. Tseng, Baicheng Zhou and Zhungzheng Pan

Institute of Oceanology, Academia Sinica, Qingdao, China

Vertical distribution of the benthic seaweeds is much more complicated than generally accepted. One of the factors influencing vertical distribution is the relationship between the intensity and quality of light and photosynthesis. Because of the obvious compensational adaptation, the red seaweeds can efficiently utilize the green light(1-3), and are dominant in the deeper parts of the sea. At Qingdao there is no obvious division of the coastal region vertically into distinctive belts each dominated by one group of seaweed. Generally, in different levels, the green, the brown and the red seaweeds grow together. In the intertidal zone, the mixed growth phenomenum is especially conspicuous. Light intensity is especially strong in the intertidal zone, and therefore the adaptation of the red seaweeds to the strong light becomes an especially prominent problem. According to Yokohama(4), the photosynthetic light curve of the photosynthesis of red algae in the intertidal and upper sublittoral zone (about 1 meter below the lowest tidal belt) belongs to the sun type and only the red algae in deeper part of the sea are really of the shade type.

There are many species of red algae in the intertidal zone at Qingdao. According to the conditions of the light factor, three groups may be distinguised: the low light intensity group consisting of a few red seaweeds,

Xth International Seaweed Symposium
Printed in Germany

growing beneath the large kelps and in places shaded by rocks; the high light intensity group consisting also of a few red seaweeds growing in the upper littoral zone; and the intermediate group of the majority of the red seaweeds growing on rocks in the middle and lower littoral zone. The present study deals with the photosynthesis of the red algae of these three groups in relation to the light intensity. Twenty species of intertidal benthic red algae were studied on their photosynthetic rate as measured by continuously recording oxygen electrode system[5] at 20°C in relation to the light intensity.

On the basis of the photosynthetic light curve (figs. 1-2) obtained, the 20 species of red algae under investigation fall into three groups: (1) those growing in the higher intertidal zone, their photosynthetic light saturation values are higher than 40,000 lux; (2) the shade forms in the intertidal zone have their photosynthetic light saturation values at about 10,000 lux; (3) the exposed intertidal forms have their photosynthetic light saturation values between 20,000 and 40,000 lux.

Belonging to the first group are Gelidium divaricatum, Gloiopeltis furcata Gymnogongrus flabelliformis and Porphyra yezoensis. The Gelidium and the Gloiopeltis grow on rocks in the upper littoral zone, and in the Qingdao region are the red seaweeds occurring in the upper most part of the intertidal zone. The Porphyra grows on raft, floating near the sea surface, receiving strong radiation all the time. There are two kinds of Gymnogongrus. Some of them are yellowish green in color, characterically sun forms, with very high photosynthetic light saturation point. The other kind grows in places shaded by rocks, appears purplish red in color and has photosynthetic light saturation value at about 18,000 lux.

In the second group are Acrosorium yendoi growing on the holdfast of Un-

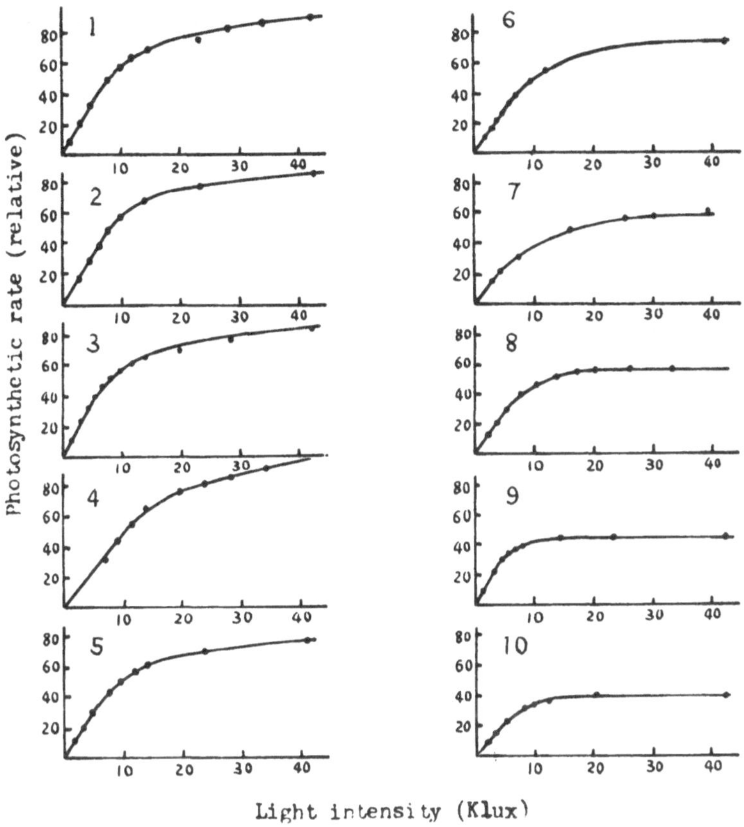

Fig. 1. The photosynthetic rates of inter-tidal red algae as a function of light intensity

1. Gloiopeltis furcata ; 2. Gelidium divaricatum ;

3. Porphyra yezoensis ; 4. Gymnogongrus flabelliformis ;

5. Gelidium amansii ; 6. Ceramium kondoi ;

7. Polysiphonia japonica ; 8. Halymenia sinensis ;

9. Heterosiphonia japonica; 10. Plocamium telfairiae .

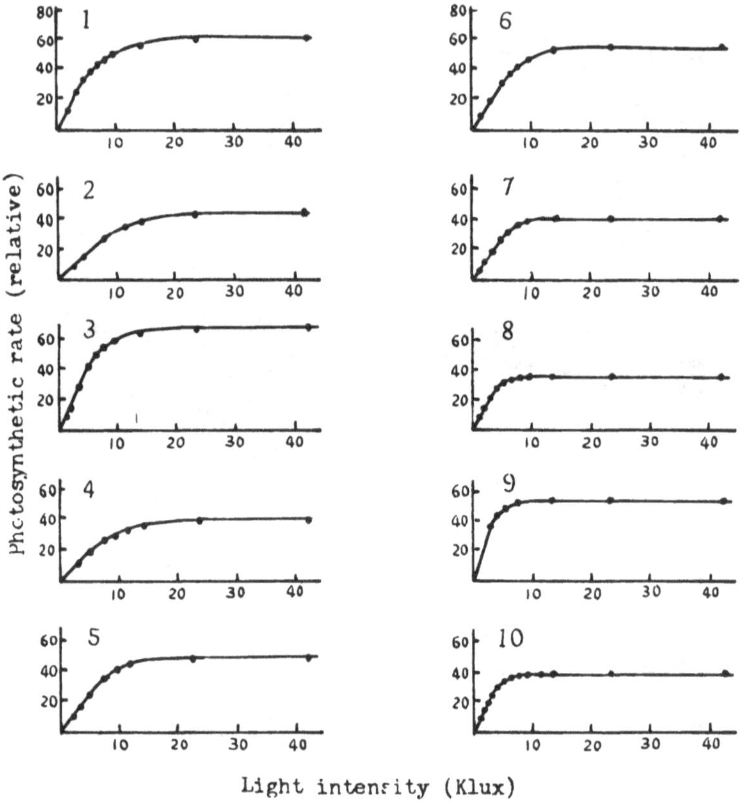

Fig. 2. The photosynthetic rates of inter-tidal red algae as a function of light intensity

1. Polysiphonia urceolata ; 2. Corallina officinalis ;
3. Heterosiphonia japonica ; 4. Gracilaria verrucosa ;
5. Chondrus sp. ; 6. Campylaephora hypnaeoides;
7. Symphyocladia latiuscula ; 8. Gracilaria taxtorii ;
9. Gigartina intermedia ; 10. Acrosorium yendoi .

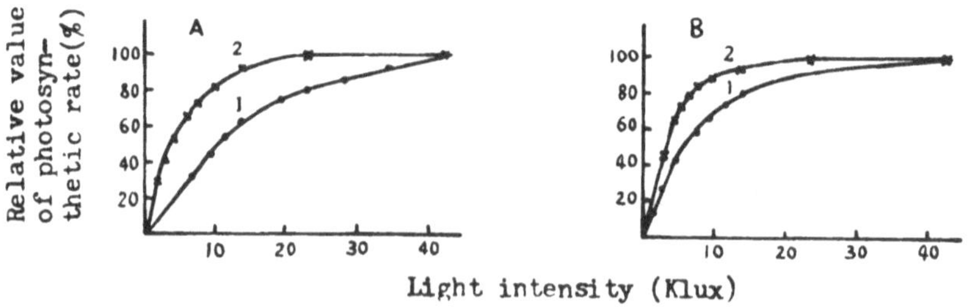

Fig. 3. Photosynthesis Rate in Relation to Light Intensity. A. Gymnogongrus flabelliformis : (1) Sunny (2) Shade. B. Gelidium amansii : (1) Lower littoral (2) After 15 days room culture

daria pinnatifida and Gigartina intermedia growing at the base of Sargassum pallidum. These two small red seaweeds were all shaded by the large seaweeds.

Most of the intertidal red seaweeds belong to the third group, including Campylaephora hypnaeoides, Ceramium kondoi, Chondrus sp., Corallina officinalis, Gracilaria textorii, G. verrucosa, Halymenia sinensis, Heterosiphonia japonica, Hyalosiphonia caespitosa, Plocamium telfariae, Polysiphonia japonica, P. urceolata and Symphyocladia latiuscula. These seaweeds are all from the middle and lower littoral zone, a few of which grows also in the sublittoral zone.

Thus, it shows clearly the close relationship between the photosynthesis of the red algae and the light intensity to which they are subjected. The only exception is Gelidium amansii, which normally grows in the sublittoral region, and appears purplish red in color. According to the report of Yokohama(6), the Gelidium growing in one meter below the low tide region has its photosynthetic light saturation value at about 20,000 lux and that growing in 7-10 meters below low tide zone lower than 10,000 lux. The Gelidium we collected at the low tide zone was light yellowish green in color and had its photosynthetic light saturation value higher than 40,000 lux. It is of interest to note that after being cultured for 15 days in the laboratory, the Gelidium had its light saturation point drop down to about 18,000 lux. This shows that the photosynthesis of the Gelidium is quite well adapted to the change of the light intensity.

The above results show that, similar to the benthic green and brown algae (6-8) the intertidal red algae, with the exception of the few shade forms, have their photosynthetic light saturation value higher than 20,000 lux. Gloiopeltis furcata and the other three species belonging to the first group acts very similarly to the green alga Ulothrix flacca, their photosynthetic light saturation value being even higher than that of the intertidal green and brown algae. Therefore the benthic red algae may also be classified into two general groups, one adapted to high light intensity and one to low light intensity. This reflects the actual condition of the vertical distribution of the benthis red algae.

References

1. Levring, T. : Göteborg K. Vetensk. Vitterh.-Samh. Handl. 5, 1-90(1947).
2. Haxo, F. T., Blinks, L. R. : J. Gen. Physiol. 33, 389-422(1950).
3. Halldal, P. : N. G. Jerlov and E. Steemann Nielsen (eds.) Optical Aspects of Oceanography, Acad. Press, London · New York 1974. p. 345-360.
4. Yokohama, Y. : Bull. Jap. Soc. Phycol. 21, 119-124 (1973).
5. Zhou, B. C., Pan, Z. Z. : Haiyang yu Huzhao Tongbao, 1979(2), 59-64 (1979). (in Chinese)
6. Yokohama, Y. : Bull. Jap. Soc. Phycol. 21, 70-75 (1973).
7. Kageyama, A., Yokohama, Y. : Bull. Jap. Soc. Phycol. 22, 119-123(1974).
8. Zeng, C. K. (Tseng, C. K.), Zhou, B. C., Pan. Z. Z. : Oceanol. et Limnol. Sinica 11, 134-140(1980).

UPTAKE AND INCORPORATION OF ^{35}S INTO CARRAGEENAN AMONG DIFFERENT STRAINS OF *CHONDRUS CRISPUS*

E. Tveter-Gallagher, D. Cheney*, and A. C. Mathieson
Jackson Estuarine Laboratory, University of New Hampshire
Durham, New Hampshire

Introduction

Chondrus crispus is the primary source of the commercially important sulfated polysaccharides kappa and lambda carrageenan. These phycocolloids, along with iota carrageenan, form the basis of a multi-million dollar carrageenan industry. Although a considerable amount of information is available on various aspects of the chemistry (1-4) and production-ecology (5-7) of carrageenan, little is known about its sulfation (8,9) or cytological localization (10,11). A preliminary report by Loewus et al. (8) demonstrated the incorporation of ^{35}S into KCL-soluble and-insoluble fractions of carrageenan. More recently, Jackson & McCandless (9) suggested that the rates of sulfate uptake in *C. crispus* were similar in haploid and diploid plants from a given area cultured for a similar time, while plants from different sources showed varying uptake and incorporation rates.

In the present account, we describe the uptake and incorporation of ^{35}S into carrageenan among several strains of *Chondrus crispus* having different growth rates and carrageenan yields (12). Although our results are preliminary in nature, they suggest that ^{35}S uptake rates may reflect the relative growth status of a plant and be potentially useful in screening fast-growing strains for mariculture.

In addition, the localization of ^{35}S incorporated into *Chondrus*

*Present address: Dept. of Biology, Northeastern University, Boston, Massachusetts.

Xth International Seaweed Symposium
Printed in Germany

tissue was determined using ultrastructural autoradiographic techniques, in order to better understand the site of carrageenan synthesis.

Materials and Methods

Five strains of gametophytic and sporophytic plants of Chondrus crispus were employed for ^{35}S uptake and incorporation. Each of the strains had been previously grown at the Environmental Systems Laboratory (ESL) of the Woods Hole Oceanographic Institute, Woods Hole, Massachusetts, where their growth rates (13) and carrageenan yields (12) were characterized. Similar uptake studies were conducted on field plants collected from Dover Point, New Hampshire (43°07'05"N, 70°49'50"W). Prior to the ^{35}S studies, the plants were washed in filtered sea water, cleaned of epiphytes, and divided into 1 g segments ("whole"). Some of the 1 g segments were then sectioned into 11 mm^2 segments to assess the effects of additional wounding on uptake. The "whole" and "segmented" plants were placed in 50 ml of sterile-filtered ASP_{12} sea water (14), containing 2.8 Ci/ml of carrier-free $Na_2{}^{35}SO_4{}^{2-}$; $MgSO_4$ was replaced by $MgCl_2$. The plants were incubated at 7°C under 8000 lx. The rate of uptake was determined by sampling the medium every 24 h over a 4-day period. All experiments were conducted with five replicates.

Carrageenan extractions were performed on oven-dried "whole" and "segmented" portions of Chondrus plants, according to the methods of Craigie and Leigh (15). The percent yields are based on solvent-washed weights, which are approximately 80% of the original plant dry weight. Levels of radioactivity were determined on a Hewlett-Packard liquid scintillation counter; corrections were made for radioactive decay and counter variations.

Sporelings and segmented portions of Chondrus plants were fixed for transmission electron microscopy (TEM) according to the methods of Tveter & Mathieson (16). The plants were prepared for ultrastructural autoradiography using Ilford L4 emulsion according to the modified techniques of Salpeter et al. (17) and Farquhar (personal communication).

Results

A comparison of the ^{35}S uptake rates amongst different cultivated strains of Chondrus is shown in Figure 1. In general, the uptake rates were greater for "whole" than "segmented" plants. Conspicuous differences in uptake rates were noted among "whole" plants which appeared to be related to Jan.-Feb. growth rates (Table I). For example, strain CH61, a sporulating tetrasporophyte, showed the lowest uptake rate (i.e., 119 X 10^{-3}cpm/mg dry wt) and no growth. In contrast, strain BH56 exhibited the fastest growth rate (.92 %/day) and the greatest uptake (241 X 10^{-3} cpm/mg dry wt). Strains BH50 and BH46 had similar growth rates (Table I) and comparable uptake rates. Plant BH59 was an exception, as it had moderate uptake rates but exhibited no growth. It should be noted that BH59 was in a relatively young stage of carposporic development. The ^{35}S uptake of four "whole" field-collected plants is also shown in Figure 1. Overall, the tetrasporic plants showed a higher uptake rate (approximately 30%) than the cystocarpic ones after 24 and 48 h.

The incorporation of ^{35}S into extractable carrageenan appears to be related to sulfate uptake (Table I). For example, 56% of the sulfate taken up by plant BH56 after 48 h was incorporated into raw carrageenan. In contrast, plant CH61 exhibited the lowest uptake (119 X 10^{-3}cpm/mg dry wt) and incorporation rates (39%). All of the plants examined had approximately the same carrageenan yields in Jan. (Table I).

Electron microscopic autoradiograms (Fig. 2-4) indicate that young sporelings and cortical cells of mature Chondrus plants contain the normal complement of organelles (e.g., 16) and that no apparent radiation damage is evident. Silver grains,

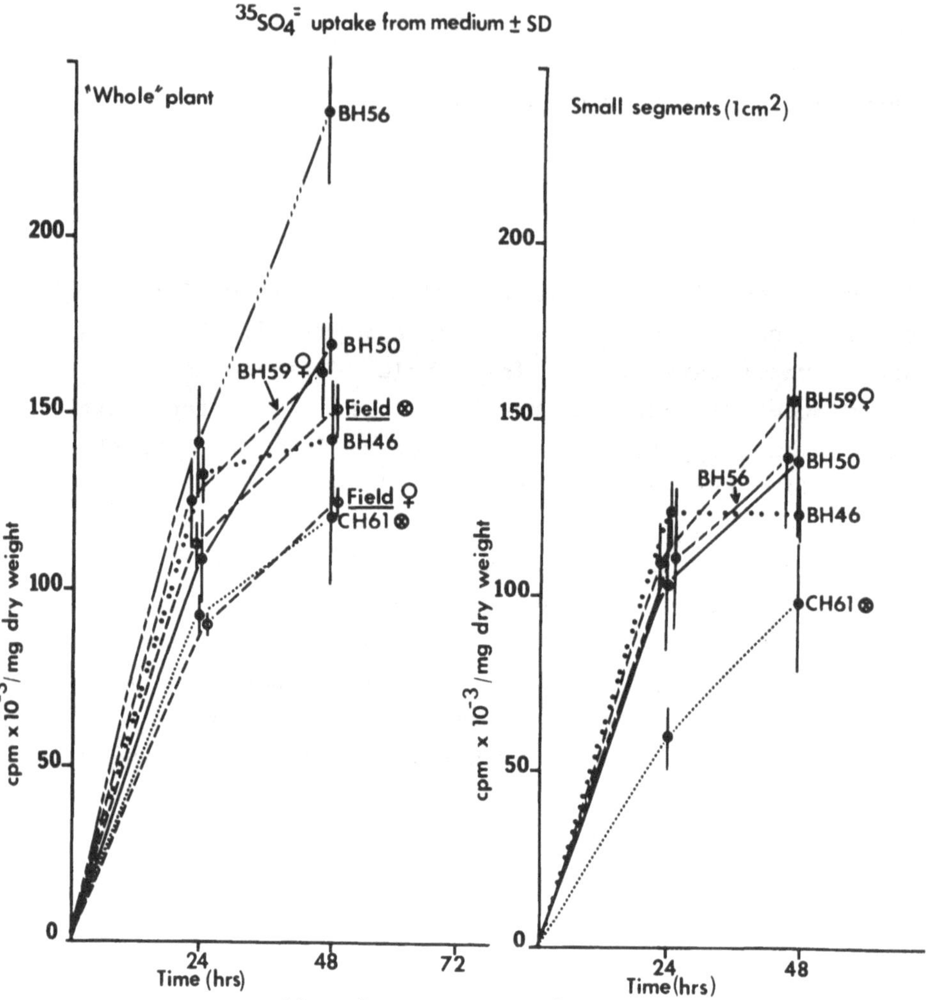

Fig. 1. Amounts of $^{35}SO_4^{2-}$ *(cpm X* 10^{-3}*/mg dry wt plant tissue) removed from the medium by "whole" (A) and "segmented" (B) plants after 24 and 48 h.*

representing disintergration patterns of beta particles from ^{35}S atoms, are evident within the cytoplasm and cell walls of both types of plants (Fig. 2-4). An analysis of grain distribution (17) indicates that much of the activity within the cytoplasm is associated with golgi bodies and Floridean starch granules (Fig. 3).

Discussion

The sulfate uptake and incorporation rates reported in the present study appear to be greater than those noted by Loewus et al. (8) and Jackson & McCandless (9). Such differences in ^{35}S assimilation may be due to experimental design (e.g., the specific activity and molarity of $^{35}SO_4^{2-}$), as well as possible biological differences between plants. In addition, plant preparation appears to be a major factor affecting uptake as "segmented" plants removed less sulfate from the medium than "whole" plants (Fig. 1), presumably due to differential wound respiration.

As noted by Mathieson and Norall (18), the reproductive status and "environmental history" of Chondrus may be critical in determining its physiological state. The present study documents extreme differences in ^{35}S uptake amongst plants of varying reproductive status (Fig. 1). For instance, the sporulating tetrasporophytic strain CH61 exhibited the lowest incorporation rates

Table I. $^{35}SO_4^{2-}$ uptake and incorporation rates (cpm X 10^{-3}/mg dry wt) of "whole" (W) and "segmented" (S) strains of Chondrus compared to their growth rates (%/day) and carrageenan yields at ESL. Activity of carrageenan is based on 2 samples except where indicated.

CULTURE DATA			$^{35}SO_4^{=}$ (cpm x 10^{-3}/mg dry wt.)					
PLANT NO.	CARRA-GEENAN YIELD(%)	GROWTH %/day	UPTAKE (MEDIUM) W	UPTAKE (MEDIUM) S	CARRAGEENAN ACTIVITY W	CARRAGEENAN ACTIVITY S	PERCENT INCORPORATION W	PERCENT INCORPORATION S
BH56	55	.92	241	141	135	62	56	44
BH50	52	.47	165	138	79	62	48	46
BH46	55	.52	143	121	no data	49*	-	41
BH59	50	no growth	161	159	77*	65	48	41
CH61	54	no growth	119	93	46	27	39	28

*based on a single sample

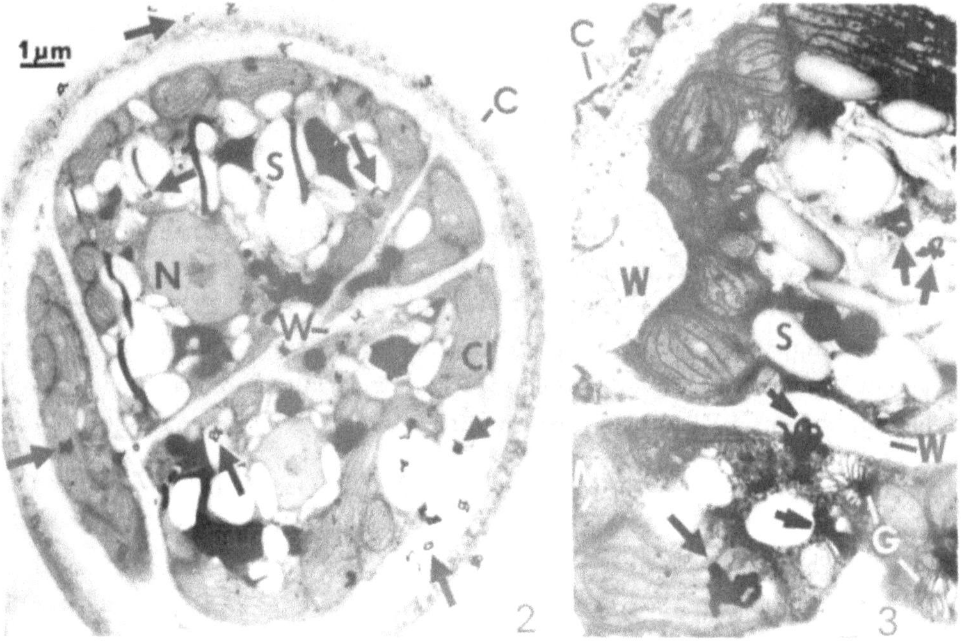

Fig. 2&3. Electron microscopic autoradiograms of 2 day old *Chondrus* *carpospores grown in* $^{35}SO_4{}^{2-}$ *for 48 h with no chase. Fig. 2. Note cell divisions and distribution of silver grains throughout the cytoplasm and cell walls (W). Fig. 3. Grains appear to be associated with golgi bodies (G) and Floridean starch (S). Cuticle (C), chloroplast (Cl), mitochondrium (M), nucleus (N).*

of all plants tested, while the younger field-collected tetrasporic plants had much higher rates of uptake. Jackson & McCandless (9) suggested that the rate of ^{35}S uptake in Chondrus is comparable for gametophytic and sporophytic plants, even though there are varying degrees of sulfation between the two generations. However, we observed a 30% differential in uptake rates (Fig. 1) between tetrasporic and cystocarpic plants collected from the same field site(Dover Point). Concurrently, we did not observe a similar differential uptake rate for the two generations in culture. Thus, a greater number of plants should be evaluated throughout the year and tested in their vegetative state to determine if differential uptake exists.

The rate of assimilation and incorporation of sulfate into

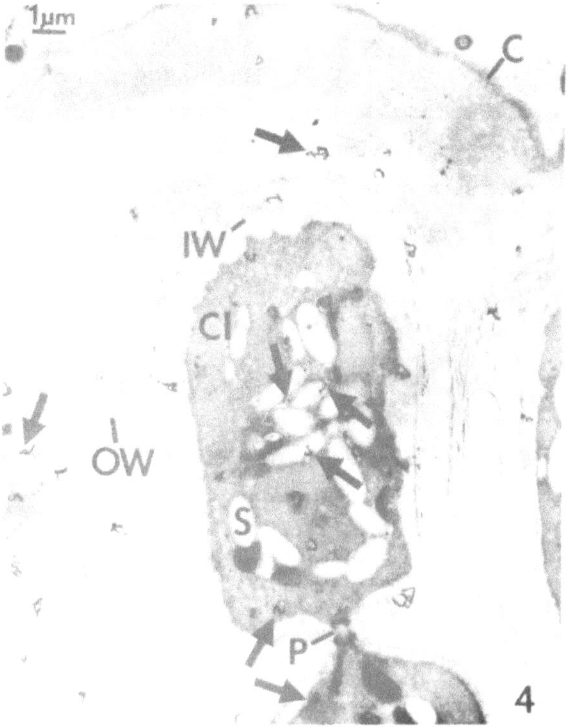

Fig. 4. Electron microscopic autoradiogram of BH59 grown in $^{35}SO_4^{2-}$ for 48 h, no chase. Silver grains are distributed throughout the cytoplasm and cell walls as in the young sporelings shown in Figures 2&3. Inner wall (IW), outer wall (OW), chloroplast (Cl) Floridean starch (S), pit connection (P), cuticle (C).

extractable carrageenan amongst the different strains appeared to be directly related to their growth rates (Table I). Such results are of considerable interest as they suggest that ^{35}S uptake rates may provide a rapid method for selecting fast-growing strains for mariculture; however, further studies should be conducted using plants with varying growth rates.

The autoradiographic studies indicated the presence of ^{35}S within both the cytoplasm and cell walls of Chondrus plants. Such results are in contrast to those reported by LaClair & Dawes (10) in the red alga Eucheuma, which indicated that radioactive ^{35}S was restricted to cell walls. The high concentration of ^{35}S within the cytoplasm of Chondrus is a good indication that the golgi is involved in synthesis of carrageenan, as has been shown for structural polysaccharides in brown algae (19-23). Further pulse-chase studies, employing ultrastructural autoradiographic techniques are presently underway to determine sulfate transport. Such studies may be helpful in determining the biosynthesis of carrageenan.

Acknowledgements

We wish to thank Dr. John Ryther, Director of ESL, Dr. Marilyn Farquhar, Yale School of Medicine, Miss Jan Belmonte, and Miss Valerie Hovey for their assistance in this project.

The investigation was supported by the Office of Sea Grant Programs of the National Oceanic Atmospheric Administration as Research Grant numbers 4-21130 and 4-21129. Published as Scientific Contribution Number 114 from the New Hampshire Agriculture Experiment Station; also issued as contribution number of the Jackson Estuarine Laboratory.

References

1. Rees, D. A. Structure, confirmation and mechanism in the formation of polysaccharide gels and networks. Adv. Carbohydr. Chem. Biochem. 24:267-332. (1969)

2. Stancioff, D.J. and Stanley, N.F. Infared and chemical studies on algal polysaccharides. Proc. 6th Intern. Seaweed Symposium, Marina Mercante, Madrid. 595-609 (1969).

3. Wong, K.F. and Craigie, J.S. Sulfohydrolase activity and carrageenan biosynthesis in *Chondrus crispus* (Rhodophyceae). Plant Physiol. 61:663-666. (1978).

4. McCandless, E.L. and Craigie, J.S. Sulfated polysaccharides in red and brown algae. Ann. Rev. Plant Physiol. 30:41-53 (1979).

5. Chen, L.C-M and McLachlan, J. The life history of *Chondrus crispus* in culture. Can. J. Bot. 50:1055-1060. (1972).

6. McCandless, E.L., Craigie, J.S. and Walter, J.A. Carrageenans in the gametophytic and sporophytic stages of Chondrus crispus. Planta (Berl.) 112:201-212. (1973)

7. Mathieson, A.C. and Tveter, E. Carrageenan ecology of Chondrus crispus Stackhouse. Aquat. Bot. 1:25-43. (1975)

8. Loewus, R., Wagner, G., Schiff, J.A., and Weistrop, J. The incorporation of ^{35}S-labeled sulfate into carrageenans in Chondrus crispus. Plant Physiol. 48:373-375. (1971).

9. Jackson, S.G., and McCandless, E.L. Incorporation of ^{35}S Sulfate and ^{14}C bicarbonate into karyotype specific polysaccharides of Chondrus crispus. Plant Physiol. 64:585-589 (1979).

10. La Clair, J.W. II and Dawes, C.J. An autoradiographic and histochemical localization of sulfated polysaccharides in Eucheuma nudum (Rhodophyta). J. Phycology, 12(4):368-375 (1976).

11. McCandless, E.L. Okada, W.T., Lott, J.N.A., Vollmer, C.M. and Gordon-Mills, E.M. Structural studies of Chondrus crispus: the effect of extraction of carrageenan. Can. J. Bot. 55(15):2053-2064. (1977).

12. Cheney, D., Mathieson, A.C. and Schubert, D. The application of genetic improvement techniques to seaweed cultivation: I. Strain selection in the carrageenophyte Chondrus crispus. Proc. Xth Intern. Seaweed Symposium, Goteborg, Sweden (accepted for publication) (1980).

13. DeBoer, J.A. and Ryther, J.H. Potential yields from a waste-recycling algal mariculture system. In: The marine plant biomass of the Pacific northwest coast. (Ed.) R. Krauss., Oregon State Univ. Press, Corvallis. 231-249. (1977).

14. McLachlan, J. Growth media-marine. In: Handbook of phycological methods. (Ed.) J. Stein. Cambridge Univ. Press N.Y. 25-51. (1973).

15. Craigie, J.S. and Leigh, C. Carrageenans and agars. In: Handbook of phycological methods. (Ed.) J.A. Hellebust and J.S. Craigie. Cambridge Univ. Press, N.Y. 109-131(1978).

16. Tveter-Gallagher, E. and Mathieson, A.C. An electronmicroscopic study of sporeling coalescence in the red alga Chondrus crispus. Scan. Elec. Micro. 3:571-580. (1980).

17. Salpeter, M.M., Bachmann, L. and Salpeter, E.E. Resolution in electron microscope radioautography. J. Cell Biol. 41: 1-20 (1969).

18. Mathieson, A.C. and Norall, T.L. Photosynthetic studies of Chondrus crispus. Mar. Biol. 33:207-213. (1975).

19. Bidwell, R.G.S., Percival,E. and Smestad, B. Photosynthesis and metabolism of marine algae. VIII. Incorporation of of ^{14}C into the polysaccharides metabolized by Fucus vesiculos during the pulse labeling experiments. Can. J. Bot. 50: 191-197. (1971).

20. Fulcher, R.G. and McCully, M.E. Histological studies on the genus Fucus V. An autoradiographic and electron microscopic study of the early stages of regeneration. Can. J. Bot. 49:161-165. (1971).

21. Evans, L.V., Simpson,M. and Callow, M.E. Sulphated polysaccharide synthesis in brown algae. Planta (Berl.) 110: 237-252. (1973).

22. Evans, L.V., Callow, M.E., and Coughlan, S.J. Sulfated polysaccharide synthesis in Fucus (Paheophyceae) zygotes. Proc. 9th Intern Seaweed Symposium, Santa Barbara, Calif. 329-335. (1977).

23. Evans, L.V. and Callow, M.E. Polysaccharide sulphation in Laminaria. Planta(Berl.). 117: 93-95. (1974).

IMMUNOELECTROPHORETIC STUDY OF CHANGES IN CELL WALL ANTIGENS DURING FUCUS EMBRYO DEVELOPMENT

V. Vreeland and W.M. Laetsch
Department of Botany, University of California
Berkeley, California 94720 U.S.A.

Introduction

Cell wall formation in Fucus zygotes has been studied by biochemical, histological and ultrastructural methods.[1-4] The unfertilized egg has no wall, and cell wall assembly begins immediately after fertilization. This primary wall contains alginate and cellulose, and traces of fucan have been detected one hour later.[1] By four hours after fertilization, the wall contains as much fucan as cellulose, and retains its shape after extraction.[1] Alginate contains sequences of mannuronic acid units (M blocks), guluronic acid units (G blocks) and mixed sequences (MG blocks).[5] Fucans contain fucose, sulfate, glucuronic acid, galactose, mannose and glucose.[6] Both types of acidic carbohydrate are variable in composition, size and sugar arrangement.

An immunological approach can provide great sensitivity and specificity for the study of these variable heteropolysaccharides. It can allow the study of specific sequences up to seven sugar units in length, which are present as a very small percentage of the carbohydrate. Antibodies can detect conformational differences in carbohydrates, which can be related to its properties, interactions and functions.[7] This report describes the first application of immunological methods to the study of initial cell wall production, and subsequent changes in wall carbohydrates during early Fucus development.

Methods

Unfertilized eggs from Fucus vesiculosus, and zygotes and embryos from

Xth International Seaweed Symposium
Printed in Germany

F. distichus were shed and cultured according to Quatrano.[8] Embryos were studied during initial wall assembly (1 and 4 hour old zygotes), after rhizoid germination and first cell division (1 day old), and germlings containing several cell types (2 weeks old: carrot-shaped, with apical hairs and interior mucilage). Eggs and embryos were extracted at 55° C for 1 hour, three times, and pooled. They were extracted with either 0.1 M EDTA or 0.2 N HCl, which was followed by extraction in 3% sodium carbonate in the same manner. Using purified 1 day embryo alginate and fucan, and an adult alginate fraction, rabbit antisera were prepared as described.[9] Immunizing preparations were screened for immunological purity by immunodiffusion and rocket-line immunoelectrophoresis using previously available antisera.[9] Embryo extracts were studied by agarose electrophoresis and crossed immunoelectrophoresis.[9] Embryo alginate proved weakly antigenic, and so an antiserum to adult alginate fraction 1 was used. For fucan studies, an antiserum to carbonate-extracted fucan from 2 cell embryos was used.

Results

Carbohydrate electrophoresis in agarose gives several kinds of qualitative information about the acidic carbohydrates in Fucus embryo extracts: 1) relative electrophoretic mobility reflects charge density and/or size differences, 2) retardation by a small amount of calcium in the buffer is interpreted as caused by intermolecular crosslinking of G blocks[10], and 3) M blocks stain purple and G blocks stain blue with toluidine blue.[9] As seen in Figure 1, unfertilized eggs have virtually no high molecular weight alginate, and have only one of the three electrophoretically distinct regions seen in the similar EDTA extract one hour after fertilization. EDTA and carbonate extracts of embryos 1 hour to 2 weeks old show a gradual change of pattern during development. Egg and early embryo samples are blue, while 2 week germling samples are purple. The proportion of high molecular weight material also increases with development. The components of fastest electrophoretic mobility differ for EDTA and carbonate extracts, and decrease during development. In Figure 2, partially

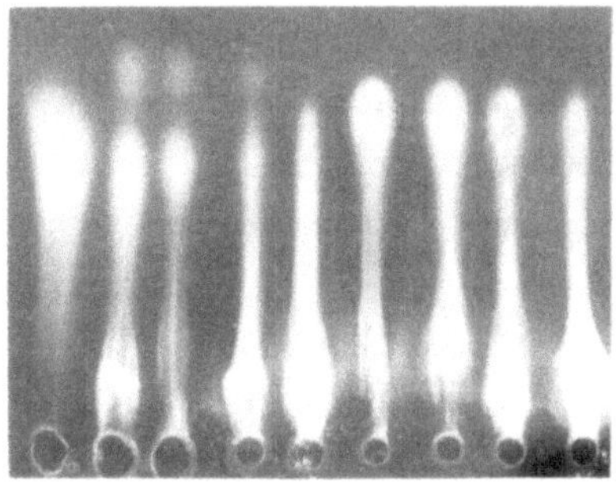

Figure 1. Agarose electrophoresis of *Fucus* egg, zygote and germling extracts in 1 mM EDTA. 1-5 are EDTA extracts, and 6-9 are sodium carbonate extracts. 1, egg; 2, 1 hour zygote; 3, 4 hour zygote; 4, 2 cell embryo; 5, 2 week germling; 6, 1 hour zygote; 7, 4 hour zygote; 8, 2 cell embryo; 9, 2 week germling. All samples are 25 μg, except 1-3 are 400 μg.

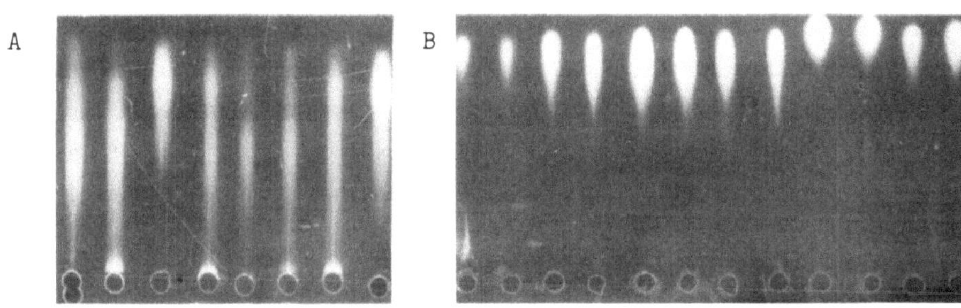

Figure 2. Agarose electrophoresis of partially hydrolyzed alginate from zygotes, germlings and adult *Fucus*, A, in 0.4 mM $CaCl_2$, and B, in 1 mM EDTA. 1-4 are EDTA extracts, and 5-12 are sodium carbonate extracts. 1, 1 hour zygote; 2, 4 hour zygote; 3; 2 week germling; 4, 2 cell embryo; 5, 1 hour zygote; 6, 4 hour zygote; 7, 2 cell embryo; 8, 2 week germling; 9, *Laminaria digitata* M blocks; 10, *F. distichus* G blocks; 11, *F. distichus* adult fraction 1; 12, *F. distichus* adult fraction 2. All samples are 5 μg, except 5 is 25 μg and 6 is 10 μg.

hydrolyzed extracts up to the two cell stage are blue, while 2 week germling alginate resembles adult alginate in being purple. Embryo alginates in EDTA (B) have not been degraded to the size of blocks, but have components of lower mobility than adult fractions prepared in the same manner. In calcium (A), embryo alginates up to the two cell stage are greatly retarded and stain blue.

In Figure 3, crossed immunoelectrophoretic patterns of fucan and alginate antigens show loss of antigenic components when the embryos have been pre-extracted with HCl compared to samples pre-extracted with EDTA. In Figure 4, alginate crossed immunoelectrophoretic patterns of acid pre-extracted samples increase in intensity of the immunoprecipitate and in completeness toward the electrophoretically fast side of the peak. The upper peak in the 4 hour zygote is completed laterally in the samples extracted with and following EDTA (see Figure 3B, left). It is largest in the 4 hour carbonate extract.

Discussion

Data presented in Figures 1, 2 and 4 show qualitative changes in carbohydrate electrophoretic patterns which are consistent with increasing similarity to adult Fucus carbohydrates as development progresses. Quantitative immunoelectrophoretic data (unpublished) also support this conclusion. Unfertilized egg EDTA and carbonate extracts gave a reaction similar to G blocks in rocket immunoelectrophoresis (unpublished result). Fucan is barely detectable in 1 hour zygotes chemically,[1] but shows a considerable and intense immunoelectrophoretic reaction. During primary wall deposition, early zygotes appear to contain guluronic-rich alginate of low molecular weight, possibly in unusually long homopolymeric sequences, and much of which may be calcium cross-linked to more tightly bound alginate. Between the two cell stage and the complex two week old germling stage, alginate becomes less retarded by calcium, richer in mannuronic acid and more antigenic, more closely resembling adult alginate.

Although acid pre-extraction is used when extracting alginate,[1,5] increasing exposure to acid caused progressive degradation of carbohydrate antigens, and should be minimized for immunochemical analysis. An alginate antigenic component is described which appears to be specific for zygotes and early embryos. The antiserum used was to adult alginate, and so the embryo specific quality must reside with the arrangement of antigenic determinants on separate molecules, or with the same determinant on "carrier" regions of alginate which differ in precipitability. Monospeci-

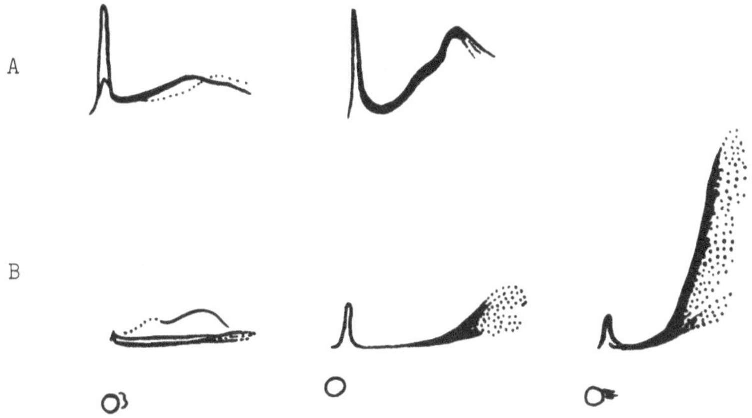

Figure 3. Effect of acid pre-extraction on crossed immunoelectrophoretic pattern of fucan (A) and alginate (B) from sodium carbonate extracts of 2 cell embryos. Left, following EDTA pre-extraction; middle, following HCl pre-extraction; right (on B only), partially hydrolyzed. The extracts were first electrophoresed in agarose to the right, and subsequently to the top into agarose containing anti-fucan serum (A) or anti-alginate serum (B).

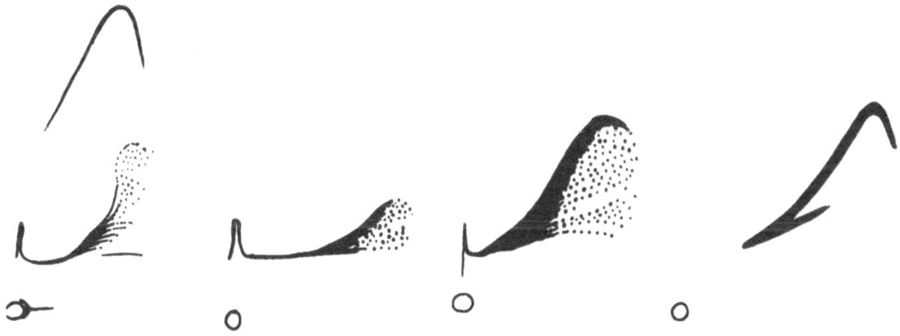

Figure 4. Crossed immunoelectrophoretic patterns of alginate in carbonate extracts which followed HCl pre-extraction. From the left: 4 hour zygote, 2 cell embryo, 2 week germling, and adult alginate fraction 1. The last sample was partially hydrolyzed.

fic antibodies are needed due to such situations, so that the nature of antigenic changes in cell wall carbohydrates during early development can be clarified.

Acknowledgements. This research was supported by NSF Grant PCM 78-15988. We thank J. West, T. DeCew, A. Nonomura, C. Tarp, E. Mastro and J. Fiore for collecting *Fucus* plants. M blocks were the gift of B. Larsen.

References

1. Quatrano, R.S., Stevens, P.T.: Plant Physiol. 58, 224-231 (1976).
2. Vreugenhil, D., Dybstra, M.L., Libbenga, K.: Protoplasma 88, 305-313 (1976).
3. Brawley, S.H., Wetherbee, R., Quatrano, R.S.: J. Cell Sci. 20, 255-271 (1976).
4. Callow, M.E., Evans, L.V., Bolwell, G.P., Callow, J.A.: J. Cell Sci. 32, 45-54 (1978).
5. Haug, A., Larsen, B., Smidsrød, O.: Acta Chem. Scand. 21, 691-704 (1967).
6. Percival, E.: Brit. Phycol. J. 14, 103-117 (1979).
7. Preston, R.D.: Ann. Rev. Plant Physiol. 30, 55-78 (1979).
8. Quatrano, R.S.: in, Handbook of Phycological Methods, Developmental and Cytological Methods, ed. Gantt, E., Cambridge University Press, New York, 59-68 (1980).
9. Vreeland, V., Chapman, D.J.: J. Immunol. Meth. 23, 227-240 (1978).
10. Morris, E.R., Rees, D.A., Thom, D., Boyd, J.: Carbohyd. Res. 66, 145-154 (1978).

AGAR FROM AN INTERTIDAL POPULATION OF GRACILARIA SP.

J.N.C. Whyte and J.R. Englar

Fisheries and Oceans Canada, Technology Research Laboratory
6640 N.W. Marine Drive, Vancouver, B.C. Canada V6T 1X2

Introduction

Agar consists of a spectrum of closely related polysaccharides which differ in the extent of substitution of the repeating disaccharide unit (composed of 1,3-linked β-D-galactopyranosyl and 1,4-linked 3,6-anhydro-α-L-galactopyranosyl units) with methoxyl-, sulphate ester-, 4,6-carboxyethylidene-, and glucuronic acid-groups (1,2). Properties of agar, reflecting degrees of substitution, differ in general with generic type; Gelidium species provide rigid gelling agars useful for bacteriological and biomedical applications while Gracilaria species afford soft pliable gelling agars suitable for the food industry (3,4). Nevertheless, quality of agar from the Gracilaria genus also appears to be dictated by season (5,6), environment (7), species (8) and morphotypes (9). This paper presents the seasonal variation in quantity and quality of agar from Gracilaria pseudoverrucosa variant (9), an assessment of the recoverability of the agar involving an additional acid treatment during extraction, and the possible role of this polymer in the alga's intercellular matrix.

Material and Methods

Gracilaria was collected at monthly intervals in 1975 from the head of Bamfield Inlet, Vancouver Island, at a site consisting of fine sand and mud flats. This intertidal population consisted of a free floating mat of vegetative fragments each 12 cm approximately in length and varying in colour from brown to yellow depending on exposure to light. Agar from monthly samples was extracted with boiling water and isolated by ethanol

Xth International Seaweed Symposium
Printed in Germany

precipitation. Details of this procedure together with chemical and physical measurements on 1% aqueous solutions have been described previously (9).

Isolation of agar before and after acid treatment. A composite sample of *Gracilaria* from June and August (26.6g) was extracted (x4) with chloroform-methanol-water (1:2:0.5, 600 ml) (10) to yield a trisolvent soluble fraction (8.5g; 32% of dry alga) and polymeric residual material (18.1g; 68%). This polymeric residue I (18g) was extracted (x4) with water (1L) at 95-100° for 1 hour then centrifuged. Combined centrifugate, reheated to 95°, was then clarified by passage through a Sharples centrifuge, concentrated, and freeze dried to yield crude agar (10.25g; 38.7% of dry alga). Crude agar (9.8g) dialyzed overnight against running water provided a freeze dried polymer Agar A (8.74g). Agar A (4.09g) in water (400ml) was precipitated by addition of ethanol (5 volumes) to yield Agar Ap (3.19g) and by evaporation of the aqueous-ethanol solution, Agar As (0.90g). Polymeric residue II from aqueous extraction was suspended in water (1.7L) and stirred at 21° for 24 hours at pH 5.0 using 1N H_2SO_4 for pH control. Stirring was continued for a further 6 hours at 80° with checks and adjustments to pH 5.0 being made when necessary. On cooling, the mixture was neutralized with 0.1N NaOH and extracted as described previously. Centrifugate was clarified by passage through a Sharples centrifuge, dialyzed and freeze dried to afford Agar B (1.54g). Agar B (0.79g) on dissolution in water (80ml) yielded Agar Bp (0.47g) on precipitation with ethanol (5 volumes) and an aqueous-ethanol soluble fraction Agar Bs (0.32g). Polymeric residue III from the second extraction was freeze dried (5.1g; 19.3% dry alga; N, 6.75%).

Results and Discussion

Yields of agars increased from the June sample of *Gracilaria* to the highest inclusion in the August specimen, thereafter a gradual decline followed as the season progressed, Table 1. In general, the 3,6-anhydrogalactose component provides a measure of gelling ability with the highest theoretical level, 47%, corresponding to the idealized agarose molecule devoid of

TABLE 1. Yield Composition and Properties of the Agar Isolated from *Gracilaria* Collected over the Summer Months

Parameter	Month Collected				
	June	July	Aug.	Sept.	Oct.
Yield, % dry alga	26.0	32.2	35.4	34.8	31.0
Anhydrogalactose, %	32.2	31.0	30.1	29.2	30.5
Sulphate, %	3.57	4.52	4.81	4.84	4.91
Gelation temperature, °C	38.8	39.5	39.6	39.8	39.5
Gel melting temperature, °C	82.8	86.7	83.7	80.5	82.4
Viscosity, cps.	27.4	63.0	240	136	173
Gel strength, g/cm^2	190	65	126	95	160
Gel deformation, mm.	3.7	3.6	6.7	5.0	6.9
Gel cohesion, mm.	2.4	2.2	3.5	3.5	3.8
Gel rigidity, $g/cm^2/mm$.	40.1	15.4	18.2	13.4	21.1

charged groups. Only a slight decline in content of this residue was noted throughout the season but when coupled to a progressive increase in sulphate ester content a marked difference in physical properties was recorded Table 1. As the gelation temperatures of the agars varied little, 39.3 ± 0.5°C, the observed 40.7 to 47.2°C hysteresis reflected the variable gel melting temperatures. This parameter which is concentration dependent is also considered dependent on molecular weight distribution (11). However, no correlation could be found between gel melting temperatures and the viscosity values which normally provide a comparison of polymer size. Although these values, Table 1, indicated that agar of higher molecular weight was formed in the latter part of the season no direct correlation existed with corresponding gel strengths suggesting that gelling ability of the agar was dependent on the extent of concurrent unsubstituted sugar residues (uninhibited agarose blocks) within the molecule rather than the molecular size.

As gelation involves association of chains of double helices having all but the 2-hydroxyl group of the D-galactose residue pointing outwards from the helix cavity (12) partial substitution at this position or incomplete anhydro-bridge formation of the L-galactose residue by substitution at the 6-hydroxyl group (13) would inhibit normal coiling and afford inferior gels. A strong rigid gel was provided by the June sample but the following months provided agars with decreased rigidities, Table 1, suggesting inhibition by the sulphate ester groups at these two critical positions in the repeating disaccharide unit. Values for agars in July and September were typical of soft, pliant and weak gels; whereas, the August and October agars provided soft, more pliant, yet stronger gels with increased degrees of deformation, considered to be a result of the higher molecular weight polymers affording greater intermolecular association. With these results, and our other studies, it has been noted that gel rigidity rather than gel strength provides a clearer indication of the type of agar molecules interacting to form the gel matrix. As the season progressed therefore, the gelation characteristics of the agars changed from a brittle, rigid system in June to a more variable softer and more elastic system.

In our continuing studies of agarophytes yields of 11-36% agar have been observed (7,9) although in apparent contradiction polymeric material in these algae constituted 65-75% of the dry weight. An accountability for the agar in the Gracilaria from Bamfield Inlet was therefore performed by systematic extraction, involving an acid treatment and subsequent analysis, Table 2. Losses occurred on dialysis (4%) and in isolation of the polymer by addition of ethanol despite observed apparent completeness of precipitation. Polymer fractionation by ethanol precipitation illustrated non-recovery of agar molecules with high contents of 6-O-methylgalactose and sulphate ester units. These polymers (Agars As and Bs) associated with slightly more protein, also exhibited reduced anhydrogalactose contents and gelation capacities. Polymers recovered on addition of ethanol (Agars Ap and Bp) retained xylose and glucose residues which together with associated protein had no apparent effects on gelling abilities. A comparison of the agars, Table 2, illustrated preferential removal of 6-O methylgalactose containing agar molecules prior to acid

TABLE 2 Yield, Composition and Properties of the Agar Fractions Isolated before (Agars A) and after (Agars B) Acid Treatment

Parameters	Agar A	Agar Ap	Agar As	Agar B	Agar Bp	Agar Bs
Yield, % dry alga	34.5	26.9	7.6	5.8	3.5	2.3
Anhydrogalactose, %	30.9	31.8	28.9	23.1	23.5	23.0
Galactose, %	43.0	44.7	35.0	40.3	44.6	32.7
6-O-Methylgalactose, %	7.4	5.5	12.6	3.8	1.8	5.8
Xylose, %	0.7	0.8	0.2	0.8	1.0	0.2
Glucose, %	5.2	5.6	2.8	10.7	13.1	5.5
Sulphate, %	3.7	3.2	5.4	6.3	5.0	7.9
Nitrogen, %	1.3	1.2	1.5	2.0	1.8	2.1
Gel strength, g/cm^2	74.4	103	21.5	53.3	62.8	22.2
Gel deformation, mm	4.3	4.3	4.4	4.5	5.0	4.0
Gel cohesion, mm	2.7	2.9	2.8	3.2	3.5	2.7
Gel rigidity, $g/cm^2/mm$	13.9	17.6	3.8	8.3	9.0	4.1
Viscosity, cps	67.4	159	-	9.8	-	-

treatment; subsequent release of agar molecules with increased sulphate ester groups and associated glucose, xylose and protein residues followed acid treatment. Removal of cations by acid therefore appeared to disrupt sulphate ester-cation bridges binding these agars in the algal intercellular matrix.

That residual agar, even after exhaustive extraction, was associated with cellulosic and proteinaceous components was demonstrated by the presence of corresponding constituent sugars in the hydrolysate of the polymeric residue III which contained 42.2% protein accounting for 67% of the total protein in the alga. Increased levels of xylose and glucose in Agars Ap and Bp had no effect on gelling abilities suggesting these adjunct molecules, if attached, must be on the hydroxyl groups facing outwards from the helix coils. As cellulose fragments were removed from the agar solutions by high speed centrifugation, the increase in glucose was attributed to soluble glycoprotein conjugates. As a matrix intercellular polymer

(14) the agar may be linked to inner cell-wall polymers (cellulose, xylan) through glycoprotein type bridges or sulphate ester-divalent cation bridges attached to any of the peripheral hydroxyl groups in the double helix coil, three of which in each disaccharide repeating unit are known to point outwards from the helix cavity.

References

1. Araki, C.: Proc. Intl. Seaweed Symp. 5, 3-19 (1966)
2. Duckworth, M., Yaphe, W.: Carbohyd. Res. 16, 189-197 (1971)
3. Tsuchiya, Y., Hong, K.C.: Proc. Intl. Seaweed Symp. 5, 315-321 (1966)
4. Yaphe, W., Duckworth, M.: Proc. Intl. Seaweed Symp. 7, 15-22 (1972)
5. Hoyle, M.D.: Bot. Mar. 21, 347-352 (1978)
6. Oza, R.M.: Bot. Mar. 21, 165-167 (1978)
7. Whyte, J.N.C., Englar, J.R.: Proc. Intl. Seaweed Symp. 9, 437-443 (1978)
8. Duckworth, M., Hong, K.C., Yaphe, W.: Carbohyd. Res. 18, 1-9 (1971)
9. Whyte, J.N.C., Englar, J.R.: Bot. Mar. 23, 277-283 (1980)
10. Whyte, J.N.C., Southcott, B.A.: Phytochem. 9, 1159-1161 (1970)
11. Selby, H.H., Wynne, W.H.: Industrial Gums, Polysaccharides and their Derivatives, p. 23, Acad. Press Inc., New York, (1973)
12. Arnott, S., Fulmer, A., Scott, W.E., Dea, I.C.M., Moorhouse, R., Rees, D.A.: J. Mol. Biol. 90, 269-284 (1974)
13. Rees, D.A.: Adv. in Carbohyd. Chem. Biochem. 24, 267-332 (1969)
14. Akatsuka, I., Iwamoto, K.: Bot. Mar. 22, 367-370 (1979).

STUDIES ON PHENOLIC CONTENT OF SOME BROWN ALGAE FROM ADRIATIC SEA

N. Zavodnik
Center for Marine Research, "Rudjer Bošković" Institute
52210 Rovinj, Yugoslavia

Introduction

The presence of phenolic compounds in brown algae has been known for some time. Their location in physodes - brown algal subcellular bodies - was observed over a century ago (1). Ragan (2) has given a very detailed review of the investigations of physodes, their chemical contents, classes, distribution, biogenesis and physiological role. Recently various methods for estimation the total polyphenols have been developed and tested (3). Chemical composition of phenolic compounds in Phaeophyta and of their exudates were studied by many authors (4). But there is little information on the seasonal variations and the distribution of phenolic compounds during the growth of the algae.

It was the aim of present research to estimate the phenolic content in some brown algae common to the littoral zone in the north Adriatic Sea, to follow their variations during the vegetative period of the investigated species, and to compare the results with comparable algae from the Norwegian Sea. Special interest was paid to the influence of the dessication of intertidal algae on exudation processes which occur after the reimersion of plants at high tide.

The investigated algae belong to two ecologically different groups:

a) Species which occur in the midlittoral zone: <u>Fucus virso-</u>

Xth International Seaweed Symposium
Printed in Germany

ides, Colpomenia sinuosa and Scytosiphon lomentaria. Due to tidal action all these algae can be subjected to periodic dessication, which can be particularly severe under high summer temperatures.

b) Species which are distributed in the upper sublittoral zone: Cystoseira barbata, C. fimbriata, Dictyota dichotoma, Taonia atomaria, Dilophus fasciola, and Halopteris scoparia. In their natural habitats in the North Adriatic, these algae are only accidentally exposed to air during maximum low tides in the winter, when low temperatures and high humidity preclude any serious dessication.

Methods

1) Variations in total phenols. Variations in total phenols were determined for all species listed above. In addition, in Fucus virsoides the phenolic contents were measured also in different parts of the thallus, e.g. in old and new parts.

The whole plants, free from epiphytes, were cut into small pieces with a scissors prior to extraction. The extraction method has been previously described (5). For determination of the phenols in the aqueus concentrate, the Echtrotsalz GG (Schmidt Chromo-Gesellsachaft, Stuttgart) was used instead of Brentamine Fast Red 2 G Salt. According to Ragan and Jensen (3) this compound has chromogenic properties virtually identical to those of Brentamine.

2) Exudation studies. Analyses of exudation were made on six species (Table II). Two experimental approaches were utilized. In the first, algae were allowed to exudation for 1 hour in 1000 ml beakers exposed to direct sunlight immediately after collecting during high tide. In the second series, the algae were exposed for 1 hour to dessication under direct sunlight, and at various temperatures. The

effects of dessication on the quantities of algal exudates were obtained after reimersion of algae in sea water. Other details of methods were given previously (5).

Results and discussion

Among the algae investigated, *Fucus virsoides* contented the highest amount of phenols (2-3 % of d.w.). Of other species, *Cystoseira barbata*, *C. fimbriata*, *Colpomenia sinuosa* and *Scytosiphon lomentaria* contained more phenols than *Dictyota dichotoma*, *Dilophus fasciola* and *Halopteris scoparia* (Table I), in which quantities were sometimes even below the level of detection. The amounts of phenolic compounds reported by Munda (6) for some of the same species, are lower, perhaps due to different experimental methods; but the ratio of phenol contents between the Fucaceae and the Dictyotaceae determined by Munda is confirmed by results reported in this paper.

Table I.: Contents of phenolic compounds in Adriatic brown algae.

Species	Dry weight (%)	Phenolic compounds (% of d.w.)
Fucus virsoides	24	2.4 - 3.3
Scytosiphon lomentaria	11	1.1 - 1.3
Colpomenia sinuosa	9	1.6
Cystoseira barbata	17	0.7 - 1.8
Cystoseira fimbriata	15	1.2 - 2.0
Dictyota dichotoma	14	0.1 - 1.0
Dilophus fasciola	14	0.3
Taonia atomaria	14	1.4
Halopteris scoparia	12	0.6

According to Haug and Larsen (7) phenol concentrations are related to salinity of water: they increase with a lower salinities. The results reported in this paper do not, however, support this hypothesis. The Adriatic algae investigated were growing at salinities of 36-38 ^{o}/oo, and in the Northern Adriatic where these algae were collected, only exceptionally, and for short period, does the salinity decrease to 35 ^{o}/oo or below. In contrast, in Trondhejm Fjord, the salinity is usualy below 30 ^{o}/oo. Yet here only Pelvetia canaliculata was lower in phenols (1-3 % of d.w.); in Ascophyllum nodosum and Fucus vesiculosus the phenolic content of 3-6 % of d.w. was previously estimated (5). Consequently, the phenolic content in brown algae is probably a specific characteristic which does not depend exclusively on environmental conditions, such as, for example, the salinity.

In Fucus virsoides the variations in total phenol content follow mostly variations in algal dry weight. The greatest amounts were found in March and April, during the time of the most luxurious growth of seaweed, followed by a decrease correlated with the riping of receptacules. Similar changes in total phenol content occured in the different parts of the thalli: phenols decreased in the upper part during the fructification, but increased in the lower part until winter. However, the concentration on phenols in the lower part of thalli never reached that found in the upper part in spring time.

Differences in the amounts of algal exudates in brown seaweeds immediately after collecting without previous dessication, and after dessication, are given in Table II. It is evident that the quantities of algal exudates varied from species to species, independently of light intensities and temperature. But after dessication an appreciable amount of exudation occured in all species, and therefore seems to be a process related to the extent of dessication. Tha analyses indicate that in immersed algae which previously have been exposed to dessication, the

Table II.: Exudation of carbohydrates, proteins and phenols (mg/100 g d.w./h) by some Adriatic brown seaweeds.

Species	Temp. (^oC)	Light (Lux)	Before Dessication			After Dessication			
			Carbo hydr.	Protein	Phenols	Carbo hydr.	Protein	Phenols	Loss of water (%)
Fucus virsoides	24	56000	95	190	47	107	1690	73	36
Fucus virsoides	19	47000	35	30	36	52	194	43	8
Scytosiphon lomentaria	23	47000	27	353	27	88	792	8	12
Cystoseira barbata	22	52000	260	110	24	300	1013	133	13
Cystoseira fimbriata	17	41000	35	1164	57	78	1158	103	25
Dictyota dichotoma	18	45000	100	278	57	163	614	55	25
Halopteris scoparia	26	63000	40	96	1.4				

phenolic content decreases more than 50 % of total phenols compared with algae which were not dessicated. The maximum exudation occurs in the first hour of immersion under experimental conditions. In this period, Fucus virsoides exude about 1.7 % of its total phenolic content when the algae were not exposed to dessication, and about 3 % if they had been previously dessicated. Similar or even higher values were found for other species: Cystoseira barbata, 1.8 and 12 %; C. fimbriata, 3 and 6 %; and not dessicated, Dictyota dichotoma, 0.12 %; Scytosiphon lomentaria, 1 %; Halopteris scoparia, 0.16 %.

According to Sieburth and Jensen (8), carbohydrate and polyphenolic materials are the usual products of exudation, but also appreciable amounts of proteinaceaous materials are exuded as the result of previous dessication. In our observations, however, all algae exuded large amounts of proteinaceous material independently of the degree of dessication. This is in agreement with our previous results on Norwegian algae (5).

References

1. Crato, E.: Bot. Ztg. 51, 157-195 (1893).
2. Ragan, M.A.: Bot. Mar. 19, 145-154 (1976).
3. Ragan, M.A., Jensen, A.: J. exp. mar. Biol. Ekol. 30, 209-221 (1977).
4. Jensen, A.J., Ragan, M.A.: Tetrahedron Letters 9, 847-850 (1978).
5. Zavodnik, N., Jensen, A.: Proc. 8th int. Seaweed Symp., Bangor (ed. G.E. Foff) (in press).
6. Munda, I.: Acta Adriat. 10, 1, 3-11 (1962).
7. Haug, A., Larsen, B.: Acta Chem. Scand. 12, 650-657 (1958).
8. Sieburth, J.McN., Jensen, A.: in: Organic Matter in Natural Waters (ed. D.W. Hood), Univ. Alaska Occas. Publ. 1, 203-223 (1970).

4. Cultures

GROWTH OF THE CYANOPHYTA *AGMENELLUM QUADRUPLICATUM* CULTIVATED ON GLYCINE OR NITRATE AT VARIOUS TEMPERATURE AND IRRADIANCE CONDITIONS

D. Bonin

Station marine d'Endoume et Centre d'Océanographie, Rue de la Batterie des Lions, 13007 Marseille, France.

N.J. Antia

Pacific Environment Institute, Environment Canada, 4160 Marine Drive, West Vancouver, British Columbia V7V 1N6, Canada.

J. Pelaez-Hudlet

Institute of Marine Resources, A-018, Scripps Institution of Oceanography, UCSD, La Jolla CA 92093, U.S.A.

The Cyanophyta *Agmenellum quadruplicatum* (Menegh.) Brébisson strain PR6 (Van Baalen's collection) has been cultivated on a medium containing nitrate or glycine as the sole nitrogen source to test its growth under different conditions of temperature and irradiance (T ranged from 11 to 40°C and R ranged from 2 to 547 $\mu E\ m^{-2}\ sec^{-1}$).

At the lowest temperature of 11°C, no growth has been observed neither on glycine nor on nitrate, for any irradiance, and even after an incubation period of 55 days. At the lowest irradiance of 2 $\mu E\ m^{-2}\ sec^{-1}$, there has been no more growth whatever the temperature was. On nitrate, the highest growth rates appear with the highest levels of temperature and irradiance. On glycine, the growth rates are always lower than those observed on nitrate for the same physical conditions. Differences between the two corresponding growth rates vary according to the acting couples of physical factors. A similar observation

Xth International Seaweed Symposium
Printed in Germany

can be made about the maximum yields reached at the end of the experiments.

Comparing the growth curves, it is possible to assert that the shorter difference between them appear for the couple of factors as follows : R = 7.6 $\mu E\ m^{-2}\ sec^{-1}$ and T = 29°C. Fair good responses on glycine are also observed with the combinations : R = 7.6 $\mu E\ m^{-2}\ sec^{-1}$ and T = 34°C; R = 98 $\mu E\ m^{-2}\ sec^{-1}$ and T = 40°C; R = 547 $\mu E\ m^{-2}\ sec^{-1}$ and T = 40°C. These data demonstrate that the growth of *Aymenellum quadruplicatum* on glycine is greatly influenced by interactions between physical factors of the environment. The authors emphasize the need for cautious examination of data, obtained by others, concerning utilization of nitrogen organic compounds under fixed physical conditions which may be different, and, then, are not necessarily comparable.

GROWTH RESPONSE OF *CHONDRUS CRISPUS* (RHODOPHYTA, GIGARTINALES) TO LIGHT AND TEMPERATURE IN LABORATORY AND OUTDOOR TANKS CULTURE

J.P. Braud
Ceca S.A., Usine de Baupte, 50500 Carentan, France

R. Delépine
Biogéographie et écologie benthiques, B.V.M., 7 Quai St Bernard
75230 Paris, France

Introduction

Among the numerous parameters conditioning the growth of cultured seaweeds, special attention must be paid to water temperature and solar radiation, these factors being uncontrollable on industrial scale. In France, for solar radiation and seawater temperature respectively, the extreme mean values with regard to season and latitude range from 300-1,800 $J.cm^{-2}.d^{-1}$ (1) and 6-18°C (2) at 51°N to 700-2,800 $J.cm^{-2}.d^{-1}$ (1) and 13-26°C (3) at 43°N. If south seems favourable as to solar radiation, temperature maximum may give rise to overheating problems. On another hand, power stations located on the Channel coast (≈50°N) would be able to deliver seawater with a Δt of 15°C and to offer the opportunity of working near optimum throughout the year. According to these data, it is all-important to determine the parameters values leading to maximum production to choose the best place for industrial scale culture, taking also into consideration the distance from the extractive plant.

Material and methods

Erected fronds of *Chondrus crispus* were collected on the Channel

Xth International Seaweed Symposium
Printed in Germany

french coast at Pleubian (48°52'N-3°08'W) and cultivated for 6 months in outdoor tanks before experiments. All the seaweeds studied in laboratory or outdoor arose from this initial harvest.

Outdoor culture. Six circular tanks (Ø = 3 m, d = 0.4 m) were inoculated under densities ranging from 0.7 to 5.7 $kg.m^{-2}$. Seaweeds were kept in suspension by blowing air at the bottom of the tank. Seawater inflow rate was of 6 $l.min^{-1}$ (3 tanks $volume.d^{-1}$). Nutrients were supplied by dosimetric pump (14 $mg.min^{-1}$ ammonium nitrate, 7.5 $mg.min^{-1}$ disodium hydrogen phosphate). Each week, seaweeds were weighed and densities reajusted at their initial values. Production is expressed in g wet $weight.m^{-2}.d^{-1}$ (dry matter = 19.7 % fresh weight) and the specific growth rate (G) calculated according to the formula : $G = 100\ Ln(Wt/Wo).t^{-1}$ (Wt = measured weight, Wo = initial weight, t = days of culture). Incident solar energy was calculated from tabulae of monthly mean sums of daily solar energy (1) corrected by the data of the meteorological station which is 3 km distant from Pleubian. Temperature was recorded daily with a minima-maxima thermometer.

Laboratory culture. Two 200 l thermostated tanks, each divided in 2 compartments were lighted by 8 Philips cool floods under a photoperiod of 16:$\overline{8}$. Photosynthetically active radiations (400-700 nm waveband) were measured at the tank bottom (0.4 m depth) with a Lambda Instruments LI 192 SB underwater quantum sensor. The energy levels lower than the maximum were obtained by mean of neutral screens. Temperature was risen by steps of 3°C from 8 to 20°C in five consecutive experiments (energies fixed for each compartment). 12 thalli weighing 10.5-11 g and 8 weighing 11-11.5 g were labelled and similarly fixed in the 4 compartments for all the experiments in order to avoid the interferences following the heterogeneity of light flux. Cultivation medium was recirculated by air lift and renewed by 1/5 every day (0.5 g NH_4NO_3-0.25 g Na_2HPO_4 in 40 l seawater). Each thallus was weighed at the end of experiment (15 d.) and mean growth calculated by the G formula. Mann and Witney test showed that 15 days were sufficient to obtain significant differences (U < 105 - 99 % confidence).

Results and discussion

Outdoor culture. For all densities, growth was maximum in August (1,730 $J.cm^{-2}.d^{-1}$ - 21.3°C) and minimum in January (380 $J.cm^{-2}.d^{-1}$ - 5.4°C). The 2.1 $kg.m^{-2}$ density was followed throughout an annual cycle (fig. 1). This density has been chosen because on the one hand it is the optimum for winter time and thus considered as basic minimum reference and on another hand it is the lowest density for which epiphytes did not become a too serious problem during summer. Fig. 1 shows that growth depends more on temperature variations than on solar radiation ones. As growth response might be delayed with regard to the solar or temperature stimulus, we have calculated the correlation growth/solar radiation (G/S.R.), growth/temperature (G/T) and T/S.R. The first term of each correlation was fixed at the t time values, the second shifted back 7, 14, 21, 28 and 31 days with regard to t time. Results are summarized in table 1. The correlation coefficients without shift indicate that growth is more closely linked to temperature than to solar radiation as observed by Neish (4). The decrease of G/T coefficient with shift shows that growth response to temperature stimulus is rather fast whereas the increase and parallelism between G/S.R. and T/S.R. confirm the temperature influence and illustrate the 4 weeks delay of temperature with regard to solar radiation. In spite of the good correla-

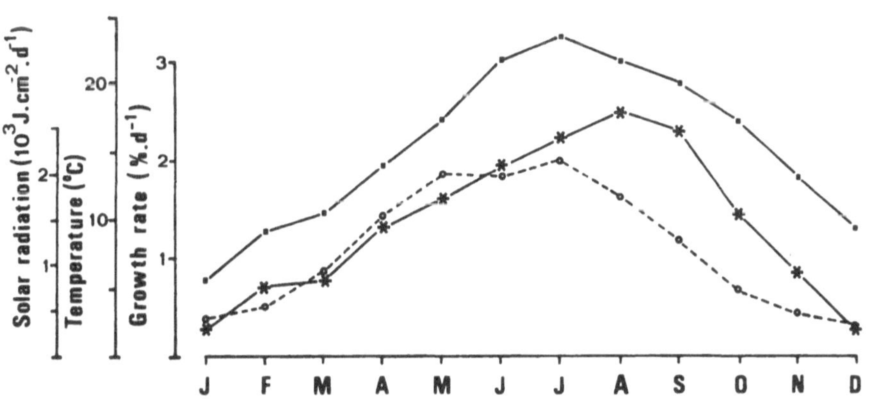

Fig. 1 - Annual variations of solar radiation (∘--∘), temperature (•—•) and growth (✱—✱) in outdoor culture tank.

Table 1 : Correlation coefficients in function of progressive time shift

Shift (d.)	Growth/solar rad. G/S.R.	Growth/temp. G/T	Temp./solar rad. T/S.R.
0	0.833	0.956	0.835
- 7	0.893	0.949	0.861
- 14	0.917	0.929	0.921
- 21	0.938	0.892	0.950
- 28	0.975	0.851	0.976
- 31	0.964	0.796	0.967

tion between temperature and growth, maximum growth occured at 21.3°C (fig. 1) and not at the temperature maximum indicating that 23°C is too high for *Chondrus*.

Figure 2 illustrates the productions obtained under different densities. January and August are respectively months of minimum and maximum production while other ones are intermediate as, for example, April. It appears clearly that the 2.1 kg.m^{-2} density is too low, except in January, for optimum utilization of solar radiation. The coefficients of the correlations maximum production/solar radiation and maximum production/temperature were respectively 0.982 and 0.834 meaning that incident energy is now predominant. The co-

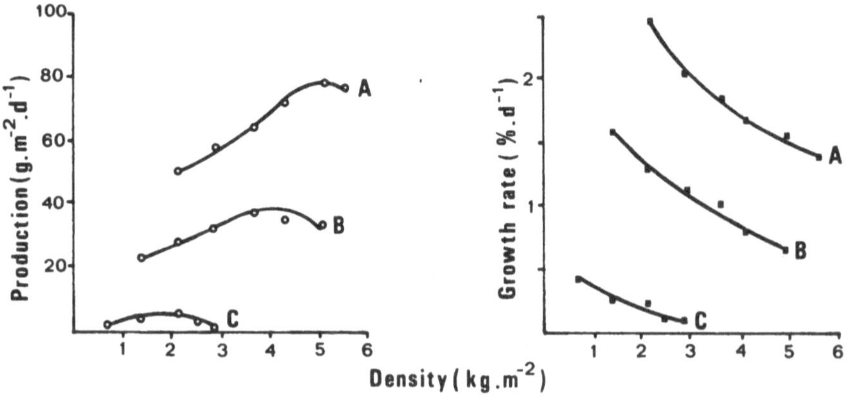

Fig. 2 - Production and specific growth rate under different densities in August (A), April (B) and January (C).

efficient G/S.R. previously calculated (0.833) traduced the misuse of energy by *Chondrus* under 2.1 $kg.m^{-2}$ density. Production curves maxima correspond to densities for which utilization of energy is optimum. Photosynthetically active radiation measurements at 0.2 m depth and under the optimum density for each considered month gave values ranging from 37 (January) to 90 $\mu E.s^{-1}.m^{-2}$ (August). So, it is necessary to modulate the density accordingly to the available energy as noted by Waaland (5). As in greenhouse experiments (4), growth rate decreases exponentially with increasing density.

Laboratory culture. Fig. 3 shows that 17°C is the optimum temperature under laboratory conditions. Bird (6) obtained nearly the same value on *Chondrus* apical segments whereas Simpson (7) noted an optimum of 20°C under greenhouse conditions, a datum close to the value measured for outdoor tanks. Growth rates were similar to those observed with apical segments (6). There is a good concordance between laboratory and outdoor tanks optimal ranges (14-20°C) giving a favourable period for *Chondrus* culture of 8 months at 50°N. Taking growth at 17°C as unity, the mean growth rate for 8, 11, 14 and 20°C were respectively : 0.71, 0.73, 0.88 and 0.76. In a similar experiment with light provided by Sylvania Grolux tubes (43 $\mu E.s^{-1}.m^{-2}$),

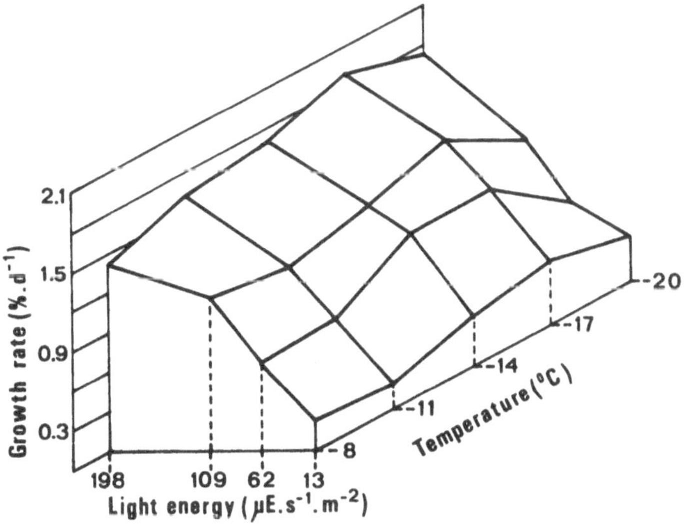

Fig. 3 - Growth rate of *Chondrus* for different temperatures and energy levels in laboratory culture.

we tested temperature down to 2°C. On the same unity basis, growth rates were 0.21 (2°C) and 0.43 (5°C), corroborating the poor growth observed outdoor during winter months. Light energy curves shapes were not sufficient to attest that seaweeds were light saturated but the highest level tested (198 $\mu E.s^{-1}.m^{-2}$) was similar to the saturation level noted by several authors (8, 9). From November to January, the mean daily incident energy in outdoor tank (0.2 m depth) was lower than the maximum studied in laboratory. Like in light/temperature studies on *Chondrus* (8), the saturation level seems to be reached more rapidly at 8°C than under the other values. With growth obtained under 198 $\mu E.s^{-1}.m^{-2}$ as unity, the mean growth rate for 109, 62 and 13 $\mu E.s^{-1}.m^{-2}$ were respectively 0.72, 0.48 and 0.19 traducing the energy influence on growth during winter time in outdoor culture.

In conclusion, as far as growth rate is concerned, i.e. culture under a fixed density, temperature appears as the prevailing factor conditioning the metabolism of *C. crispus*. On another hand, maximum production for a given area can only be reached by adjusting density to available solar radiation.

References

1. Perrin de Brichambaut, C., Lamboley, G. : Cahiers A.F.E.D.E.S. 1, 11-109 (1968).
2. Kling, R. : Thèse 3e cycle Fac. Lille (1978).
3. Bougis, P., Corre, M.C. : Cahiers océanographiques, 23 (8), 733-735 (1971).
4. Neish, A.C., Shacklock, P.F., Fox, C.H., Simpson, F.J. : Can. J. Bot. 55, 2263-2271 (1977).
5. Waaland, J.R. : *In* the marine plant biomass of the pacific northwest coast. R.W. Krauss Edit., Oregon state Univers. press, 1977.
6. Bird, N.L., Chen, L.C.M., Mc Lachlan, J. : Bot. Mar. 22, 521-527 (1979).
7. Simpson, F.J., Shacklock, P.F. : Bot. Mar. 22, 295-298 (1979).
8. Mathieson, A.C., Norall, T.L. : Mar. Biol. 33, 207-213 (1975).
9. Enright, C.T. : Proc. Int. Seaweed Symp. 9, 209-218 (1979).

THE APPLICATION OF GENETIC IMPROVEMENT TECHNIQUES TO SEAWEED CULTIVATION: I. STRAIN SELECTION IN THE CARRAGEENOPHYTE *CHONDRUS CRISPUS*

Donald Cheney*, Arthur Mathieson and David Schubert
Jackson Estuarine Laboratory, University of New Hampshire, Durham, New Hampshire, 03824, USA

Introduction

The majority of the world's current supply of kappa and lambda carrageenan comes from harvested populations of *Chondrus crispus* from Canada's Maritime Provinces, and to a lesser extent from populations in New England and Northern Europe. Although productivity from harvests of wild populations appears to have stabilized in recent years (1), cultivation programs are being developed by the phycocolloid extraction industry as a long range solution for providing reliable and economic sources of raw material, as well as a method for increasing product control (2). Current methods for culturing *Chondrus* have largely evolved from the early studies of A.C. Neish and coworkers at the Atlantic Regional Laboratory in Halifax, Nova Scotia, which demonstrated that detached plants could be successfully grown in tanks, using the fast-growing T4 strain (3,4). In comparison to the success of these and more recent Canadian studies (5-7), previous efforts to culture *Chondrus* in the temperate waters of southern New England have been unsuccessful (8), presumably due to the species' intolerance to high summer water temperatures.

---

*Present address: Biology Department, Northeastern University, Boston, MA, 02115, USA

Xth International Seaweed Symposium
Printed in Germany

As with other types of cultivation, the genetic improvement and/or strain selection of Chondrus will ultimately prove crucial to the development of cost-effective cultivation programs. To date, there have been very few studies concerned with strain selection and/or improvement in seaweeds (9-13). The present study is part of a larger project ultimately concerned with developing improved strains of Chondrus for cultivation in New England. In the present account, we summarize preliminary results of a strain selection program currently being conducted in Woods Hole, Massachusetts. Growth rate and carrageenan yield data over a 6 month period are presented for selected strains.

Materials and Methods

Approximately 100 individual plants (i.e. different strains) of Chondrus crispus from 5 wild populations (Isles of Shoals, New Hampshire; Miminegash, Charlottetown, and Basin Head, Prince Edward Island; Pubnico, Nova Scotia), as well as of the NRC strain T4, were grown at the culture facilities of the Environmental Systems Laboratory (Woods Hole Oceanographic Institution) in Woods Hole, Massachusetts, from July - Jan., 1979-1980. Plants were individually tagged and grown in large rectangular culture tanks (described in 14), with a flow rate of approximately 6 tank vol (10,950 l)/day. The water was enriched twice per week at a concentration of 1mM No_3 (as $NaNo_3$) from July to mid-Sept. (= N1 in Fig. 1), and at 0.66 mM No_3 from mid-Sept. to Dec. (= N2 in Fig. 1); during Dec. and Jan., the water was unenriched (= N3 in Fig. 1). The salinity was constant (35 o/oo) throughout the study, however water temperatures varied seasonally (for comparable data, see 14). Initial weights were determined after a 7 day acclimation period. The initial stocking density was approximately 500 g/tank; the density in Sept. and thereafter varied from 1300-2000 g. All inocula of strains initially weighed between 5-10 g; after obtaining 25 g or more, they were broken into smaller pieces and retagged. Growth rates were based upon damp dry fresh wts and expressed as both biomass increase /g inoculum (i.e. total final wt of strain - initial wt/initial wt) and % growth/day according to the method described by DeBoer and Ryther (14). The carrageenan yields of individual plants were determined periodically, using the microanalytical techniques of Craigie and Leigh (15). All of the analyses were conducted on similar portions (i.e. primary and secondary branches) of plants, weighing 3-5 g fresh wt. Yields were reproducible within ± 2%. A standard of known yield (provided by Dr. J.S. Craigie, Atlantic Regional Laboratory) was periodically run with strain samples. Carrageenan yields are expressed as the % of oven-dried, solvent-washed dry wt; solvent-washed dry wts equalled approcimately 80% of dry wts. Carrageenan production refers to the g wt of carrageenan produced per g fresh wt of inoculum.

Results and Discussion

The majority of strains tested died after 3 months or less in culture; less than 10% survived longer than 6 months. The growth rate and carrageenan yield data presented below describes the performance of the fastest-growing, longest-surviving strains. All but one of these strains were from a single wild population in Eastern Canada (Basin Head).

The temporal variation of carrageenan yields amongst strains is illustrated in Figs. 1 and 2. The initial yields of these plants (55-75%) approximated the range observed in 7 populations from New Hampshire and Eastern Canada (16 and in prep.) Thus nearly as great a range in yields existed initially amongst the strains tested as we've observed collectively in wild populations. After 6 months in culture, however, the yields of the strains became markedly lower and more uniform (50-56%), suggesting that most of the variation initially observed amongst strains was not genetically determined.

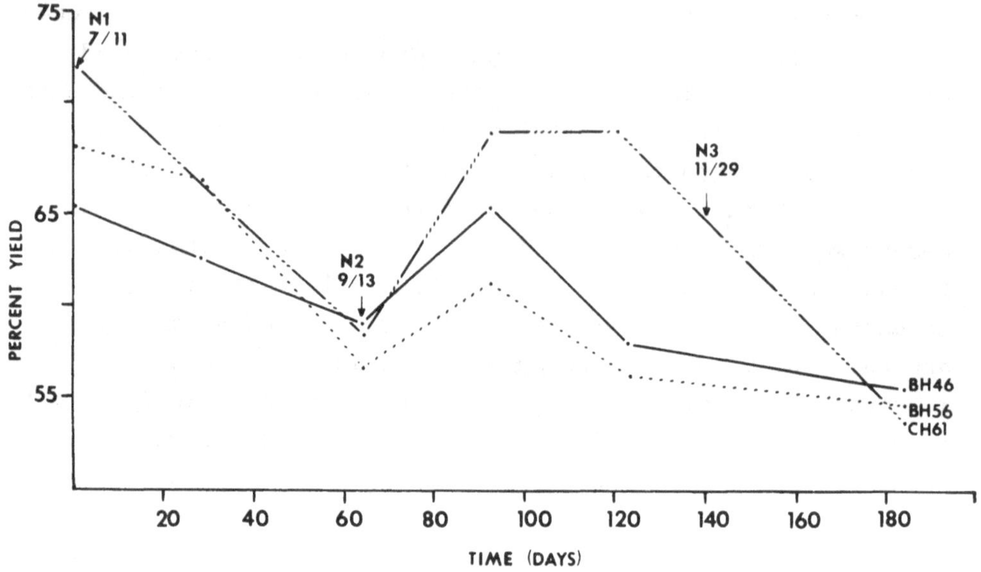

Fig. 1. Changes in percent yield from July 11 - Jan. 11 for strains BH 46, BH 56, and CH 61

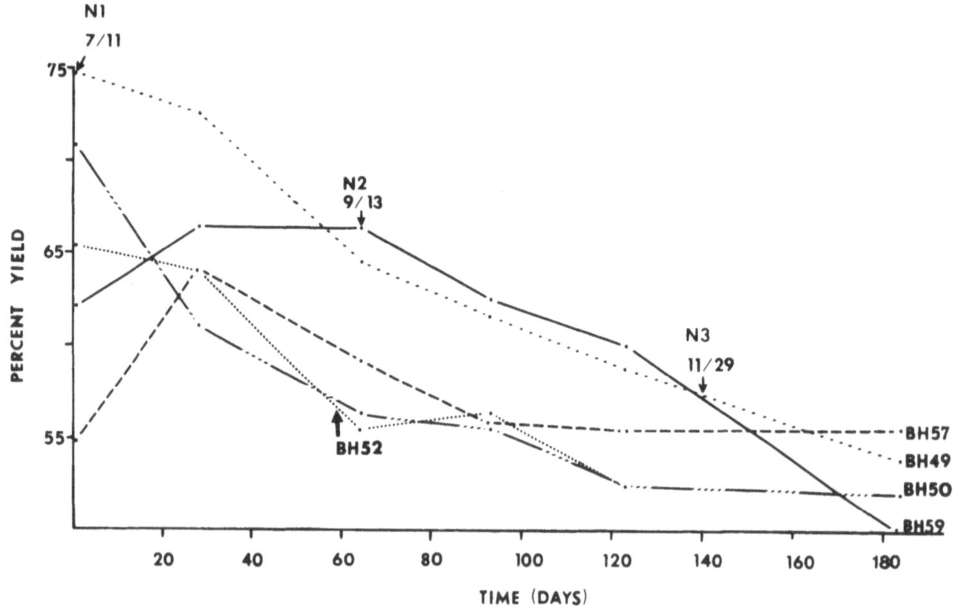

Fig. 2. Changes in percent yield from July 11 - Jan. 11 for strains BH 57, BH 49, BH 50 and BH 59

Several studies have evaluated the effects of various environmental factors (e.g. temperature, nutrient concentration, pH, and salinity) on carrageenan yield in T4 (5-7, 17). In the present study, not all strains responded similarly in their carrageenan yields to changes in culture conditions; rather, two major patterns were evident. In the first (Fig. 1), yields initially decreased in culture and then showed a marked increase during periods of reduced nutrient concentration. Such a pattern is similar to what has been described as the "Neish effect" (6). In the second group of plants (Fig. 2), however, yields either continuously decreased (e.g. BH49, 50 and 52), or initially increased and then decreased (e.g. BH57 and 59). A comparison between an individual strain's relative ranking in carrageenan yield and growth rate (compare Figs. 1 and 2 with 3 and 4), does not show a consistent relationship between the two. Thus, the carrageenan yield of a plant would appear to be determined by current, as well as antecedent environmental (i.e. growth determining) conditions. Hence, the variation observed in the initial yields of the strains, as well as in their subsequent response patterns to changes in culture conditions, may be explained by differences in microhabitat conditions and/or growth rates at the time of collection.

The rate of biomass production differed substantially amongst strains. The differences were most evident after the 64 day period from July to Sept. (Fig. 3), in which there was as much as a 4 fold differential in the amount of biomass produced between the fastest and slowest growing strains. Since all plants were grown under identical conditions (and all but one came from the same original population), most of this discrepancy would appear to reflect genetically determined differences. Although heritability studies are needed to say more about the genetic control of growth rates, it would appear from the present data that growth rate, rather than carrageenan yield, should be the major parameter used in strain selection.

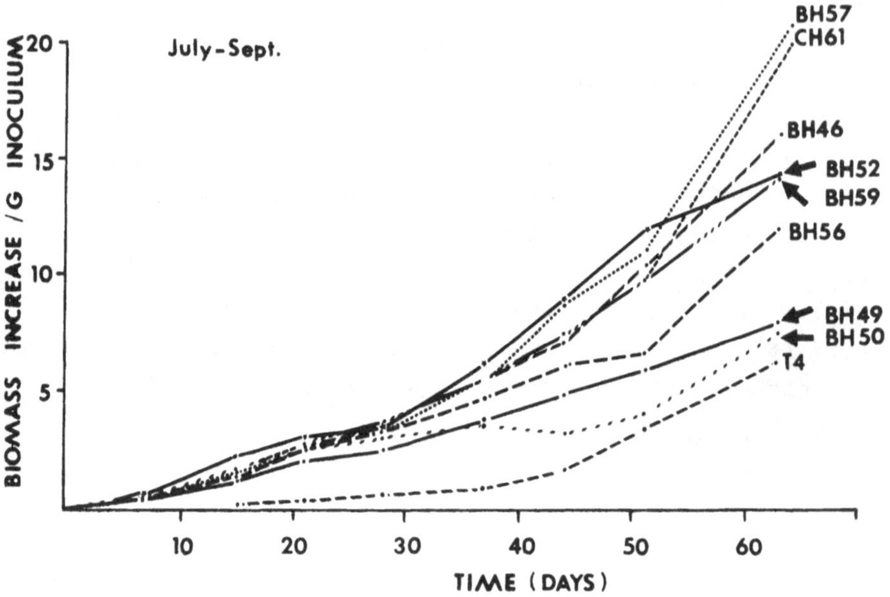

Fig. 3. Biomass production rates amongst strains for the period July - Sept.

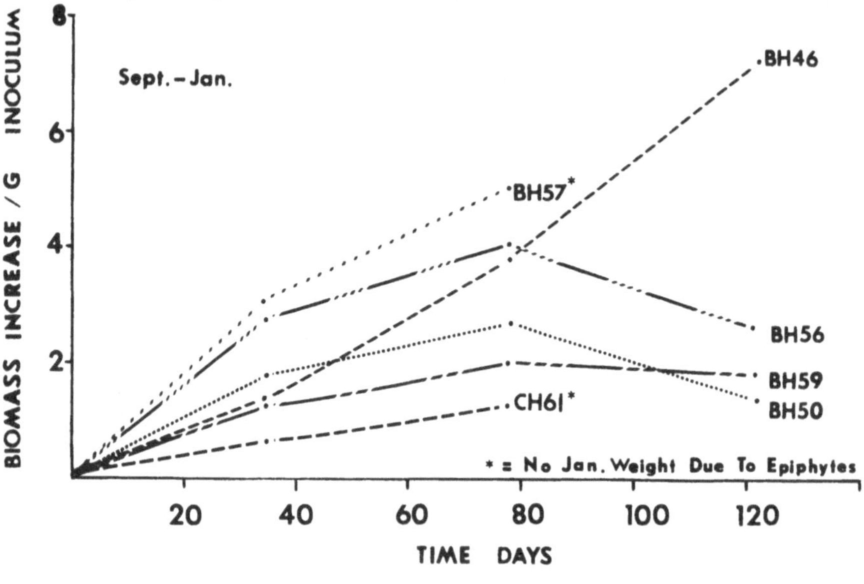

Fig. 4. Biomass production rates amongst strains for the period Sept. - Jan. Strains BH49 & 52 showed similar rates to BH56.

As shown in Figs. 3 and 4, biomass production rates were highly variable, depending upon culture conditions. The present study confirms previous findings (3-7) demonstrating the importance of temperature and nutrient concentration on biomass production. In addition, the relative reproductive status of a plant can have a major effect on its growth rate. For example, strain CH61 exhibited a much lower growth rate prior to and in Dec. (Fig. 4), when it became reproductively mature (tetrasporic), than during July to Sept. (Fig. 3). It should also be noted, that the comparative growth rates of different strains depended upon the specific culture conditions. This is an important consideration in regards to strain selection, as an "optimal" strain may only be optimal under rather specific culture conditions. For example, BH46 demonstrated the greatest biomass production during the Sept. to Jan. period when N concentrations and water temperatures were reduced, while it ranked third in biomass production for July to Sept.

Table 1. A comparison of carrageenan yields, growth rates and net carrageenan production amongst strains grown under enriched (1mM No_3) culture conditions for a 64 day period (July-Sept.).

PLANT NO.	YIELD ORIG	YIELD FINAL	GROWTH %/DAY	BIOMASS PROD.	% DRY WEIGHT	CARRA PROD.
BH 46	65	59	4.4	16.1	17.5	1.65
BH 49	75	64	3.5	8.1	17.9	0.93
BH 50	71	56	3.4	7.6	16.1	0.69
BH 52	65	55	4.3	14.5	17.6	1.41
BH 56	67	65	4.0	12.1	17.4	1.37
BH 57	55	60	4.8	20.9	19.6	2.45
BH 59	62	62	4.3	14.4	16.6	1.48
CH 61	72	58	4.6	18.2	19.1	2.02
T 4	nd	63	3.9	7.9	16.4	0.82

For a Chondrus cultivation program to be successful in New England, comparable or greater carrageenan production rates must be achieved than those reported for Eastern Canada. Table 1 compares the growth and carrageenan production rates of all strains during the period July to Sept. Although it is difficult to make direct comparisons between our studies and those in Canada, due to varying culture techniques and methods of calculating growth rates, it would appear that the biomass production rates of strain BH57 (i.e. 20.9 g/g inoculum) are equal to or greater than those reported for T4 in Canada (3-7, 17). For example, the culture data presented in Simpson and Shacklock (7) for Chondrus growing at 20°C and with an initial stocking density of 500 g, yields a growth rate of 3.8 %/day (calculated as in the present study). Similarly, the data presented in Table 6 of Neish et al. (6), yields a growth rate of 4.6 %/day; based upon a 64 day growth period (as in Table 1), the latter data would result in a biomass increase of 18.1 g/g inoculum.

In conclusion, the present study suggests that Chondrus may be successfully grown in southern New England, particularly if an active strain selection program is incorporated.

Acknowledgements: We wish to express our special appreciation to John Ryther and the staff of ESL for their support of this project. Our special thanks also to John Pringle, Louis Hanic, Jim Murchison and Valarie Hovey for their help. This investigation was supported by the Office of Sea Grant Programs of the National Oceanic & Atmospheric Administration as Research Grant No. 4-21129. Published as Scientific Contribution No. from the New Hampshire Agriculture Experiment Station; also issued as Contribution No. 115 of the Jackson Estuarine Laboratory.

References

1. Pringle, J.: Proc. Inter. Seaweed Symp. 9, 225-232 (1979).
2. Neish, I.: In: Actas primer symposium sobre algas marina Chilenas. (Ed.) B. Santelices, Santiago, 59-74 (1979).

3. Neish, A. and Fox, C.: Technical Report No. 12, Atlantic Regional Laboratory, National Research Council of Canada (NRCC) No. 12034, 35 pp (1971).

4. Neish, A. and Shacklock, P.: Technical Report No. 14, Atlantic Regional Laboratory, NRCC No. 12253, 25 pp (1971).

5. Shacklock, P., Robson, D. and Simpson, F.: Technical Report No. 21, Atlantic Regional Labora-ory, NRCC No. 14735, 27 pp (1975).

6. Neish, A., Shacklock, P., Fox, C. and Simpson, F.: Can. J. Bot. 55, 2263-2271 (1977).

7. Simpson, F. and Shacklock, P.: Bot. Marina 22, 295-298 (1979).

8. Ryther, J., DeBoer, J. and Lapointe, B.: Proc. Inter. Seaweed Symp. 9, 1-16 (1979).

9. Academica Sinica (Anonymous):Proc. 3rd Inter. Ocean Dev. Conf., Tokyo, 1-12 (1975).

10. Miura, A.: In: Advances of phycology in Japan. (Ed.) J. Tokida and H. Hirose, Junk, The Hague, 273-304 (1975).

11. Hasegawa, Y.: J. Fish. Res. Board Can. 33, 1002-1006 (1976)

12. Doty, M.: Proc. Inter. Seaweed Symp. 9, 35-58 (1979).

13. Waaland, J.: Proc. Inter. Seaweed Symp. 9, 241-247 (1979).

14. DeBoer, J. and Ryther, J.: In: The marine plant biomass of the Pacific northwest coast. (Ed.) R. Krauss, Oregon State Univ. Press, Corvallis, 231-249 (1978).

15. Craigie, J. and Leigh, C.: In: Handbook of phycological methods. Vol II, Physiological and biochemical methods. (Ed.) J. Hellebust and J. Craigie, Cambridge Univ. Press, New York, 109-131 (1978).

16. Cheney, D., Mathieson, A. and Ragonese, A.: J. Phycol. 15 (Suppl.), 18 (1979).

17. Simpson, F., Neish, A., Shacklock, P. and Robson, D.: Bot. Marina 21, 229-235 (1978).

CULTIVATION OF *GRACILARIA* (RHODOPHYCOPHYTA, GIGARTINALES) IN TAIWAN

Young-Meng Chiang

Institute of Oceanography, National Taiwan University, Taipei, Taiwan, Republic of China

Abstract

Cultivation of *Gracilaria* in brackish water ponds is extensively carried out in southern parts of Taiwan. *Gracilaria verrucosa* is the species most commonly cultivated, but *G. gigas* and *G. lichenoides* are also propagated usually mixed with *G. verrucosa* in some areas. *Gracilaria* cultivated in Taiwan is used mainly for the domestic agar industry, export, and for feeding stuff for small abalone, *Haliotis diversiolor supertexta*, mariculture. The most recent farming method which is practiced by most farmers is the polyculture system. Fronds of the seaweed are torn in pieces and are evenly thrown on the bottom of the pond in which milkfish, *Chanos chanos*, grass shrimps, *Penaeus monodon*, or crabs, *Scylla serrata* are cultivated. These animals free the seaweed from harmful epiphytes growing on them and increase the economic efficiency of the pond. Continuing studies, especially on the development of high-yielding and fast growing strains of *Gracilaria*, are being conducted to develop further the potential of this important commodity. There are now over 300 hectares of culture area in Taiwan producing about 12,000 tons/year of dry seaweeds for the market. It is possible to increase this production for the domestic agar industry and for export by raising the quality of the seaweed.

Introduction

Xth International Seaweed Symposium
Printed in Germany

In 1961 the Taiwan Fisheries Bureau initiated a program to develop an economically feasible seaweed aquaculture industry in the water of southern Taiwan. To this end, the native agarophyte Gracilaria verrucosa (Hudson) Papenfuss was selected. At the early stage of Gracilaria cultivation, the method used was inserting fronds of seaweed into the twist of ropes and cultivated at shallow coastal areas, or ditches of brackish water ponds used for milkfish cultivation (Kuan, 1962; Hsu, 1963). Subsequent experiments paved the way for the commercial propagation of Gracilaria in ponds. This paper presents a discussion on the farming method practiced by most farmers.

Commercial production pattern

The most recent Gracilaria farming method practiced in Taiwan is the polyculture system. That is, fronds of Gracilaria are freely cultivated with milkfish, Chanos chanos Forskal, in brackish water ponds. In addition to milkfish, sometimes grass shrimps, Penaeus monodon Fabricius and crabs, Scylla serrata Forskal are also introduced to increase the economic efficiency of the pond.

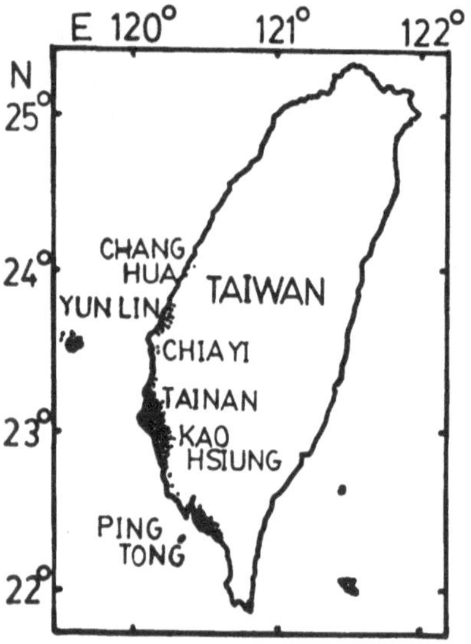

Fig. 1. Map showing distributions of Gracilaria cultivation in Taiwan.

According to the Taiwan Fisheries Year Book, the total culture area in 1979 was more than 300 ha. with a production of about 12,000 tons of dry seaweed. Most of the Gracilaria ponds are located in the Anping, Tainan, and Pingtong areas and to a lesser extent in the Changhua, Yunlin and Chiayi areas (Fig. 1). The seaweed produced is used mainly for the domestic agar industry, export and feeding stuff for small abalone, *Haliotis diversiolor supertexta* Lischke, culture. With the rapid development of abalone mariculture in the north-eastern part of Taiwan, the demand for fresh *Gracilaria* for abalone has increased tremendously. In 1979, about 120 tons of fresh *Gracilaria* went to abalone farms, and it is expected to increase in the coming years.

The pond

The *Gracilaria* ponds are all located at places where no strong wind prevails and where both sea and fresh water can be obtained easily. Most of them are about the size of one hectare or smaller. Small ponds are preferable to large ones, because the thalli of *Gracilaria* in large ponds are easier removed and accumulated at a corner of the pond by the wind than those in small ones. Small ponds are also easier to manage when shrimps and/or crabs are additionally cultivated.

Ponds are usually about 50 to 80 cm deep and with one or two places for the exchange of water. The bottom is flat and formed of clay loam, sand, silt loam, sandy loam or loamy sand (Huang, 1978). It was found that the seaweed in ponds with sandy bottoms are easier buried in sands by strong wind (Huang, 1978). Therefore, the size of the pond with sandy bottom should be smaller and the water should be deeper than usual during windy days.

In the medial portion of a large pond, such as those larger than one hectare, there usually one or two rows of windbreaks made of bamboo pieces are installed perpendicular to the direction of the wind to protect the seaweed removed by the wind.

Seeding

Gracilaria verrucosa is the species most extensively cultivated, but G. gigas Harvey and G. lichenoides (L.) Harvey are also propagated, usually mixed with the former in some areas.

The procedures of cultivation are as follows: (a) drying the pond in the sun for several days. Once the cultivation is started and goes well, the seaweed can be grown continuously in the same pond for many years; (b) introduction of water; (c) seeding. Cultured strains are used as planting stocks. About 5,000 to 6,000 or more kg/ha of fresh thalli are seeded by tearing them randomly into pieces and throwing them evenly into the pond; (d) introduction of milkfish to control epiphytes. Usually about 1,000 milkfishes (size of about 15 cm long) per hectare are introduced prior to the appearance of epiphytes.

Pond management

The depth of pond water (the distance between the top of the algal thallus and the water surface) is usually kept at 30 to 40 cm. From June to August, however the air temperature in southern Taiwan very often reaches 32°C or more. Therefore, during the summer pond water is usually increased to 50 to 60 cm, to prevent an undue increase of water temperature. On the other hand, when the air temperature is lower than 10°C or so, the depth of the pond water is lowered to about 20 to 30 cm deep.

Frequent change of water is necessary to maintain a proper range of salinity and to provide nutrients for the growth of the seaweed. One change of water in two or three days is the usual practice, and only one half to two-thirds of the water is changed each time. The salinity of the pond water is usually kept at 10 to 20‰, but sometimes it may be as low as 4‰ or as high as 40‰ (Lin, 1970). This range of salinity coincides with the results of recent study (Huang, 1980), that the cultured strain of G. verrucosa can grow under the salinity from 5 to 50‰ and the maximum growth was at salinity 15‰.

Sometimes fertilization of the pond water is needed. Usually about 3 kg/ha of urea or 120 to 180 kg/ha of fermented manure from pigs or chickens is used.

The main pests of Gracilaria are species of Enteromorpha, Chaetomorpha, and Ectocarpus, with the former two most harmful and most difficult to control. At present, control of epiphytes with milkfish seems very satisfactory. Milkfish will consume Gracilaria after the epiphytes are gone. Therefore, when this happens large fishes must be removed and small ones introduced again. Besides epiphytes, Tilapia mossambica Peters may become a pest of Gracilaria. Tilapia are widely distributed in southern Taiwan and get into Gracilaria ponds during the change of water. Due to very rapid propagation, their population usually becomes very large in a short time to cause great damage to Gracilaria by their strong grazing power and by digging many various size of holes into the bottom of the pond. Therefore, before they causing damages, efforts should be made to catch as many as Tilapia every two or three days to keep down their population.

The seaweed is harvested once every 30 to 35 days, or 45 days during the winter. Generally, about one third to one half of the seaweed is harvested each time. After harvesting, the thalli remaining in the pond are torn into pieces and are seeded again. Generally about 16,000 to 43,000 kg/ha/yr. of dried Gracilaria (with 1 to 6-7 drying ratio) are produced making the profit gained from polyculture higher than that gained from the cultivation of milkfish alone.

In conclusion, we have succeeded in cultivating Gracilaria by vegetative propagation in brackish-water ponds with milkfish.

It seems that the present technique is adequate and profitable for the farmers. But for further development of this important commodity, studies on the high-yielding and fast growing strains of *Gracilaria* must be continued.

References

1. Kuan, W.H.: China Fish. Monthly No. 111, 3-9 (1962). (in Chinese)
2. Hsu, C.W.: China Fish. Monthly No. 121, 2-8 (1963). (in Chinese)
3. Huang, C.H.: M.S. Thesis, Natl. Taiwan University, Taiwan. 61 pp. (1978). (in Chinese)
4. Huang, S.P.: M.S. Thesis, Natl. Taiwan University, Taiwan. 76 pp. (1980). (in Chinese)
5. Lin, M.N.: Fish. Series, Joint Commission on Rural Reconst., Taiwan. No. 9, 43-53 (1970).

MANIPULATION OF THE LAMINARIALEAN LIFE-CYCLE AND ITS CONSEQUENCES FOR KOMBU MARICULTURE

L.D. Druehl and R. Boal
Bamfield Marine Sta., Bamfield, B.C., VOR 1BO, Canada

Introduction

For the mariculture of Laminariaceae (e.g. Kombu) to achieve the sophistication of terrestrial agronomy, genetic control is required, as well as the ability to delay some stage of the life-cycle. Genetic control is necessary for product quality and uniformity, and a delayed phase allows for outplanting of young laboratory-grown sporophytes at any time.

Two discoveries facilitate the application of genetic control and the delay of the laminarialean life-cycle. Schreiber (1) demonstrated the potential of cloning gametophytes, and Saga et al. (2) have cloned sporophytes of *Laminaria angustata* Kjell. by induction of callus-like tissue and subsequent fragmentation. Cloning results, in the case of gametophytes, in numerous male and female plants producing genetically identical eggs and sperm which will, through syngamy, produce identical sporophytes. The difficulty with this approach is that the resulting sporophytes are of unknown quality until fully grown. However, cloning of sporophytes of proven quality may overcome this problem.

Lüning and Dring (3) demonstrated that blue light is required for gametogenesis in many Laminariales, and that in the absence of blue light, the gametophytes only develop vegetatively. This is a possible mechanism for prolonging the haploid life-history stage.

Xth International Seaweed Symposium
Printed in Germany

The purpose of this study was to expand earlier studies on cloning, and to explore growth in red light and survival in darkness as means of prolonging different laminarialean life-history stages.

Materials and Methods

The response of gametophytes to the absence of blue light was tested with mixed male and female cultures under continuous red light conditions ($8\mu E \cdot m^{-2} \cdot sec^{-1}$, General Electric F20T12-Red fluorescent tubes at 10C).

Cultures of meiospores, gametophytes, and micro- and macroscopic sporophytes of *Laminaria groenlandica* Rosenvinge were maintained in darkness at 10C. Viability was evaluated by observing the plants after reintroduction to light.

Isolated male and female gametophytes for cloning were cultured and fragmented using the methods of Lüning and Dring (3, 4). Gametogenesis was induced in white light ($20\mu E \cdot m^{-2} \cdot sec^{-1}$, Sylvania F20T12-CW fluorescent tubes) or natural light (up to $1000\mu E \cdot m^{-2} \cdot sec^{-1}$ in a greenhouse).

Sporophyte cloning was attempted on large wild plants of *Laminaria setchellii* Silva and *L. groenlandica* having callus-like outgrowths on the laminae, and small (ca. 4cm long) laboratory reared *L. groenlandica*. Plant tissue (callus, meristematic tissue and whole small plants) was cleaned and placed in liquid and agar (0.8%) phases of culture medium with sucrose ($20g \cdot l^{-1}$) and IAA ($2.0mg \cdot l^{-1}$), and with *L. groenlandica* extract ($5ml \cdot l^{-1}$). Cultures were maintained under continuous white light ($15\mu E \cdot m^{-2} \cdot sec^{-1}$, light source as above) at 10C.

Results

This study emphasized British Columbia species having kombu potential (L. groenlandica, Cymathere triplicata (Post. & Rupr.) J. Ag. and Pleurophycus gardneri (Setch. and Saund.). These species produced sori at different times of the year (Table 1). The longest period through which cultures could not be established from wild plants was 243 days (C. triplicata) followed by 225 days (L. groenlandica) and 182 days (P. gardneri). Gametophytes of all three species have been grown under red light conditions for over one year and these plants, upon exposure to blue light, have become fertile.

The experiment on dark survival of various life-history stages continues. At the time of writing (June, 1980) most of the cultures were still viable. Meiospores of L. groenlandica have survived up to 350 days. The deaths recorded are for sporophytes over 2mm long. Microscopic gametophytes, and sporophytes 2mm long and smaller were still viable up to 150 days after initiation of the experiment.

We have grown gametophytes of 13 species under red light conditions to determine whether or not they would remain asexual (Table 2). Gametangia were not observed on 6 species, were infrequently observed on 4 species, and predominantly on 3 species.

```
Laminaria            *****----        ----            ---**********
 groenlandica
Pleurophycus             ----*********-----************
 gardneri
Cymathere                                         ----***********--
 triplicata

                    |___|___|___|___|___|___|___|___|___|___|___|
                   Jan  F   M   A   M   J   J   A   S   O   N  Dec
```

Table 1. The seasonal appearance of sori (--) and of good meiospore release (**) of three species of Laminariales in Barkley Sound.

Species	Response	Intertidal Position
Alaria marginata Post. & Rupr.	-	I/S
Costaria costata (Turn.) Saunders	+	I/S
Cymathere triplicata (Post. & Rupr.) J. Ag.	+	I/S
Eisenia arborea Areschoug	-	S
Hedophyllum sessile (C. Agardh) Setchell	++	I
Laminaria ephemera Setchell	-	I/S
L. groenlandica Rosenvinge	+	I/S
L. saccharina (L.) Lamour.	-	I/S
Lessoniopsis littoralis (Farl. & Setch.) Reinke	++	I
Macrocystis integrifolia Bory	-	I/S
Pleurophycus gardneri Setchell & Saunders	+	I/S
Postelsia palmaeformis Rupr.	+++	I
Pterygophora californica Rupr.	-	S

Table 2. The production of gametangia by gametophytes in the absence of blue light. (I) intertidal, (S) subtidal, (-) vegetative growth only, (+) infrequent gametogenesis, (++) frequent gametogenesis, (+++) predominant gametogenesis.

Red-light grown gametophytes of Pleurophycus gardneri, Laminaria groenlandica, and L. saccharina (L.) Lamour. have produced presumably genetically uniform sporophytes following fragmentation and subjection to blue light.

Treatment of sporophyte tissue with IAA, sucrose, L. groenlandica extract, and no additional substance (control) resulted in the production of uniseriate filaments from meristoderm cells. These filaments had the general appearance of laminarialean gametophytes. At the time of the last observation (26 weeks after initiation of cultures), there was no indication of a sporophyte morphology.

Discussion

The production of young sporophytes for outplanting can now be achieved at any time of the year by growing gametophytes in the absence of blue light or maintaining cultures in darkness.

Prior to our study, the blue light response had been tested on nine species of Laminariales (4, 5). We have tested two species that had been investigated earlier and obtained the same results (*L. saccharina* (4), *Pterygophora californica* Rupr. (5)). Thus far, four species are known to produce gametangia in the absence of blue light: *Hedophyllum sessile* (C. Ag.) Setch., *Lessoniopsis littoralis* (Farl. & Setch.) Reinke, *Postelsia palmaeformis* Rupr. (present study) and *Egregia menziesii* (Turn.) Aresch. (5). The three former species are essentially intertidal and the latter species extends from mid-intertidal to subtidal. We are presently attempting to relate the blue light response to vertical distributions.

The field culture of presumably identical sporophytes is possible for those species requiring blue light for gametogenesis. This results from the successful cloning of gametophytes. A second advantage of this method is that once a desirable male x female cross has been identified, the same genetic material can be used for successive cultures.

We attempted to clone sporophytes using the general techniques of Saga *et al.* (2). However, rather than obtaining callus-like tissue, we obtained filamentous plants resembling gametophytes. These filamentous plants were produced by both species investigated, all of the tissues cultured, and for all treatments, including the control. The resulting plants closely resembled the diploid monoecious gametophytes of *Alaria crassifolia* Kjell. derived through apospory by Nakahara and Nakamura (6, Fig. 2-I). Plants resembling microscopic sporophytes have not yet been noted in our cultures.

Acknowledgements

Ms. K. Lloyd conducted the sporophyte cloning experiments, and Ms. S. Smith-Pakula assisted in all stages of the study. Funding was provided by M.R.B., Department of Environment, British Columbia, and NSERC Canada.

References

1. Schreiber, E.: Planta 12, 331-353 (1930)
2. Saga, N., Uchida, T., Sakai, Y.: Bull. Jap. Soc. Sci. Fish., 44, 87 (1978).
3. Lüning, K., Dring, M.J.: Planta 104, 252-256 (1972).
4. Lüning, K., Dring, M.J.: Mar. Biol. 29, 195-200 (1975).
5. Lüning, K., Neushul, M.: Mar. Biol. 45, 297-309 (1978).
6. Nakahara, H., Nakamura, Y.: Mar. Biol. 18, 327-332 (1973).

KELP GROWTH ON AN OCEAN FARM IN RELATION TO FERTILIZING

V. Gerard and W. North

Kerckhoff Marine Laboratory, California Institute of Technology, Corona del Mar, California, USA 92625

Introduction

Selection of the giant kelp, Macrocystis pyrifera, as a candidate for large-scale production of marine biomass was based on a number of desirable characteristics (1). Primary among these was the potential high productivity of this species because harvestable yield is an important factor determining economic feasibility of ocean farming. However, early attempts to grow Macrocystis at offshore sites showed that growth was limited by low macronutrient concentrations in offshore surface water, especially by low nitrogen availability (2,3). Use of artificially upwelled deep water to fertilize kelp on ocean farms may provide a solution to this problem: below depths of about 300 m, macronutrient levels are consistently high (e.g., 25-35 μM NO_3 and 1-3 μM PO_4 in the Southern California Bight). Deep water supported good growth of juvenile Macrocystis sporophytes in laboratory culture (2). One of the major objectives of the Marine Farm Project is to assess deep water as a nutrient source for adult kelp plants. If artificially upwelled water can be used to maintain optimal nutrient conditions for kelp growth and production, Macrocystis may provide a significant future energy resource.

Fertilizing Studies on the Test Farm

The test farm structure was designed to hold 100 adult kelp

Xth International Seaweed Symposium
Printed in Germany

plants in an offshore environment with continual exposure to nutrient-rich water pumped into the surface canopy from 450 m depth (4). The structure was installed at the 550 m depth contour 7 km offshore from Laguna Beach, California, on 14 September 1978. During November and early December, we transplanted 103 adult kelp plants, averaging 30 fronds/plant, from an inshore forest to the test farm. We measured elongation rates of meter-long and canopy fronds on the transplants and followed changes in N content of blade tissue as an indication of nutritional status (5). On 12 December 1978, a curtain was hung around three sides of the structure, extending from the surface to 11 m depth, to reduce current speed and increase residence time of upwelled water within the farm. The curtain remained intact for only one week. Strong currents then caused the transplants to tangle and abrade on various parts of the test farm structure and, by early February 1979, plant mortality reached 100%.

In spring 1979, juvenile Macrocystis sporophytes appeared on all parts of the test farm structure, presumably recruited from spores released by the original adult transplants. By May, kelp population on the planting substrate alone was ca. 40,000 plants between 2 and 60 cm in length. These plants received little exposure to the artificially upwelled water which was being dispersed at 2 m depth. We fertilized different groups of juvenile plants by three methods and compared their net elongation rates and tissue N contents to those of unfertilized controls. One group of plants received a few minutes of exposure to 1 M ammonium sulfate dispersed semiweekly by hose from a surface supply. Two sacks containing 22 kg of a controlled release fertilizer, Osmocote (19% N, 6% P), were located adjacent to a second group of plants. The third group was supplied with artificially upwelled deep water via an extension from one of the test farm dispersal hoses.

Effects of Fertilizing on Kelp Growth and Nutritional Status

Response of adult kelp transplants to varying nutrient supply was reflected by changes in tissue N content. Mean N content of canopy blades remained greater than 2% of dry weight during December 1978, despite low ambient NO_3 concentrations (Figure 1). The week of exposure to deep water while the curtain remained intact may have actually resulted in N contents averaging almost 4%. Following loss of the curtain, mean N content dropped to 1%. Subcanopy blades also showed a decrease in N content during January, but later and less extreme than the canopy blades. In contrast, N content of basal blades on canopy fronds remained around 1.5%. These differences may be explained by slower growth of older, deeper tissues and by higher ambient NO_3 concentrations at greater depths. Although the canopy fronds showed measurable growth, insufficient data were collected to determine changes over time. However, mean elongation rates of meter-long fronds on the transplants did not fluctuate significantly during the 2 mo period.

Juvenile kelp plants on the test farm showed significant changes in growth rate and tissue N content in response to varying nutrient supply. After the last upwelling event in early May, ambient NO_3 concentrations at 18 m depth remained below 1 µM (Figure 2). Mean net elongation rate of unfertilized plants dropped from 1 to 0 cm/day, then became negative as tissue loss overtook growth. Mean N content dropped as low as 0.5% of dry weight. Plants fertilized with deep water grew significantly faster than the control plants. The fertilized tissues maintained mean N contents over 2%, indicating that growth was not N-limited (5). Osmocote was equally effective in stimulating growth but only for plants within 1-2 m of the fertilizer. Brief exposures to ammonium caused a significant increase in N content over controls but did not enhance growth. Exposure times may have been too short for sufficient uptake, or low phosphorous availability may have become limiting to growth.

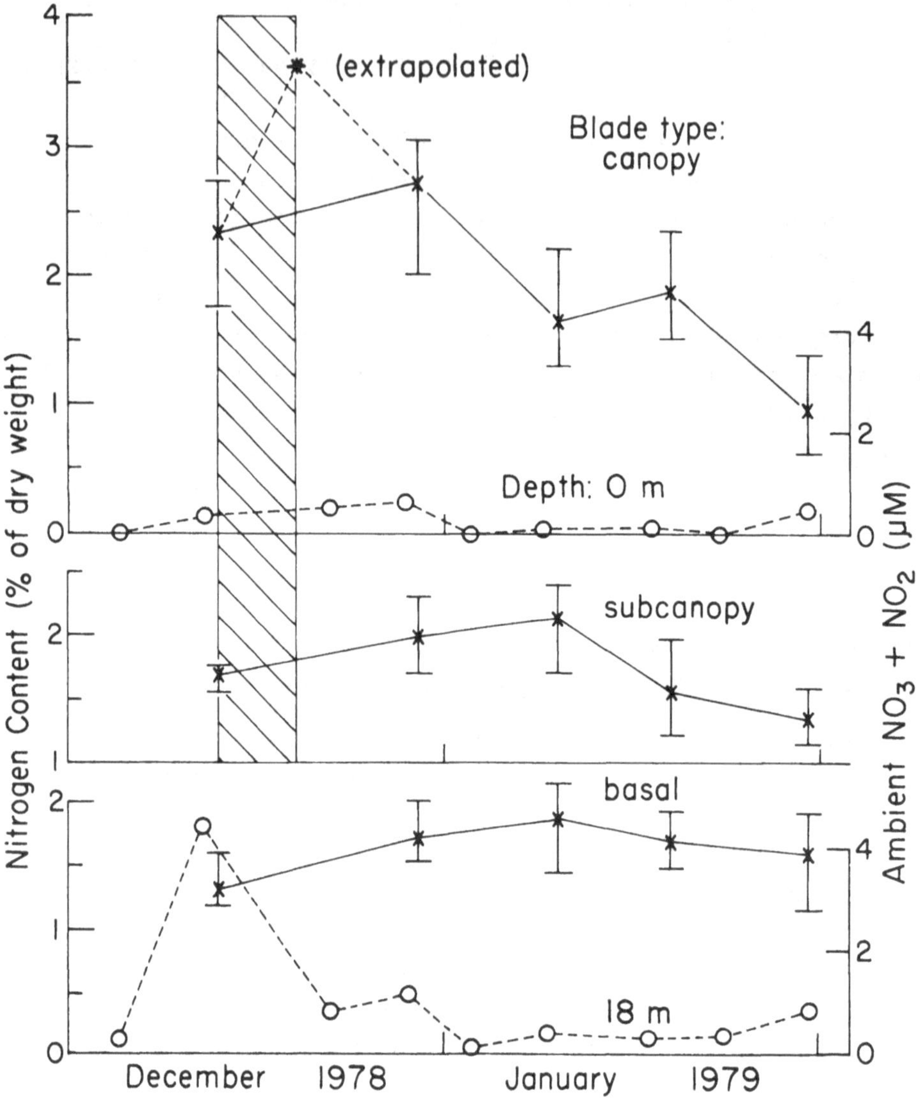

Figure 1. Mean N content (x——x) of canopy, subcanopy, and basal blade tissues collected from adult kelp transplants on the test farm (n=6, range shown) and ambient nitrate concentrations (o---o) at 0 and 18 m depths. Hatch-marked bar designates tenure of nutrient retention curtain.

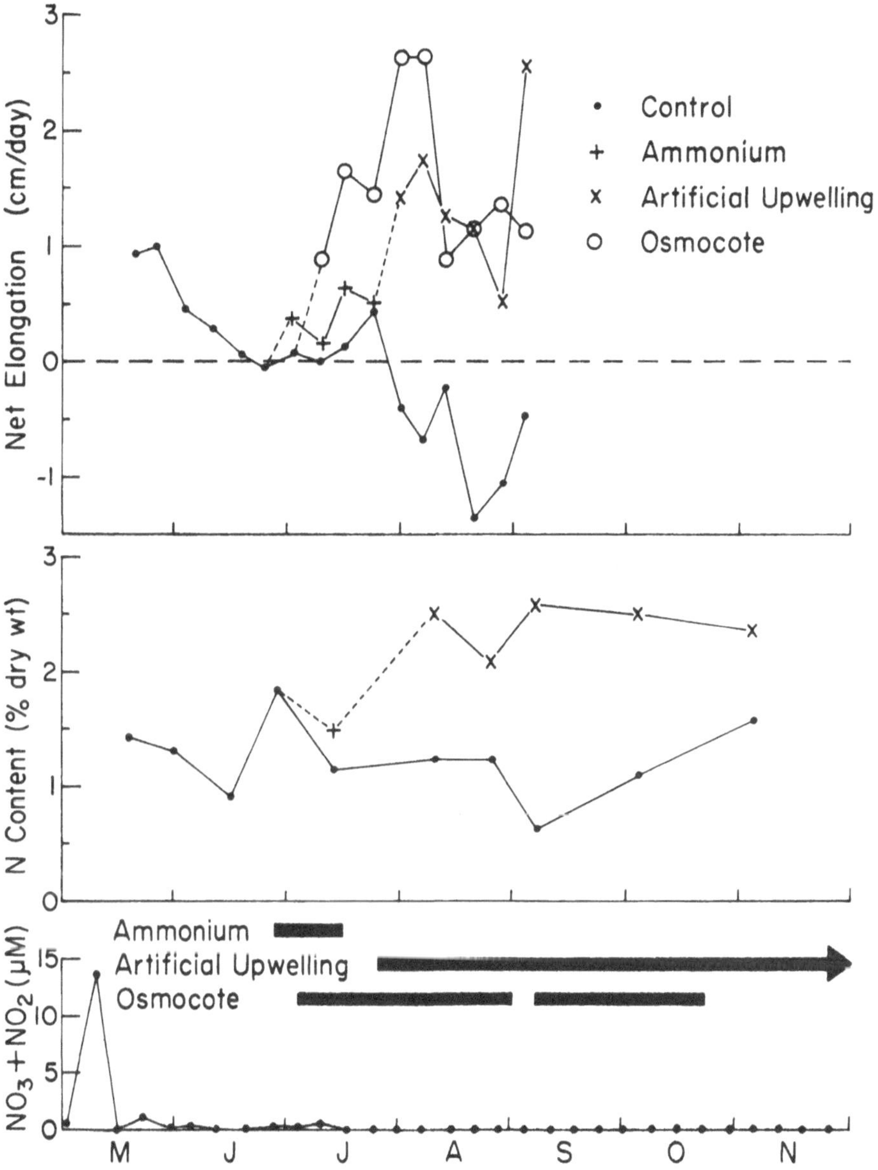

Figure 2. Mean net elongation rates (n=4-28) and tissue N contents (n=6) compared for four groups of juvenile kelp plants on the test farm. Each group either remained as an untreated control or was fertilized by one of three methods as indicated. Duration of fertilizing treatments and ambient nitrate concentrations at 18 m depth are shown.

Conclusions

Results of fertilizing experiments on the test farm supported previous evidence that low macronutrient supplies limit kelp growth in offshore surface water. Enhanced N contents of blade tissues from adult and juvenile plants and enhanced growth of juvenile plants indicated that artificially upwelled deep water could provide a suitable source of nutrients. However, measurement of harvestable production depends on our ability to prevent damage to plants by currents and abrasion, which would be edge effects on a large-scale ocean farm. We have now successfully held three adult *Macrocystis* plants on the test farm for one year by locating them away from major structural elements. At least ten times that many plants, suitably protected and fertilized with deep water, are necessary for us to accomplish our immediate goal of determining yield. The test farm is now being redesigned to meet these requirements.

References

1. Jackson, G.A., North, W.J.: U.S. Naval Weapons Cntr, Contract No. N60530-73-MV176, Final Rpt. (1973).
2. North, W.J.: Biological Solar Energy Conversion, Academic Press, New York 1977, p. 347-361.
3. North, W.J.: Proc. Symp. Fuels from Biomass 2, 999-1040 (1978).
4. Barcelona, M., Gerard, V., Kuwabara, J., Lieberman, S., Manley, S., North, W.: Proc. Biomass Energy Systems Conf. 3, 33-37 (1979).
5. North, W.J., Gerard, V.A., Kuwabara, J.S.: Synthetic and Degradative Processes in Marine Macrophytes, Bamfield, B.C. 1980 (proceedings in press).

EXPLORATORY STUDIES OF VEGETATIVE PROPOGATION OF MARINE AGLAE: PROCEDURE FOR OBTAINING AXENIC TISSUES

A. Gibor, M. Polne, M. Biniaminov and M. Neushul

Marine Science Institute, University of California Santa Barbara, Santa Barbara, California 93106

Introduction

We started a research project with a long range objective of developing techniques for vegetative propogation and breeding of economically interesting marine algae. We hope to devise procedures which will not only facilitate the propogation of these plants but will also permit a program of breeding and selection aimed at improving their properties. We are thinking of tissue cultures, culture of cell suspensions, cultivation of protoplasts and hybridization by protoplast fusion. These modern plant breeding techniques are also applicable to algae as is evident from a number of recent publications.

Schiff and coworkers (1972) mechanically disrupted the thallus of the green alga Prasiola and grew the isolated cells under defined culture conditions. They also defined the conditions which led to the attachment of the isolated cells to solid substratum (Bingham and Schiff, 1972). Chen and Taylor (1977) reported the development of techniques for isolation and axenic culture of medullary tissues of female gametophytic fronds of a commercially harvested strain of Chondrus cruspus. Enriched sea water with added "growth-regulators" was used as the culture medium. They obtained fronds with normal morphology from tissue cubes 2 mm in dimension, cut from the colorless medulla of Chondrus plants. A Japanese patent was issued to the Kureha Chemical Company in 1974 for a method of agar production employing in vitro callus cultures of algae such as Gelidium and Gracilaria. From a small piece of a Gelidium thallus, callus was induced after growth for two days at 15° on an agar medium containing 2mg/l IAA and 0.2mg/l kinetin. The medium also contained

Xth International Seaweed Symposium
Printed in Germany

sucrose, ammonium nitrate, coconut milk, yeast extract, metal-ion supplements and sea water. Pieces of callus were transferred to fresh medium and it is claimed that they increased in mass eleven-fold in twenty days. The callus tissue was subsequently extracted and found to yield high amounts of good quality agar (Misawa, 1977). A number of studies were reported on the isolation and maintaining in culture of algal protoplasts. Gabriel (1970) isolated protoplasts from the green alga Uronema gigas. Treatment of the cells with snail gut enzymes caused the liberation of living protoplasts. Regeneration of cell walls was achieved when the protoplasts were cultured in a pectin containing medium. By a similar treatment Braun and Aach (1975) obtained protoplasts from Chlorella. These authors also reported that they observed paired protoplasts suggesting the possibility that protoplast fusion has occurred. No regeneration of cell walls was reported in this study. Protoplasts were prepared from Spirogyra and Zygnema by Ohiwa (1977) and from Microasterias and Cosmarium by Berliner and Wenc (1976). These protoplasts could be maintained in a medium of high osmotic pressure. The regeneration of normal appearing filaments of Spirogyra and Zygnema was achieved by a gradual reduction of the osmotic pressure in the culture medium. Marchant and Fowke (1977) and Marchant (1979) isolated protoplasts of Mougeotia and studied their regeneration. The role of microtubules in the regeneration of the cell walls and the characteristic cell shape was studied by electron microscopy of the membranes of the protoplasts during the regeneration.

The protoplast isolations referred to above were all accomplished by the use of mixtures of digestive enzymes obtained from either snails, fungi or bacteria. Another approach for obtaining protoplasts utilized the specific cell wall digesting enzymes produced by an organism at some point in the life cycle. Schlosser, Sachs and Robinson (1976), Robinson and Schlosser (1978), and Matsuda et al. (1978) obtained protoplasts of Chlamydomonas by use of a specific "autolysine" enzyme which is released by mating gametes of this organism, these protoplasts readily regenerated their cell walls.

A different approach for obtaining protoplasts or even drops of cytoplasm without nuclei (cytoplasts) utilizes giant coenocytic algae such as Nitella or Acetabularia. From such cells the living protoplasm can be squeezed out mechanically. Rounded-droplets of the protoplasm are capable of regenerating walls and in some cases further growth (Gibor, 1965, Werz, 1970; and Kobayashi and Kanaizuka, 1977).

Other examples of experimental studies on algal protoplasts are the attempts of Ohiwa (1978) to produce hybrids of the protoplasts of Zygnema and Spirogyra. Waaland (1978) used vegetative filaments of Griffithsia tenuis, and by culturing them in the confined space of an empty Nitella cell succeeded in producing hybrids of female and male plants. It is evident from the list of experimental work on algal cells and protoplasts that algal cells are amenable to the same manipulations which proved to be productive in the propagation and hybridization of higher plants.

Major problems must be solved before successful routine vegetative cultivation of seaweeds could be undertaken; one of these is getting pure, bacteria free, tissues of the desired plants. Several successful isolations of sterile macroalgae were reported. To obtain such cultures mixtures of antibiotics were used by Provasoli (1957, 1958), Fries (1963) and Boalch (1961). A combination of antibiotics and mechanical dragging of tissues through soft agar was used by Fries (1961), Iwasaki (1967) and Chen and Taylor (1978). A mild antiseptic followed by antibiotic treatment was used by Gibor and Izawa (1963). The published procedures were not very satisfactory, they are quite laborious and resulted in a small percentage of the treated material remaining alive and sterile. We undertook therefore as a first objective of our project to develop a reliable procedure for obtaining bacteria free tissues from algal material which arc collected from nature.

Experimental Studies

Determination of viability: A prerequisite of all experimental treatments of algal tissues with germicidal agents is that the treatment does not kill the algal cells as well as their contaminating bacteria. It is essential, therefore, to have good criteria by which the viability of algal cells could be determined. It is important to note that some of the germicidal agents tend to act as fixatives and result in a life like appearance of the fixed tissues. The criteria which we adapted include the following:

Use of vital dyes. a. Neutral-red uptake and accumulation by live cells

b. Evans-Blue exclusion from living cells

Florescence microscopy. The chatacteristic fluorescence of live cells can be distinguished from that of damaged cells. We found the change in fluorescence to be a most useful and reliable method for immediate evaluation of the effects of chemicals on algal tissues.

Removal of macroepiphytes. Algal material collected in nature is invariably infested by many epiphytes. Commonly, these can be removed by vigorous brushing of the surfaces manually. This procedure is tedious and laborious. We experimented, therefore, with several mechanical alternatives to the manual brushing. Vigorous mechanical shaking in sea water removes many epiphytes. Shaking in a suspension of abrasive materials such as fine sand in sea water increases the efficiency of epiphyte removal. Best results were obtained, however, by the use of an ultrasound treatment. We found that most algal tissues were not damaged by treating them in an ultrasound cleaning bath for several minutes. The treatment can be repeated several times, with a change of sea water between the treatments. Older, strongly cemented epiphytes were cropped close to the holdfast but the holdfast itself was not dislodged. This procedure does not eliminate all the epiphytes but it is very efficient as a preliminary cleaning of the algal surfaces.(Polne 1980). Before adapting the ultrasound procedure for treatment of delicate algal cells it is advisable to establish the limits of tolerance of the specific organism to this treatment. We were successful in cleaning the giant cells of Acetabularia, by sonication for periods of less than one minute. Longer treatments killed these delicate cells.

Sterilizing treatments. The following is a list of germicidal agents which we tried. U. V. light, H_2O_2, Zephiran, Argyrol, Furazone-green, Hexachlophene, Merthiolate, Nalidixic acid, Betadine. We experimented with these at different concentrations and time of exposure. In general, the most efficient bactericidal agents were also fatal to the algal cells. The best treatment which we found was a short exposure, up to 5 minutes, to a solution of 1% Betadine. This is an iodinated polyvinyl pyrrollidone (I-PVP) which maintains a very low concentration of free iodine in its fresh solution. After the Betadine treatment the algal material was washed with sterile sea water and then submerged in a sterile solution of antibiotics in enriched sea water for one week. After the antibiotic treatment the algal fronds were washed and subjected to sterility tests.

Sterility tests. The sterility of the treated algal tissues was checked by inoculating the treated tissues into diluted nutrient broth solution (0.1% Nutrient broth (Difco) in sea water). Some samples were inoculated into the same broth which was further enriched to 0.1% Glucose and 0.1% yeast extract. A third test was the placing of algal tissue samples on the nutrient broth solution which was solidified with 0.6% Agar. In some cases the same algal pieces which passed the nutrient broth test were then transferred to the solidified agar plates for further testing.

It is important to stress that our claim to sterility of algal materials is defined by these sterility tests. We consider that for the experimental purposes for which we are isolating these algal tissues the above tests are adequate. It is also to be noted that remaining algal epiphytes are likely to also survive the sterilization procedures, thus we obtained in some cases bacteria free but not unialgal material.

Conclusions.

By the use of these procedures we have now obtained sterile tissues from the following algae. *Porphyra lanceolata*, *Gelidium nudifrons*, *Gracilaria sp.*, *Eucheuma uncinatum*, *Antithamnion sp.*, *Prasinocladus sp.*, *Macrocystis pyrifers* The efficiency of the sterilizing procedure depends on the state of the starting material. In some cases we obtained 90% of the tissue-pieces in a sterile and viable state. We are now using this general procedure routinely with modification, especially in the time of exposure to the Betadine solution appropriate to specific algal species.

References.

1. Berliner, M.D., and Wenc, K.A.: Protoplasma 97, 185-200 (1977).
2. Bingham, S.E., and Schiff, J.A.: Biol. Bull. 154, 425 (1973)
3. Chen, L.C.M., and Taylor, A.R.A.: Can. J. Bot. 56, 883-886 (1978)
4. Fries, L.: Physiol. Planat. 16, 19-31 (1973)
5. Gibor, A., and Izawa, M.: Proc. Natl. Acad. Science (Wash) 50, 1164-1169 (1963)
6. Iwasaki, H.: J. Phycolog. 3, 30-34 (1967)
7. Kobayashi, K. and Kanaizuka, Y.: Plant and Cell Phys. 18, 1373-1377 (1977)
8. Matsuda, Y., Tamaki, S., and Tsubo, Y.: Plant Cell Phys. 19, 1253-1261 (1978)
9. Misawa, M., in Barz, W. et al., eds. Plant Tissue Culture and its Biotech., Springer (1977)
10. Polne, M.A., Gibor, A., and Neushul, M.: Botanica Marina (1980) (In press)

11. Provasoli, L.: Biol. Bull. 113, 321,(1957)

12. Robinson, D.G., and Schlosser, U.G.: Planta 141, 83-92 (1978)

13. Schiff, J.A., Quaterno, R.S., Harris, G.C., Leggand, M., and Staley, J.: Biol. Bull. 143, 476 (1972)

14. Waaland, S.D.: Planta 138, 65-68 (1978)

15. Werz, G.: In. J. Brachet Med. Biology of Acetabularia. Acad. Press (1970)

Acknowledgment

This work was partially supported by the NOAA Office of Sea Grant, Department of Commerce, under Grant # NOAA-448668-22557.

REVOLUTIONARY IMPACT OF EUCHEUMA CULTIVATION IN THE SOUTH CHINA SEA ON THE CARRAGEENAN INDUSTRY

P. Laite

FMC - Marine Colloids Division
Rockland, Maine, U.S.A.

M. Ricohermoso

Marine Colloids (Phil.), Inc.
Ouano Compound, Looc, Mandaue City

Abstract

The improvement in cultivation technology of Eucheuma in the South China Sea, particularly in the Sulu and Central Visayas areas increased farm production dramatically. In 1979, an over-supply situation developed, especially for E. spinosum species. From 1975 to 1979, E. cottonii production increased ten fold from 1,680 M.T. to 16,480 M.T. and a supply-demand balance has been developed. In the case of the E. spinosum, the increase was 54 times from 60 M.T. in 1975 to 3,251 M.T. in 1979. This was an incredible achievement by the seaweed farmers which caused an over-supply situation for the major carrageenan processors.

The revolutionary impact of this development was the emergence of new production process especially alkali modification of cottonii as well as the emergence of Japan as a major importer of this product.

The price of Eucheuma has been maintained at a reasonable level which should enable processors to discover more applications as well as expand the market for new products.

This paper presents the capability of the Philippines as a reliable raw material resource for Eucheuma-derived carrageenan requirements for the

Xth International Seaweed Symposium
Printed in Germany

world market.

Rewiew

During the Seaweed Symposium held in Santa Barbara, California, U.S.A. last August 26, 1977, M.A. Ricohermoso and L.E. Deveau discussed a paper entitled; "Commercial Propagation of *Eucheuma* spp. Clones in the South China Sea. A discussion of trends in cultivation technology and commercial production pattern." In this paper, the authors reported the development of a seaweed cultivation technology which enabled farmers to produce commercial quantities of *Eucheuma* particularly *Eucheuma cottonii* hence, putting an end to the indiscriminate gathering of wild *Eucheuma* in the Philippines.

This paper deals with the further development which took place during the last two years not only in the field of cultivation but also in the development of new product and new market for both *Eucheuma cottonii* and *Eucheuma spinosum*.

Revolutionary Impact

Increased Farm Production

The data on the Philippine farm production of *E. cottonii* is shown in Table 1. The total farm harvests increased from 480 M.T. in 1973 to 16,480 M.T. in 1979. This shows that the farm production increased 34 times in six (6) years or about 5.7 times annually. In 1979, surplus *E. cottonii* production was achieved which indicated that the Philippines had satisfied the world market requirement for kappa carrageenan derived from this type of seaweed.

The production of *E. spinosum* which is a source of iota carrageenan is shown in Table 2. In 1973, the farm production was only 25 M.T. which gradually increased to 284 M.T. in 1977. In 1978, the production increased to almost three (3) times that of the 1977 production but in 1979, the production increased by more than four (4) times that of the 1978 production. At the end of 1979, there was a surplus production of about 720 M.T. This over-production was unexpected for a crop that took so long to develop. As a result of this problem, the E. spinosum price declined dramatically from December, 1978 to June, 1980.

Increased Philippine Export of Eucheuma

The export data for E. cottonii and E. spinosum are shown in Table 1 & 2, respectively. The raw E. cottonii exported in 1978 was more than twice that of 1977, but the 1979 export was only 1.38 times that of 1978. It appeared that the volume of export of E. cottonii had begun to level off in 1979, especially the raw Cottonii.

The export of E. spinosum also increased in 1978 by almost three (3) times compared to 1977 and then 3.27 times in 1979 over that of the 1978 export. However, unlike E. cottonii, the volume of E. spinosum export was way below the total Philippine harvest thus, creating an over-supply of this species of Eucheuma. This supply-demand problem is discussed in succeeding section.

Emergence of New Market

The consumers of Philippine-grown Eucheuma are listed in Table 3 . In 1978, the big importers were Japan, Europe and U.S.A. Japan became a major buyer by sharing 20% of the Philippine Cottonii export. However, in 1979, Japan bought only 16%, but European processors purchased 56% while the U.S. purchased 24%.

In the case of E. spinosum, the U.S. processors dominated the field by importing 60%, 64%, and 88% in 1977, 1978, and 1979, respectively. A European company imported 28% in 1977 but stopped buying in 1979. Another European company shared 7%, 13%, and 7% in 1977, 1978, and 1979, respectively. Thus, in 1979, the Philippine E. spinosum was imported only by the U.S. (88%), Europe (8%) and Japan (4%).

Emergence of New Products

In 1978, a semi-processed product derived from E. cottonii was exported from the Philippines mostly to the U.S.A. and Japan. This product is exported in chopped or coarse ground form and usually possesses high gel strength and high viscosity water gels. As shown in Table 1, 313 M.T. was exported from the Philippines in 1978 which increased to 689 M.T. in 1979. The exportation of the semi-processed Cottonii will at least double in 1980 and continuously increase in the future as more applications are derived from the new product.

Need for More Application Research

The development of technology in the cultivation of Eucheuma as well as in the production of semi-processed E. cottonii in the Philippines was indicated by the demand for this type of seaweed by the developed countries like U.S.A., Japan and most countries in Europe. This is an example wherein a developing country can supply the requirements of the developed countries if provided the technical and financial resources.

Aided by the technology developed over the years in cultivating E. cottonii, the cultivation of E. spinosum was more rapid than projected and over-supply was quickly developed as new iota application had not progressed in the market place.

Since 1975, the major importers have been urging the Filipinos to produce as much E. spinosum as possible. Now, E. spinosum cultivation technology has developed to the extent that it can supply to meet existing and projected future demand. The price has now declined temporarily due to the over-supply situation but the technology and technique are available given adequate lead time when new applications and uses are developed and demand increases.

Table 1. Farm Production and Export of E. Cottonii from the Philippines, 1973-1979 in Metric Tons

YEAR	TOTAL PHIL. HARVEST	PHIL. EXPORT RAW MATERIAL	PHIL. EXPORT ALKALI-TREATED
1973	480	480	-
1974	12000	6000	-
1975	1680	2640	-
1976	1440	3300	-
1977	5760	4800	-
1978	9720	11102[x]	313
1979	16480	15334[xx]	689

x - includes the raw material usage of 313 M.T. alkali-treated product

xx - includes the equivalent raw materials of 597 M.T. washed product and 689 M.T. alkali-treated product

Table 2. Farm Production and Export of E. spinosum from the Philippines, 1973-1979 in Metric Tons

YEAR	TOTAL PHIL. HARVEST	TOTAL PHIL. EXPORT
1973	25	25
1974	40	40
1975	60	60
1976	250	250
1977	284	284
1978	775	775
1979	3251	2531

Table 3. Importers of Philippine-Grown E. Cottonii & E. Spinosum

IMPORTER	1977		1978		1979	
	COT	SPN	COT	SPN	COT	SPN
1. U.S.A.	33%	60%	24%	64%	24%	88%
2. Europe	43%	37%	47%	36%	56%	8%
3. Japan	9%	2%	20%	-	16%	4%
4. Australia/New Zealand	3%	-	3%	-	1%	-
5. Taiwan	-	-	2%	-	-	-
6. Argentina	2%	-	-	-	0.7%	-
7. Hong Kong	1%	1%	1%	-	-	-
8. Korea	9%	-	3%	-	2%	-

BREAKTHROUGH IN THE COMMERCIAL CULTURE OF *EUCHEUMA SPINOSUM* IN NORTHERN BOHOL, PHILIPPINES

Jose Rodolfo Lim and Hans Porse

Genu Products Philippines, Inc., Cebu City, Philippines and
The Copenhagen Pectin Factory Ltd., Supreme House,
Penang Road, Singapore

The high world demand for *Eucheuma spinosum* during the early 1970's had encouraged the production of this species in Southern Philippines. However, due to many ecological problems such as lack of knowledge of requirements, slow growing strains, occurrence of diseases, seasonality and the unstable political conditions, production lagged behind. Production was highly unreliable and an alternate site became a must.

In 1975, Genu Products, Philippines, Inc., an affiliate of Copenhagen Pectin Ltd. in Denmark seriously considered the development of an alternate site for "spinosum" production. In November 1976, GENU started an experimental farm in Danajon Reef. A year later, the farm became semi-commercial and to date (June 1980) had produced 215 MT of dried "spinosum". This modest success attracted many people to venture into farming.

Research and Development

Initial research efforts were confined principally to field surveys for potential sites, field testing with *E. spinosum*, growth rate studies, monitoring of ecological parameters, multiplication of seedstocks and determination of feasible culture method. Motorized outrigger banca was used for field survey works. Field testing with "spinosum" cuttings

Xth International Seaweed Symposium
Printed in Germany

was undertaken in the first experimental site in Gilutongan Is., an hour-ride from Mactan, Cebu. Activities centered on growth studies using the raft and monoline methods of culture. Monitoring of ecological parameters such as salinity, water and air temperature, current and rainfall were also made. Fertilization studies as well as screening of "spinosum" strains for seedstock development were also done. Preliminary results showed that the site and the local seedstocks were unsuitable. Although better growth rates were obtained using the raft, this method was hardly commercially profitable. This site was abandoned after almost one and one-half years of operation.

Fig. 1. MAP OF NORTHERN PORTION OF BOHOL

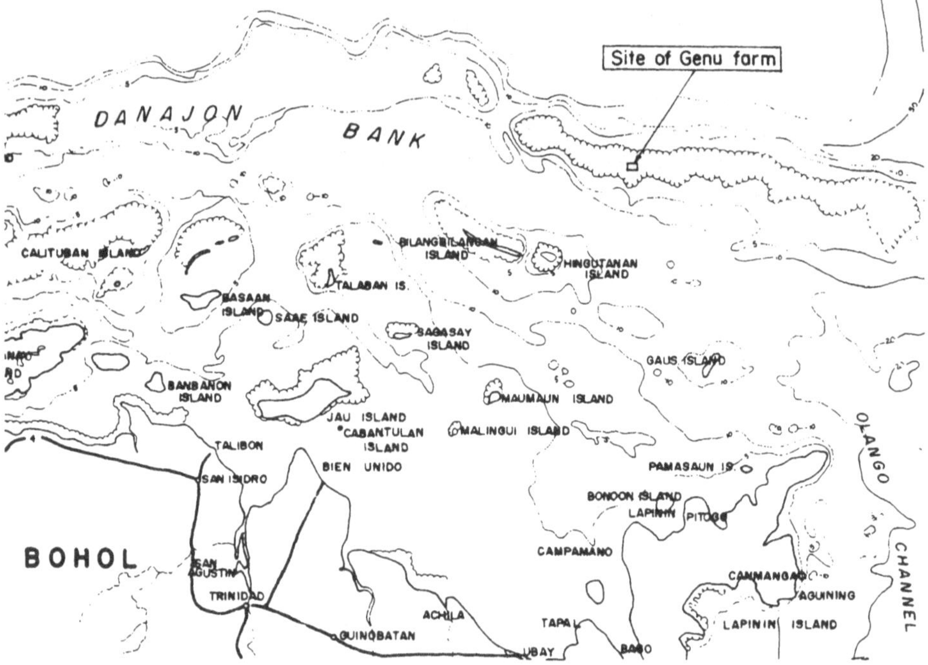

The search for a new site was extended farther north of Bohol towards the Danajon Reef (Fig. 1). Results of surveys were very encouraging as the site supports natural crops of both

E. spinosum and *E. striatum*. The tidally induced current is strong, continuous and uni-directional with depth of one foot even during the lowest spring tides. The site, however, is exposed to strong NE winds, pouching and drying of the produce were problems and the site is far from Cebu City. The site was test-planted with *E. spinosum* and *E. striatum* seed-stocks from Tawi-Tawi, Southern Philippines. After 66 days, *E. striatum* plants averaged 1.5 kgs each while *E. spinosum*, 1.0 kg. Efforts to multiply and screen promising varieties of *E. spinosum* continued. By April 1977, 3 months after the site was test-planted, the seedstocks grew to more than 100 kgs. The original seedstocks had been multiplied more than 50 times in about a 3-month period. Daily growth rates showed an average of 5.9% and 5.7% for *E. spinosum* and *E. striatum*, respectively. The monoline method of culture proved to be efficient and economically feasible method of culturing *E. spinosum* in Danajon Reef.

The first six months in Danajon Reef was very successful. The farm started to produce dried *Eucheuma* and became semi-commercial hacienda-type farm before the end of the year.

The various farm activities follow strict schedules based on the prevailing oceanographic conditions. Working hours are limited by water depth and would last 4 to 5 hours during spring low tides. An average of 10 days of good low spring tides and 20 days of low neap tides are realized in a month. Area preparation, building of the support system, maintenance, etc., are priority activities during very low tides. The schedule of activities is posted for guidance.

For ease in management, the farm is divided into units called modules which consist of 28 monolines. Each monoline is planted with 35 to 36 plants. One module contains about 1000 plants. Thirty-five modules make a hectare containing approximately 35,000 to 36,000 plants. A five-foot space is maintained between modules. The following activities are done when building a module: clearing, staking, installation of monolines, preparation of seedlings and planting. The gathering and preparation of planting materials are done prior to low tides. Six to eight week old plants are selected and harvested, placed in baskets, brought to the drying platform but are left submerged in water until they are to be used. One hundred to 150 gram seedlings are tied with plastic straws and bundled in fives, returned to the baskets submerged in water. Seedlings are transported to the site for planting in floating baskets. Seedlings are tied to the monofilament at 8 to 10 inches intervals. Maintenance is a very important activity a few weeks after planting. A team of workers would go through the modules repairing posts, weeding, replacing lost plants, etc. Plants are harvested when these are a kilo or so in weight. Harvest is transported to the drying platform on a flat bottom boat. Fresh seaweeds are spread thinly on the drying platform at approximately 10 kgs per m^2. The harvest dries up in one and one-half days during sunny days. Dried seaweeds are placed in sacks, each weighing approximately 70 kgs.

Some agronomic techniques have been developed and/or modified from previous techniques to adapt to the conditions in the site. Monolines are placed at right angle to the current about six to eight inches from the substratum. This prevents the plants from sliding to one end of the line and avoid the full force of the current. The use of rattan baskets with floats and scoop nets as container for seedstocks or harvest has proven to be quite efficient. The line cleaner, made of an iron hook with a handle is useful in ridding the monolines of weeds, old tie-ties and silt. Staking in hard rocky bottom is done using pointed steel bar and sledge hammer in making holes into which the stakes are finally planted during building and repair of the support system. The flat bottom dinghy is best adapted in hauling seedstocks and harvest in shallow waters. The drying platform, though contributing to the increase in capitalization, has proven to be more economical than transporting the produce to the islands several kilometers away. The building of a catchment net at the leeward side of the farm has offset the loss due to breakage by strong currents. Harvest from the catchment net is 400 kgs per day.

Socio-Economic Impacts

Most of the people living in the vicinity of the farm site (Danajon Reef) consisting of approximately 1,500 families may be considered as belonging to the low income strata with per capita income of approximately $510 per annum. Most of them are sustenance fishermen whose additional income comes from the gathering of wild seaweeds and production of copra. Thus, the introduction of seaweed farming in the area had paved the way for a labour-intensive but productive employment opportunities for these people.

The development of seaweed farming in the area took approximately 3.5 years. Except for the hacienda type farm of the Company, the others are individual (using hired labor) or family type farms where labor is supplied by the members of the family. Family type of farming was introduced in the area at the beginning of 1978. Farming manual both in English and local dialect, was prepared and distributed to potential farmers who were trained in the company farm. Seedlings were given free and towards the end of 1978, commercial quantities of *Eucheuma* were produced in these family farms. Analysis of the cost of production for the three types of farms showed that the hacienda type would cost U$0.34 or ₱2.49 per kilo, the individual type U$0.37 or ₱2.71 per kilo while the family type only U$0.20 or ₱1.44 per kilo.

Many more were encouraged by the high price initially paid for the produce. By the middle of 1979, Danajon Reef has totally changed to a large thriving community. There are

Fig. 2. Total E.spinosum produced per month by family farms
Total accumulated production Oct. 1978 - May 1980 1,980 m/ts.

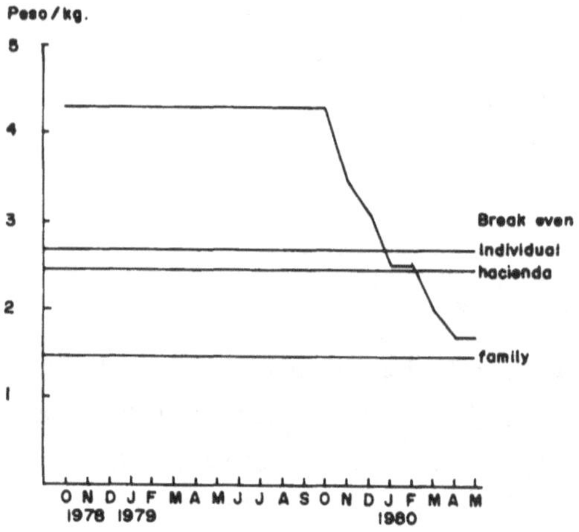

Fig. 3. E. spinosum prices paid to fisherman
Break even point different farming set-ups.

200 farm houses/platforms constructed, 500 hectares planted, 15 million plants in water, 2000 people working daily in the farms and 8000 people or 1000 families depending fully or partly on income from farming. The average monthly production of <u>E. spinosum</u> in family farms and the corresponding price paid to fishermen are shown in Figures 2 and 3. Unfortunately, the carrageenan industry was not able to absorb all the produce and by the end of 1979 the industry suffered another slump experienced in 1974.

We believe that farming will pick up at the start of 1981 when prices should stabilize at a fair price of ₱2.00-₱2.50 ($0.27-$0.34) per kilo. Our farm will continue to act as catalyzer until such time when supply and demand has been balanced through moderate pricing policy.

Total production in Danajon Reef is projected at 1000 MT in 1981 which may more than double in 1982. Our project intends to undertake the improvement of the existing culture system and the genetic improvement of seedstocks (high yielding and disease-resistant variety).

GROWTH RATES OF CORALLINA (RHODOPHYTA, CORALLINACEAE) IN JAPAN

T. Masaki and M. Miyata

Laboratory of Marine Botany, Faculty of Fisheries, Hokkaido University, Hakodate 041, Japan

H. Akioka

Laboratory of Biology, Hokkaido, Kyoiku University, Hakodate Branch, Hakodate 040, Japan

H. W. Johansen

Department of Biology, Clark University, Worcester, Massachusetts 01610, USA

Introduction

Growth rate data for coralline algae reveal that environmental factors affect the rates (1, 2, 3), but the results are suggestive at best. Much of the growth in a coralline is due to the activity of meristems at branch apices and crust margins. The terminal meristems in articulated coralline algae result in elongation and are more localized in action than the marginal meristems by which crustose algae spread over a substrate.

More than 1,400 branches of three species of *Corallina* L., *C. pilulifera* Post. & Rupr., *C. officinalis* L., and *C. squamata* L., were cultured in the Laboratory under controlled conditions and measured for growth. The name for *C. squamata* in Japan is unknown; we shall call it *Corallina* X. The type locality is in Europe, and *C. squamata*, as recognized in that part of the world, has been transferred to *Haliptilon squamatum* Johans., Irv. & Webst. (4). Four forms of *C. pilulifera* were designated A, B, C, and D, and their elongation data were kept separate (Figs. 1, 2).

Our hypotheses were that (1) light intensity and water temperature affect branch elongation in *Corallina*, (2) the three species respond differently

Xth International Seaweed Symposium
Printed in Germany

to environmental conditions, (3) plant sensitivity to environment varies with season and (4) differential growth rates are related to plant habitat and taxonomic status.

Materials and Methods

The plants used in these experiments were collected from known populations in intertidal and upper subtidal regions at Shinari, Hakodate, and at Usujiri, Minami-Kayabe in Hokkaido. The fronds were washed and brushed in sterile seawater and cut into pieces 2-5 cm long. They were planted in plexiglass holders and placed in glass aquaria each containing 2 liters of media. The medium consisted of seawater from Point Anama, Hakodate, to which was added inorganic and organic chemicals (5). The cultures were incubated in chambers at 10, 15 and 20°C. Two light regimes supplied by white fluorescent bulbs provided 500 or 1000 lux on a 12/12/L/D photoperiod. The cultures were maintained for two or three months, and the branches were traced at weekly or monthly intervals from enlarged close-up photographs or by using a Nikon Profile Projector, type V-16A. With this instrument, tracings at standard magnifications between 10 and 20 X were made. The branches were measured from the tracings.

Two sets of identical experiments were carried out. The first was with plants collected and planted in October, 1978 (fall-grown) and the second set involved fronds collected and planted in May, 1979 (summer-grown).

Results

The rates of elongation ranged from about 1.5 mm/mo to more than 5 mm/mo (Figs. 1, 2) with the slowest rates recorded from *C. pilulifera*, the intertidal species. At 10°C, the summer-grown branches of all species grew more slowly than the fall-grown branches (Fig. 1). At 20°C, a similar comparison revealed that in *C. officinalis* and *Corallina* X the reverse was true; that is, the summer-grown plants grew faster than the fall-grown plants (Fig. 2). In C. pilulifera, the most rapid growth occurred in branches collected in the fall, except in one population grown in the summer (population A).

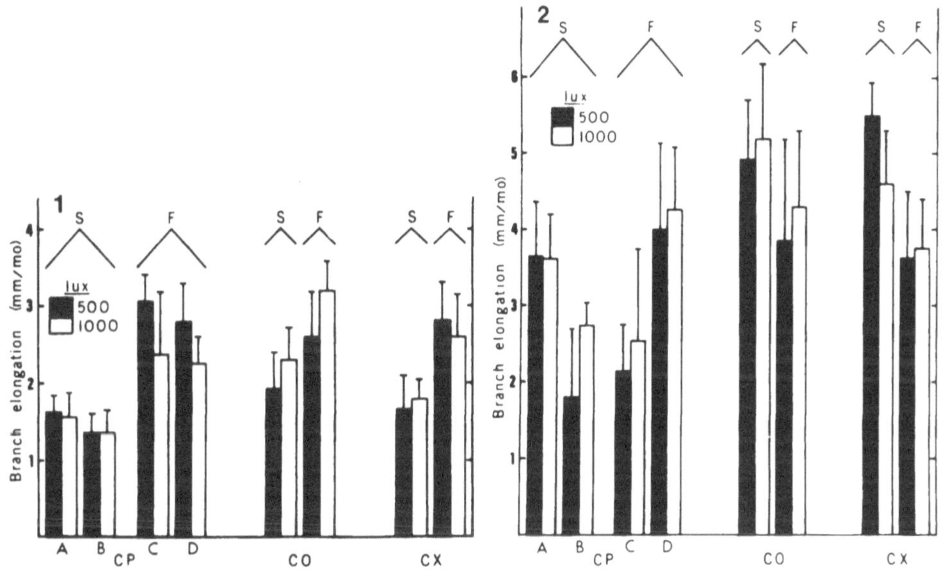

Figures 1-2. Mean rates and standard deviations of branch elongation in all populations of Corallina pilulifera (CP, populations A-D), Corallina officinalis (CO), and Corallina X (CX).

Figure 1. At 10°C. In all species rates in the Fall (F) were greater than those in the Summer (S).

Figure 2. At 20°C. In Corallina officinalis and Corallina X the rates were greater in the Summer (S) than in the Fall (F).

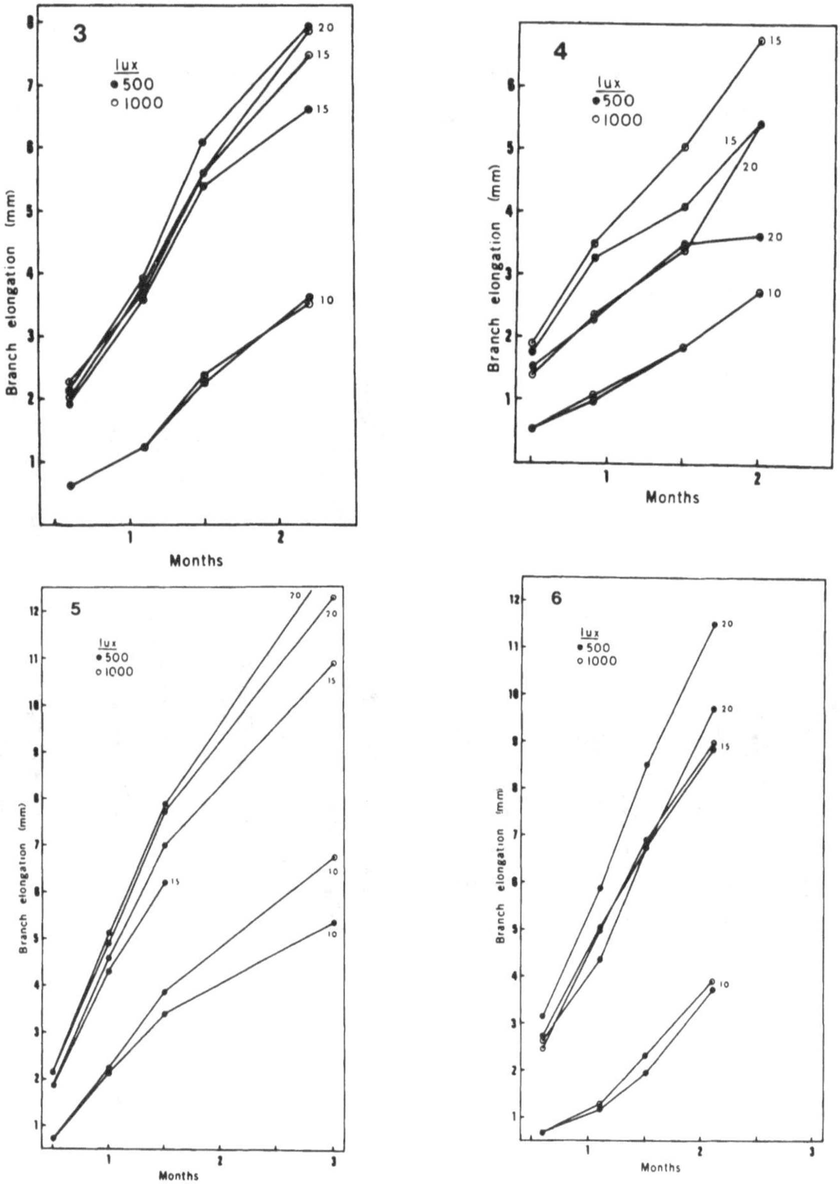

Figures 3-6. Branch elongation in plants collected and planted in May (summer-grown). Numbers to the right are degrees Celsius.
Figure 3. *Corallina pilulifera*, population A.
Figure 4. *Corallina pilulifera*, population B.
Figure 5. *Corallina officinalis*.
Figure 6. *Corallina* X.

All species exhibited a relative insensitivity to the two light intensities used, and temperature was more important in regulating their growth rates (Figs. 3-6). In all experiments the plants at 10°C grew significantly more slowly than those at higher temperatures. The differences in rates between those at 15°C and 20°C were not as great as between 10°C and 15°C although in C. officinalis and Corallina X those at 20°C grew consistently fastest (Figs. 5, 6).

Conclusions

The results substantiated our hypotheses, but the wide range of growth rates was surprising. The conclusions may be summarized as follows: (1) Branches at 15°C and, especially, 20°C grow markedly faster than those at 10°C, in all three species. Seawater temperatures in summer (August) in the collecting sites are 20°C or higher, whereas in winter they are less than 10°C. Although other factors may modify growth rates, on the basis of temperature growth is probably fastest in the summer. (2) The two light intensities used had little effect on growth rates. Possibly intensities lower than 500 lux would result in etiolation, but further experimentation is needed to support or refute this. (3) Corallina X resembles C. pilulifera in appearance, but is closer to C. officinalis in growth rate. It is sometimes difficult to distinguish the first two species in the field; they are both densely branched, in contrast to C. officinalis, which is lax. Possibly the slower growth rate in C. pilulifera is associated with its intertidal habitat. (4) Plants collected at different times of the year responded differently in culture. This differential response was modified by water temperature and suggests that, like many plants, a seasonal cycle of growth is inherent in the branches.

References

1. Adey, W. H.: Phycologia 12, 111-118 (1973).

2. Adey, W. H. and Vassar, M. J.: Phycologia 14, 55-69 (1975).

3. Colthart, B. C. and Johansen, H. W.: Mar. Biol. 18, 46-49 (1973).

4. Johansen, H. W., Irvine, L. M. and Webster, A. M.: Brit. Phycol. J. 8, 212 (1973).

5. Brown, V., Ducker, S. C. and Rowan, K. S.: Phycologia 16, 125-131 (1977).

We gratefully acknowledge the help of R. E. Levenbaum.

EXPERIMENTAL FERTILIZING OF COASTAL *MACROCYSTIS* BEDS

W. North, V. Gerard, R. McPeak
California Institute of Technology, Pasadena, California 91125
and Kelco Company, San Diego, California 92123

Introduction

Our studies seek to determine whether harvestable productivity of *Macrocystis* beds in southern California could be increased significantly by nutrient enhancement. Upwelling and runoff are the two primary sources of nutrient (nitrogen and phosphorus) renewal for kelp beds of southern California. Data on kelp composition and growth indicate that productivity very likely is not nutrient-limited during intense upwelling or runoff events (1). Periods of weeks or months occur, however, when neither significant runoff nor upwelling occurs. Kelp canopies may remain static or even decline. Fertilizing operations would almost certainly be beneficial at such times. Our efforts have been aimed at determining potentially limiting nutrients, selecting kinds and quantities of fertilizers, designing an efficient fertilizing method, and evaluating effectiveness in terms of yield and cost.

Methods and Materials

Laboratory research on *Macrocystis* nutritional requirements was conducted using 5-7 preconditioned juvenile sporophytes (each 0.1 to 10 gm wet weight) held in 40 l Plexiglas tanks containing recirculating seawater (2). Surface seawater collected about 6 km offshore was typically low in ammonium, nitrate, and phosphate and was used as a basal medium to which known amounts

Xth International Seaweed Symposium
Printed in Germany

of nutrients were added. Field work involved measuring various physical, chemical, and biological parameters in kelp beds and conducting fertilizing experiments. Qualitative and semi-quantitative estimates of currents within and around kelp canopies were made by dye dispersal techniques combined with aerial photography. Plant growth was measured as elongation rates of canopy and meter-long fronds. Nitrogen content of plant tissues was employed as an indicator of nutritional status (1). Drawings from aerial photographs were used to estimate canopy areas of experimental kelp beds. Canopies were harvested with a commercial vessel operated by Kelco Company. Yield was computed as harvestable biomass per harvested canopy area. Our most recently employed fertilizing method involved dispersal of ammonium sulfate and phosphate into the kelp canopy by commercial crop-dusting helicopters.

Summary of Results

Several lines of evidence indicate that nitrogen most frequently limits kelp growth in southern California waters. Traces of phosphate (ca. 0.2 μM) almost always occur in sea-water samples, even when nitrate falls to undetectable levels (<0.05 μM) (1,3). The "saturation" concentration for nitrate (i.e., a value where further concentration increases cause no significant rise in growth rates of juvenile Macrocystis sporophytes) varies with light intensity. At 53,000 ergs/cm^2 sec, growth saturates at about 15 μM NO_3 (Figure 1). Ambient nitrate concentrations within and near kelp beds are usually much lower than this (1,3). Juvenile Macrocystis sporophytes exhibit "luxury uptake" of nitrogen, but N reserves can deplete after one week of rapid growth. Thus, Figure 1 illustrates significantly lower growth rates after two weeks of culturing at high illuminations in 3 and 6 μM than in 15 and 30 μM nitrate enrichments. No significant differences occurred among these cultures during the first week. For phosphate,

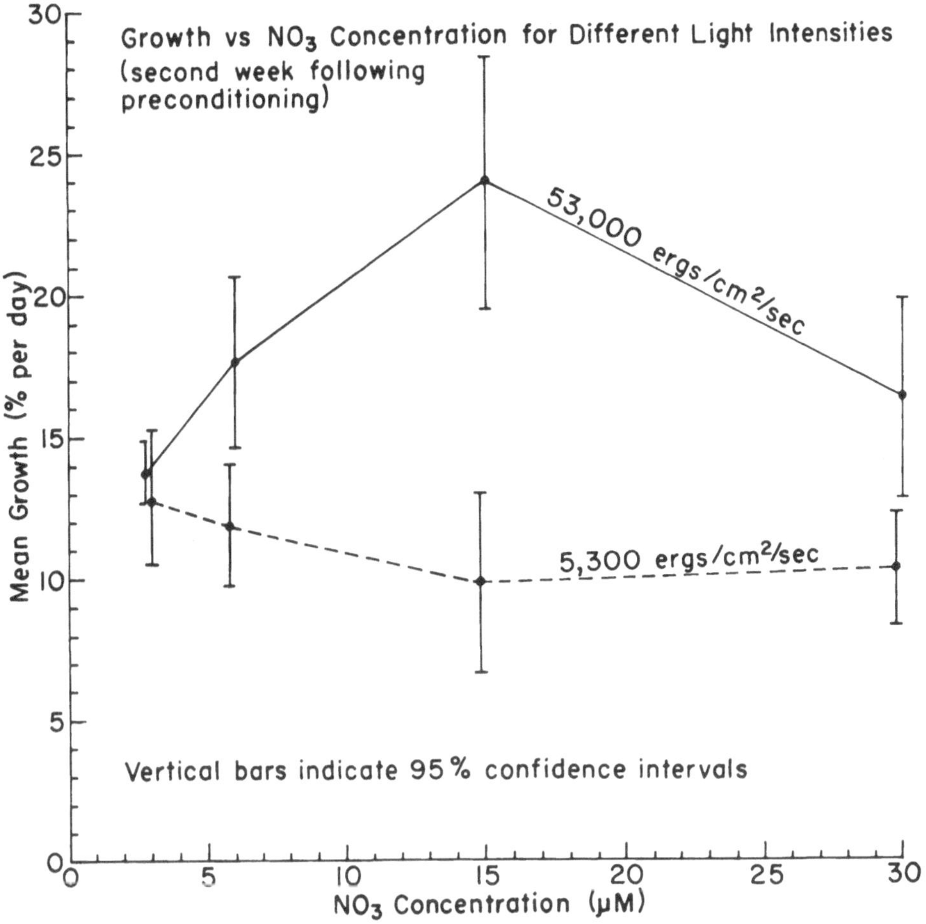

Figure 1. Relationships between illumination, nitrate concentration, and mean specific growth rates (by weight) shown by groups of six juvenile Macrocystis sporophytes held in offshore surface seawater enriched with 1 μM PO_4 and with NO_3 as indicated. All plants were preconditioned in 30 μM NO_3 and 2 μM PO_4 for three weeks before initiation of the experiment.

growth saturates at about 1 µM at 24,000 ergs/cm^2 sec. We concluded that the optimal fertilizer for Macrocystis was some inorganic form of nitrogen with an admixture of phosphate at an N/P ratio of about 15.

Dye dispersal studies showed that current velocities in channels and open spaces of Macrocystis beds ranged up to 9 m/min. Interstitial water movements within the canopies, however, were much slower. In the absence of strong winds, fertilizer solutions might reasonably be expected to linger for 1-3 hr if they were introduced into the canopy interstices. Rapid uptake of the dispersed material was thus desirable for high fertilizing efficiency. Macrocystis accumulates ammonium at high concentrations (>20 µM) more rapidly than nitrate (4). Ammonium sulfate was slightly toxic but less damaging to small Macrocystis sporophytes than the chloride or nitrate salts during short exposures to 1 M solutions. Crystals smaller than 1 mg dissolved in a minute or less in seawater, within the top meter of the water column where the greatest kelp biomass concentration occurs. In our most recent experiments, ammonium sulfate and phosphate crystals were scattered over a kelp bed by conventional crop-dusting techniques. Initial ammonium concentrations in surface water were usually about 100 µM but fell to around 20 µM within an hour or less (1).

We have assessed effects of fertilizing various kelp populations since 1976. Our largest and most recent experiment was conducted from November 1978 through March 1980 at a relatively small isolated bed off Leucadia, California, (33°05'N, 117°19'W) with a total canopy area of 0.1 to 0.2 km^2. Computations indicated that 5-10 M.T. of commercial grade ammonium sulfate would increase N contents of the kelp tissues by about 1% of dry weight. One ton was dispersed daily for 10 to 15 days during each of four fertilizing periods in the 15-month study (Table 1). N contents of the plants rose by roughly 1% during each fertilizing period (1). Unfortunately,

Table 1. Summary of events during the Leucadia kelp bed fertilizing experiment.

Dates	Event	Remarks
Nov 29-30 1978	First harvesting	375 wet tons from 0.08 km^2
Jan 8-24 1979	First fertilizing	13 tons dispersed
Feb 28 1979	Second harvesting	203 wet tons from 0.11 km^2
April 9-24 1979	Second fertilizing	13 tons dispersed
May 23-June 1 1979	Third harvesting	399 wet tons from ca. 0.1 km^2
Sept 17-Oct 3 1979	Third fertilizing	10.4 tons dispersed
Nov 29-Dec 10 1979	Fourth harvesting	1104 wet tons from 0.10 km^2
Jan 25-Feb 28 1980	Fourth fertilizing	15.6 tons dispersed

the study coincided with a year of unusually large nutrient inputs from upwelling and runoff, and control kelp beds also showed high N contents, growth rates, and harvest yield. Results, nonetheless, were encouraging. The computed annual yield from the Leucadia kelp bed was about 36 dry tons/ha/yr. This is the highest yield ever reported for a southern California kelp bed.

Fertilizing costs as of March 1980 were: ammonium sulfate, \$150/ton; premixed sulfate and phosphate at a 1/20 ratio, \$336/ton; price of application, \$360/day. The costs assume fertilizer is purchased in 25 ton lots and 1 ton is dispersed daily. If further research verifies that ammonium uptake efficiencies (calculated to be 30-60%) and harvest yields as high as observed at our Leucadia site can be reliably achieved, it seems probable that fertilizing *Macrocystis* beds during periods of low natural inputs of nutrients will be economically justifiable.

Acknowledgements

Research support from the Office of Sea Grants, Kelco Company, Union Oil Company, the U.S. Department of Energy, the U.S. Navy, and the National Science Foundation is gratefully acknowledged. We thank the staff of Kerckhoff Marine Laboratory for assistance in all aspects of the work.

References

1. North, W.J., Gerard, V.A., Kuwabara, J.S.: Synthetic and Degradative Processes in Marine Macrophytes, Bamfield, B.C. 1980 (proceedings in press).
2. North, W.J.: Clean Fuels from Biomass Symp., Inst. Gas Tech., Chicago, 1977, p. 128-140.
3. Jackson, G.A.: Limn. & Oceanog. 22, 979-995 (1977).
4. Haines, K.C., Wheeler, P.A.: J. Phycol. 14, 319-324 (1978).

STUDIES IN DOMESTICATION OF *EUCHEUMA UNCINATUM*

M. Polne, M. Neushul, and A. Gibor
University of California, Santa Barbara, California 93106

Introduction

An important requirement for the cultivation of any organism is knowledge of its growth patterns, growth rates, and its productivity. Such knowledge has been obtained for the red alga, *Eucheuma uncinatum*, an iota-carrageenan producer. In nature it is an annual endemic plant of the Gulf of California. It is commonly found in Feb. - July, (Norris 1975). Natural populations have been experimentally harvested and good quality iota-carrageenan has been extracted (Dawes et al. 1977). From our observations of natural populations the plants are quite sparse, and are attached to mid- and low- intertidal rocks by a small and relatively weak holdfast.

Our investigation of *Eucheuma uncinatum* as a suitable alga for cultivation has included the following basic topics:

1. Conditions for optimum growth
2. Maximum potential for growth
3. Effectiveness of various nutrient application methods
4. Manipulation of stages in the life history
5. Selection of superior quality plants
6. Storage of seed stock

Results and Conclusions

Growth

In April 1977 a small branch (2 gr) of *E. uncinatum* was transplanted from Guaymas, Mexico (water temperature about 28°C) to the cool 17°C waters of

Xth International Seaweed Symposium
Printed in Germany

a sea water supplied greenhouse in Santa Barbara (Charters and Neushul, 1979). The plant seemed to have no trouble adapting to the lower temperature environment. Further research demonstrated that the best growth occured at 17-24°C, with a maximum of 11% wet weight increase per day. Low temperatures of 9-15°C and high temperatures of 28-30°C suppressed growth. 9-10% growth per day was common among the over 200 plants studied. Within the optimum temperature range the plants resembled the original plants collected from nature: They were light brown to reddish brown, large (30-70 cm long), and flexible. When the temperature was raised gradually to 30°C (about 1/2°C per day) the plants survived, but growth dropped down to 1-2% per day. Plants grown at 29-30°C were dark, deep red, very brittle, and developed a thin, short branching pattern which gave the thallus a thick, bushy appearance. No plants survived temperatures of over 30°C, even though they were exposed to a gradual increase in temperature of one degree every week.

Light

E. uncinatum's growth response to light was as expected (Fig. 1). Optimum growth occurred at light levels of 500 to 1500 uE m^{-2} sec^{-1}. Full exposure to sunlight, 2000 uE m^{-2} sec^{-1}, however, caused bleaching and eventual mortality. Once this light damage occured it was irreversible. Bleached plants which were moved to low light never recovered.

Nutrients

The availability of nutrients was a critical factor affecting the growth rates of E. uncinatum. Growth in natural ocean water (3-5 ugat/l NO_3, 1 ugat/l PO_4) was slow. The addition of nutrients boosted growth ten fold within one or two days of nutrient application (Fig. 2). Three different methods of applying the nutrients to the plants were studied. 1) The total body of medium was enriched with nutrients (up to 30 ugat/l NO_3, 5 ugat/l PO_4) using either a slow release commercial source of fertilizer, or by the addition of pure salts of $NaNO_3$ and KH_2PO_4. This provided continuous availability of nutrients. 2) The plants were dipped

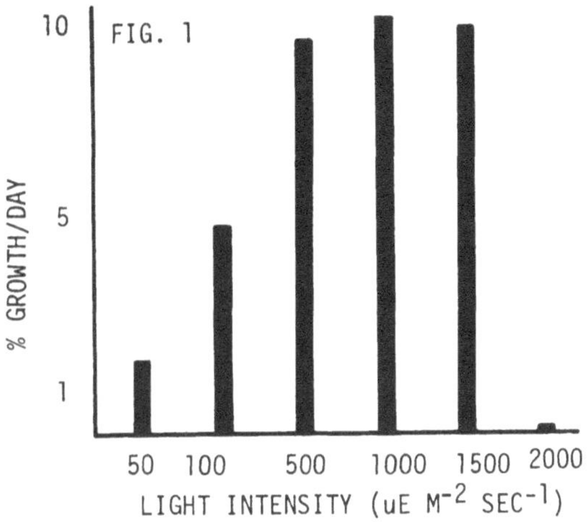

FIG. 1. Effect of light intensities on the growth of Eucheuma uncinatum

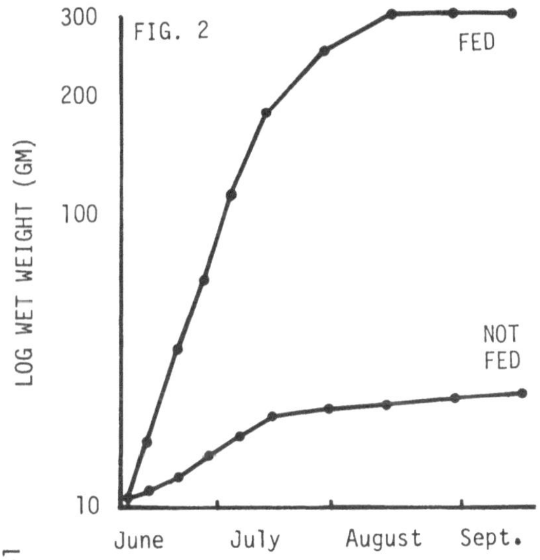

FIG. 2. Effect of "feeding" on the growth of Euchemia uncinatum

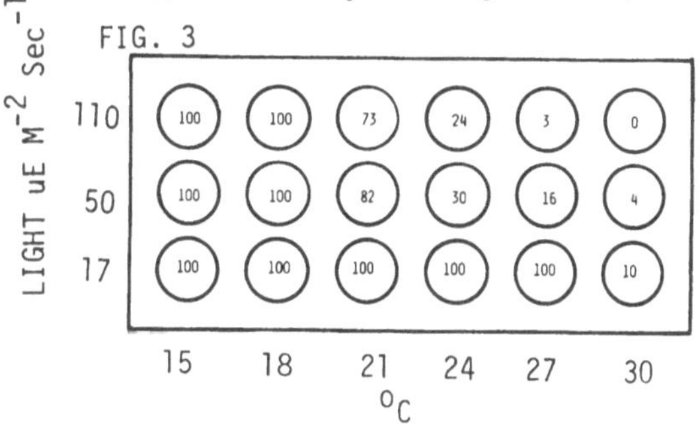

FIG. 3. Percent survival after one month on a temperature/light cross gradient.

in nutrient rich medium once every week, for about an hour in the light. (Polne et al. a). With this dipping method, occasional stirring was necessary to ensure that there was active uptake. 3) A source of concentrated nutrient solution was applied to one part of the thallus, preferentially to a thick older stem. With this point-application method, it was assumed that the part of the plant which was in contact with the fertilizer source would take up nutrients, and would transport nutrients to the entire plant.

Both continuous-availability, and the weekly-dip methods of applying nutrients were successful and supported good growth rates. However, the continuous-availability method also supported the active growth of epiphytes, and thus created several undesired blooms of diatoms and *Ectocarpus*.

To see if the point-application method worked, we looked for the existence of translocation in *E. uncinatum* by following labelled isotopes via autoradiography and scintillation counts. $^{33}P-H_3PO_4$ and ^{14}C-Bicarbonate (1 uCi/ml) were applied to a branch for periods as short as 30 minutes. Traces of the label were detected in other parts of the plant (Polne et al. b). These results show a clear translocation of both ^{33}P and ^{14}C throughout the *E. uncinatum* plants. The label seemed to move efficiently towards young, actively growing branches and to accumulate there. These results also suggest the possible use of point application of nutrients in farming not only brown algae, in which translocation is known to occur (Parker, 1965; Penot, 1979), but red algae as well.

Manipulation of stages in life-history

Male and female plants:

All stages of the life-history of *E. uncinatum* were grown in the laboratory. Light levels, nutrient availability, and water motion were the main factors affecting the morphology of cultured plants.

Female plants developed cystocarps in culture vessels which were held under low light conditions (30-50 uE m^{-2} sec^{-1}), low water motion and a regular supply of nutrients. Both male and female plants were identifiable when they reached 10 cm in length. They were very branched, and weighed between 73-150 gr. Cystocarps released normal size carpospores which developed into normal looking germlings.

Induction of tetraspores:

Tetraspore release was induced on a light/temperature cross gradient table like those built by Halldal (1958) and Edwards (1970). Spores developed over a 9-30°C temperature range and 17-110 uE m^{-2} sec^{-1} light range. However, the largest number of spores and best survival and development occurred at 21-27°C and 17-50 uE m^{-2} sec^{-1}. Figure 3 shows that cool temperatures (below 18°C to 9 C) supported 100% survival of spores and germlings, but growth rates were directly proportional to the temperature. No active growth occurred below 12°C. Best growth occurred under the combination of 17 uE m^{-2} sec^{-1} and 19-29°C. The ability to induce the release of a large number of spores makes it possible to start a culture of identically aged plants, grow large numbers of them in the laboratory under controlled conditions, and select for desirable characteristics such as high growth rates, disease resistance, temperature tolerance, etc.

Storage of seed stock:

A safe method for storing selected seed stock is a necessity for an ongoing cultivation plan. The seaweed farmers in the Orient, working with plants having a heteromorphic life-history, have developed the conchocelis shell storage system for Porphyra, and can store gametophytes of kelps in culture.

Several experiments were run, where we studied Eucheuma spores and germlings (<1 cm) for long time periods (over three years). We found that germlings in dish cultures could be stored successfully at low light (3-5 uE m^{-2} sec^{-1}), 16-22°C, salinity of 30-41 ppt and with nutrient change once every two months, for three years at the time of writing. These stored, dormant "algal seeds" would grow normally on transfer to better growth conditions (30-100 uE m^{-2} sec^{-1}, regular nutrient supply, and water motion).

References

Charters, A. C. and M. Neushul, (1979). Aquatic Botany, 6: 67-78.

Dawes, C. J., N. F. Stanley, and R. E. Moon, (1977). Bot. Mar. 20: 437-442.

Edwards, P. and C. Van Baalen, (1970). Bot. Mar. 13: 42-43.

Halldal, P. and C. S. French, (1958). Plant Physio. 33(4): 249-252.

Norris, J. N., (1975). Unpublished Ph.D. Dissertation, Univ. of Calif.

Parker, B. C., (1965). J. Phycol. 1: 41-46.

Penot, M. and M. Penot, (1979). Z. Pflanzenphysiol. Bd. 95: 265-273.

Polne, M., M. Neushul, and A. Gibor, in preparation.

Polne, M., K. Eckhart, and A. Gibor, in preparation.

THE "ICE-ICE" PROBLEM IN SEAWEED FARMING

F. R. Uyenco, L. S. Saniel and G. S. Jacinto
Marine Sciences Center, University of the Philippines, Diliman, Quezon City, Philippines

Introduction

The problem of 'ice-ice' in Eucheuma farms still prevails today, six years since its first report (Trono, 1974). Since then, attempts have been made to explain the occurrence of this ailment - manifested by the prescence of a white powdery growth on the thallus which causes the loss of pigments, gradual consumption and fragmentation of the alga. Doty (1973) believes that the problem is due to senescence and terms it "aging effect". Trono (1974) suggests that "ice-ice" is brought about by unfavorable ecological conditions that lower the resistance of the algae. The presence of bacteria isolated from affected thalli of Eucheuma (Uyenco, 1977) brought in the possibility that 'ice-ice' may actually be a disease caused by pathogenic microorganisms.

This study was initiated to obtain empirical data and determine which factors causes the problem. One aspect involved the isolation, identification and cultivation of microorganisms from Eucheuma samples obtained from the field. At the same time, field studies were undertaken in seaweed habitats to look into abiotic and biotic factors and possibly correlate these with the incidence of "ice-ice".

Xth International Seaweed Symposium
Printed in Germany

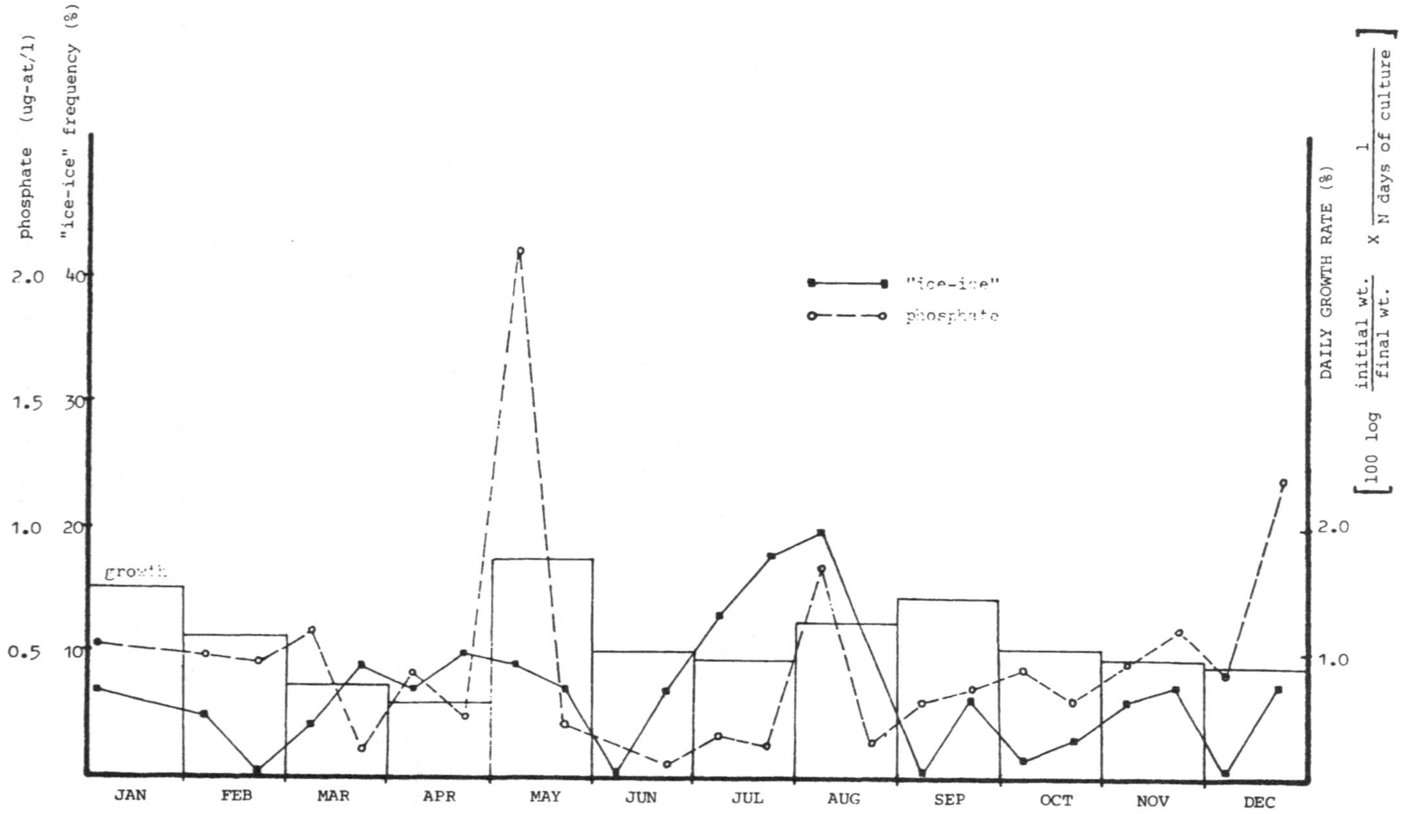

Fig. 1 Variation of phosphate concentration and growth rates with incidence of "ice-ice".

Methods and Materials

Regular field studies (fortnightly) were undertaken at the GENU Products farm off Hingutanan Island, Bohol ($10^{o}17'$ N - $124^{o}24'$ E) starting January, 1979. The algae cultivated in the farm is mainly E. spinosum with only a small percentage of E. striatum. Three sampling stations approximately 250 meters away from each other were arbitrarily chosen to be monitored in the study area.

Diurnal water movement, temperature (minimum-maximum) and nutrients (nitrate-nitrogen, nitrite-nitrogen, and reactive phosphate-phosphorus) were determined from bottom-water samples in each sampling station. Tagged plants were also established in each station in order to evaluate growth rates. In addition, random collections of about 100 harvested plants were made monthly to determine the mean weight of the algae.

Incidence of "ice-ice" as well as epiphytes in each 10-meter quadrat established at the sampling stations were determined. This was done by dividing the number of plants in each quadrat.

Bacteria obtained from sections of unhealthy and healthy Eucheuma were identified using standard techniques. Algae epiphytic on Eucheuma were also collected and identified.

Results and Discussion

The values of the various parameters measured in the three sampling stations have been found essentially the same. Thus, the choice of one station may be considered representative of the other two.

The average current velocity in the area was found to be 50 cm/sec, currents being primarily tidal. This magnitude is considered fairly strong and insures substantial diffusion of

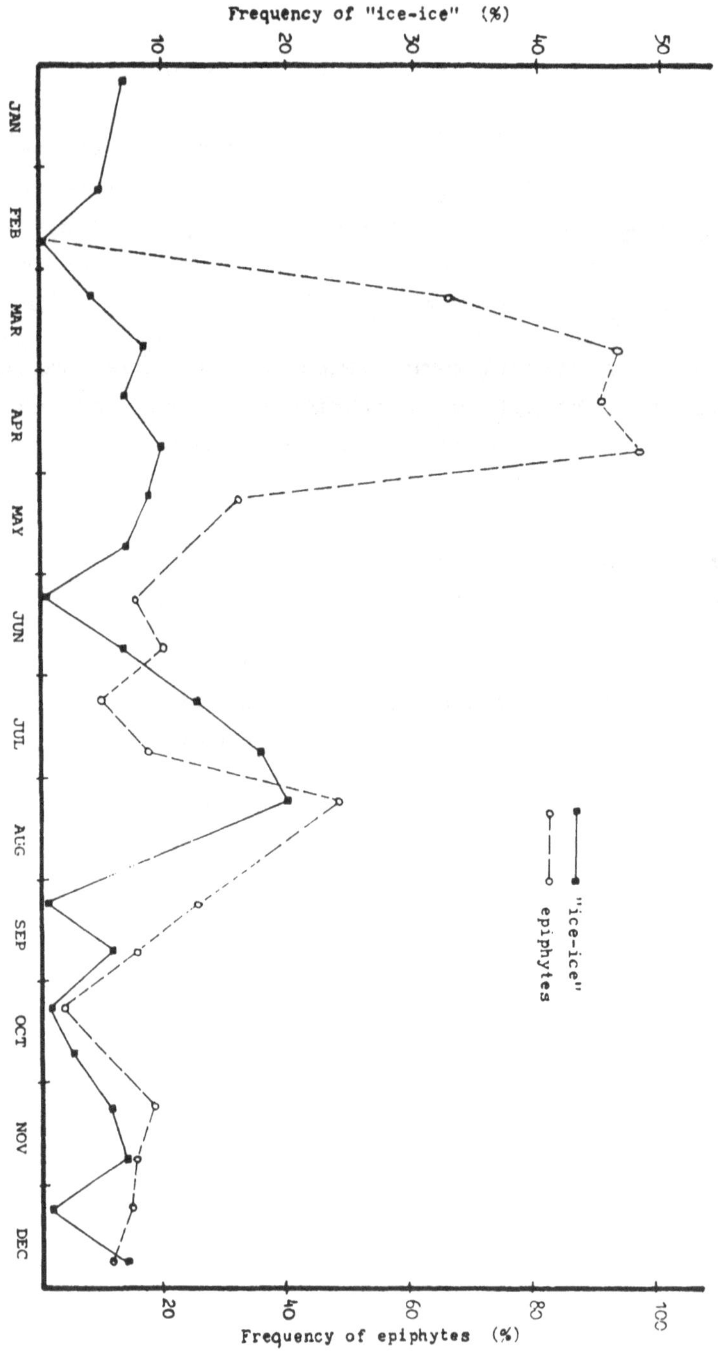

Fig. 2 Variation of "ice-ice" incidence with occurrence of epiphytes.

whatever nutrients are available in the seawater of the algae. Since water velocity has been found rather constant during the year, this parameter is not considered to be a source of stress to the Eucheuma in the area.

A possible correlation has been found between phosphate concentration in the seaweed habitat and the incidence of "ice-ice". Periods of relatively high phosphate seem to show a lowering of the number of propagules found with the problem (Fig. 1.) This has not been found true with variations in nitrate concentrations, however. The input of phosphate is likely to be of terrestrial origin as high values were obtained during rainy periods of the year experienced by surrounding islands.

During months of high "ice-ice" incidence, high occurrence of epiphytes have also been generally found. The relationship of epiphytism to "ice-ice" incidence, however, remains unclear. To be considered still is whether epiphytism causes a substantial stress on the Eucheuma making it more susceptible to "ice-ice" or if the red alga at periods of high incidence of the problem somehow become favorable substrates for the epiphytes to thrive on.

Bacteria obtained from sections of unhealthy Eucheuma thalli at various months of the year were from the genera Pseudomonas, Flavobacterium, Vibrio, Xanthomonas and Achromobacter. Like the earlier findings of Uyenco et al. (1977) no specific bacteria was found associated with each incidence of "ice-ice". This strengthens the belief that the role of microorganisms in the onset of the problem is secondary. It is likely that the bacteria are able to effectively invade the algae only at periods when the Eucheuma is less able to produce extracellular products that may be inhibitory to the microorganisms due to stress it may be experiencing or during periods of non-active algal growth. In Fig. 2, high growth rates particularly during the months of May and September, show a decrease in "ice-ice" incidence. As indicated by Sieburth (1968), there appears to be a direct correlation of inhibitory activity and also concentration of inhibitory substances with active growth of seaweeds.

Thus, the interplay of ecological or abiotic factors with the physiological state of the Eucheuma during its growth stages seem to be the more plausible explanation to the "ice-ice" problem.

References

1. Doty, M.S. Farming red seaweed, Eucheuma for carrageenans, Micronesica 9(1), 59-73 (1973).
2. Dawes, C., Mathieson, A., Chenny, D. Ecological Studies of Floridean Eucheuma (Rhodophyta, Gigartinales) I. Seasonal growth & reproduction. Bull. Mar. Sci. 24(2) (1974).
3. Rodina, A.G. Methods in Aquatic Microbiology. University Park Press, Baltimore (1972).
4. Sieburth, J.McN. The influence of algal antibiosis on the ecology of marine microorganisms. Advan. Microbiol. Sea 1, 63-94 (1968).
5. Strickland, J.P.H., Parsons, T.R. A Practical Handbook of Seawater Analysis. Fisheries Research Board of Canada (1968).
6. Trono, G.C. Jr. Eucheuma Farming in the Philippines. U. P. Natural Science Research Center, Quezon City (1974).
7. Uyenco, F.R. Microbiological Studies of Diseased Eucheuma sp. and Other Seaweeds. National Seaweeds Symposium. Metro Manila, Philippines (1977).
8. Uyenco, F.R., Saniel, L.S., Gomez, E.D. Microbiology of Diseased Eucheuma striatum Schmitz. Proceedings Ninth International Seaweed Symposium. Sta. Barbara, California (1977).

DEVELOPMENT OF A COASTAL MARINE FARM AND ITS ASSOCIATED PROBLEMS

W.N. Wheeler, M. Neushul and B.W.W. Harger
Neushul Mariculture Inc., 275 Orange Ave., Goleta, CA 93017, USA

Introduction

Marine farmers must be able to provide selected seed stock, suitable substratum and adequate nutrients for crops grown on their farm structures. This paper describes farm structures and growth of macrophytes on them. This experimental farming program has been under way for the past year at two sites near Goleta, California, USA.

Methods and Materials

Farming is carried out shoreward from the Goleta Bay kelp bed, in Goleta Bay, and at Goleta Point at the west end of the bay. The point site is exposed to longshore currents and wave action, while the bay site is more protected.

To characterize these environments, daily and weekly environmental measurements have been made. Nitrate levels, current speed and secchi depth are discussed here. Nitrates were determined according to (1) on filtered, frozen samples. Current speed was measured using a modified TSK flow meter, and a 25 cm diameter secchi disk was used.

In order to measure the growth response of individual plants, these are grown on planting units consisting of two vertical spar buoys with a horizontal substrate suspended between them (Fig 1). The unit parts are made of two-inch diameter PVC (polyvinyl chloride) pipe. The system is anchored with concrete-filled scrap tires on the sea floor. The spar

Xth International Seaweed Symposium
Printed in Germany

buoys are attached to the anchors so that they do not twist in the current. Individual plants are attached to short pieces of rope that are in turn attached to a slit collar that is snapped over the substrate when the plants are grown and can be easily removed when they are being measured. On a commercial farm, the measurement of individual plants would not be necessary. We have modified the scientific farm design shown in Fig 1 to be more like a commercial farm. Plants are inserted on 15 and 30 cm centers along 12 meter-long ropes, which are attached to modified substrates (Fig 2).

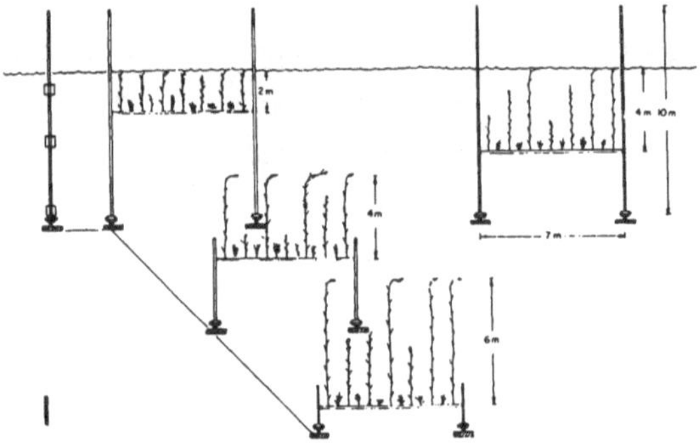

1

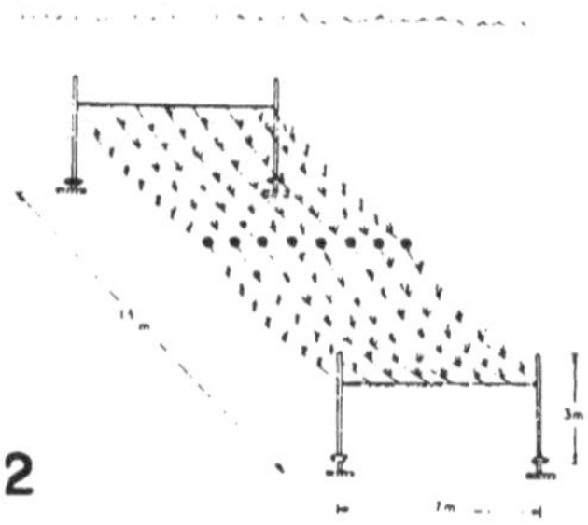

2

Fig 1: A modular test farm structure, consisting of 2 spar buoys anchored to at the bottom, with a planting substrate tied at a fixed depth between them.
Fig 2: A rope farm, consisting of the basic structures shown in Fig 1, with 12 m rope planting lines strung between them.

Seed stock plants have been collected from kelp forests near Santa Barbara and off Santa Cruz Island. Macrocystis pyrifera (L) CA Agardh and Sargassum muticum (Yendo) Fensholt were grown first in the laboratory and then mounted on collars or rope lines. The culture methods are described elsewhere in detail (2,3). Gelidium robustum (Gardner) Hollenberg & Abbott and G. nudifrons Gardner were vegetatively divided and put directly onto collars and ropes.

Results

A body of background information is provided by measurements of water transparency with the secchi disk (Fig 3a), of current speed (Fig 3b) and of nitrate concentration (Fig 4) at both the bay and point sites. Low secchi depth and high current speeds are good indicators of storm conditions, which were present in late December, January and February. Calm conditions were generally present from October through early December and March through June. Upwelling, as indicated by high nitrate concentrations, occurred infrequently, only once in late November and again in May.

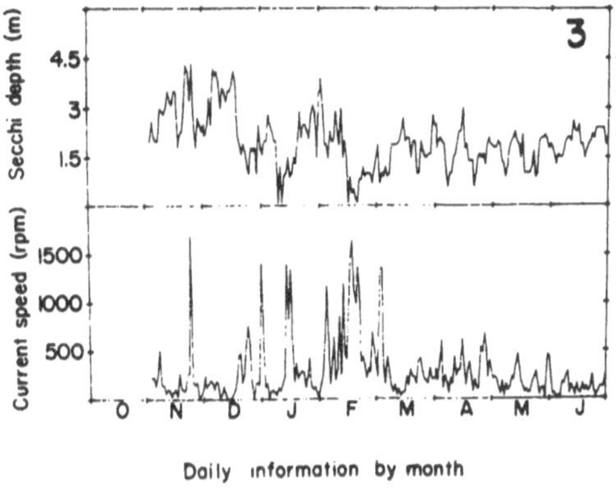

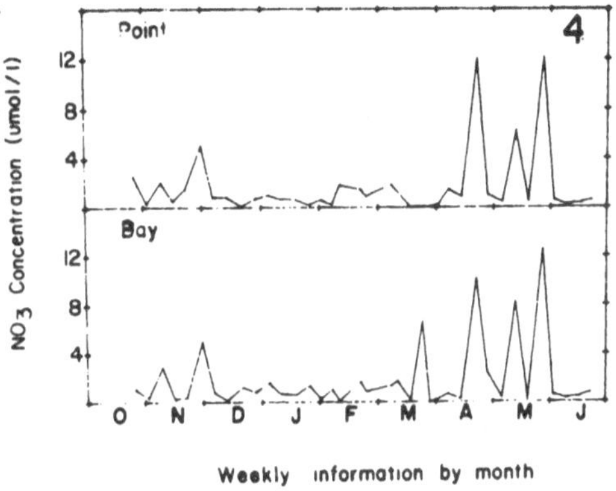

Fig 3: Daily secchi disc measurements (top) and current speed measurements (bottom) made on a daily basis from the end of Goleta Pier.
Fig 4: Weekly sampling of nitrate concentrations at the point (top) and bay (bottom) sites. Samples taken from the top 2 m of the water column.

Growth experiments were performed from October to December, January to March and March to May. These dates correspond to a calm, nutrient-poor period, a stormy period and an upwelling period.

The fall growth rates are presented in Table 1. Along with these rates, the starting size of the cultivated plants, and ending size as shown. The weight of epiphytes on the plants was also measured. Initial observations indicated that starting size may influence growth rates, since regression analyses of the dependence of growth rate on starting weight showed significant correlations in some cases. Further, regression analysis of growth rate versus epiphyte load showed no significant correlation, even though some species had rather high epiphyte loads. <u>Sargassum muticum</u>

had the highest mean growth rate, approaching 8%/day calculated on a logarithm basis. Macrocystis pyrifera had the second highest growth rate, 4.6% per day, and was also the cleanest, having only 1.8 g of epiphytes per plant. Of the two Gelidium species, G. nudifrons had the highest growth rate, possibly because of its greater number of apices.

Table 1: Growth comparison farmed macrophytes - Fall Experiment

Species (number)	Start wt	End wt	Epiph.	GR (%/d)
G. nudifrons (222)	6.0	14.8	8.8	1.4
G. robustum (169)	5.6	6.7	4.7	0.8
M. pyrifera (123)	0.9	28.5	1.8	4.6
S. muticum (57)	0.2	5.2	3.4	7.9

Table 2: Growth comparison of the four species - Winter Experiment

Species (number)	GR (%/d)	% (+) GR
G. nudifrons (120)	$0.97 \pm 0.08^{*}$	62
G. robustum (50)	0.37 ± 0.05	52
M. pyrifera (11)	2.48 ± 0.60	100
S. muticum (6)	2.48 ± 1.02	100

* $\pm$ standard error

During the winter experiment (Jan-Mar) the farms were subjected to storms that destroyed many substrates. Many of the plants were lost. Those that survived, were, in most cases badly damaged. The growth rates for this period reflect the generally poor growing conditions. Growth rates are calculated only on plants showing positive growth. In spite of the poor conditions, the percentage of plants showing some growth is still fairly high.

From April to June, there were 3 small upwelling events (Figure 4). This natural fertilization along with the stabilized weather and other factors provided better growth conditions than before. All growth rates, with the exception of that of *Sargassum*, were highest during this period. The drop in the growth rates of *Sargassum* is probably due to the much greater initial size of the plants.

Table 3: Growth comparison of the four species - Spring Experiment

Species (number)	Starting wt	Ending wt	GR (%/d)
G. nudifrons (74)	9.1	45.7	2.22 $\pm$ 0.19
G. robustum (90)	6.7	15.1	1.26 $\pm$ 0.10
M. pyrifera (5)	6.5	450.2	5.84 $\pm$ 1.18
S. muticum (35)	4.0	19.3	2.19 $\pm$ 0.20

Rope line outplantings were started in May, during the last upwelling event of the season. This took place over about one week after which nitrogen levels remained low for the rest of the experiment. *G. nudifrons* on ropes grew at a rate of 1.8 %/d, while *G. robustum* grew at 1.0 %/d. This corresponds to an increase in biomass from 7.5 Kg to 17.0 Kg for *G. nudifrons* and an increase in biomass from 3.0 Kg to 4.6 Kg for *G. robustum*.

Discussion

Although the growth rates of these plants under natural conditions is quite good, it is hoped that by: 1) providing supplemental nitrogen, 2) adjusting the depth of the cultivation so as to provide optimum light levels, and 3) devising some method for controlling epiphytization; even better growth rates will be possible.

Because the height of the planting substrates can be varied, it is possible to optimize the light level for best growth of the crop plants, and worst growth of photosynthetic epiphytes. Some preliminary experiments have been carried out with dramatic results, but not all of the experiments showed positive results. Much of the observed variation in growth rate could be due seasonal and site-specific effects. The more sheltered bay site had significantly lower growth rates during the fall experiment. However, during the winter storms, those plants in the relatively calmer waters of the bay were the only ones to survive. During the spring experiment, differences in epiphytization with site were noticable, plants at the point site were epiphytized more heavily by hydrozoans and Ectocarpus-like plants which may have masked some of their increase in growth. Epiphytization made it difficult to assess the growth of the crop. Control of epiphytes and the provision of nutrients are two major problems for the would-be macroalgal farmer.

This material is based upon work supported by the National Science Foundation under grant number PFR-7911715.

References

1. Strickland, J.D.H., Parsons, T.R.: Bull. Fish. Res. Bd. (rev ed) 167, 1-310 (1972).
2. Charters, A.C., Neushul, M.: Aquatic Bot 6, 67-78 (1979).
3. Sanbonsuga, Y., Neushul, M.: J. Phycol. 14, 214-224 (1978).

TIP-CUTTING OF FRONDS AS A MEANS OF INCREASING PRODUCTION IN <u>LAMINARIA</u> AQUACULTURE

Wu, C. Y., S. Q. Zheng and C. K. Tseng
Institute of Oceanology, Academia Sinica, Qingdao, China

Our early work on the translocation and accumulation of assimilates in the blade of <u>Laminaria japonica</u> started in the late fifties* and the tip-cutting method of the blade based on these studies to enhance the production of the cultivated <u>Laminaria</u> was thereafter introduced to the cultivation industry.

1. <u>Frond cutting experiment</u>. In order to detect if translocation and accumulation of assimilates occur in the blade, and if so, their direction, the blades were cut transversely and longitudinally in different positions and small holes were punched at the corresponding place of the blades. In Fig. 1, the corresponding region of the blade No. 2 is 6.1 cm longer than that of No. 1; the blade No. 3 on which holes are punched at 15 cm and 30 cm from the base increased 8.6 and 9.3 cm more than that of the corresponding regions of No. 2. All these indicate the close dependence of the growth at the intercalary growing region upon the length of tip cut, i.e. the more the transverse cut, the less the growth. Comparison of Nos. 3, 4 and 5, shows that there is no wound effect on the growth in length. A possible point of view for explaining the results can be suggested as follows: A certain amount of assimilates which is partly formed in and translocated downward from middle-upper part of the blade is needed for the intercalary

* Part of the data has been quoted in the Manual of Haidai (Laminaria japonica Aresch.) Cultivation. Tseng, C. K. and C. Y. Wu (ed.), 1962.

Xth International Seaweed Symposium
Printed in Germany

growing region in the rapid growing stage. The decreased rate of growth is accompanied by the reduced translocation of assimilates owing to the greater cutting area. In Fig. 2, No. 1 is 2.9 g less than the corresponding part of No. 2. Likewise, the weight of the base part of No. 2 represents 76% of the corresponding part of No. 3. This means that the greater the cutting area, the less the accumulation of assimilates. It is suggested that the base part of the blade is supplied with certain amount of assimilates from the upper part and there must be a translocation from tip to the base of the blade.

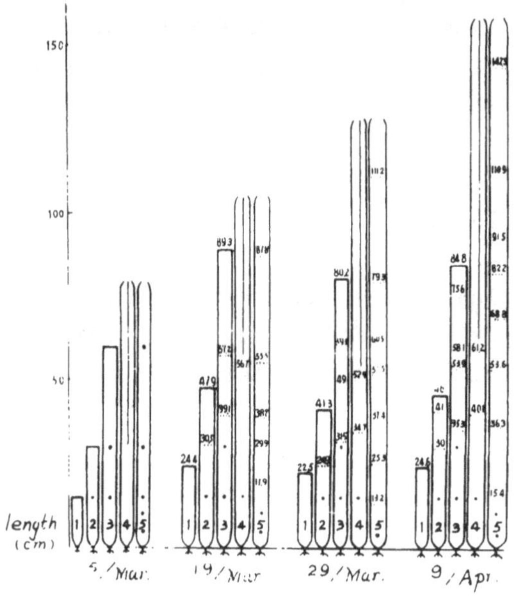

Fig. 1. Growth of the frond after cutting

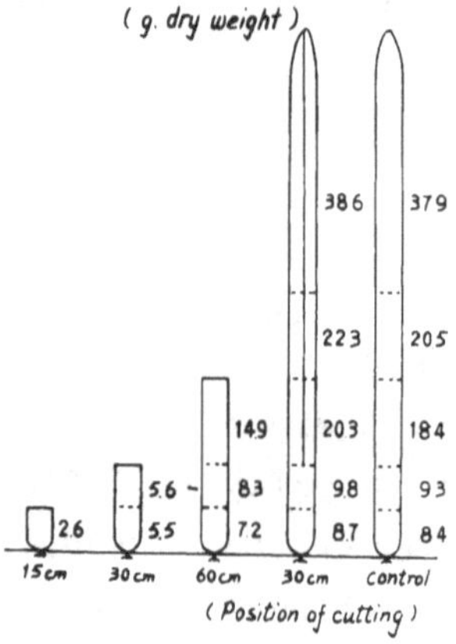

Fig. 2. Increment of the frond after cutting (g dry wt.)

2. Experiment on accumulation of assimilates. In early May while the water temperature reached to 13°C, the rate of elongation of the blade diminished rapidly, whereas the average dry weight of the blade as a whole enhanced without interruption, and water content reduced continuously. These indicate that accumulation of assimilates in the blade oc-

curred.

Experimental data show that there is no significant difference in the water content in different parts of the blade before early May. As soon as the temperature raises to 13°C, the relative water content reduces rapidly, especially in the two base parts of the blade. On July 2, the relative dry weight of these two parts reaches to 21.1-21.6%, whereas the one in the tip only 13.7%. This result elucidates that a great deal of assimilates is accumulated in the base parts of the blade which act as a sink.

3. Rate of photosynthesis at different parts of the blade.
Four 25 cm^2 pieces of tissue were punched out from different parts of the blade and cultivated in the sea for 2-3 days. The rate of photosynthesis was measured by dark-bottle method and the following conclusions may be derived:

1) In table 1, the rate of photosynthesis is low near the intercalary growing region which undergoes cell division and enlargement vigorously and much higher in the middle and upper parts of the blade.

2) Table 2, shows that the average daily increment in dry weight of the whole blade is 1.07 g, whereas the net photosynthesis calculated as $C_6H_{12}O_6$ is merely 0.8 g. That is to say the actual increment is 34% more than that of the net photosynthesis. According to the chemical analysis, the blade contains more than 30% of ash by dry weight. By adding the ash to the products of net photosynthesis, the total amount would be approximately equivalent to the actual daily increment.

3) The theoretical daily net photosynthesis calculaded as glucose is 0.1 g in the intercalary growing region, whereas

that in the middle-upper parts is 0.7 g. On the contrary, the average actual daily increment in dry weight is much higher in the intercalary growing region than that of the middle-upper parts. By adding the 34% of ash to 0.1 g of the daily net photosynthesis in the intercalary growing

Position	Oxygen evolved (ml $O_2 \cdot g(dw)^{-1} \cdot h^{-1}$)
	2.00
	1.25
	1.14
	0.25

Table 1. Rate of photosynthesis of the Laminaria frond

Frond parts	% of the total frond length	Average dry weight	Average daily increment in dry weight (g.)	Excess of average photosynthesis over respiration calculated as $C_6H_{12}O_6$ (g/24hr.)	Excess of average photosynthesis over respiration calculated as $C_6H_{12}O_6$ plus 34% ash	Rate of export (g/day)
Total frond	100	96.8	1.07	0.80	1.07	0
	23.4	7.6	0.03	0.15	0.20	0.17
	16.7	11.3	0.07	0.14	0.19	0.12
	31.8	34.6	0.39	0.41	0.55	0.16
	28.1	43.3	0.58	0.10	0.13	-0.45

Table 2. Export of assimilates in different parts of the frond

region, it makes 0.13 g which is less than 0.58 g of the actual daily increment. Where does the excessive assimilates come from? In regard to the middle-upper parts, it is found that 0.70 g of theoretical daily net photosynthesis is 0.21 g more than 0.49 g of the actual daily increment in dry weight. By adding 34% of the ash to the 0.70 g, the excess would be 0.45 g. The point in question is where does the excessive assimilates go? We presume that the excessive assimilates in the intercalary growing region ought to be translocated downward from the middle-upper parts, and this is in agreemtnt with the inference that there is more production and less increase in the middle-upper part.

4. Main chemical components in different parts of the blade. In order to prove that the accumulated assimilates at some part of the blade exceeding the amount produced is due to the translocation from other parts rather than the absorption of inorganic salts by the tissue from sea water. The analysis of ash, crude fiber and crude protein was made. Analytical date show that in the course of experiment there is no significant change in the relative content of crude fiber. The relative content of crude protein is a little higher in the rapid growth period, but slightly lowered since early May. Variation in the relative content of ash is rather great. Before early May there is no significant change; however, a remakable decrease is found when the rapid growth in length stopped, especially in the several base parts. These results elucidate that the increment in weight of the intercalary growing region is not caused by the absorption of inorganic salts from sea water. It is concluded that the increment in weight of the intercalary growing region which acts as a sink is proved to be partly translocated from the surplus-producing middle-upper part of the blade.

5. P^{32} trace experiment. Different parts of the blades were immersed in P^{32}- phosphate sea water. After labeling, tissues from different parts of the blade were assayed for radioactivity. Data showed that there was only one distinct downward

translocation of the assimilates from the upper part to the base part of the blade. The rate of translocation averaged 12 cm/hr in situ was observed. After quick freezing, it was found that the radioactivity was confined to medulla where sieve elements occurred. It is therefore inferred that sieve elements are connective elements which translocate the assimilates in Laminaria japonica.

6. Tip-cutting experiment. Frond cutting and accumulation experiments showed that growth in length and in weight were not obviously diminished if 1/3 of the frond were cut from the tip. In late March to April in North China the Laminaria under cultivation reach three meters in length and the crowdiness led to serious rot of the tip about one meter of the old tissue weighing more than 10 g by dry weight per plant was eventually cast off and lost to the sea. In case of tip-cutting, this part may serve as one reason to enhance the production. Experiments had been conducted by cutting 1/3 of the fronds in early April in Qingdao. It was found that the resistance of the blades to the tidal current was decreased, resulting in a change of the floating angle of the blade from 8° to 27° with a current speed at 5 cm/sec. This resulted in a better lighting condition of the blades and naturally in a higher photosynthesis. Production experiments of tip-cutting of the blades showed an increased production of 12-15%. This method has been widely accepted by the Laminaria aquaculturists as a normal practice in Laminaria production in North China.

Reference

Schmitz, K. and C. S. Lobban : A survey of translocation in Laminariales (Phaeophyceae). Mar. Biol. 36, 207-216 (1976).

PRELIMINARY STUDIES ON MASS CULTURE OF GRACILARIA SP USING DIFFERENT NUTRIENT MEDIA.

Yocie Yoneshigue-Braga and Maria Helena C. Baeta Neves.
Instituto de Pesquisas da Marinha
Projeto Cabo Frio
28.910 - Arraial do Cabo - Rio de Janeiro - Brasil.

Abstract - *Gracilaria* sp. was cultured in five different concentrations of nitrogen and phosphorus in addition to unenriched seawater. The alga grew well when the seawater was fortified with ammonium. These experiments were done at Cabo Frio Island, Brasil, using concrete tanks outdoors. The results indicate that Cabo Frio, Island is potentially suitable for the cultivation of seaweeds.

Introduction

Studies on *Gracilaria* are of special interest because the agar produced by these algae. Commercially, agar is in high demand and species of *Gracilaria* are the major industrial seaweed resource in Brasil.

The unnamed species of *Gracilaria* (= *Gracilaria* sp)used in our study by widely distributed along the coast of Brasil. Recently this alga was found in hypersaline waters (45‰) in Araruama Lagoon (1), State of Rio de Janeiro.

Our objective was to study the growth of *Gracilaria* sp in outdoor tanks under different concentrations of nitrogen and phosphorus. This work is preliminary to evaluating the potential of the alga and available facilities for mariculture of seaweeds, especially agarophytes.

Material and Methods

All experiments were done at Cabo Frio Island, about 5 km distant from the laboratory at Arraial do Cabo, Rio

Xth International Seaweed Symposium
Printed in Germany

de Janeiro.

The tanks, located outdoors, were of concrete with a capacity of 1000 l (1,38 m long x 1,18 m wide x 0,69 m height). Vigorous aeration was provided by compressed air pumped through perforated "tygon" tubing laid diagonally across the bottom of the tank. Seawater was pumped on the seaward side of the island from a depth of 50 m; if this system failed, surface water was automatically pumped from the leeward side of the island. The water was not filtered or otherwise treated. Water was pumped through the tanks at a rate of 8.5 tank volumes (8640 l) per day. Temperature of the water in the tanks was monitored daily between 0700 and 0900 h.

In addition to unfortified seawater five nutrient conditions were established as indicated and added directly into the tanks (batch culture). The concentrations are per litre of seawater, and the nutrients were renewed weekly.

1- (VST) - Von Stosch medium modified (2), lacking vitamins.
2- ("A) - $(NH_4)_2 SO_4$, 14 mg; $NaH_2PO_4 \cdot 12 H_2O$, 0,5 mg
3- ("B) - $NaNO_3$, 14 mg; $NaH_2PO_4 \cdot 12 H_2O$, 0,5 mg
4- (SCH) - "Salitre do Chile" commercial preparation for higher plants, 0,1 g giving 0,0 067 mg $N\text{-}NH_4^+$
5- (SOV) - "Super Ouro Verde", commercial preparation as above, (Reg. nº SPO5372), 0,1 g, giving 56 mg N as NH_4^+

In our initial experiment, done during October 1979, we evaluated Hypnea cornuta, Gracilaria sp and Acanthophora spicifera from Araruama Lagoon and Pterocladia capillacea from the upwelling area. In subsequent experiments we concentrated on Gracilaria with an initial inoculum of 75 g wet weight. A 30-day period was sufficient to evaluate the results of an experiment.

The plants were removed weekly, drained for 10 min. in a nylon-mesh bag, weighed, and returned to the tank. Experiments with Gracilaria were conducted from November to April 1980.

Physical and Chemical Conditions

The mean water temperature in the tanks was 27°C from November through February, and 25.5°C during March and April. Irradiance averaged 456 ly. d^{-1} during October-February and 270 ly. d^{-1} throughout the rest of the year (3). During the course of our experiments, the range in phosphate values in the incoming seawater was 0.07 - 0.9 µg.at. l^{-1} and the corresponding nitrate values were 0.04 - 10.4 µg.at.l^{-1}.

Results

Species Selection

Results of Screening experiment using the four species indicated are shown in Table I in which the initial weight of the inoculum is compared to the final weight after 30 day's of incubation. Clearly the best results were obtained with *Gracilaria* sp and *P. capillacea*. On the basis of these data we concentrated on the former species.

Table I: Variation in fresh weight after 30 day's incubation (running seawater)

Species	Inoculum(g)	Final weight(g)
Hypnea cornuta	200	Deteriorated
Gracilaria sp	100	140
Acanthophora spicifera	200	Deteriorated
Pterocladia capillacea	200	250

Effect of Nutrients on Growth of *Gracilaria sp.*

The results of these experiments are summarised and displayed in Table II. Maximum growth was obtained with "A" (5,84% $d.^{-1}$) in November, "VST" (6.14% $d.^{-1}$) in December, and "SOV" (5.37% $d.^{-1}$) in March. Our upwelling occurs during November, December, and January and sporadically during March. In unfortified seawater growth was poor, and the plants lost red pigmentation, becoming yellow-green. This suggests possible nitrogen deficiency. In media "B" and "SCH" the plants grew only during the first two weeks of November, there after loosing weight and deteriorating. Similar results were obtained in "VST" during March and April.

As a result of a number of factors, undoubtedly including water quality, temperature and irradiation, *Gracilaria* became covered with epiphytes. The epiphytes were removed by hand each week before weighing. The most commom epiphytes were species of *Enteromorpha*.

These epiphytes presented such a serious problem that it was necessary to discard the *Gracilaria* each month and reinoculated the tanks with fresh collected material.

Discussion

Our best growth was obtained with seawater enriched during the upwelling season. Although the Von Stosch medium was developed for culture of macrophytes, we were unable to sustain satisfactory growth using this formulation. Possibly important substances such as trace metals and organic compounds are in excess in our waters (5).

Our findings indicate that the ammonium in media "A" (5.84% $d.^{-1}$) and "SOV" (5.37% $d.^{-1}$) stimulated growth of *Gracilaria* sp. This is not surprising as the habitat from which the alga was taken has a high concentration of ammonium, 9 µg $at.l^{-1}$ (Jacob, in verb.). In a study using both *Gracilaria* and *Neoagardhiella*, De Boer and Ryther (6) reported that the growth rates of these species were more rapid with ammonium than with nitrate. Moreover, ammonium was assimilated preferentialy. In our experiments, Gracilaria sp incubated in unenriched seawater grew very poorly indeed as did Acetabularia calyculus (7), indicating symptons of nitrogen deficiency (i.e loss of red pigmentation). This was in spite of a flow rate of 8.5 tank - volumes per day of tropical surface waters. Much better growth was obtained with Gr. tikvahiae in unenriched Canadian waters (8).

We have experienced many problems with our cultivation experiments. Nevertheless, these preliminary data suggest that the mariculture of seaweeds in land-based tanks at Cabo Frio has considerable potential.

TABLE II – Growth of *Gracilaria* sp. using different nitrogen concentration, expressed

$$\text{in \% Day}^{-1} = \left[\frac{wt}{wo}\right]^{\frac{1}{t}} - 1 \times 100, \quad (4)$$

Tx = % Increase in fresh weight per day

Time / Media	NOVEMBER						DECEMBER					
	07	14	21	28	06	Tx	07	13	20	27	03	Tx
SCH	75	90	106	100	60	1.36	–	–	–	–	–	–
SOV	75	110	155	205	225	3.60	75	130	180	205	256	4.65
A	75	215	235	315	436	5.84	75	125	210	275	315	3.48
VSt	75	140	170	205	225	3.60	75	145	336	356	375	6.14
B	75	100	123	90	50	0.87	–	–	–	–	–	–
Control	75	80	82	90	100	0.93	75	88	89	98	105	1.25
Time / Media	JANUARY						FEBRUARY					
	04	11	18	25	01	Tx	04	11	18	25	03	Tx
SOV	75	121	164	208	246	4.33	75	98	95	108	114	1.50
A	75	140	190	262	317	5.28	75	83	100	105	156	2.65
VSt	75	121	169	232	285	4.88	75	60	45	–	–	–
Control	75	77	89	92	105	1.20	75	84	93	102	11	1.47
Time / Media	MARCH						APRIL					
	04	11	18	25	01	Tx	02	09	16	23	30	Tx
SOV	75	200	245	300	325	5.37	75	90	110	118	135	2.12
A	75	125	160	220	285	4.88	75	145	198	223	265	4.61
VSt	75	115	130	149	165	2.85	75	80	70	50	–	–
Control	75	99	104	121	135	2.12	75	83	85	92	99	0.99

Acknowledgements

The authors thank to Drs: Jack Mc Lachlan and Eurico C. Oliveira Filho for their helpful comments and suggestions in review of the manuscript. Special thanks to Mr. Frederico Koenig Neto for field assistance.

This research was financied by FINEP (Finanças, Estudos e Projetos). For the first author this work was partially supported by Grant 1111.2326/76 from the National Research Council of Brasil.

References

01 - YONESHIGUE-BRAGA, Y. and COUTINHO, R. - 1980 Phycologia Latino Americana (In press).

02 - EDWARD, P. Contributions in Marine Science. Supplement to Vol. 15:1-127. (1970).

03 - YONESHIGUE-BRAGA, Y. and DRUEHL, L. Aquaculture, 14:1-12. (1978).

04 - BIRD, N. L., CHEN, C. M. and MC LACHLAN, J. Bot. Mar. Vol. XXII:521-527. (1979).

05 - YONESHIGUE-BRAGA, Y.; MAESTRINI, SERGE, Y.; GONZALEZ RODRIGUEZ, E. C. r. hebd. Séanc. Acad. Sci. Paris 288: 135-138. (1979).

06 - DE BOER, J. A. and RYTHER, J. H. Plant Biomass of the Pacific Northwest Coast. Ed. by Robert W. Krauss. pp.: 231-249. (1977).

07 - YONESHIGUE-BRAGA, Y. Marine Ecology - In Press. (1980).

08 - EDELSTEIN, T., BIRD, C. J.; and MC LACHLAN, J. Can. J. Bot. Vol. 54 (15):2275-2290. (1976).

Key words:

Gracilaria sp, culture, nutrients.

5. Application, Biomass, Technology, Utilization

THE COMPOSITION OF MAËRL BEDS OF ECONOMIC INTEREST IN NORTHERN BRITTANY, CORNWALL AND IRELAND

G. Blunden*, W.F. Farnham**, N. Jephson**, C.J. Barwell*, R.H. Fenn+ and B.A. Plunkett++

Departments of Pharmacy*, Biological Sciences**, Physics+ and Chemistry++, Portsmouth Polytechnic, Portsmouth, Hants., PO1 2DZ, U.K.

Introduction

Certain calcareous red algae of the family Corallinaceae, known collectively as maërl, are harvested and sold primarily for use on acidic soils to increase their pH values (1). Maërl is collected commercially from the Gulf of St. Malo and around the Glenan Islands in Brittany, France, and from Falmouth Harbour, Cornwall, in the U.K., although commercial harvesting has been in progress at Falmouth for only the past few years. The composition of the maërl beds of the Glenan Islands has been described (2) and in this paper other commercially-exploited beds in the Gulf of St. Malo and in Falmouth Harbour are discussed, as well as the composition of some of the deposits found off the west coast of Ireland, which have been considered for economic utilisation.

Results

Gulf of St. Malo. In Paimpol Bay 13 stations were examined, most of them near to where boats were dredging. At a number of the stations the maërl formed a ridge and furrow system, the wave heights being about 1.5 m and the distance between adjacent ridges about 5-8 m. The maërl particles were, on average, larger in the furrows than on the ridges. Few living plants were observed. At other stations the sea bed was

Xth International Seaweed Symposium
Printed in Germany

comparatively flat and there was a higher proportion of living plants. At one station, away from the dredging areas, the top 1-2 cm was composed of living maërl. The predominant calcareous species was Phymatolithon calcareum (Pall.) Adey et McKibbin. At all the locations studied the maërl deposits were characterised by very sparse epiphyte cover. Bivalve shells were abundant in certain areas.

Falmouth Harbour. Large quantities of living P.calcareum were found on the St. Mawes Bank. Smaller, more delicate plants, identified as Lithothamnium corallioides Crouan frat. (3) predominated at depths down to 6 m, resting on a muddy bottom. Larger rhodoliths of more typical P.calcareum were observed at depths of 6-10 m. This material supported more epiphytes, in particular flocculent, ephemeral algae, such as Antithamnion spirographidis Schiffner. On the west side of the outer harbour, fragmented dead maërl was found in beds of Zostera marina L. Elsewhere in the outer harbour and in Falmouth Bay, dead maërl fragments have aggregated and are present in the characteristic ridge and furrow system described for Paimpol Bay and the Glenan Islands. The dead maërl has few epiphytes.

The algal flora associated with the Falmouth maërl beds is similar to that of the Glenan Islands (2) and consists of seasonably variable species, with maximum diversity in the summer. Common species include A.spirographidis, Chylocladia verticillata (Lightf.) Bliding and Polyneura hilliae (Grev.) Kylin. Noteworthy species found were Dudresnaya verticillata (With.) Le Jol., Carpomitra costata (Stackh.) Batt., Vaucheria piloboloides Thur. and Seirospora seirosperma (Harv.) Dixon. Mature tetrasporangial plants of P.calcareum were found in December; these were reported to be rare in Spanish beds (4).

Ireland. Three sites were examined: Finavarra Bay, Co. Clare, Mweenish , near Carna, Co. Galway, and Kilkieran Bay, Co. Galway. In Finavarra Bay, 7 stations were studied. Maërl was found at depths ranging from 10-18 m. No ridge and furrow system was observed at any of the stations, but at many of them the sea bed had a continuous cover of maërl overlying mud; many of the plants were living. At some stations the maërl cover was patchy. The depth of maërl was usually thin ranging from 10-20 cm, although it was considerably deeper in other places. Variable amounts of shell debris and rocks were observed.

The algal epiphytes associated with the maërl were generally similar to those of Falmouth Harbour and the Glenan Islands. Characteristic species were A.spirographidis, Brongniartella byssoides (Good. et Woodw.) Schmitz, Dasya corymbifera J.Ag., Halarachnion ligulatum (Woodw.) Kütz. and Myriogramme bonnemaisonii (C.Ag.) Kylin.

The site at Mweenish was a deep, steep sided hole approximately 100 m across and 14 m deep located at the entrance of a nearly land-locked bay. The sea bed at the base of the slope is made up of fragments of P.calcareum. In addition there are many large rhodoliths, up to 20 cm in diameter, which appear to have developed around stones. These living plants are considered to be P.polymorphum (L.) Foslie. Interesting algal epiphytes of the maërl include Monosporus pedicellatus (Sm.) Solier and Pterosiphonia parasitica (Huds.) Falkenb.

In Kilkieran Bay 8 stations were examined and maërl was found at depths ranging from 4-18 m. At some stations maërl fragments, some alive, were found in shallow water in beds of Zostera marina. At other stations the sea bed had up to about 60% of its surface covered with a thin layer of live maërl, usually in a ridge and furrow arrangement. The thickness of the maërl was variable and it usually overlaid mud, but sometimes either rock or sand. The ridge and furrow system was well developed in deeper waters, the ridges being about 40 cm high and the distance between adjacent ridges about 2 m.

There were more living plants in the furrows than on the ridges and a greater abundance of living plants in the shallower waters. The maërl deposits were accompanied by considerable quantities of shell debris. The major maërl species was P.calcareum, with some material resembling Falmouth L. corallioides. Both species have been reported from Mannin Bay, Co. Galway (5). A further species was found which has been tentatively identified as Lithothamnium dentatum (Kütz.) Foslie. The algal epiphytes of the maërl were similar to those found at Falmouth and the Glenan Islands.

Maërl beds generally support rich algal associations. Our survey of the beds in Falmouth and of the west coast of Ireland has resulted in the discovery of previously unrecorded species for the British Isles. Solieria chordalis (C.Ag.) J.Ag. was found for the first time in the British Isles at Falmouth (3). Only tetrasporangial plants have been found (April, June and December) and it is likely that S.chordalis reproduces mainly by vegetative fragmentation and subsequent re-attachment.

An encrusting red alga, Cruoria cruoriaeformis (Crouan frat.) Denizot, was found growing on maërl at Falmouth. This is the first record of this species for the British Isles, although the likelihood of its occurrence on British maërl has been pointed out (6). C.cruoriaeformis was located later in Finavarra Bay. Tetrasporangial plants were found in December and April at Falmouth and in June in Finavarra Bay. C. cruoriaeformis specimens often contained a species of the endophytic green alga Codiolum.

A foliose red alga unlike any known foliose British species was found growing on maërl in Falmouth Harbour (3). Characteristic features of the alga are that the plants are dioecious, with well-developed stellate cells containing highly refringent cytoplasm, and with simple carpogonial branches giving rise to inwardly developing carposporophytes. In July, 1976, no further specimens could be found.

A small red alga, looking like a diminutive Callophyllis, but anatomically different from it, was found growing on maërl in Falmouth Harbour and in Finavarra Bay; it has been found since in the River Yealm in south Devon. Only tetrasporangial plants have been found and the species has not been identified.

Another foliose red alga was found at the Mweenish station, near Carna, in May 1977. The structure of the alga is similar to that of Halymenia; the occurrence of cruciately divided tetrasporangia (c. 17 x 14 µm) supports this idea, which needs to be confirmed by examination of cystocarpic material. If it is a Halymenia species then it is not H.latifolia Crouan frat., which had been erroneously recorded for the British Isles (7). In the absence of cystocarpic material the species may be in another genus, such as Sebdenia.

From each station examined a representative sample of maërl, free from other materials, was assayed for its content of calcium and magnesium (2). The calcium content of the samples from Paimpol Bay, north Brittany, varied from 29.5-31.7% of the dry weight (mean value 30.8%), from Falmouth Harbour from 29.2-33.1%, (mean value 32.1%), and from Finavarra Bay from 31.4-32.9% (mean value 32.0%). The magnesium content of the samples from Paimpol Bay varied from 2.7-3.5% of the dry weight (mean value 3.1%), from Falmouth Harbour from 2.2-3.4% (mean value 3.1%), and from Finavarra Bay from 3.2-3.9% (mean value 3.5%). These values are in agreement with those recorded for the material from the Glenan Islands (2). At each station examined a 20 cm^2 quadrat/core sampler was used and the top 5 cm of material collected was dried, powdered and a sample taken to determine its acid-insoluble ash content in order to estimate the silica present. For the samples from Paimpol Bay values varied from 7.1-31.7% of the dry weight (mean value 15.7%), from Falmouth Harbour from 12.8-28.2%, and from Finavarra Bay from 2.6-43.0% (mean value 11.5%). Material collected by using grabs at such sites would have these quantities of silica present, which would significantly reduce the calcium and magnesium contents of the dried commercial material.

As with the maërl from the Glenan Islands (2), that from the Gulf of St. Malo, Falmouth Harbour and Ireland had as its major component calcite with aragonite present as a minor constituent. The percentage of aragonite to calcite was calculated as being 15% for the material from northern Brittany and approximately 10% for the samples from Falmouth Harbour and Ireland. These estimations are, however, only approximate and no significance should be put on the differences obtained. The X-ray photographic powder diffraction data obtained in this present study and that which was published earlier for maërl collected from the Glenan Islands (2), give strong grounds for the identity of aragonite, although Dixon (8) has stated that there is no firm evidence for its presence in red algae.

References

1. Blunden, G., Binns, W.W., Perks, W.: Econ.Bot. 29, 140-145 (1975).
2. Blunden, G., Farnham, W.F., Jephson, N., Fenn, R.H., Plunkett, B.A.: Botanica Mar. 20, 121-125 (1977).
3. Farnham, W.F., Jephson, N.A.: Br.phycol.J. 12, 119 (1977).
4. Adey, W.H., McKibbin, D.L.: Botanica Mar. 13, 100-106 (1970).
5. Bosence, D.: Palaeontology 19, 365-395 (1976).
6. Dixon, P.S., Irvine, L.M.: Seaweeds of the British Isles, Vol.1. Rhodophyta. Part 1. Introduction, Nemaliales, Gigartinales. British Museum (Natural History), London 1977.
7. Dixon, P.S., Irvine, L.M.: Bot.Notiser. 123, 474-487 (1970).
8. Dixon, P.S. Biology of the Rhodophyta. Oliver and Boyd, Edinburgh, 1973.

DO WE REALLY NEED TO GROW MACROCYSTIS IN EUROPE?

Gerald T. Boalch

Marine Biological Association of the U.K., The Laboratory, Citadel Hill, Plymouth, England

Macrocystis is a genus of large brown algae which is mainly confined to the southern hemisphere and particularly to the southern Pacific, although it does extend into the southern Atlantic as far as the Cape of Good Hope and spreads up the western coast of America as far as the U.S.A./Canadian border. The two main species are Macrocystis integrifolia and M. pyrifera. Individuals of these species can be very large and may reach 150 m in length. They can grow in waters up to 25 m deep or as shallow as the low tide level. In some areas they form extensive forests offshore and in other areas they form dense cover from the low tide level down to deep water. Macrocystis is of economic value as a source of alginates. On the other hand it can present considerable problems to the navigation of small boats by impeding progress, tangling around propellers and blocking cooling water intakes. A considerable amount of information on Macrocystis is contained in a book edited by W.J. North (1).

For a number of years Macrocystis has been harvested on the Pacific coast of U.S.A. for the extraction of alginic acid. In California the commercial harvesting is carried out using specially constructed vessels which cut and collect the weed from the offshore beds.

Alginic acid, probably as various alginates, is an important constituent of the cell walls of brown algae. The percentage alginate composition of brown algae differs from species to species and also varies throughout the year for any one species. Further, the actual composition of the alginic acid varies from one species to another. Thus the best alginate for a specific purpose may be most advantageously extracted from one species at perhaps only one period in the year. Macrocystis has proved to produce alginate that has wide commercial applications, the content is high and the

Xth International Seaweed Symposium
Printed in Germany

weed can be harvested over a long period of the year.

In the U.K. most alginate is produced from the knotted wrack, Ascophyllum, which is harvested mainly in the summer and almost entirely from north-west Scotland. France and Spain harvest the oar weeds, Laminaria, for their alginates and the harvesting is mainly a 'cottage' industry in that it is carried out as a part-time job in the spring and summer by fishermen and coastal farmers.

In 1950 the Scottish Seaweed Research Association wanted to introduce Macrocystis integrifolia, to Scottish coasts to increase the production of alginates. On this occasion the British Advisory Committee on Scientific Policy set up a special committee which considered evidence from the seaweed industry, the Ministry of Agriculture, Fisheries and Food, the Admiralty and a number of marine biologists. The committee recommended that, "It might be practicable, but it would not be desirable to introduce foreign seaweed of industrial value into British inshore waters", and the proposed introduction was forbidden by Government Order. Many of the arguments put forward against the introduction in 1950 are still valid today.

Early in 1973 it came to the notice of British phycologists that French scientists from the Institut Scientifique et Technique des Pêche Maritimes (ISTPM) were carrying out experiments which involved growing Macrocystis pyrifera in the sea at Roscoff on the north coast of Brittany. Further enquiries revealed that ISTPM had plans for introducing Macrocystis pyrifera to the coast of Brittany as a source of alginates (2, 3). This proposal caused alarm to a number of British phycologists as they considered that a wild population of this giant alga on the coast of Brittany would mean that it would very soon spread to the coast of the U.K. The physical, hydrographical and chemical conditions appear to be suitable for the growth of this alga from the north coast of Spain to about half way up the Norwegian coast. This alga is far larger than any native species in the North Atlantic and it could probably grow on submerged reefs of rock on which at present there is no growth of large algae.

Due to its great length and tendency to form dense beds on offshore reefs it would constitute a new hazard to inshore navigation. This was one of the main reasons for opposition by the Admiralty in 1950. In their present habitat the beds of *Macrocystis* are charted and thus can be avoided. During a natural spread in European waters it would be many years before the beds were stabilized and could be charted. Offshore forests of this alga extending from the sea bed to the surface and spreading along the surface would have considerable effects on local water movements and would thus change the dispersal patterns of sewage and other waste outflows. Rocky areas at the moment colonised by the brown seaweeds of the genus *Laminaria* are good fishing grounds, especially for shellfish. If these areas were taken over by *Macrocystis* the massive growths of this alga would make the present fishing methods difficult if not impossible. The *Macrocystis* beds off California do support fish which are mainly used for sport fishing.

The commercial harvesting of large beds of *Macrocystis* on the west coast of U.S.A. is much easier than the present system for harvesting *Laminaria* in Europe. However ISTPM has stated that at the moment only about half the *Laminaria* which could be used is harvested.

The experiments of ISTPM were under way before any news reached British phycologists. The first news came via an American colleague who wrote to the Conservation Committee of the British Phycological Society (BPS) in February 1973 pointing out that scientists from ISTPM had visited California and had discussed their experiments which involved growing *Macrocystis* in the sea in European waters. British phycologists immediately started investigations and discovered that *Macrocystis pyrifera* plants had been grown in the sea near the Roscoff Marine Biological Laboratory in northern Brittany (4). In December 1972 the Société Phycologique de France discussed the *Macrocystis* experiments and expressed anxiety and questioned their wisdom. Further, they had asked that a group of their members be given the opportunity to discuss this project with the scientists from ISTPM.

In March 1973 Dr Boalch informed Dr H.A. Cole, MAFF, Lowestoft and at that time Chairman of the Working Group on the Introduction of Non-Indigenous Marine Species of the International Conseil for the Exploration of the Sea

(ICES), of the potential danger of the introduction of Macrocystis to European waters and asked him to put the proposal before his working group. The Working Group discussed the French proposal in Lisbon in October 1973. The general Council of ICES passed a resolution (1973/2:18), "It was decided that:- a) the Working Group on the Introduction of Non-Indigenous Marine Organisms should meet as soon as practicable to consider further the proposed introduction of Macrocystis to France in the light of the detailed assessment prepared by Peres et al., taking into account the discussion of this question by the 61st Statutory Meeting of ICES. b) in view of the importance of the matter all potentially affected member countries be requested to appoint members to the Working Group."

The next detailed account of the French experiments to introduce Macrocystis to European waters was given at the meeting of the Société Phycologique de France at Roscoff 8-10 November 1973. Several British phycologists attended this meeting and heard a paper read by Dr R. Perez of ISTPM (5). Dr Perez discussed his laboratory experiments with Macrocystis and also described how he had planted young sporophytes in the sea near Roscoff in 1972 and how they had grown to 13 metres during the summer. By the time they had reached this length the sporophylls were developing and the Macrocystis was removed from the sea. This paper was followed by considerable discussion.

At the annual meeting of the British Phycological Society at University College, London, 3-4 January 1974 a report was given on the French proposals. On 4 January 1974 at the Annual General Meeting of the British Phycological Society the following resolution was passed (nem. con.).

Resolution

"The British Phycological Society strongly opposes any proposal to introduce the giant kelp Macrocystis into European waters. The British Phycological Society would appreciate being officially represented on any national and international committee or working group that may be set up to consider this matter in detail". H.T. Powell and G.T. Boalch were appointed representatives of the British Phycological Society in matters concerning Macrocystis. Copies of the resolution were sent to ICES and ISTPM.

On 31 January - 1 February 1974 the ICES Working Group on the Introduction of Non-Indigenous Marine Organisms held a meeting at the Headquarters of the Ministry of Agriculture, Fisheries and Food (MAFF) in London.
Dr Boalch was invited to attend as an observer for the British Phycological Society. At this meeting there was discussion for and against the proposed introduction of Macrocystis to Europe (6). This working group presented a report to the meeting of ICES in the summer of 1974 and the report concluded:

"The majority of the Working Group members were convinced that the spread of Macrocystis to their coasts would be followed by serious disadvantages".

The VIII International Seaweed Symposium was held at Bangor, North Wales in August 1974 and a session was devoted to the Biology of Macrocystis. Once again there was considerable discussion on the possible effects of the introduction of this alga to European waters. This time there was the added advantage that a number of American speakers had worked on Macrocystis on the California coast. There were about 350 seaweed experts from all over the world at this symposium and at the final session two resolutions were passed with one vote against the first resolution and no-one voting against the second (7). The Resolutions were: 1) (To be sent particularly to the General Secretary, ICES, ISTPM and other organisations). "At the closing plenary meeting of the VIIIth International Seaweed Symposium held at the University College of North Wales, Bangor, U.K., 23 August 1974, great concern was expressed at the proposal of the Institut Scientifique et Technique des Pêches Maritimes to carry out further experiments with Macrocystis in French coastal waters".
2) (To be sent particularly to the Secretary of the International Phycological Society and other organisations and individuals). "At the closing plenary meeting of the VIIIth International Seaweed Symposium held at the University College of North Wales, Bangor, U.K., 23 August 1974, great concern was expressed at the proposals of the Institut Scientifique et Technique des Pêches Maritimes to carry out further experiments with Macrocystis in French coastal waters. It was further recommended that the

International Phycological Society be asked to convene rapidly a Working Group of experts in the Biology of *Macrocystis* forest and European *Laminaria*-forest to evaluate all aspects of the French proposal to introduce *Macrocystis* to Europe".

ICES again considered the proposals of ISTPM to introduce *Macrocystis* to European waters at their meeting in Copenhagen in October 1974. Mr H.T. Powell represented the BPS at this meeting. By this time there had been considerable international discussion of the proposals (8, 9) and there was clearly considerable opposition from a number of European countries. As a result ICES expressed disapproval of the proposed introduction and because of this international pressure the scientists at ISTPM agreed not to carry out any further experiments with *Macrocystis* in European waters for the time being.

At the joint meeting of the IX International Seaweed Symposium, the International Phycological Society and the American Phycological Society at Santa Barbara, August 1977, there was a special one day Symposium on *Macrocystis*, an evening with a discussion panel, kelp harvesting demonstrations and a visit to a kelp processing plant but no organised discussion on the proposals to introduce *Macrocystis* to Europe (10).

Early in 1979 the author was informed that ISTPM were planning another experiment with *Macrocystis* in the Baie of Douarnez. It was understood that ISTPM proposed to present details of their planned experiments to ICES in the summer of 1979. At the ICES meeting in Warsaw in September 1979 the matter of the introduction of *Macrocystis* to Europe was raised but there was not a full discussion as the scientists from ISTPM said they did not have full approval for the experiments from their government. However, the delegates did claim that in future experiments they would use effective methods to prevent the spread of *Macrocystis*, but did not give details.

In reply to a question in the European Parliament in June 1979, the French authorities said in Nov. 1979 that their Ministry of the Environment had still not given permission for any experiments with *Macrocystis* in the sea and in any case they "do not intend to authorize any experiments unless they are totally convinced that they will in no way affect live marine resources and/or shipping in coastal waters".

At a meeting of the ICES Introduction Working Group in Nantes in April 1980, scientists from ISTPM said they were still keen to carry out experiments with Macrocystis on the coast of southern Brittany but they still did not have permission from their Ministry of the Environment. No doubt the matter will be raised again at the ICES meeting at Sete in September 1980.

The records of where man has introduced aquatic plants from one area to another all indicate that the immigrant has got out of control and caused problems. In fresh water the Canadian pondweed Elodea canadensis and the water hyacinth Eichhornia crassipes are striking examples. Codium fragile and Sargassum muticum are examples of recent transplantations of marine algae. In the latter case efforts to control the spread of the weed on the south coast of England have proved totally ineffective.

Many phycologists must realize that the introduction of Macrocystis into the North Atlantic will cause problems. At the moment the world demand for alginic acid is declining as the world recession reduces the demand for textiles and paper - these two industries use about 60% of the alginate produced. Further, in the U.K. the harvesting of Ascophyllum is now proving so expensive due to the high cost of fuel for harvesting, drying and transportation that it is cheaper to collect naturally caste and dried Macrocystis from Chile and bring it back to Europe. With a decline in the demand for alginates this is certainly the wrong time to be thinking of introducing Macrocystis to Europe. Instead, efforts should be concentrated on ways of increasing the production of natural populations of Laminaria and Ascophyllum in Europe, and Macrocystis on the Pacific coasts of North and South America. Further research should be aimed at less energy consuming ways of harvesting the native seaweeds rather than introducing species which may disturb the natural populations and have adverse effects on coastal navigation.

References

1. North, W.J.: Beih. Nova Hedwigia. 32 (1971).
2. Perez, R.: Science et Pêche. 216, 1-9 (1972).
3. Anon.: Science et Pêche. 228, 15 (1972).
4. Perez, R., Mesni, J.C. du., Colin, Y., Fur, Le. le., Didou, H.: Rev. Trav. Inst. Pêches marit. 37, 307-361 (1973).
5. Braud, J.-P., Etcheverry, H., Perez, R.: Science et Pêche. 233, 1-15 (1974).
6. Perez, R., Braud, J.-P.: I.C.E.S., C.M., E47 (1974).
7. Anon.: Phycologia. 14, 123 (1975).
8. Franklin, A.: New Scientist. (1974).
9. Chassé, C.: Penn ar Bed. 9, 416-428 (1974).
10. Jensen, A., Stein, J.R. (Editors): Proc. Int. Seaweed Symp. 9, (1978).

DISCUSSION

J.D. Pringle (Canada)

Yesterday we heard an interesting talk by Professor Tseng on the introduction of *Laminaria* to southern China. I wonder if he can comment on the effect it had on the environment?

C.K. Tseng (China)

I do not know much about *Macrocystis*. I do think the make up of the ecosystem is very important and the introduction of a new species may go either way - it may be good or it may be bad. I do not know what the effect of *Macrocystis* in western Europe will be. It will certainly modify the ecosystem, perhaps very much, and this is probably why some scientists are very much against the introduction. Once *Macrocystis* is established in Europe I think it will stay for ever because the conditions are suitable. I do not know if I am for or against this introduction but I believe we should be very cautious before changing the ecosystem.

J. Dugoujon (France)

I represent CECA. You ask if we need to grow *Macrocystis* in Europe and I answer yes if we want an active alginate industry. The cost of producing alginate from brown algae has increased and there are now other compounds competing against alginates. In my opinion the growth of *Macrocystis* is the only way to get enough alginate. The first experiment was only a small one and I feel it is not a good scientific attitude to oppose such experiments.

R. Delepine (France)

If we could be sure that we could master the experiment then we should do it, but can be we sure? My second point is, will *Macrocystis* save the European alginate industry? I don't think we can be sure of this either.

K. Nielsen (Denmark)

The question is do we really need *Macrocystis* at present? There are many other suitable brown seaweeds in Europe and we should develop and exploit these first. At the same time we could try a controlled experiment with *Macrocystis*. However if we really need *Macrocystis* we can collect plenty from where it grows naturally.

S. Bonotto (Belgium)

I would like to ask whether any country in Europe has the right to carry out experiments which cannot be controlled and which will affect other countries?

J.-P. Braud (France)

I am rather astonished that everytime there is discussion about the spread of *Macrocystis* there is reference to *Sargassum*. The life cycles in the two genera are quite different so *Sargassum* is not a good example to use.

G.T. Boalch (United Kingdom)

I agree that the two have different life histories but *Macrocystis* produces many more spores than *Sargassum* does eggs. It is the spread of these microscopic spores by tidal currents that we are worried about. We have heard of efforts to produce a sterile Macrocystis but is it possible to guarantee that it will always remain sterile?

J.-P. Braud (France)

In the experiments carried out by Professor North in California they found a maximum spread of 80 m from the parent plant.

G.T. Boalch (United Kingdom)

I understand that this experiment was with a plant that had been lifted from its normal habitat and placed in a sandy bay, hardly a plant under normal conditions. This experiment does not give any indication of how far the spores may be carried by strong tidal currents in Brittany.

R.J. King (Australia)

The original question was 'Do we really need Macrocystis in Europe'. Our French colleagues suggest that we do. Can someone from the industry tell us whether we really need something like Macrocystis to keep the alginate industry going?

R.J. King (Australia)

If no one can show there is a real need for Macrocystis then there is no justification for further experiments.

J. Sanderson (United Kingdom)

I would like to correct one error in that the amount of Macrocystis imported into Britain by the alginate industry has been very small as it was only for experimental purposes.

E.K. Ganesan (Venezuela)

I am against the transplanting from one ecosystem to another. I suggest this problem should be considered by the various national phycological societies and a referendum taken from phycologists all over the world as the principle involved has world-wide implications.

J.M. Jones [Kain] (United Kingdom)

I would like to make two points. Firstly going back to the spread of Macrocystis this has exactly the same life history as Laminaria and we have heard at this Symposium that Laminaria is now growing on the North Sea oil platforms so I do not think there is much doubt about the spread. Secondly, if we need more plants - more biomass - for alginates or other reasons let us first look at the species we have already. The Chinese have shown us how they produce vast amounts under controlled conditions so let us consider something like this first.

J.P. Cullinane (Ireland)

The Irish Government has shown interest in the French plans but we had heard that the experiments had been suspended. Can anyone tell us the present state of the experiments?

J.-P. Braud (France).

There is no change in the situation. The experiments are still suspended.

G.T. Boalch (United Kingdom)

Please note they are suspended, not completely abandoned. Some of us feel certain that if the Minister of the Environment in France gave clearance the planned experiments would go ahead.

AGAR FROM *GRACILARIA VERRUCOSA* (HUDSON) PAPENFUSS AND *GRACILARIA SJOESTEDTII* KYLIN FROM NORTHEAST BRASIL

Muttutamby Durairatnam and Neuson de Queirõz Santos
Departamento de Oceanografia e Limnologia, Universidade Federal do Rio Grande do Norte, Praia de Mãe Luiza, Natal-RN 59000, Brazil

Introduction

Since World War II there has been an increase in demand for seaweeds for the extraction of agar for various uses in brewing, manufacture of jams, meat preservation, confectionery food, pharmaceutical, bacterial media and other uses. Many investigations on the extraction of agar have beem carried out throughout the world on different species of *Gracilaria* especially in Japan, Korea, Philippines, Thailand, Singapore, Malaysia, Sri Lanka, India, Australia, Denmark, Chile, Peru, Argentina, Brazil, United States of America and South Africa. Of the different species of *Gracilaria* used for the extraction of agar the most important is *Gracilaria verrucosa* (Hudson) Papenfuss which was formerly known as *Gracilaria confervoides* (Linnaeus) Greville. In India Mohanty (1956) has shown that a good quality agar yields from 30-45% was obtained from *Gracilaria verrucosa*. In Sri Lanka Durairatnam and Medcoff (1954) and Durairatnam (1979) has showed that the yield of pure agar varied from 28 to 32% in *Gracilaria verrucosa*. In Australia Wood (1946) has shown that agar of excellent quality and superior in some respects to Japanese agar can be produced from *Gracilaria confervoides* in yields of 27.25 to 32.75%. Dias-Piferrer and Perez (1964) has shown that the

Xth International Seaweed Symposium
Printed in Germany

yield of agar varied from 23 to 37% for plants from Puerto Rico. In Chile Tagawa, Kojima and Kono (1963) treated Gracilaria verrucosa with different concentrations of sodium hydroxide at various temperatures. Treatment with 2% sodium hydroxide at 90°C for 3 hours and autoclaving at 125 to 130°C for 3 hours gave a yield of 17 to 18% and 27 to 28% using crude and refined material respectively. In Japan Ohmi and Kuroda (1955) reported that the yield of agar from Gracilaria verrucosa from various localities in Japan varied from 38.1 to 42% and the yield of agar from Gracilaria sjoestedtii from Carpentaria, California was 49.91%. He pointed out that Gracilaria verrucosa was the only source of agar in 9 out of 16 factories operating in Japan in 1957. Kojima and Funaki (1951) showed that Gracilaria confervoides treated with alkaline solution containing calcium was far superior to the others in quality and far exceeds the standard U.S. Pharmacopoeia with a yield of crude agar 54.6% and pure agar 31.6% with gel strengths of 233.3 and 500 respectively. Durairatnam (1980) has shown that there are 19 species of Gracilaria in Northeast Brazil of which 4 species Gracilaria verrucosa, Gracilaria sjoestedtii, Gracilaria debilis and Gracilaria cylindrica produce agar comparable to Japanese grade I agar. Production of agar is very important for the development of Northeast Brazil where Gracilaria verrucosa and Gracilaria sjoestedtii are found in reasonable large quantities but no detail work has been done on the extraction of agar. It was decided to carry out investigations on the extraction of agar from collections made quarterly to determine its suitability for industrialisation.

Material and Methods

The material for this study was from seaweeds cast ashore,

handpicked from rocks, corals and from sandy bottoms at low tide as well as in deeper areas collected by using divers. The seaweeds collected were washed in sea water and brought to the laboratory where they were washed several times in fresh water and after removal of the impurities were dried until completely bleached. 60 grams of the bleached seaweeds on an anhydrous basis were placed in a 1-3% aqueous NaOH solution made up to 1.5 litres and boiled at 90°C for 3 hours. The seaweeds were then thoroughly washed in fresh water and placed in a vessel containing 1.5 litres of distilled water at PH at 130°C. After boiling for 3 hours the acidity of the hot solution was adjusted to 6.5, strained through several layers of cheese cloth and filtered through filter paper in a pyrex funnel on a tripod with a collecting tray below. The agar was heated when the filter paper became clogged. The filtered agar was collected in stainless trays and allowed to stand for 24 hours. They were then placed in a freezer at 0°C to -5°C for 24 hours and then at -5°C to -10°C for a further 24 hours. This is done to avoid gel freezing only at the surface. The trays were then removed and the frozen gel allowed to thaw. During the process of thawing 85% of the water was removed with the soluble impurities such as sugars, salts and pigments. It was then allowed to dry on cloth in the sun and when completely dry the percentage yield of agar was calculated. The procedure was repeated for unbleached seaweeds. The gel strength, setting point and melting points were determined by the method of Thomas and Krishnamurthy (1976).

Results

Table 1. Physical properties of pure agar from *Gracilaria verrucosa* and *Gracilaria sjoestedtii*.

	Jelly		1.5% solution	
	strength	PH	setting temp.	melting temp.
G. verrucosa	501	6	40.1	91
G. sjoestedtii	438	6	36.6	82

Table 2. Yield of pure agar and gel strengths obtained from unrefined *Gracilaria verrucosa* and *Gracilaria sjoestedtii* on an anhydrous basis.

Date of harvest	Percentage yield without alkaline treatment	Percentage yield with alkaline treatment	Gel strength 1.5% agar (g/cm^2)	
			without alkaline treatment	with alkaline treatment
16-4-79				
G. verrucosa	50.2	17.8	340	510
G. sjoestedtii	48.3	16.8	321	496
19-7-79				
G. verrucosa	48.8	17.2	291	480
G. sjoestedtii	47.8	16.6	285	448
18-10-79				
G. verrucosa	51.2	17.4	324	514
G. sjoestedtii	49.7	17.5	316	520
14-12-80				
G. verrucosa	54.2	19.3	345	562
G. sjoestedtii	49.7	17.3	305	485

Table 3. Yield of pure agar and gel strengths obtained from refined *Gracilaria verrucosa* and *Gracilaria sjoestedtii* on an anhydrous basis.

Date of harvest	Percentage yield without alkaline treatment	Percentage yield with alkaline treatment	Gel strength 1.5% agar (g/cm^2)	
			without alkaline treatment	with alkaline treatment
16-4-79				
G. verrucosa	42.0	26.1	285	480
G. sjoestedtii	43.4	27.2	260	432
19-7-79				
G. verrucosa	41.8	24.3	278	458
G. sjoestedtii	40.2	23.5	261	412
18-10-79				
G. verrucosa	42.4	27.2	305	520
G. sjoestedtii	41.5	25.8	289	476
14-2-80				
G. verrucosa	45.6	28.2	320	548
G. sjoestedtii	42.6	25.3	294	492

Table 4. Effect of concentration of sodium hydroxide solution during alkali treatment of raw seaweeds at 90°C for 3 hours on the yield of agar extracted.

	1%	1.5%	2%	2.5%
G. verrucosa	14.6	12.8	17.9	12.8
G. sjoestedtii	13.4	12.0	17.0	11.9

Table 5. Effect of concentration of sodium hydroxide solution during alkali treatment of refined seaweeds on the yield of agar prepared by extraction at 130°C for 3 hours.

	1%	1.5%	2%	2.5%
G. verrucosa	24.5	27.2	26.4	21.4
G. sjoestedtii	23.8	26.1	25.2	20.3

Table 6. Effect of concentration of sodium hydroxide solution during alkali treatment of raw seaweeds at 90°C for 3 hours on the gel strength of agar extracted.

	1%	1.5%	2%	2.5%
G. verrucosa	412	605	548	430
G. sjoestedtii	381	524	490	385

Table 7. Effect of concentration of sodium hydroxide solution during alkali treatment of refined seaweeds on the gel strength of agar prepared by extraction at 130°C for 3 hours.

	1%	1.5%	2%	2.5%
G. verrucosa	260	410	501	481
G. sjoestedtii	242	368	448	415

Conclusion

Experiments carried out showed that good quality agar comparable to grade I Japanese agar can be produced from

Gracilaria verrucosa and *Gracilaria sjoestedtii* by treatment with 2% sodium hydroxide at 90°C for 3 hours followed by autoclaving at 130°C for 3 hours as suggested bu Tagawa, Kojima and Kono (1963).

The yield of agar without treatment with sodium hydroxide using refined seaweeds gave a yield of 42% compared to 23 to 26% when sodium hidroxide was used. For the domestic market it would be economical to prepare agar without alkali treatment, but for export purposes it is necessary to treat the seaweeds with alkali as the gel strength is considerably increased, although the yield is reduced by 50%. The cost of a factory in Brazil to produce 3 tons of agar per month is estimated at US$ 1,000,000.00.

Reference

1. Dias-Piferrer, M., Perez, C.C.,: Inst. Biol. Marina, C.A.A.M; Univ. Puerto Rico 1-145, (1964).

2. Durairatnam,M., Medcoff J.C.: Ceylon Trade Journal 19, 1-6 Ceylon (1954).

3. Durairatnam,M.: Proceedings of the seaweed symposium on marine algae of the Indian Ocean Region. India(1979) (in press).

4. Durairatnam, M.: Ciência e Cultura, Brazil(1980) (in press).

5. Kojima, Y., Funaki, K.: Bull. Jap. Soc. Sci. Fish. 16, 43-46 (1951).

6. Mohanty, G.B.: G.B.: J. sci. indust. Res., 14A, 36 (1955).

7. Ohmi, H.: Kuroda, K.; Bull. Jap. Soc. Phycol., 3(1) 19-22 (1955).

8. Ohmi, H.: Fac. Fish. Mem. Hokkaido Univ. 9(1), 1-66. (1955).

9. Tagawa S., Kojima. Y, Kono M.: J. Shimonoseki Univ. Fish. 13(1), (1963).

10. Thomas, PlC., Krishnamurthy.: Bot. Mar. 19, 115-117 (1976).

11. Wood, E.J.F.: Coun. Sci. Ind. Res. Bull. 203 1-43 (1943).

THE COMMERCIAL COLLECTION AND UTILISATION OF SEAWEEDS IN IRELAND.

M.D. Guiry
Department of Botany, University College, Galway, Ireland

and G. Blunden
School of Pharmacy, Portsmouth Polytechnic, King Henry I Street, Portsmouth, PO1 2DZ, U.K.

Introduction

Seaweed has been gathered in Ireland for hundreds of years; the earliest printed reference would appear to be in 1726 (1). Until the present century seaweed was collected either as a fertilizer or as a food. Kelp burning was practiced on the west coast during the 19th century, mainly as a source of fertilizer; drift seaweed was, and still is, collected in areas where soil is either absent or very poor and mixed with sand to grow crops such as potatoes. Some information on commercial utilization of seaweeds in Ireland has appeared in other publications (2-6) but no overall review has been undertaken. In comparatively recent years a number of seaweed drying plants have been established but little processing beyond drying and milling is carried out in Ireland.

A. *Ascophyllum nodosum* (Phaeophyta)

This is the most important seaweed collected in Ireland. It is gathered all the year around and is cut by hand from the intertidal ensuring that not less than 15 cm. of weed is left for regeneration. In areas not accessible by road the fronds are made into floating rafts and towed to collection points. Most is now transported wet by road to factories at Kilkieran, Co. Galway, Newport, Co. Mayo and Meenmore Quay, Co. Donegal. Here

Xth International Seaweed Symposium
Printed in Germany

the plants are dried and exported for alginate production in Scotland. The principal areas where the species is collected are: the Connemara area (Co. Galway), Co. Mayo, Co. Donegal, the Shannon Estuary (mainly Co. Clare), with smaller quantities from Co. Cork, Co. Kerry and Co. Sligo. The weed is dried mainly using rotary driers and the product is milled prior to export. Some wet weed is similarly dried at a factory in Kilrush, Co. Clare and exported for use in animal feedstuffs and for the preparation of liquid seaweed extracts for horticultural and agricultural uses. Quantities collected are shown in Table 1.

B. *Laminaria* species (Phaeophyta)

L. digitata is collected mainly as cast weed on Inishiar (Aran Islands) and Quilty, Co. Clare. The plants are mostly air dried *in situ*. Most is processed at the Kilrush factory, but the quantity used is very small (Table 1). Good quality weed is used for the production of iodine tablets. Some weed is used to prepare liquid seaweed extracts.

L. hyperborea is collected mainly in April-June as cast weed. The old laminae ('Mayweed') are cast off in spring (2) and large quantities are thrown up during storms. Both stipes and laminae are air dried separately *in situ* and transported by road; stipes are processed at Kilkieran and the laminae at Kilrush (Table 1). The laminae are oven dried and milled prior to being exported mainly for animal feed with some being used for producing liquid seaweed extracts. Stipes are exported whole for sodium alginate production without further drying; in very recent years stipes have not been used to the same extent as formerly. Collection is mainly on the coast of Co. Clare, particularly in the Kilkee and Quilty areas.

	1973 (t)		1979 (t)	
Ascophyllum nodosum	39,000	(wet)	62,000	(wet)
Laminaria hyperborea				
Laminae	500	(dry)	225	(dry)
Stipes	1,200	(dry)		
Laminaria digitata	300	(dry)	100	(dry)
Fucus vesiculosus	20	(dry)	5	(dry)
Fucus serratus	25	(dry)	20	(dry)
Chondrus crispus	?		165	(dry)
Palmaria palmata	?		5	(dry)
'Blackweed'	8-10,000	(dry)	8,250	(dry)

Table 1. Quantities of seaweeds being collected and utilised commercially in Ireland, 1973 and 1979.

C. *Fucus* species (Phaeophyta)

Small amounts (Table 1) of *F. vesiculosus* are collected in the Shannon Estuary and on the Dingle Peninsula, Co. Kerry all year around. The weed is air dried *in situ* and is mainly exported for use in the health food industry. *F. serratus* is also collected all year around in small quantities (Table 1), again mainly in the Shannon Estuary. The weed is air dried and exported; some is used in the manufacture of seaweed extracts.

D. *Chondrus crispus* (Rhodophyta) - Irish Moss

This is collected by hand picking, mainly in June-September on spring tides. Mostly mature fronds are gathered but many collections are of poor quality with up to 30 species of other seaweeds being present. The alga is collected principally on the coasts of Co. Clare (Castlegregory, Quilty, Kilkee and the Burren), Co. Kerry (Fenit, Dingle Peninsula) and Co. Mayo (Clew Bay) with smaller quantities being obtained from Co. Donegal,

Co. Sligo and Co. Galway. Plants are air dried and bleached *in situ* and most is exported whole and unpackaged. A firm at Moycullen, Co. Galway packages whole and ground *Chondrus* for sale as food. Annual tonnages of this species peaked in 1942 when 625 t were gathered (5) (see Table 1). Dried seaweed is sold to factories at Kilrush, Maam Cross, Moycullen and Newport. Exported material is used in the brewing and printing industries and for culinary purposes.

E. *Palmaria palmata* (Rhodophyta) - Dulse

Small quantities (Table 1) are being collected by one firm (Moycullen) for drying and sale, ground or whole, as a condiment and for human consumption.

F. 'Blackweed' (Mainly Phaeophyta)

Blackweed is the commercial name given to total drift weed. It usually comprises *Fucus* spp, *Ascophyllum nodosum*, variable quantities of *Laminaria* spp and smaller quantities of numerous other species. It is collected mainly in Co. Clare and Co. Galway and the drift weed is raked initially into small piles and gradually transferred up along the shore where it is air dried in larger piles. It is then transported by road to factories at Maam Cross, Roundstone (Co. Galway) and Kilrush where it is dried either in tunnel or rotary driers. After drying the product is milled and is used either for animal feed or for the manufacture of liquid seaweed extracts. Approximate quantities are given in Table 1.

Discussion

Ascophyllum nodosum is the most intensively harvested and profitable seaweed currently being exploited in Ireland. It is cut from areas of shore regarded as being the property by tradition of an individual; the other species are generally regarded as being available to anybody who wishes to harvest them (although in general it is the adjoining landowners). *A. nodosum* is sold wet and there is a well organized transport and purchasing system. As a result the average age of the gatherers is relatively low and although the annual quantities harvested seem to be somewhat cyclical (depending on weather, unemployment levels, market price etc.), the ready market assures the delivery of adequate material for the maintenance of the industry. It seems likely that there is not a consistently high enough tonnage available to justify the processing of the weed for sodium alginate production in Ireland; also there are problems in obtaining a sufficient freshwater supply and disposing of effluent in areas where the weed is collected.

Most of the 'blackweed' processed in Ireland is milled for sale as animal feed. Because of the low calorific value and high price of the product demand has fallen. Also, recent increased fuel costs have meant that more air drying is needed to cut down on transport and drying costs. The method of air drying currently being used is detrimental to the quality of the end product; the stacking of the material in large piles and the wet, warm Irish weather results in very high bacteriological activity. The method of collection is labour-intensive and the average age of the collectors is high. The method of stacking also leads to the leaching and decomposition of the active compounds necessary for growth promotion making the product unsuitable for the preparation of seaweed extracts. In the case of the food species, the principal one, *Chondrus crispus*, is often of poor quality and, although there is a ready market and a good price, there is not much expansion likely to take place due largely to a lack of suitable and easily-gathered stocks. The average age of *Chondrus* collectors is also increasing and production seems to be cyclical for the same reasons as outlined for *A. nodosum*. The demand and quantities available of the other food species are generally low.

The future prospects for the Irish seaweed industry are generally poor due largely to increasing fuel costs, increasing average age of the collectors, lower demand for certain seaweed products, and increasing affluence in Ireland. Better quality control, improved packaging and marketing, and initiation of research into other aspects of the uses of seaweeds may improve the situation.

Acknowledgements

We would like to thank: the National Board for Science and Technology (Ireland) for providing financial aid for this study; Mr. J.P. McCormack and Mr. C. Glynn (Irish Marine Products Ltd); Messrs Casburn (Seaweed Ltd); and Senator M. Staunton (Clew Bay Seaweeds Ltd) for information on the Irish industry.

References

1. Threlkeld, C.: Synopsis Stirpium Hibernicarum...., Dublin (1726).
2. Brennan, A.T.: Notes on Some Common Irish Seaweeds, The Stationery Office, Dublin (1950).
3. Chapman, V.J.: Seaweeds and their uses, Second Edition, Methuen, London (1970).
4. Cotton, A.D.: Proc. R. Ir. Acad., 31, sect. 1(15), 1-178 (1912).
5. Kirby, R.H.: Seaweeds in Commerce, H.M.S.O., London (1953).
6. Michanek, G.: Fish. tech. Pap. F.A.O. no. 138 (1975).

METHANE PRODUCTION FROM THE RED SEAWEED *GRACILARIA TIKVAHIAE*

M. Dennis Hanisak
Harbor Branch Foundation, RR 1, Box 196, Fort Pierce, Florida, 33450, USA

Introduction

Research has been conducted for several years at the Harbor Branch Foundation on the cultivation of seaweeds as a possible source of biomass that can be converted to methane or other fuels (1,2). Of over 40 species of seaweeds examined, the rhodophyte *Gracilaria tikvahiae* had the highest sustained yield and can be vegetatively propagated indefinitely in an aquaculture system. Its productivity can be as high as any terrestrial crop on earth (3). This communication reports on the sustained anaerobic digestion of this species to produce methane and on how the loading rate affects gas production.

Materials and Methods

Two types of digesters were used in this study. The first (with a functional volume of 120 l) have been previously described (4). Digestion was started by adding the following into the digester: 20 l of anaerobic sediments collected from a nearby estuary, 20 l of seawater, and 5-25 kg wet weight of *Gracilaria*. Gas production began within 5 days. The seaweed was not pretreated (e.g. shredded) in any way prior to being loaded into the digesters. Loadings were made from October 1978 to June 1980, at approximately weekly intervals, usually with 5 kg wet weight; an equal volume of residue was removed. The contents of the digester were mixed manually at each

Xth International Seaweed Symposium
Printed in Germany

loading; this was the only agitation employed. The digesters were kept at ambient temperature (30 C ± 5 C) during most of the year. During the cooler months of the first year (November-March), the digesters were kept partially submerged in a 3800 l circular water tank which was kept at approximately 30 C by an immersion heater.

The second type of digester was a sealed 3.8 l Nalgene wide-mouth bottle. An exhaust line from each digester was connected in series to two 3.8 l glass jugs; the first of these contained a 0.05 N sulfuric acid solution and the second one was empty. Biogas production was monitored as the displacement of liquid into the second jug. Each digester was started with 2 kg wet weight of *Gracilaria* and inoculated with 0.2 l of liquid from the larger digesters. After 7 weeks in a batch mode, the digesters were loaded at rates corresponding to retention times (RT) of 10-60 days. The data reported here are for a 60 day period commencing 3 months after the initiation of the differential loading, a length of time adequate for steady-state conditions to be achieved.

Gracilaria fermented in this study was cultivated in outdoor cultures (5). Digesters were loaded once a week, when samples for gas analysis, volatile solids, and pH were taken. Biogas was measured 3 times a week. Gas data are reported for standard conditions of temperature and pressure (STP). Methane levels in the biogas were measured with a MSA Total Hydrocarbon Analyzer, calibrated against known amounts of methane. The volatile solids (vs) in plants or digester residue were determined after ashing dried samples (dried at 60 C for 48 hrs) for 4 hrs at 550 C in a muffle furnace.

Results

Stable, continuous anaerobic digestion of *Gracilaria* was

TABLE 1. Characteristics of stable, continuous anaerobic digestion of *Gracilaria tikvahiae*.

Temperature	30 C ± 5 C
Agitation	None
Load Composition	*Gracilaria* at 11% total solids, of which 58% is volatile solids
Mean Loading Rate	0.54 g volatile solids/l of digester/day
Normal pH Range	6.8-7.5
Gas Production	0.4 l/g volatile solids, at 60% methane
Bioconversion Efficiency	48%

maintained in the large (120 l) digesters for over 20 months, with an average gas production of 0.4 l/g vs, at 60% methane (Table 1). The heat of combustion of *Gracilaria* is 4.5 kcal/g ash-free dry weight (3) or 18.8 kJ/g vs. Since pure methane has an energy content of approximately 37.7 kJ/l, the above methane production represents an average bioconversion efficiency of about 48%.

As the retention time (RT) increased (i.e. loading rate decreased) from 10 to 60 days, the total production of biogas and methane (Table 2) as well as the percent methane and the reduction of total vs (Table 3) increased to maxima at a 30 day RT and then decreased at a 60 day RT. Biogas production and methane production on the basis of vs added increased up to a 60 day RT (Table 2) as did the percent vs reduction (Table 3). The amount of biogas and methane produced on the basis of vs destroyed (Table 2) was remarkably constant, except at a 10 day RT. Gas production at a 10 day RT was erratic and considerably less than all other RT's, and low pH (<6.0) was common at this high loading rate. The pH in the digesters increased as the RT increased. Bioconversion efficiency was maximal at the longest RT and was closely correlated with percent vs reduction.

Table 2. The effect of retention time (RT) on gas production in anaerobic digesters of *Gracilaria tikvahiae*.

RT	Biogas Production			Methane Production		
(days)	(l)	(l/g vs added)	(l/g vs destroyed)	(l)	(l/g vs added)	(l/g vs destroyed)
10	28.9	0.04	0.67	8.8	0.01	0.20
15	35.2	0.07	0.76	20.7	0.04	0.45
20	41.0	0.11	0.73	24.0	0.06	0.43
30	48.9	0.19	0.73	32.2	0.13	0.48
60	44.6	0.35	0.73	25.4	0.20	0.42

Table 3. The effect of retention time (RT) on biogas composition, pH, volatile solids reduction (VSR), and bioconversion efficiency in anaerobic digesters of *Gracilaria tikvahiae*.

RT	Biogas Composition	pH	VSR		Bioconversion Efficiency
(days)	(% Methane)		(g)	(%)	(%)
10	30.4	5.9	42.9	5.6	2.3
15	58.9	6.2	46.4	9.1	8.1
20	58.6	6.4	56.3	14.7	12.6
30	65.8	6.7	66.9	26.2	25.3
60	57.0	7.1	60.9	47.7	40.0

Discussion

This study demonstrates that *Gracilaria* can be fermented to produce methane for a sustained period of time. Both the gas production and bioconversion efficiency compare favorably with those of other biomass substrates (e.g. 6,7). The emphasis in this study has been on non-intensive digestion (i.e. energy inputs into the digesters are minimized). Procedures such as continuous agitation or heating would increase gross methane production; however, the net energy yield could decrease as

Jewell (8) has observed for the digestion of cow manure.

This study has shown how loading rate affects gas production and other digestion characteristics. This information will be helpful in designing large-scale digesters with Gracilaria as a substrate. The rate of gas production is maximal at a 30 day RT, but, if the digesters were relatively inexpensive to construct, a longer RT (e.g. 60 days or more) might be justified and would increase the bioconversion efficiency.

Little comparison can be made between the digestion of Gracilaria and other seaweeds due to a paucity of research in this area. Preliminary studies indicate that Gracilaria may not be as readily digested as some other seaweeds, particularly chlorophytes (Hanisak, unpublished data). The rates measured for Gracilaria at longer RT's are similar to those reported for the phaeophyte Macrocystis (9), but Gracilaria is considerably less digested at shorter RT's. This reduced digestibility is probably a function of differences in experimental conditions and the greater resistance of the agar in Gracilaria to microbial breakdown.

Gracilaria has been proposed as a possible "energy crop" (1,2). It is now technically feasible to cultivate this species in an aquaculture system for this purpose, but this is probably not economical to do without considering the value of its agar, which is worth far more than the methane produced from its biomass (Hanisak, in preparation). Since, at a 30 day RT, little of the agar is digested, it could be extracted before or after the Gracilaria is loaded into the digester. Under these conditions, both agar and methane could be extracted with little reduction in yield for either product. The energy derived from the methane could be used in processing agar or in cultivating the agarophyte.

Aquaculture is a new, energy-intensive industry that will be developed only if its energy needs can be met at a reasonable cost. Seaweeds, grown as part of an aquaculture system, can serve as an alternative energy source and replace the need for conventional fossil fuels. Any energy conserved in this manner would, of course, reduce overall energy demands. Thus, the use of seaweeds as an "energy crop", may play a role as part of man's solution to the present energy crisis.

Acknowledgements

I thank Craig Tucker and Dan Young for performing the methane analyses, Coddy Williams for technical assistance, and Linda Hanisak for assistance with manuscript preparation. This research was supported in part by Contract No. XR-9-8133-1 from the Solar Energy Research Institute. This paper is Contribution No. 195 from the Harbor Branch Foundation, Inc.

References

1. Ryther, J.H., DeBoer, J.A., Lapointe, B.E.: Proc. Int. Seaweed Symp. 9, 1-16 (1978).

2. Ryther, J.H., Williams, L.D., Hanisak, M.D., Stenberg, R.W., DeBusk, T.A.: Proc. Second Fuels from Biomass Symp. 2, 978-989 (1978).

3. Lapointe, B.E., Ryther, J.H.: Aquaculture 15, 185-193 (1978).

4. Hanisak, M.D., Williams, L.D., Ryther, J.H.: Resource Recovery and Conservation 4, 313-323 (1980).

5. Hanisak, M.D.: International Council for the Exploration of the Sea, Code Number C.M. 1979/F:24 (1979).

6. Jewell, W.J. (ed.): Energy, Agriculture, and Waste Management, Ann Arbor Science Publishers, Ann Arbor, Michegan 1975

7. National Academy of Sciences: Special Report No. 19, National Academy of Sciences, Washington, D.C. 1977

8. Jewell, W.J.: Report No. COO-EY-S-02-2981-11, U. S. Department of Energy, Washington, D.C. 1979

9. Klass, D.L., Ghosh, S., Chynoweth, D.P.: Paper presented at 175th National Meeting Am. Chem. Soc. (1978).

THE BIOMASS OF SEAWEEDS AT THE KULLEN, S. SWEDEN

R.-A. Kornfeldt

Dept. of marine botany, University of Lund,
Box 740, S-220 07 Lund, Sweden

Introduction

The species composition and biomass of seaweeds at the Kullen, S Sweden was investigated. The locality is situated in the transition area between the North Sea and the Baltic. The salinity at the Kullen fluctuates between 8 and 34 o/oo, median at 5 m depth 14.8 o/oo. These salinity fluctuations combined with the absence of tide and increased pollution have a profound influence on the flora at the locality (1). Compared to the North Sea there are fewer species, the Fucus belt is broader and the dense sublittoral forests of Laminaria species are thinner (1,2). Good opportunities for red algal development exist among the scattered Laminarias.

Material and methods

Vegetation samples of 0.25 m^2 were collected from 2,4,6,8 and 10 m depth at Ransvik, the SW slope of the Kullen. A total of 302 samples were taken on 32 occations from February 1974 to November 1978. At each depth the samples were taken at random. The material was frozen within 3 h and later thawed, sorted into species or species-groups and dried on Al-foil at 65°C to constant weight. All values given are related to the sampling area, 0.25 m^2. The material was divided into groups based on depths and seasons (=quaters) for calculations. The Brillouin index (3) was chosen as a measure of the diversity. The weight

Xth International Seaweed Symposium
Printed in Germany

of each species was used as its abundance. Species with very small biomass, which were not weighed were arbitrarily set to 0.1 g. Where several species were weighed together, the weights were split according to the number of species.

Results

The observed biomasses at different depths and seasons for red, brown, perennial and annual algae, and the total weights are shown in Fig. 1. Table 1 gives adjusted means for depths and seasons respectively.

The green algae contribute very little to the biomass at all depths investigated. No sample contains more than 0.8 g and the mean weight is less than 0.1 g.

The brown algae show maximal biomass at 2 m and decline with depth (Tab.1). Fucus serratus dominates 2 and 4 m depth but is less important further down, although specimens may grow even at 10 m depth. The Laminaria species constitute most of the brown algal biomass from 6 m depth and below but they are few and unevenly distributed (Fig.1a). The brown algal biomass shows a seasonal variation ($p<0.001$) with a drop during winter and a maximum in mid-summer.

The red algal biomass increases with depth to a maximum at 6 m and with high values also at 8 and 10 m (Tab.1, Fig.1b). At 4 and 6 m Furcellaria lumbricalis dominates but further down Phyllophora spp are the dominating perennials. The composition of the red algae changes from mainly perennial at shallow depths to much more annual at 10 m. This group shows seasonal variation ($p<0.001$) with low winter and high summer values.

The biomass of the perennials decreases with depth and is at 10 m only one ninth of that at 2 m (Tab.1, Fig.1c). The shallower parts are dominated by Fucus serratus, but from 6 m and down-

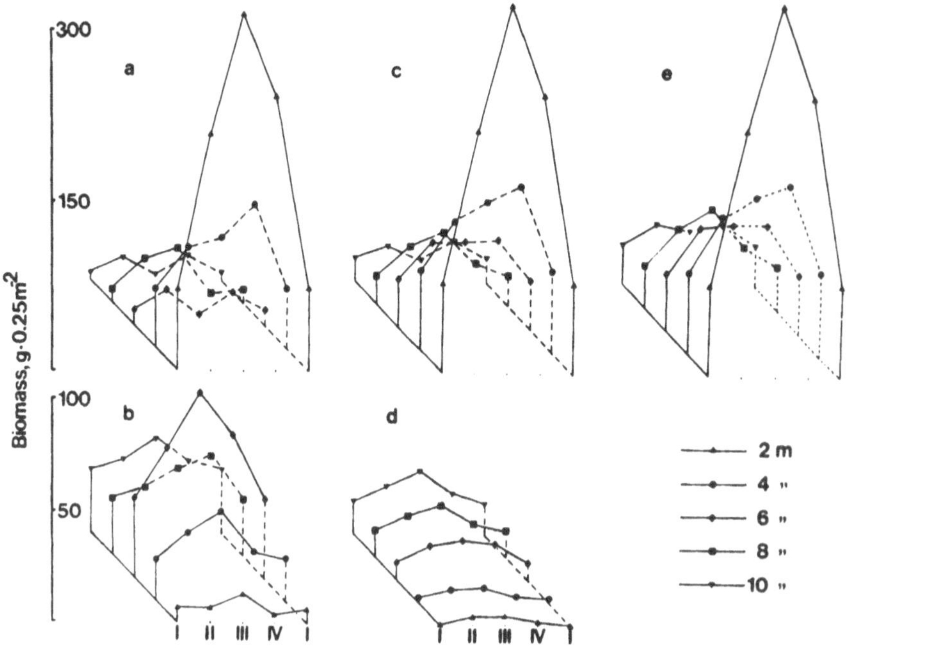

Figure 1. Observed biomasses at each depth and season, a. brown b. red c. perennial d. annual algae and e. total weight.

Table 1. Adjusted weights of red, brown, perennial and annual algae, total biomass (g.0.25 m^{-2}), mean number of species and of Brillouin index ± s.d. for the five depths and four seasons.

	n	brown	red	perennial	annual	total	number of species	Brillouin index
2 m	55	191.0	8.7	196.9	2.8	199.8	7.6±4.2	0.13±0.1
4 m	68	92.0	28.1	115.6	4.5	120.2	13.6±4.0	0.34±0.2
6 m	65	19.8	60.4	66.7	13.4	80.2	17.0±3.4	0.58±0.2
8 m	58	26.2	33.4	42.2	17.3	59.6	18.8±3.2	0.71±0.2
10 m	56	9.7	33.3	22.2	20.9	43.0	19.6±4.2	0.69±0.1
I	71	20.3	23.5	36.4	7.5	43.9	13.9±5.5	0.44±0.3
II	107	76.9	32.5	97.0	12.4	109.4	16.2±5.6	0.50±0.3
III	66	84.8	45.5	114.5	15.8	130.3	17.0±5.4	0.56±0.3
IV	58	83.5	34.0	107.2	10.2	117.5	13.7±5.7	0.45±0.3

wards the red perennials predominate. Although these species are abundant the whole year, a single specimen having a life-span of several years, the biomass shows a marked seasonal variation ($p<0.001$).

The annuals consists almost entirely of red algae, i.e. *Delesseriaceae* and filamentous species. Only a small part is brown algae e.g. *Ectocarpus siliqulosus*, *Pilayella litoralis*, *Desmarestia viridis*, which are all most common during spring in the shallower part. Some of the red algae, e.g. *Phycodrys rubens*, *Polysiphonia nigrescens*, *Ceramium rubrum*, are usually considered perennials (4), but at Kullen show a distinct seasonality in occurence (1). The biomass of this group increases with depth (Tab.1). There are, as expected, seasonal variations (p 0.001) with a winter minimum and a summer maximum (Fig.1d). In this group there were also differences ($p=0.002$) between the years 1976-78. The adjusted mean value for 1977 is 13.87 g, 3.44 g more than in 1978. Such differences were not seen in any of the other groups.

The total biomass decreased with depth (Tab.1) and the mean at 10 m depth is only one fifth of that found at 2 m. The biomass is totally dominated by Fucus at 2 and 4 m. At 6 m *Furcellaria* constitutes half the biomass. With increased depth the annual group becomes more important. There is a marked decrease in the biomass during winter, with a maximum in summer, resulting in a strongly significant seasonal variation ($p<0.001$) (Fig.1e).

The mean number of species per sample increases with depth ($p<0.001$) and there is also seasonal variation (p 0.001) with minimum in autumn and maximum in summer (Tab.1). When the number of species are combined with their abundance into the Brillouin index of diversity the trends are slightly different (Tab.1). The index value increases with depth down to 8 m where it seems to stabilize ($p<0.001$) and the seasonal variation is less pronounced ($p=0.006$). The low index value at 2 m proves that one species, i.e. *Fucus serratus*, dominates the samples, whereas at the deeper parts several species are important contributors to the biomass of the sample.

Discussion

The decrease in salinity from the North Sea to the Baltic proper strongly influences the flora. The *Fucus* belt becomes broader, from 0.5 m at Helgoland to 5 m at Askö and the Laminaria species disappears (5,6). At the Kullen, in the transition area, the *Fucus* belt is well developed, ca 4 m at Ransvik, and some *Laminaria digitata* and *L. saccharina* grow below 5 m. The *Fucus* biomasses are larger than those reported from the Askö area (6) but are similar to those found at the west coast of Sweden (7). The red algae dominate the zone below the *Fucales* at the Kullen and the *Laminarias* contribute only 22 o/o of the biomass. The amounts of algae below the *Fucus* zone are intermediate between biomasses from fruther north the Swedish west coast (2,7) and the east coast (6).

Seasonal variations are seen not only in the annual group but also in the perennials. The two main factors changing over the year are, available nutrients and energy input, both temperature and light. In January the energy reaching the surface is less than one tenth of the energy influx in June (1). *Fucus serratus* and *Laminaria saccharina*, the two most abundant brown algae, have increased photosynthetic rate per light unit during winter. *Furcellaria lumbricalis* and most of the red annuals have increased growth rates during spring (8). Brown algae increases markedly in late winter and early spring whereas the red algal biomass increases successively from winter to summer (Tab.1). Highest production rates were found in sheet-like and finely branched forms (8). These correspond to the annuals in this study. At 8 and 10 m depths, where this group is most important, the weight increase during spring and summer is most marked whereas the red perennials, i.e. *Phyllo-*

phora spp, show less seasonal variation. These thin forms also have the largest surface area per volume and thus better opportunities to absorb needed nutrients. The supply of inorganic nitrogen is very low during summer (1) and the cessation in growth in the big brown algae may be caused by nutrient deficiency (9). This is indicated by the nitrogen contents in Laminaria saccharina which diminish from 28 $mg \cdot g^{-1}$ in late winter to 8 in summer whereas the filamentous red algae varies only between 49 and 36 $mg \cdot g^{-1}$ (10).

The Brillouin index and the mean number of species show that the flora in deeper water is more diverse than in shallower water in spite of the lower biomasses.

References

1. von Wachenfeldt, T.: Marine benthic algae and the environment in the Öresund. Thesis, Lund 1975.
2. Kornfeldt, R.-A.: Svensk Bot. Tidskr. 73, 131-138 (1979).
3. Brillouin, L.: Science and Information Theory, Academic Press, New York 1960.
4. Mathieson, A. C.: Botanica Mar. 22, 511-520 (1979).
5. Lüning, K.: Helgoländer wiss. Meeresunters. 21, 271-291 (1970).
6. Jansson, A.-M., Kautsky, N.: in Keegan, B. F.,O'Céidigh, P.,Boaden, P. J. S. (eds.): Biology of benthic organisms, Pergamon Press, Oxford 1977.
7. Gislén, T.: Kristineberg Zoologiska Station 1877-1927. Skriftser. utg. av Kungl. Svenska Vetenskapsakad. 4, 1-380 (1930).
8. King, R. J., Schramm, W.: Mar. Biol. 37, 215-222 (1976).
9. Chapman, A. R. O., Craigie, J. S.: Mar Biol. 40, 197-205 (1977).
10. Kornfeldt, R.-A.: 1980 in preparation.

EXPERIMENTAL HARVESTING OF GRACILARIA IN NEW ZEALAND.

D.M. Luxton
Davis Gelatine (N.Z.) Ltd., Woolston, Christchurch, 2,
New Zealand.

Introduction

The dominant benthic alga of the Manukau Harbour, Auckland, is the agarophyte Gracilaria secundata forma pseudoflagellifera. Specimens from the Harbour were assigned to G. secundata Harv. by May (5), and the infra-specific taxon is distinguished by the pattern of external branching (5). The species had been previously referred to G. confervoides (6).

G. secundata exhibits three distinct modes of growth within the Manukau Harbour. Individual plants, common on intertidal mud banks, are attached by discoid holdfasts to hard substrates, which are most frequently the shells of the cockle, Chione stutchburyi. The Gracilaria also forms mono-specific algal mats composed of entangled thalli anchored down by fine sediments. The Gracilaria mats have recently been mapped by aerial photography, and their occurrence and development has been associated with the discharge of sewage effluent and land run-off (3). The mats give rise to a third growth form, unattached free-living fragments, which occur throughout the Harbour. The action of wind and tide frequently accumulates unattached Gracilaria into drift masses which may remain in the intertidal zone of sheltered bays for several weeks.

An agar industry has operated in New Zealand since 1943 (6) based on the Gelidioid algae Pterocladia lucida and P. capillacea. Early on, indigenous Gracilaria from Paremata was considered inferior and disregarded (6), but subsequently,

Xth International Seaweed Symposium
Printed in Germany

with modern extraction methods, an agar of commercial quality has been produced from Manukau Gracilaria (Luxton unpublished). The purpose of the study reported here was to determine the standing crop and regeneration rate of G. secundata, and to ascertain the feasibility of commercial harvesting.

The harvesting of wild Gracilaria crops is generally achieved by labour intensive hand gathering, frequently with the aid of hooks, graphnels, (4) or with long-handled rakes (1). In the past, unattached plants have been gathered in anchored stop nets and by trawls (2). Attempts have also been made to develop a Gracilaria trawl from a modified Agazzis trawl for gathering deeper water benthic plants (8).

Materials and Method

Standing crop and hand clearing. Biomass data on intertidal Gracilaria mats was obtained by random sampling with $0.25m^2$ quadrats. Samples were washed free of mud and oven dried to constant weight. At Little Muddy Creek (36° 58.0'S, 174° 38.7'E) seasonal changes in mat density were monitored over 12 months. Areas of $16m^2$ were hand cleared in both winter and summer and regrowth was measured at approximately monthly intervals by harvesting 3m x 0.5m strips from within previously cleared areas, using a $0.25m^2$ quadrat. The mean relative growth rate (RGR) was calculated for each density increase between samplings.

Manual raking. A small Gracilaria mat situated at Blockhouse Bay (36° 55.8'S, 174° 42.4'E) was measured for area, and sampled by random $0.25m^2$ quadrats for biomass. The whole mat was harvested in one day by six men using rakes with 0.4m wide heads and 70mm long tines. After raking, the area was sampled at regular intervals to assess recovery.

Boat harvesting. A 10.3m x 5.5m catamaran was used to harvest the extensive *Gracilaria* mats in areas on the north and south sides of Puketutu Island (36° 58.0'S, 174° 44.7'E). For harvesting in shallow water the catamaran was more suitable than conventional mono-hulls, having a draft of only 0.38m, high stability and a large deck capacity for seaweed. Initial trials involved two methods of harvesting. Firstly raking the *Gracilaria* from the mud substrate by means of a hydraulically driven 4.6m x 0.9m rake-conveyor pivoted between the hulls. Secondly, netting free-living *Gracilaria* moving with the tidal flow, and winching the net up a retractable ramp supported between the hulls. Raking was carried out at depths of 0.5m-3.0m, and netting at 1m-12m depths.

Results and Discussion

Gracilaria mats occur from 0.1m below to 1.6m above mean low water spring tides. Growth was continuous throughout the year, with maximum densities recorded during winter and spring;

Table 1 - Natural Density Changes (dry wt) in the *Gracilaria* mat in control area, Little Muddy Creek.

Date Day	Month	Mean density $g\ 0.25m^{-2}$	Variance Sd$\pm$	Mean RGR $\%\ day^{-1}$
6	8	86.6	11.4	
4	9	83.2	8.7	Loss
10	10	54.9	7.1	Loss
5	11	62.0	5.3	0.5
7	12	68.8	18.8	0.3
10	1	73.0	8.1	0.2
16	2	67.0	9.8	Loss
2	4	61.7	4.8	Loss
6	5	73.8	8.8	0.6
21	6	88.4	10.25	0.4

346g m^{-2} Little Muddy Cr.; 352g m^{-2} Blockhouse Bay; and 345g m^{-2} north side of Puketutu Is. Mean mat density over twelve months at Little Muddy Cr. was 290g m^{-2}. During September, storms reduced mat density by ca 34% (Table 1), and resulted in temporary masses of drift plants close to high water level.

The area of the Harbour covered by Gracilaria mats in 1976 was 187 ha, of which 130 ha was adjacent to sewage effluent outfalls on the north and south sides of Puketutu Is., (3). At densities of 2.90-3.45t ha^{-1} the standing crop for all the harbour mats is estimated to be 542-645t dry weight.

Table 2 - Regeneration of Gracilaria (dry wt) from hand cleared areas, Little Muddy Creek.

Period Day/Month	Days after clearing	Mean density g $0.25m^{-2}$	Variance Sd±	Mean RGR % day^{-1}
6.08- 4.09	28	18.6	2.8	4.6
4.09-10.10	64	32.0	5.9	1.5
10.10- 5.11	90	60.0	13.4	2.4
5.11- 7.12	121	67.4	9.0	0.4
7.12-10.01	155	71.2	6.7	0.2
10.01-16.02	192	65.8	6.2	Loss
16.02- 2.04	46	12.7	4.9	2.2
2.04- 6.05	81	19.0	6.4	1.1
2.04- 6.05	35	13.2	0.5	3.0
6.05-21.06	46	9.3	3.1	1.5

Regeneration in cleared areas was by vegetative growth from residual fragments held by the soft substrate. 29 days after winter clearing (Table 2) regrowth was approximately 100% cover and 21.5% of the original density. After 90 days, regrowth density was the same as that in the adjacent control area. The mean relative growth rate was greatest during the period immediately after clearing, but maximum mean productivity, 4.3g m^{-2} day^{-1}, was between densities of 128g m^{-2} and 240g m^{-2}. At Blockhouse Bay the greatest biomass increase, 5.0g m^{-2} day^{-1}, occurred between similar densities of 83g m^{-2} and 224g m^{-2}.

The area of the Blockhouse Bay mat was 0.209 ha with a standing crop of 720kg ± 68kg before harvesting. Quadrat analysis showed that 40 man hours of raking removed 540kg, but the sun dried harvest weight (680kg) was greater, because of 9-12% moisture and quantities of dried mud. The mat density recovered in eight weeks (Table 3) during which time the mean relative growth rate was 2.3% day^{-1}. Treading by collectors caused some substrate disturbance which, if continually repeated, may lead to erosion, as reported for a Gracilaria mat under similar conditions in British Columbia (7). Regeneration rates indicate that a yield of 5.8-6.9t ha^{-1} could be obtained from four harvests per year, taking 50% of the standing crop at each harvest. This represents a sustainable yield of 1084-1304t $year^{-1}$ from all the Gracilaria mats in the Manukau Harbour.

Initial yields per man hour from mechanical raking of Gracilaria mats during tidal submersion were lower than those from hand raking. Difficulties in maintaining the correct

Table 3 - Regeneration of Gracilaria from raked area, Blockhouse Bay.

Date Day	Month		Density (dry wt) g $0.25m^{-2}$	Variance Sd±(n=6)
26	9	Before harvesting	85.6	8.2
29	9	After "	20.8	9.6
27	10	4 Wks after harvesting	56.0	11.6
24	11	8 Wks " "	70.7	20.5

tine height above the substrate gave mixtures of mud and Gracilaria, which required prolonged washing. In contrast, netting trials collected clean free-living Gracilaria. Experimental nets of 50m length and 2m drop produced 1.2t-1.3t (wet wt) in four hauls over five hours. Netting also trapped various commercial fish species. The polyzoan Zoobotryon was common amongst the unattached Gracilaria.

In conclusion, as the major Gracilaria mats grow on soft deep mud, hand raking during low tide is not practical, and tramping may lead to excessive substrate disturbance and erosion. Mechanical raking and netting from a boat causes less substrate disruption and netting in particular is the most efficient harvesting method.

Acknowledgment: Grateful thanks to Professor V.J. Chapman and Dr P.R. Henriques for helpful discussions.

References

1. Cable, E.D.: Bot. Mar. 17, 60-62 (1974).
2. De Loach, W.S. et al.: Duke Univ. Mar. Sta. Bull. 3, 25-30 (1946).
3. Henriques, P.R.: Selected Ecological Aspects of the Manukau Harbour. Unpub. PhD thesis. Dept of Bot., University of Auckland, New Zealand (1977).
4. Kim, D.M.: Bot. Mar. 13, 140-162 (1970).
5. May, V.: C.S.I.R. Bull. Australia, 235, 1-63 (1948).
6. Moore, L.B.: N.Z. J. Sci. Tech. 25 Sec. B 183-209 (1944).
7. Saunders, R.G., Lindsay, J.G.: Proc. Int. Seaweed Symp. 9, 249-255 (1977).
8. Wood, E.J.F., Peddie, H.M.: Jour. C.S.I.R. Australia, 14, 315-317 (1941).

PROBLEMS OF CONVERTING LINEAR GROWTH INCREMENTS OF KELPS TO ESTIMATES OF BIOMASS PRODUCTION.

Kenneth H. Mann and Colin Mann

Department of Biology, Dalhousie University,
Halifax, Nova Scotia, Canada. B3H 4J1.

Introduction

It is relatively easy to measure growth increments at the meristems of kelps, using the hole-punching method developed by Parke (1), but much more difficult to convert these measurements to reliable estimates of biomass production. The senior author has tackled this problem on *Laminaria* in Nova Scotia (2), on *Ecklonia maxima* in South Africa (3) and on *Ecklonia radiata* in Western Australia (4). On each occasion, improvements have been made and it is now possible to look back and assess errors in previous estimates.

The Method Used with *Laminaria* in Nova Scotia.

Laminaria longicruris and *L. digitata* add to their blade tissue by growth at the meristem, and simultaneously lose tissue by erosion at the tip. By punching a hole at a known distance from the base of the blade and remeasuring after an interval of time it is possible to determine the rate of linear growth, (which is often of the order of 1 cm/day). As the plants get older, the blades get wider and thicker, so that 1 cm of growth in an older plant adds more biomass than 1 cm of growth in a young plant. A plot of biomass (W) versus length (L) for a large number of plants shows a relationship of the form: $W = aL^b$ Equ.1

Xth International Seaweed Symposium
Printed in Germany

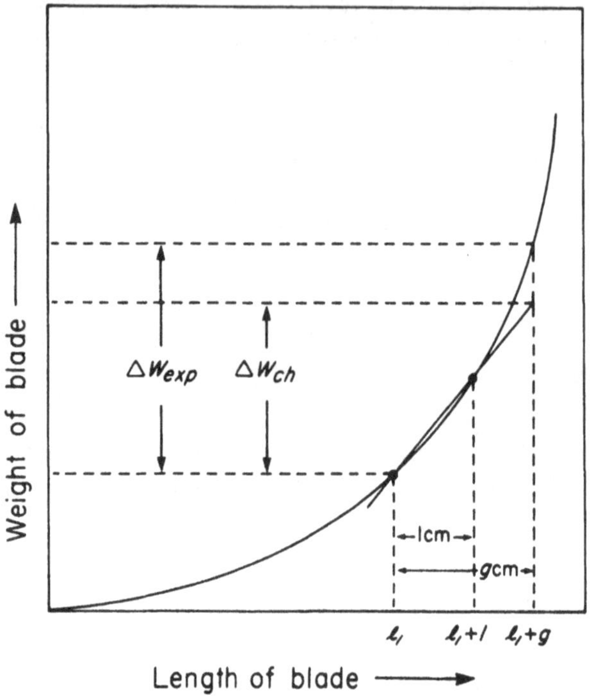

Fig. 1. Diagram showing exponential relationship between blade length and weight for a large sample of *Laminaria* sporophytes, and basis for calculating weight increase by equations 2 and 3.

in which b is of the order of 2, i.e. weight is proportional to the square of the length (Fig.1).

To calculate the biomass increase of *Laminaria longicruris* in t days, the initial length l_1 was recorded, the growth increment g was determined by the hole-punching method, and the length after time t was calculated as if there had been no erosion, as l_1+g. The biomass corresponding to l_1+g was determined from the regression of biomass on length (Fig.1). We may refer to this as the exponential model and the biomass production in time t as $\Delta W_{exp.}$ Then:

$$\Delta W_{exp} = a(l_1 + g)^b - al_1^b \qquad \text{Equ.2}$$

The Method Used with *Ecklonia maxima* in South Africa

The weakness with the exponential model is that biomass after time t is calculated as if there had been no erosion, whereas the relationship $W = aL^b$ was determined empirically, by measuring a large number of plants that had been subjected to the normal processes of erosion. This leads to an inflated value for b, and hence for biomass production. In an attempt to improve on this estimate when calculating biomass production of *Ecklonia maxima* in South Africa, a new model was devised which assumes a constant relation between biomass and length approximately equal to that occurring at the time of initial measurement. It may be called the chordal model. If the initial length is l_1 and the growth increment is g, the biomass increase is given by:

$$\Delta W_{ch} = g \times \left[a(l_1+1)^b - al_1^b\right] \qquad \text{Equ.3.}$$

Fig.1 shows that the new relationship is a chord of the original curve.

CALCULATION OF ERROR INVOLVED IN THE EXPONENTIAL MODEL

If the chordal model is reasonably accurate, what is the error inherent in the exponential model? It may be assessed by calculating the error as

$$E = \frac{\Delta W_{exp} - \Delta W_{ch}}{\Delta W_{ch}} \times 100\% \qquad \text{Equ.4.}$$

From equations 2 and 3 this is:

$$E = \frac{a(l_1+g)^b - al_1^b - g\left[a(l_1+1)^b - al_1^b\right]}{g\left[a(l_1+1)^b - al_1^b\right]} \qquad \text{Equ.5.}$$

Since, as we have seen, b is often close to 2, substitution of this value, and simplification gives

$$E = \frac{g - 1}{2l_1 + 1} \qquad \text{Equ.6.}$$

As a slight variant on the chordal model, the calculation was repeated for a tangential model, i.e. one in which the relationship between biomass and length was assumed to be constant at the value given by the tangent to the line at length l_1. If this model is taken as the true value, the error inherent in the exponential model is given by:

$$E = \frac{g}{2l_1} \qquad \text{Equ.7.}$$

which gives values very close to those in equation 6. For a *Laminaria* blade of 100 cm, and a growth of 5 cm, the error involved in the use of the exponential model is less than 2%. The largest increase used in the calculations of productivity of *Laminaria* by Mann (2) was 35 cm on a 100 cm blade, and the error involved here would be close to 17%. Hence, the range of error is 2-17%; taking 10% as a mean value, we may revise the estimate of kelp bed production in St. Margaret's Bay (Mann 2) from the original value of 1750gC $m^{-2}yr^{-1}$ to 1675gC $m^{-2}yr^{-1}$.

A TOTALLY DIFFERENT APPROACH

In calculating the productivity of *Ecklonia radiata* in Western Australia (Mann and Kirkman, 4) an approach was devised which avoids the necessity of extrapolating from a generalized relationship between weight and biomass. The only requirement is that plants are sufficiently abundant to be sacrificed after each sampling interval.

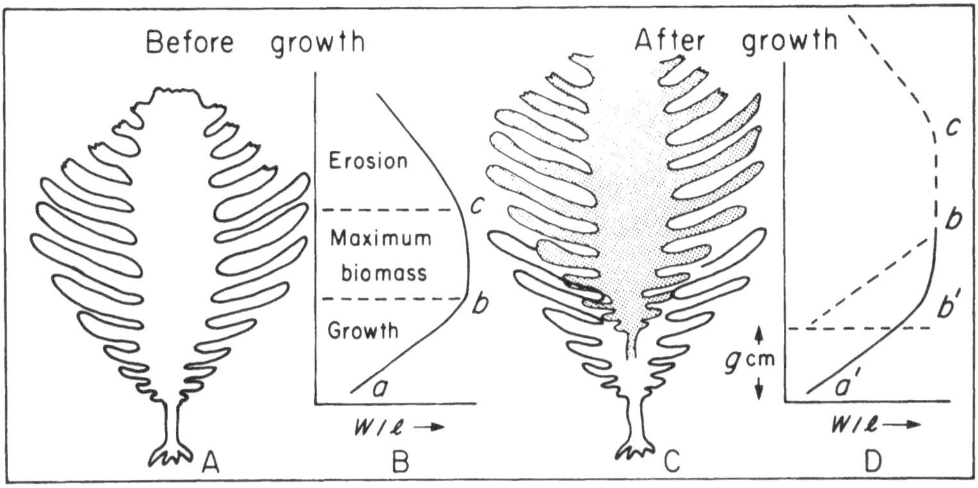

Fig. 2. Diagram showing (A) morphology of *Ecklonia radiata*; (B) a plot of the distribution of biomass per unit length along the axis of the blade; (C) the original plant (shaded) with new growth at the base (unshaded) and (D) a plot of the new distribution of biomass per unit length.

Ecklonia maxima has the morphology shown in Fig.2A. We cut plants transversely into segments, determined their biomass and dimension along the axis of the plant. From this we calculated biomass per cm, W/l. A plot of W/l against distance from base of blade (Fig.2B) shows a zone of growth, in which W/l is increasing, a zone of maximum biomass, and a zone of erosion in which tissue is being broken off and carried away in suspension. This pattern was repeated in all plants. Growth at the base of the blade can be interpreted as a shifting of the line *a b c* in Fig. 2B to a new position *a'b'b c*, as in Fig. 2D, the distance from *a* to *a'* and from *b* to *b'* being the amount of growth, *g* cm. It was shown (Mann and Kirkman, 4) both by algebra and by geometry, that provided the proportions of the plant do not change in the sampling interval, the amount of new tissue added is given by

$$\Delta W = g \times W/l \qquad \text{Equ.8}$$

where W/l is determined in the region of maximum biomass. An empirical test of the hypothesis confirmed its reasonableness. The new method has the advantage of avoiding any generalizations about the relationship of biomass to length in a population of algae. Instead, the relationship is determined for each plant individually, and the only assumption is that the relationship of length to weight remains constant between the initial marking and the final harvesting. If this interval is kept comparatively short (a few weeks) the assumption seems reasonable. The method has been tested on *Laminaria longicruris*. Although the blades of these plants appear to be parallel-sided, and one would expect to find an almost constant value for W/l along the length, we found that the distribution of biomass was very similar to that shown in Fig.2B. Hence, the arguments used for *Ecklonia radiata* (4) apply equally to *Laminaria longicruris*. We suggest that the method probably has universal applicability in the Laminariales.

References

1. Parke, M.: J. mar. biol. Ass. U.K. 27, 651-709 (1948).
2. Mann, K.H.: Mar. Biol. 14, 199-209 (1972).
3. Mann, K.H., Jarman, N. and Dieckmann, G.: Trans. Roy. Soc. S. Afr. 44, 27-41 (1949).
4. Mann, K.H. and Kirkman, H.: Austr. J. mar. freshwat. Sci. (in press).

STUDIES ON GRACILARIA SP. IN BAHIA ARREDONDO, CHUBUT PROVINCE, ARGENTINA.

Alejandro M.S. Mayer
Patagonia Comercial S.R.L., Junin 558, 1026 Buenos Aires, Argentina.

Introduction

Despite the fact that the genus Gracilaria sp. has long been known as a source of the phycocolloid agar (1) and that it has been harvested commercially in Argentina for already 25 years, little or no research has been undertaken on the taxonomy and production ecology of this genus in general (2). Only in the last decade, very preliminary data on the distribution of this genus along the patagonic coast has been made available by a qualitative survey (3). Thus Gracilaria sp. was observed restricted in its distribution to the Chubut Province and localized in the following places: Golfo Nuevo, Bahía Camarones, Puerto Melo and Bahía Bustamante. However personal observations of the author indicate the presence of Gracilaria sp. in Isla Tova (Lat. 45°05´ and Long. 65°59´), Punta Tafor (Lat. 45°03´ and Long. 66°20´) and Bahía Arredondo (Lat. 45°02´23´´and Long. 65°49´25´´) also in the Province of Chubut (Figures 1 and 2).

Of these areas the Gracilaria sp. population in Bahía Arredondo was selected to initiate quantitative biomass and seasonal fluctuation studies, as well as to relate these fluctuations to environmental parameters.

In this paper we report the results of our preliminary investigations over a two-year period: 1979-1980.

Xth International Seaweed Symposium
Printed in Germany

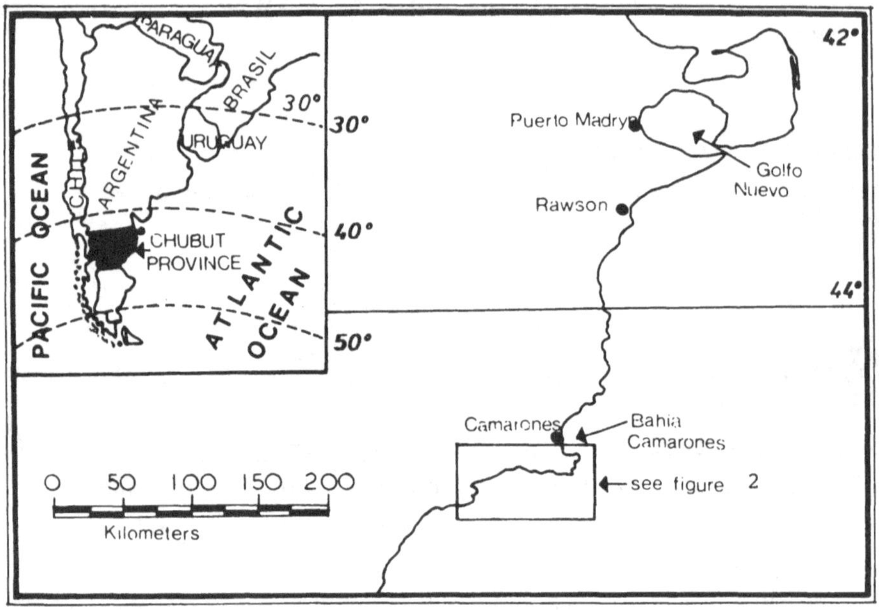

Figure 1: Relative situation of Chubut Province in Argentina. Locations of <u>Gracilaria sp</u>. are indicated.

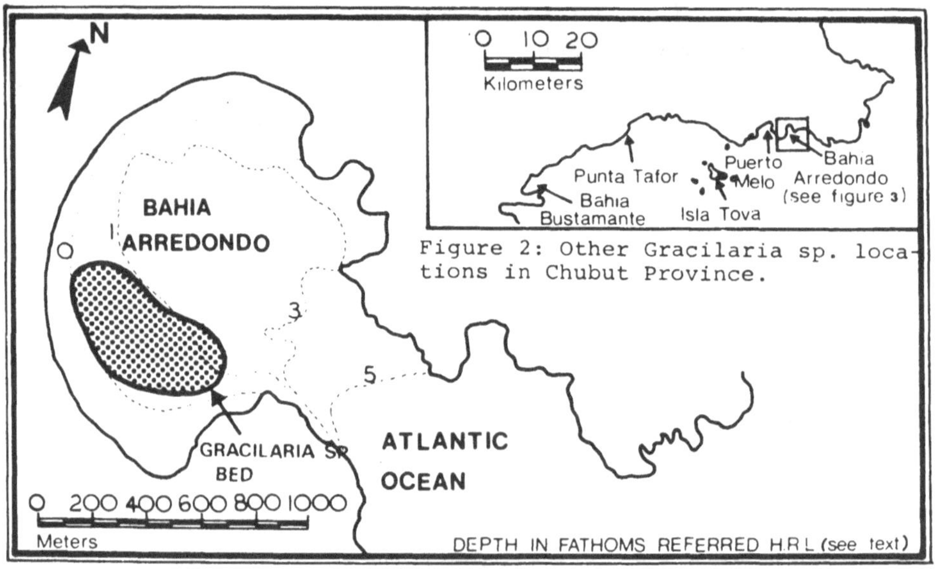

Figure 2: Other Gracilaria sp. locations in Chubut Province.

Figure 3: Gracilaria sp. bed in Bahia Arredondo, Chubut Province, Argentina.

Materials and Methods

The attached population of Gracilaria sp. in Bahía Arredondo was studied at regular intervals - weather and sea conditions permitting - using snorkel and SCUBA. Field work was performed by the author, helped by a professional diver.

The first object of the work was to determine the Gracilaria sp. bed size and shape; and thereafter to make regular biomass assesment. Accordingly, semipermanent markers were established every 50 m along the shores of the bay as reference points, slightly above the highest tide level. Then systematic observations of the sea-bottom were performed proceeding across the bay in a N-S or E-W direction, using a 50-m long polypropylene study line for measuring distances in the water. One permanent transect was established in the N-S orientation using fixed buoys for continuous observations.

The second object was to record the following data simultaneously with the mapping procedures: depth referred to the Hidrographic Reference Level (HRL) as established by the Argentine Hydrographic Service for the area; distance to the shore (to the highest tide levels); presence or absence of Gracilaria sp. ($>$ 0,3 kg per 1-m^2 wet-weight); estimated algal cover in percentage (using photography with a Nikonos III underwater camera when turbidity was low); bottom and surface temperature ; description of the substratum; and wet biomass of Gracilaria sp. per 1-m^2. To obtain these data random sampling every 50 m was performed, positioning a 0,5 m^2 (0,7m x 0,7m) stainless steel quadrat on the sea bottom, bagging Gracilaria sp. in a polypropylene bag and then on land weighing on a Triple Beam Balance, Ohaus Corporation, U.S.A. Data was expressed as wet-weight per 1-m^2. All samples were then air-dried and kept for later analysis.

Results

These preliminary observations have shown that Gracilaria sp. grows mainly as an attached population in Bahía Arredondo, in a bed approximately sized 350m (N-S) by 500m (E-W), situated in the southern part of the bay (Figure 3). The size of the attached bed varied, with a maximum in January (summer) and a minimum in June (winter) of ± 50m. Free living plants were observed in different periods of the year though mainly in summer and autumn in variable quantities, difficult to estimate quantitatively. The data collected in each period of observation are shown in Table 1.

Table 1: Survey of Gracilaria sp. in Bahía Arredondo, Chubut Province, Argentina.

Date	Number of quadrats sampled	Mean density[a] ($kg\text{-}1m^2$)	Minimum and maximum density ($kg\text{-}1m^2$)	Bottom cover (%)	Temp. (°C)
17-25/ 1/79	13	3,11	1,10 - 5,10	100	-
22-24/ 6/79	5	0,78	0,71 - 0,93	< 30	6
13-19/ 9/79	32	2,13	0,72 - 4,96	80	8
30/1- 1/2/80	12	3,12	1,74 - 4,65	100	17
10/3/80	7	2,76[b]	1,30 - 3,87	90	16-17
16/6/80	4	0,62	0,16 - 0,57	< 30	8

(a) Mean yield from all sampled quadrats.

(b) Approximately 20% wet weight corresponded to epiphytes.

Gracilaria sp. was found in abundance (> $0{,}3kg\text{-}1m^2$ wet weight) between 0m and 2,3 m referred to the HRL as a maximum. The observed mean density of the Gracilaria sp. biomass fluctuated between 0,62kg-1m^2 wet weight in June and 3,12 kg-1m^2 in January. Algal cover was lowest in June ($<$ 30%) and highest in January (100%). The temperature ranged from 6-8°C in June to 16-17°C in January-March. Practically no difference was observed between surface and bottom temperatures.

Discussion

The bay´s size (1000m x 1400 m) and the presence of a sandy-muddy-stony bottom in a greater area than the one actually occupied by the Gracilaria sp. could at least theoretically permit the existence of a greater bed than the one actually found, but the great tide differences in the area(2,5-6m) which cause daily currents might explain if too strong in those other areas this observed size limitation. Work is underway to evaluate this possibility.

The Gracilaria sp. bed size fluctuations are still not completely clear, and will however be better understood and studied using fixed buoys anchored to concrete blocks in the field limits and permanent transects following recent suggestions (4). Further work is also necessary to evaluate the possible importance of a free living stage of Gracilaria sp. also recently observed (5). However its permanent existence as an important stage of Gracilaria sp. life history in Bahía Arredondo seems doubtful due to the special oceanographic conditions prevailing in the area with strong daily tide differences and the corresponding currents as well as the direct action of the normally occuring autumn and winter storms which are responsible for the annual commercial harvest in this bay.

The observation of Gracilaria sp. in the depths indicated as well as the observed mean density per 1-m^2 seems to be well in agreement to other data in the literature (6,7,8,9). More work actually in progress is needed to evaluate the role of the different environmental parameters on the ecology of *Gracilaria sp.* in Bahía Arredondo.

Acknowledgement

Technical support from Engineer Folmer V.T.J. Fauring, the diving assistance from Mr. Pablo Ljunberg for the collection of samples, and help from many other people of Patagonia Comercial S.R.L. who directly or indirectly participated in this work is gratefully acknowledged. I am also grateful for the useful comments and ideas submitted by Dr. Dong Ho Kim (1977) and Dr. Jack McLachlan (1980) in their individual visits to Bahía Arredondo. Recommendations and cooperation by Lic. Alicia Boraso de Zaixso and Lic. Hector Zaixso are also much acknowledged.

References

1. Levring, T., Hoppe, H.A., Schmid, O.J.: Cram, de Gruyter and Co., Hamburg. 421 p. (1969).
2. Dr. Carmen Pujals, Museo Argentino de Ciencias Naturales "Bernardino Rivadavia". Buenos Aires. Argentina. Personal communication. (1980).
3. Halperin, D., Asensi, A., Boraso, A.: CIBIMA, Contr.Tec.14, 1-33 (1973).
4. Dr. Jack McLachlan, Atlantic Regional Laboratory, Halifax, Canada. Personal Communication. (1980).
5. Lic. Alicia Boraso de Zaixso. Centro Nacional Patagónico, Puerto Madryn, Chubut, Argentina. Personal Communication. (1980).
6. Kim, D.H. Botanica Marina XIII, 140-162 (1970).
7. Bird, C.J., Edelstein, T., McLachlan, J.: Naturaliste can., 104, 257-266 (1977).
8. Bird, N.: Proc. N.S. Inst. of Sci.: 27, 144-158 (1977).
9. Santelices, B., Fonck, E.: Actas I Symp. Algas Marinas Chilenas 165-200 (1979).

APPLICATIONS OF AGARCSE IN BIOMEDICAL TECHNIQUES IN RELATION TO PHYSICAL AND CHEMICAL PROPERTIES

K. Pluzek

The Protein Laboratory, University of Copenhagen,
34 Sigurdsgade, 2200 Copenhagen N, Denmark.

Introduction

Agarose is a linear polysaccharide consisting of alternating D-galactose and 3.6-anhydro-L-galactose units (1). It is purified from agar, which is separated into fractions with varying amounts of charged groups (sulphate, carboxyl) (2,3). Agarose is the fraction with lowest content of charges, and has now replaced the raw agar, and other anticonvection mediums, in many biomedical techniques, due to its special properties. Thus it is used as a porous stabilizing medium in analytical and preparative methods for investigating or separating molecules, (for example as a thin gel layer for electrophoresis, immunochemical, immunological and microbiological methods, or as beaded grains for chromatographic methods). Agarose is commercially available with well-defined properties (e.g. Litex, Marine Colloids Division, L'industrie Biologique Française, Pharmacia Fine Chemicals) as a non-toxic material.

Agarose properties

Controlled gelling temperature. Agaroses are available with defined low gelling temperatures (35-42°C) and with a rapid and uniform gel formation (4). The gelling is reversible, but once formed, the gel is stable up to 90°C. This prevents melting of the gel during electrophoresis, where the temperature

Xth International Seaweed Symposium
Printed in Germany

may increase due to insufficient cooling. Normally the gel must be boiled, before it is melted completely. The low gelling temperature allows incorporation of heat labile products without denaturation (e.g. antigens, antibodies, enzymes, ampholytes, cells). A special agarose (Sea Plaque, Marine Colloids) has been produced for use in the Jerne plaquing assay, where functions of cells are studied (5,6). The gelling temperature is below 30°C, and the agarose will remain fluid at 37°C for at least 24 hours, which simplifies the procedures, and ensures greater cell viability. This agarose remelts, when the temperature exceeds 65°C.

Large pore size. If the concentration of agarose is varied, gels are formed with different pore sizes. Agarose types are available to form a stable gel down to about 0.15%, due to the high gel strength. In plaquing techniques (5-7) and migration inhibition techniques (8), concentrations about 0.5-2.0% are used. Normally a 1% agarose is used for electrophoresis (9), immunoelectrophoresis (10-13) and immunodiffusion methods (14,15). The large pores allow a rapid penetration of macromolecules ($MW<20 \cdot 10^6$ daltons) through the pores due to diffusion or electric forces without sieving effect. The molecules have space to form macromolecular reaction products in the gel. Also the pores permit small immunocomplexes to move in the gel. Because of the open gel structure, agarose is preferred to for example polyacrylamide in assays, where especially macromolecules are investigated (16). Also the open gel structure makes a rapid fixation, staining and destaining procedure or buffershift possible.

For chromatographic techniques agarose is prepared as beaded grains (17,18). This form, combined with the open structure, improves the chromatographic procedure (flow of liquid through the medium). Varying the concentration of agarose (e.g. up to 6%), different types of gel filtration mediums are obtained to allow molecules up to $150 \cdot 10^6$ daltons to be fractionated

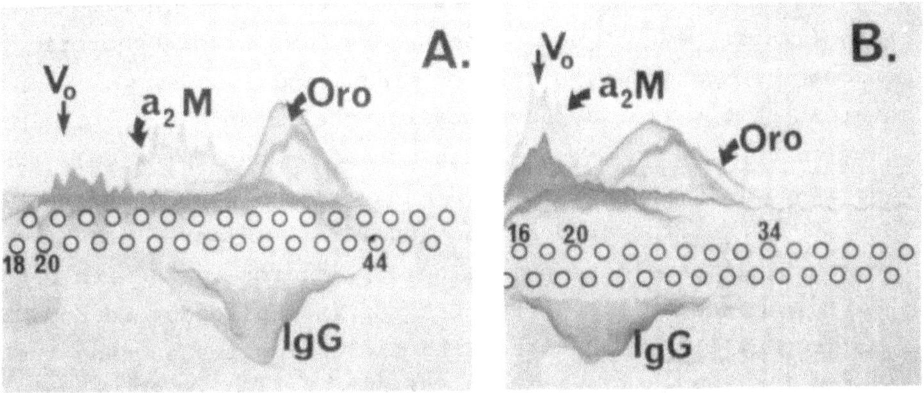

Fig. 1. Fused rocket immunoelectrophoresis (11)(in 1% agarose, Litex HSA) of human serum protein fractions from gel filtrations (17-19), where agarose medium A (Sepharose CL-6B) or dextran medium B (Sephadex G-200) have been used. Alpha-macroglobulin (MW $9 \cdot 10^5$ daltons) is fractionated in Sepharose, indicating a penetration through the pores, but are eluted with the void volume (V_o) in Sephadex.

according to size (19). Therefore agarose is one of the mediums of choice, when macromolecules or fragments of cells are separated (Fig. 1).

High gel strength. Even in low concentrations agarose exhibits high gel strength. This makes the agarose easy to handle. For example many immunoelectrophoretic methods involve transfer of gel slabs (1.5 mm thick) from one plate to another (11), or agarose gels are used for gel bridges (9). The high gel strength combined with the open gel structure allows a successive pressing and swelling of the gel, thereby removing unreacted soluble proteins (11). The chromatographic agarose matrix can stand a high gel pressure (high liquid flow) compared to other mediums (e.g. dextran (17)).

Controlled amount of charged groups. The different types of agaroses contain different amounts of negatively charged groups. These may influence the assays. In electrophoretic methods the charges give rise to electroendosmosis, which is movement of the liquid phase towards the cathode (20). In immunoelectrophoretic methods with antibody-containing gel, agarose is normally chosen to give an average migration of nearly zero of the incorporated antibodies (10-13)(Figs. 2 and 3). For counterimmunoelectrophoresis (21), agarose gels with high electroendosmosis (e.g. Litex LSA, Marine Colloids SeaKem HE) is preferentially used, directing the antibodies towards the cathode and the antigen. Isoelectric focusing (16,22) requires an agarose with negligible electroendosmosis (e.g. Litex HSIF, Marine Colloids Isogel, Pharmacia Agarose IEF), in order to avoid a drift of the pH gradient during the electrofocusing. Generally the relatively low amount of anions in agarose gives rise to only minimal adsorption effects, but basic proteins (e.g. C-reactive protein) may bind to the agarose or cause trailing in electrophoresis (9). Agaroses are now available

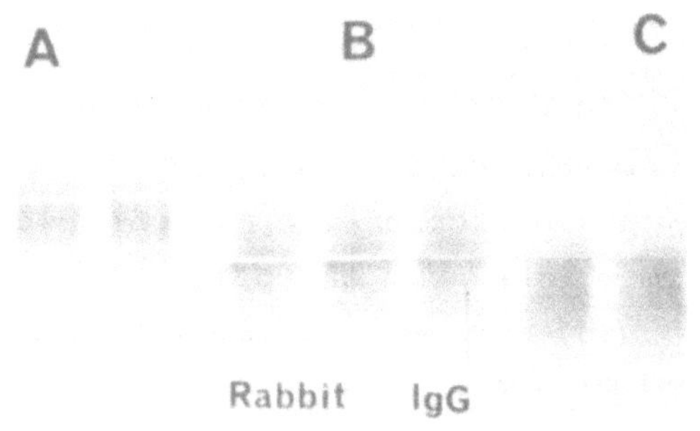

Fig.2. Electrophoresis(9) of rabbit antihuman albumin, DAKO 10-001, in 1% Agarose, Litex HSIF Mr-0.00(A), Litex HSA Mr-0.13 (B), Litex LSA Mr-0.25 (C), at buffer pH 8.6. Agarose with electroendosmosis about Mr-0.13 gives an average migration of nearly zero of the antibodies.

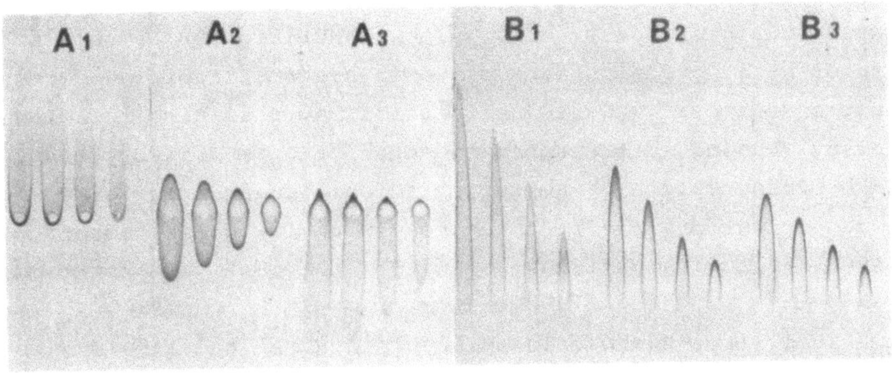

Fig. 3. Rocket immunoelectrophoresis (10,11) in 1% agarose, Litex HSIF Mr-0.00 (A1+B1), Litex HSA Mr-0.13 (A2+B2), Litex LSA Mr-0.25 (A3+B3). Human serum run against anti-human IgG DAKO 10-090 (A1-3), and anti-human albumin DAKO 10-001 (B1-3). Normally agarose with electroendosmosis about Mr-0.13 is chosen for the assay, when using a buffer pH 8.6.

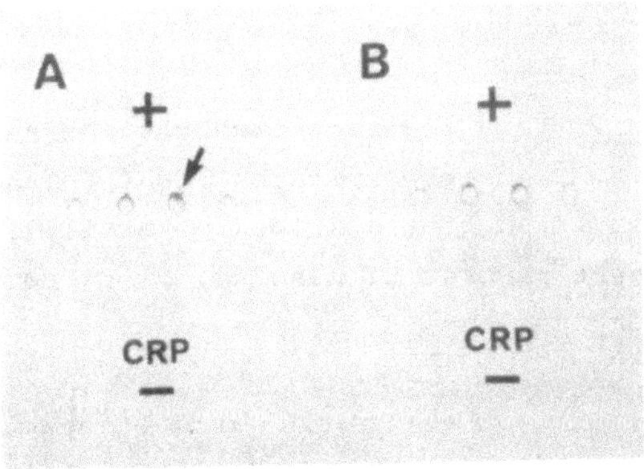

Fig. 4. Rocket immunoelectrophoresis (10,11) in 1% agarose. Litex HSHE Mr-0.35 (A), Litex LSHE Mr-0.4 (B). C-reactive protein (CRP) rich human plasma is run against anti-human CRP DAKO A 073.
Agarose A adsorbs CRP due to high sulphate amount (arrow), whereas agarose B shows reduced adsorption effect and measurable rockets, due to reduced sulphate amounts, but the same electroendosmosis.

with a high electroendosmosis,but with low sulphate amounts (Marine Colloids SeaKem HEEO, Litex LSHE)(Fig. 4). Thus adsorption of basic proteins is avoided in systems requiring high electroendosmosis. Basic antibiotics may also bind to anions, thereby causing interference in antibiotic sensitivity tests (23). Furthermore high amount of anionic charges inhibits many plaque formations (5-7). An agarose with low anion amounts may prevent such disadvantages.

High stability. Agarose is stable in neutral and alkaline mediums, but hydrolysis takes place below pH 5. However, assays in agarose can be performed at lower pH, using modifications of the procedure (24). Agarose is insoluble in nearly all organic solvents (exceptions e.g. dimethylsulphoxide). It may be degraded by strong oxidizing agents and forms complexes with borate salts. The physical and chemical stability of the beaded agarose can be improved by cross-linking the agarose chains (18).

High transparency. Many reaction products can be examined in the wet gel (e.g. immunoprecipitates), due to the transparency of the gel. When dried, the agarose is totally transparent and provides permanent records of results. It is optically clear for scanning purposes.

Derivatives. Derivatives of the hydroxy groups in the agarose can easily be prepared with different chemical groups, for example for use in chromatography. Important examples are ion exchange,hydrophobic and affinity mediums (19). The open structure of agarose makes the interior of the matrix available for ligand attachment and ensures good binding capacities.

Mixing and combination with other gels. In analytical techniques agarose in gel form has been mixed with derivatives of beaded agarose(25)(Fig.5), and combined with polyacrylamide

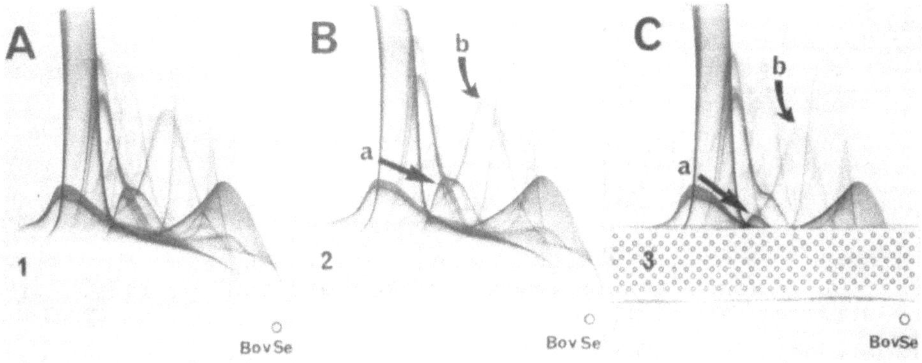

Fig. 5. Crossed immuno-affinoelectrophoresis (11,25) (in 1% agarose, Litex HSA) of bovine serum proteins run against anti-bovine serum proteins, DAKO 132 SX.
(A) intermediate gel 1, 1% agarose.
(B) intermediate gel 2, 1% agarose + 0.4% Sepharose 4 B.
(C) intermediate gel 3, 1% agarose + 0.4% Con A Sepharose 4 B.
No change can be seen of the pattern after incorporation of beaded agarose into the intermediate gel (B compared to A). Changes are first seen, when proteins (a and b) react by a specific interaction with the immobilized concanavalin A (C compared to B).

gel (26). For chromatography agarose has been mixed with polyacrylamide in different concentrations (19). Thus the properties of both gels are utilized.

References

1. Araki, C.: in Proceedings of the 4th International Congress of Biochemistry (M.L. Wolfram, ed.) Vol.I, pp. 15-30,1959.
2. Duckworth, M., Yaphe, W.: Carbohydr.Res. 16, 189-197 (1971).
3. Hjertén, S.: J.Chromatogr. 61, 73-80 (1971).
4. Arnott, S., Fulmer, A., Scott, W.E., Dea, I.C.M., Moorhouse R., Rees, D.A.: J.Mol.Biol. 90, 269-284 (1974).
5. Jerne, N.K., Henry, C., Nordin, A.A., Fuji, H., Koros, M.C. Lefkovits, I.: Transplant.Rev. 18, 130-191 (1974).
6. Kaliss, N.: Transplantation 12 (2), 146-147 (1971).

7. Borden, E.C., Gary, G.W., Murphy, F.A.: Appl.Microbiol. 20 (2), 289-291 (1970).

8. Clausen, J.E.: The agarose migration inhibition technique for in vitro demonstration of cell-mediated immunity in man (Thesis) Copenhagen, 1975.

9. Johansson, B.G.: Scand.J.Clin.Lab.Invest. 29, Suppl. 124, 7-19 (1972).

10. Laurell, C.-B.: Scand.J.Clin.Lab.Invest. 29, Suppl. 124, 21-37 (1972).

11. Axelsen, N.H., Krøll, J., Weeke, B.: A manual of quantitative immunoelectrophoresis. Methods and applications. Universitetsforlaget, Oslo, 1973.

12. Axelsen, N.H.: Quantitative immunoelectrophoresis. New developments and applications. Universitetsforlaget, Oslo, 1975.

13. Axelsen, N.H.: Handbook of immunoprecipitation-in-gel techniques. Scand.J.Immunol., Suppl. 10, in press.

14. Ouchterlony, Ö.: in Progress in Allergy (P.Kallós, ed.), Vol V, pp. 1-78, Basel, New York, 1958.

15. Mancini, G., Vaerman, J.-P., Carbonara, A.O., Heremans, J.F.: in Prot.biol.Fluids 11th Coll. Bruges, (Peeters, H., ed.), pp. 370-373, 1964.

16. Rosén, A., Ek, K., Åman, P.: J.Immunol.Methods 28, 1-11 (1979).

17. Hjertén, S.: Biochim.Biophys.Acta 79, 393-398 (1964).

18. Porath, J., Låås, T., Janson, J.-C.: J.Chromatogr. 103, 49-62 (1975).

19. Lowe, C.R.: in Laboratory Techniques in Biochemistry and Molecular Biology (T.S.Work, E.Work, eds.), pp. 274-518, North-Holland, Amsterdam, 1979.

20. Morris, C.J.O.R., Morris, P.: Separation methods in biochemistry, Sir Isaac Pitman and Sons, London, 1963.

21. Hierholzer, J.C.: J.Immunol.Methods 11, 63-76 (1976).

22. Saravis, C.A., Zamcheck, N.: J.Immunol.Methods 29, 91-100 (1979).

23. Hanus, F.J., Sands, J.G., Bennett, E.O.: Appl.Microbiol. 15 (1), 31-34 (1967).

24. Løwenstein, H., Ingild, A.: Anal.Biochem. 71, 204-208 (1976).

25. Bøg-Hansen, T.C.: Anal.Biochem. 56, 480-488 (1973).

26. Lundahl, P., Liljas, L.: Anal.Biochem. 65, 50-59 (1975).

THE RELATIONSHIP BETWEEN ANNUAL LANDINGS OF CHONDRUS DRAGRAKERS, EFFORT, AND STANDING CROP IN THE SOUTHERN GULF OF ST. LAWRENCE

J.D. Pringle

Marine Plants Section, Resource Branch, Fisheries and Oceans Canada, P.O. Box 550, Halifax, Nova Scotia, Canada B3J 2S7

Introduction

In Atlantic Canada a traditional inshore fishery is Irish moss (*Chondrus crispus* Stackh.) harvesting, which began during the last world war. Historical data typically employed to manage fisheries resources are lacking for *Chondrus*; the only long-term data available are total annual landings, along with price paid per kg to the harvesters (Statistics Branch, Fisheries and Oceans Canada). Harvester effort (number of boats harvesting da^{-1}) and CPUE data are available for the years 1975 through 1979 in Marine Plant Harvesting District 1 (for geographical location of districts, see reference 1). An attempt is made, using the available data, to interpret the system in this district.

Materials and Methods

The annual landings for each year from 1947 through 1979 are presented in Figure 1. An average harvest for each of three distinct periods (based on size of crop) 1947 to 1965 (Period I), 1966 to 1971 (Period II), and 1972 to 1979 (Period III), was calculated (Fig. 1). Approximately three times per week the number of boats (harvesters) actively

Xth International Seaweed Symposium
Printed in Germany

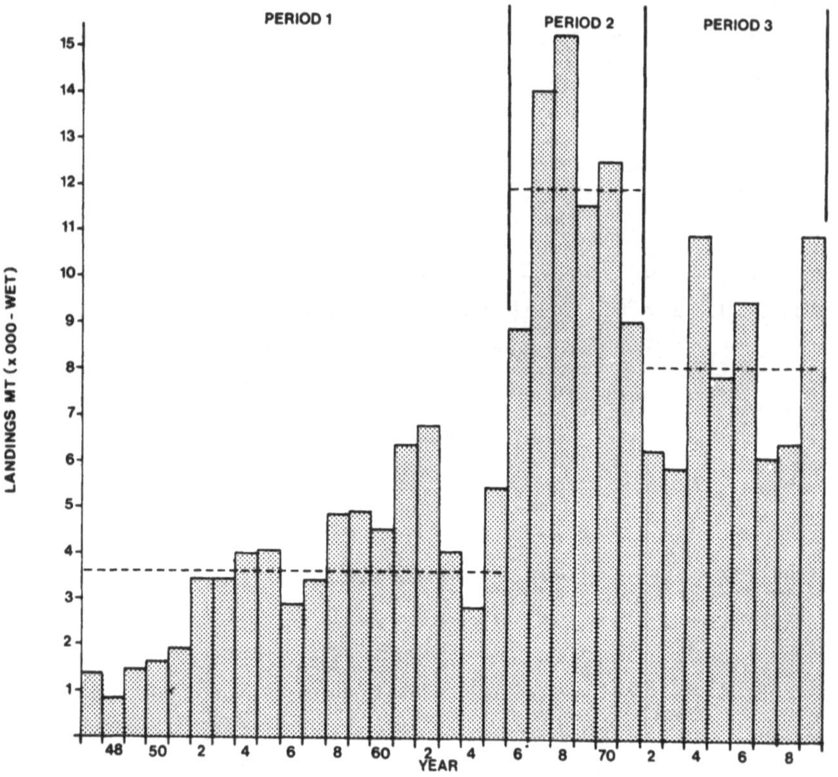

Fig. 1. District 1 annual Chondrus landings. Broken lines represent the mean annual landing for each period.

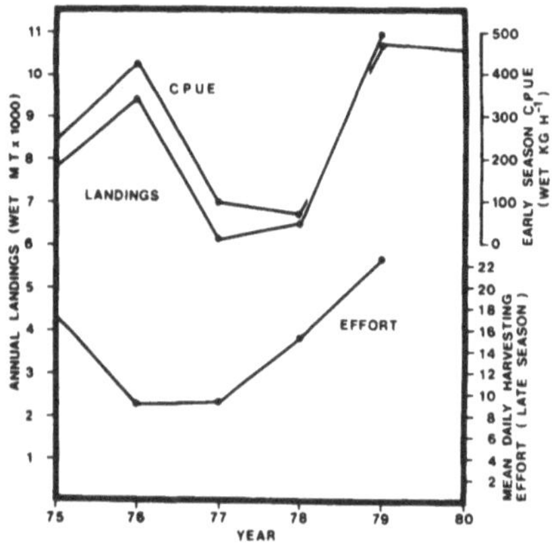

Fig. 2. Annual landings, CPUE, and effort between 1975 and 1980.

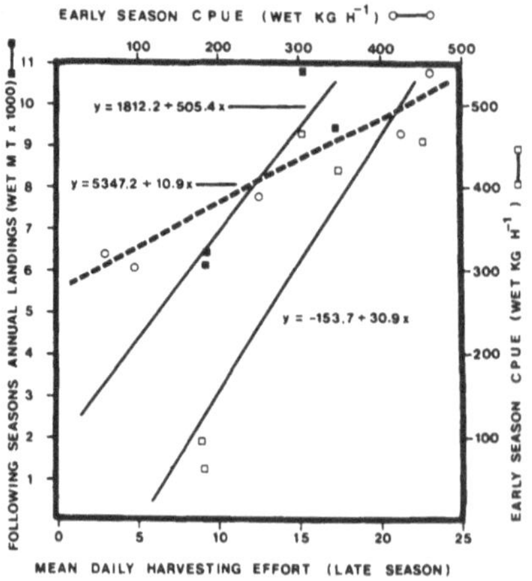

Fig. 3. Following season's annual landings and early-season CPUE as a function of late-season mean daily harvesting effort.

harvesting each bed was determined through direct observation (Table 1). The mean hourly harvest for the first three days of each season was determined in 1976, 1977, and 1979

The CPUE in 1977 and 1978 was derived from an experimental, preseason harvesting program. The regression analyses and correlation coefficients (Fig. 3 and Table 2) were done on a Monroe 325 Scientist computer (Monroe Litton, U.S.A.) with packaged programs (Alpha 325 Regression Analysis Pak).

Results and Discussion

According to anecdotal evidence, there was a steady but low demand for Chondrus between 1947 and 1965 (Fig. 1). Demand increased markedly in 1966 due to an increase in number of buyers. During Period I it is likely the harvest was below maximum sustainable levels; during Period II the mean annual harvest increased by a factor of 3.3. During Period III (Fig. 1) the mean landings decreased by a factor of 1.5 in spite of high demand. The cause of this decrease is of concern; but unfortunately data such as CPUE, effort, standing crop, etc. that would assist in interpreting the system are lacking.

Chondrus data were used in the Schaeffer Model (see reference 2 - an equilibrium bioeconomic model) (3)*. There was scatter around supply curves drawn which contradict the equilibrium concept. However, two tentative conclusions were made: 1) that MSY was between 12 000 and 14 000 MT, and 2) that because present harvests are on the upper limit of the supply curve the resource is being overharvested. Possibly by exploring that portion of Period III, where more data are available, these conclusions can be verified.

*Under contract to the Marine Plants Section, Fisheries and Oceans Canada.

Table 1. The mean number of harvesters per day per month in District 1 for the years 1975 to 1979.

Month	Mean number of Harvesters						Total	Grand Mean (Std. Dev.)
	1975	1976	1977	1978	1979	1980		
June	74.8	93.7	58.8	45.3	60.0	65.6	398.2	66.4 ± 16.5
July	84.1	38.0	45.0	33.0	54.7	-	254.8	51.0 ± 20.3
August	17.3	9.2	9.3	15.4	22.8	-	74.0	14.8 ± 5.7
Mean	58.7	46.9	37.7	31.2	45.8			45.4

Table 2. Statistical treatment of Chondrus harvesting data for Marine Plant Harvesting District 1.

Parameters	Linear regression	Correlation coefficient (r)	Standard error of estimate (S_{yx})
Late-season effort vs Following season's total harvest	y = 1812.2 + 505.4x	0.90	1253.0
Late-season effort vs Following early-season CPUE	y = -153.7 + 30.9x	0.88	111.1
Early-season CPUE vs Season's total harvest	y = 5347.2 + 10.9x	0.98	500.7
Early-season CPUE vs Early-season effort	y = 51.9 + 0.05x	0.54	15.6
Late-season effort vs Following season's early effort	y = 40.4 + 1.6x	0.52	17.5
Previous season's total harvest vs Following season's early effort	y = 45.0 + 0.0024x	0.32	17.5
Season's mean daily effort vs Season's total harvest	y = 4218.9 + 90.4x	0.46	2087.7
Price paid kg^{1} vs Season's total landings	y = 5.1 + 0.00008x	0.19	1.5

A fisherman's expected revenue is proportional to the product of the resource abundance times the price minus operating costs.

Price paid harvesters per unit of harvest does not correlate (r = 0.19) with total landings for Periods II and III; the regression was also non linear (Table 2). The correlation coefficient (r) was 0.46 between mean daily effort expended per season and total landings for the years 1975 to 1979. These suggest that price and effort fluctuations are not directly responsible for the fluctuations in annual landings. The factor remaining that likely accounts for the fluctuations in annual landings is resource abundance or catchability - see reference 4).

Change in abundance will be the difference between biomass entering and leaving the system. Accurate biomass data are not available but CPUE is an accepted index of abundance (4). Effort may also be used as an index of abundance if the following assumptions are made: 1) fishermen are opportunistic and have a number of alternative sources of income, and 2) there is a large number of potential "mossers." As these assumptions are satisfied, the following hypothesis can be made: the number of active boats reflects the expected revenue. For simplicity, it is assumed the number of active boats are linearly related to expected revenue. The lobster fishery begins August 10; consequently, *Chondrus* effort is reduced markedly by August 1 if not before, depending on revenue. For example, in 1976 the effort between June and July decreased by 59.4%; in 1979 it decreased by only 8.8% during the same period (Table 1). The assumption is thus made that the effort then, expended late in the season, is likely a function of abundance. Substantiation of this is the 1978 season where the overall landings were low but late-season effort was high compared to the preceding two years. Harvesters noted (pers. comm., R. Doucette) that CPUE had improved markedly through the season. Thus, late-season effort was plotted against total landings (Fig. 2). It can be seen that if late-season effort was high the following season's early-season CPUE and total landings were high. The regressions are linear (Fig. 3) and the correlation coefficients lie between $r = 0.88$ and 0.98 (Table 2). It appears the marked increase in 1979 landings over 1978 (69.5%) was due to an increase in abundance, not harvester effort. It may then be concluded that *in situ*, intrinsic growth rates of *Chondrus* are highly variable from season to season, and that it can be considerably higher than previously suggested (5).

Given that the population studied is heavily harvested, it is suggested that the causal factor(s) is density independent and likely operating in the late spring/early summer, influencing biomass in August and setting the "physiological stage" for a large biomass increment during the fall and winter as in Laminaria longicruris (6). Chondrus growth data verify this model. Outplants of Chondrus set out in 1975 grew markedly

The information gained from this assessment suggests that the annual landings increased/decreased in Periods II/III because of levels of standing crop. The high abundance during Period II was either due to good growing conditions or the harvesting of virgin beds, or both. The decrease in standing crop during Period III was due to poor growing conditions or overharvesting. Only further studies will confirm these hypotheses.

References

1. Pringle, J.D., Sharp, G.J.: Helgolander wiss. Merres. 33, 220-229 (in press).
2. Clark, C.W.: Mathematical bioeconomics, the optimal Management of renewable resources, John Wiley (1976).
3. Loucks, R.H.: Preliminary modelling of the dynamics of some Prince Edward Island seaweed populations. R.H. Loucks Oceanology Ltd. (1980).
4. Ricker, W.E.: Computation and interpretation of biological statistics of fish populations. Bull. 191, Env. Canada, Fish. Mar. Serv. (1975).
5. Taylor, A.R.A.: Proceedings, meeting on the Canadian Atlantic seaweeds industry. Env. Canada, 29-36 (1971).
6. Chapman, A.R.O., Craigie, J.S.: Mar. Biol. 40, 197-205.

CHILEAN SEAWEED RESOURCES: A QUANTITATIVE REVIEW OF POTENTIAL AND PRESENT UTILIZATION

B. Santelices

Departamento de Biología Ambiental y de Poblaciones, Instituto de Ciencias Biológicas, Pontificia Universidad Católica de Chile, Casilla 114-D, Santiago, Chile.

J. Lopehandía

Servicio Nacional de Pesca; Ministerio de Economía, Fomento y Reconstrucción; Santiago, Chile.

Introduction

Seaweed production in Chile has shown an overall increasing trend over the last decade which has stimulated cultivation efforts of economically important species and forced to new quantitative assesments of seaweed resources. This report provides an up to date review of chilean seaweed potential, their present utilization and major problems forseen in their cultivation.

Present Knowledge on Chilean Seaweed Resources

In Chile, seaweeds are exported as raw material, used in domestic agar and alginate industries and, to a lesser extent, consumed as a food supplement. Over the last decade, seaweed production has significantly increased (Fig. 1). Annual seaweed exports rose from 1,800 tons in 1969 to over 17,000 tons in 1979 (1)(2). Domestic agar production went up in the same period from 91 to 387 tons. Exports of chilean seaweed and seaweed-derived products were worth over US$ 17 million in 1979 (2). As seaweed extraction increased in the country, several government agencies became aware of the necessity to formulate adequate conservation regulation. Since 1976,

Xth International Seaweed Symposium
Printed in Germany

legislation requests among other requisites, that all persons who wish to harvest seaweeds must obtain prior authorization for the specific area and time interval of the year. The authorization is dependent upon a quantitative prospection of the bed which determines the algal quantity that can be safely extracted without endangering the seaweed bed. This regulation together with research efforts by various government agencies and local universities have produced close to 60 unpublished quantitative evaluations of different algal beds in the country. Taken all together, they include six of the ten most economically important chilean algal groups. An analysis of these results should provide a more realistic approximation to the chilean seaweed potential. Recent views on the subject (3)(4) are derived mostly from qualitative estimations (5)(6)(7).

Methods

Data on quantitative assessments of algal beds are taken from reports produced mainly by Instituto de Fomento Pesquero (IFOP, Chile) and a few private consulting companies and universities. The studies normally include determination of the algal bed limits and standing stock measurements in a number of stations along one or several transects. In some cases, length and wet weight of individuals as well as the wet/dry weight ratio of the species studied were also established.

Results

Quantitative measurements of the five economically most important chilean seaweed groups are shown in Fig. 2. Their relative importance as export material and destination is shown in Table 1. *Gracilaria* spp. are the most important seaweeds from the economic stand point. So far, 17 quantitative evaluations of these species have been carried out at 14 commercially important beds between 30°S and 41°S. Data indicate total standing stock values of 73,000 to 83,000

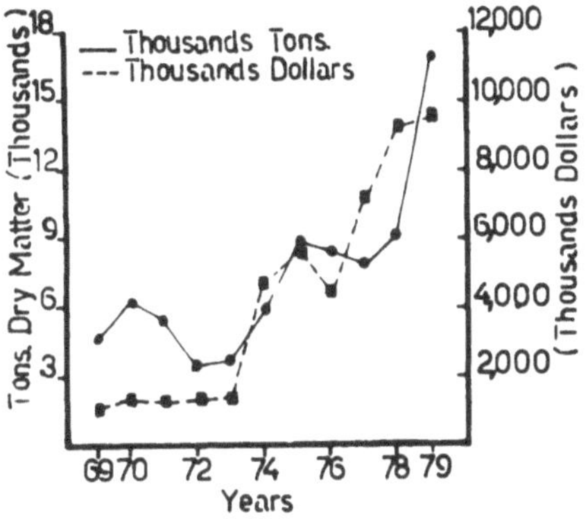

Fig. 1.- Trend of chilean seaweed export (1969 - 1979).

Table 1.- Relative importance and destination of different algal groups exported from Chile in 1979.

Country	Gelidium	Gracilaria	Iridaea	Lessonia
Canada			380.2	500.0
China		25.0		
Denmark		91.6	86.1	
EE.UU.	25.0	29.5	2,043.7	2,949.0
Scotland				499.9
Spain		475.1		
France		837.5	198.7	
Japan	93.8	6,250.1		2,079.5
Portugal		380.0		
Total	118.8	8,088.8	2,708.7	6,028.4

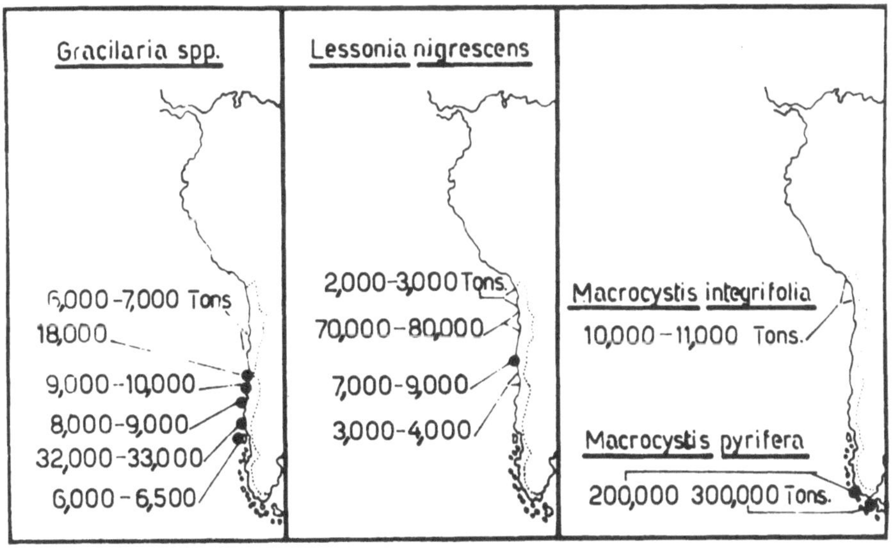

Fig. 2.- Estimated seaweed potential of different economically important algal species in Chile.

tons (wet matter). Field (wet) stock values normally range between 3 and 7 Kg/m^2. Total annual exports of chilean Gracilaria in 1979 were close to 80% the estimated total stock. Only material washed ashore is normally collected although ongoing research (8)(9) is showing that summer standing stock increases up to 8 times the winter values. Summer production in at least two of the most important beds in the country (Coquimbo and Isla Santa María) could be increased 3 to 5 times by adequate harvesting policies. Efforts are been conducted to farm species of the genus in northern Chile but so far none has been successful on a commercial scale. Although 5 species of the genus have been reported for Chile (7),indications are (10) that at least three species require further taxonomic and population biology research.

The species of Iridaea are the second most important commercial group among chilean Rhodophyta. The only survey of these species in Chile was done in 1978 around Quellón (in Chiloé). Stock values of 23.10 Kg/m^2 were reported. In 1979 close to 2,700 dry tons of Iridaea spp. were exported to the United States, Canada, Denmark and France. Most of the material was illegaly harvested by hand from unsurveyed areas. Among the economically important species of chilean algae, this group is the least studied. Even the taxonomy of the group is in need of critical analysis (10)(11).

Close to 108 tons of Gelidium spp. were exported from Chile during 1979; 94 tons were sent to Japan, while the remainer went to the USA. As in the case of Iridaea, no previous qualitative survey of Gelidium spp. have ever been performed in the country. Field (wet) stock values of Gelidium spp. range between 20 to 100 gr/m^2 in central Chile (12).

Species of Ahnfeltia are not being exported at present but a standing stock survey performed at Seno Skyring, in the proximity of Magellan Strait, reported values of 0.73 Kg/m^2.

The bulk of brown algae biomass being exported from Chile corresponds to the intertidal Lessonia nigrescens. About 6,000 tons of dry seaweeds were exported from Chile in 1979. Quantitative studies in Chile from Arica (17°S) to Caldera (28°S) yield total field stocks of 82,000 to 96,000 tons. This amount is equivalent to 2-3 years export at the present rate. Biological knowledge is necessary for the management of the species as it plays an important ecological role as a foundation species (13) in intertidal communities of most of temperate Pacific South America (14).

Two species of *Macrocystis* are reported to occur in Chile (15). Measurements of *M. integrifolia* stocks have been carried out between Arica (17°S) and Iquique (22°S) with average stock values of 5 Kg/m^2, totalling 10,000 - 11,000 tons in the 12 beds measured. Values of *M. pyrifera* have been measured in the southern tip of Chile with stock values found of 13 - 17 Kg/m^2. About 200,000 tons of dry *M. pyrifera* in Magellan Strait and close to 300,000 tons in the Beagle area have been estimated. Studies on growth rates of these beds after harvesting are just starting.

Several other chilean algal groups are reputed to be of some economic importance (present or potential) but only fragmentary quantitative information has been gathered so far.

Discussion and Conclusions

Quantitative estimates of chilean seaweed resources have accumulated to the extent that some major patterns are now detectable. Quantitative assessments of the three economically most important algal groups (*Gracilaria* spp.; *L. nigrescens* and *Macrocystis* spp.) have covered extensive parts of the chilean coastline. Future research will probably point to the formulation of biologically oriented resource management programs for these species and to further more accurate and complete assessments of other species.

Seaweed farming and mass cultivation in enclosed systems in Chile faces several problems. Although some effort has been devoted to farming of *Gracilaria*, the steep slope of the chilean coastline, the intense water movement affecting most of the highly productive coast line of northern and central Chile (high radiant energy and presence of upwelling areas), the absence of species field biological knowledge and the limited budget so far devoted to such studies have rendered many such efforts unsuccessful. Mass cultivation in enclosed systems is limited by the level of technological development in the country and the limited knowledge on production ecology of the most economically important species. In view of increasing seaweed utilization, resource management programs for these species is an urgent need in Chile.

Acknowledgments: Financial support from International Foundation for Science, Grant G 112, and OEA, Program PRDCYT, 1979, Grant C 15 is acknowledged with gratitude.

References

1. Joyce, L., Santelices, B.: Biol. Pesquera, Chile 10, 3-26 (1978).
2. Anonymous: Anuario Estadístico de Pesca. Servicio Nacional de Pesca, Santiago, Chile, 90 pp. (1979).
3. Michanek, G.: FAO Fisheries Tech. Paper 138, v + 127 pp. (1975).
4. Naylor, J.: FAO Fisheries Tech. Paper 159, v + 73 pp. (1976).
5. Llaña, A.: Rev. Biol. Mar. 1, 124-131 (1948).
6. Etcheverry, H.: Proc. Int. Seaweed Symp. 1, 108 (1953).
7. Kim, D. H.: Bot. marina 8, 140-162 (1970).
8. Santelices, B., Fonck, E.: Actas Symp. Algas Mar. Chilenas 1, 165-200 (1979).
9. Fonck, E., Santelices, B.: Biol. Pesquera, Chile (In press).
10. Abbott, I.: Actas Symp. Algas Marinas Chilenas 1, 51-58 (1979).
11. Levring, T.: Lunds Univ. Arsskr. N. F. Avd 2, 56, 1-83 (1960).
12. Santelices, B., Oliger, P., Montalva, S.: Proc. Int. Seaweed Symp. 10, (In press).
13. Dayton, P.: Ecological monogr. 45, 137-159 (1975).
14. Cancino, J., Santelices, B.: Proc. Int. Seaweed Symp. 10, (In press).
15. Neushul, M.: Nova Hedwigia 32, 211-222 (1971).

SUBCOMMERCIAL HARVESTING OF A KELP ON A SOUTH AFRICAN SHORE

R. H. Simons and N. G. Jarman

Sea Fisheries Institute, Department of Agriculture and Fisheries,
Private Bag, Sea Point, 8060, Cape Town, South Africa

Introduction

For the first time, direct exploitation of the standing stock of large South African kelps is being allowed. This concession, granted by the State for the processing of fresh weed into an agricultural fertilizer, has given us the opportunity to further assess the exploitability of local kelps. An appropriate strategy, started in September 1979, was devized and named CHEKS-1 (Commercial Harvesting of Kelp - Experiment Number One). A cutting-wire or bar was envisaged for cropping.

Here, we account for our proposals and cite the main consequences.

Materials and their Siting

Ecklonia maxima (Fig. 1), a member of the Alariaceae (Laminariales), is the subject. Its sporophyte, with trunk-like stipe, as much as 8 m long

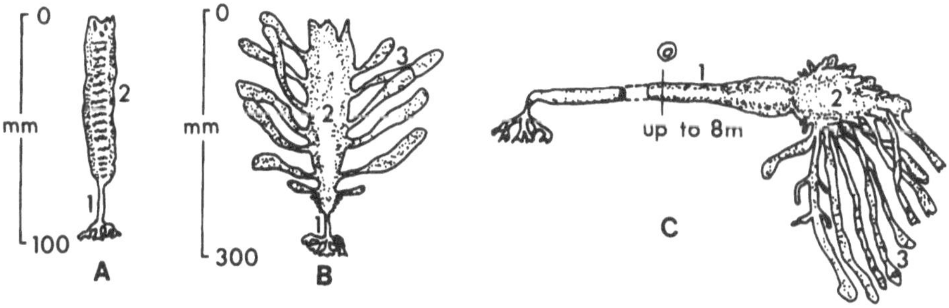

Fig. 1. Ecklonia maxima: sporophyte: A. infant; B. juvenile; C. adult: 1. stipe; 2. primary blade; 3. secondary and tertiary blades: up to 1,5 m long and 150 mm wide - blades on upper margin missing.

Xth International Seaweed Symposium
Printed in Germany

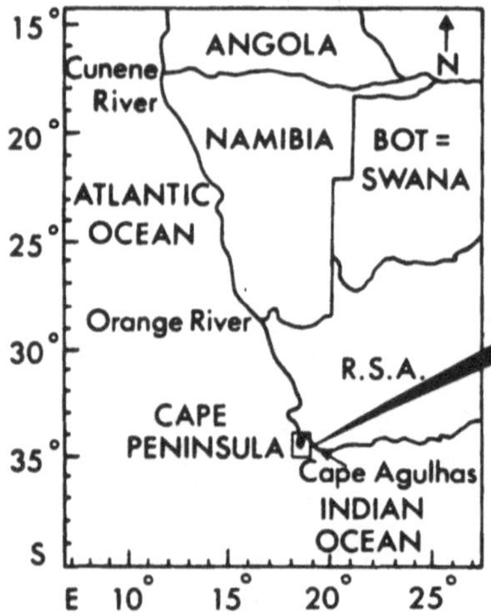

Fig. 2. Southern Africa: South-western segment

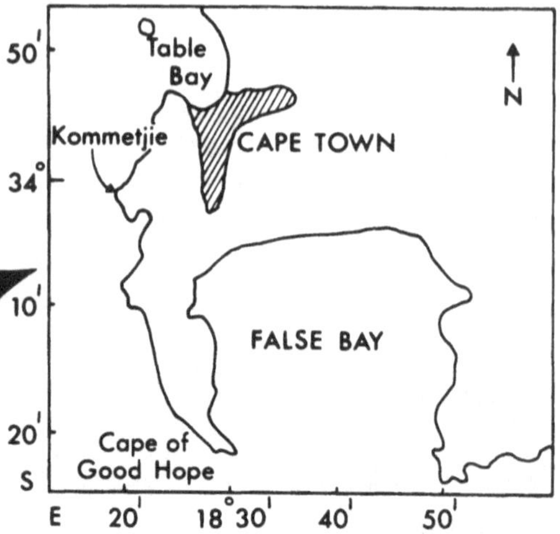

Fig. 3. Southern Africa: Cape Peninsula and vicinity

when mature, and spade-like primary blade with numerous laterals, is locally abundant between Cape Agulhas, the southern-most point of Africa, and somewhere on the Namibian coast - in all about 1200 km of shore (Fig. 2). Flourishing in the immediate sublittoral where surf runs strongly, it is ousted elsewhere by other Laminariales: Laminaria pallida on the Cape Peninsula (Fig. 3) and eastwards; L. schinzii along most of the west coast. Ecklonia, at least in the southern regions, dominates its niche and its biomass equals that of the less bulky, but commoner, L. pallida. The apex of the longest stipes (c. 8 m) of attached E. maxima reaches the surface at mean tide-level, and, though water-depth largely dictates ultimate stipe-length, plants with stipes of 1 or 2 m grow in the sublittoral fringe; the basal articulation of the buoyant, hollow stipe allows the fronds to remain submerged even when the tide is out (tidal range is at most c. 1,7 m). A steeply sloping shore supports a narrow belt of Ecklonia, whereas the terraced floor at Kommetjie has better coverage (1). All mature ecklonias are visible at the surface at low tides, making easy an aerial survey of this species.

Methods and Data for the Development of Harvesting Strategy

Resource inventory. The kelp stock at various localities on the Cape coast (Fig. 2) has been assessed through in-water studies and aerial photographs. SCUBA-equipped divers, having worked several 1 m wide transects through the kelp, acknowledge six types of kelp-community, each having a particular water-depth, kelp-biomass, species-composition, range of stipe-lengths and density. Onto colour aerial-survey photographs (1:12000) of Kommetjie, in particular, these community-types were impressed by overlaying an appropriate micro-dot grid (2. p.186) and referring simultaneously to a nautical chart of the area. Transposed to a map, these impressions allowed us to assess total biomass and coverage: standing
crop (i.e. excluding holdfasts) of *Ecklonia* was 33000 tons (wet- (I)
weight), and, of all kelps was 66000 tons; total area: of kelp- (II)
field, 700 ha; of *Ecklonia*, 250 ha; of *Laminaria*, 450 ha. (III) (IV) & (V)

Productivity. Mann et al. (3) assessed that fronds of *Ecklonia maxima* increase their biomass by 1% per day, or, produce about seven times their own weight in a year; total standing-crop turns over about twice a year.

Growth. Stipe-elongation is relatively slow, and, before this experiment began, *Ecklonia's* stipe-growth was unknown so we used data obtained from *L. pallida* (4) that unlike *L. digitata* (5), has no growth-rings. Dieckmann
deduced a growth-pattern expressed as $L=235(0{,}008)^{0{,}348^{A}}$, where L (VI)
is length (cm) and A is age (years), for 82 plants attached at -8 m; most rapid elongation takes place in the second year, though slightly greater biomass-change occurs in the third year from sporophyte-establishment; optimum length of about 2 m is attained in three or four years.

Reproduction and mortality. Because they float and because they grow in the surf-zone, ecklonias, when dislodged, often come ashore. From past records, we know that about 400 tons (air-dry) were annually taken from the beach at Kommetjie. About 90% of this (i.e. 360 tons or, 2400 tons wet-weight) would have been *E. maxima*, for it was preferred to *L. pallida*.

This 7,2% of the standing crop of Ecklonia at Kommetjie could represent natural mortality, but because (a) the sea reclaims some, and, (b) some of the cast lands in inaccessible places, the true figure is nearer 10% or about 3300 tons (wet) per year. (VII)

Stock restoration. We always found reproductive sporophylls on Ecklonia, so that expectations of de novo recolonization are, at all times, high. Yet, tests showed that sporophyte recruitment is best through early summer - the varying fortunes of the alternate micro-gametophyte generation may be responsible for this. But, cyclic reproduction has proved of less importance to the recovery of kelp-stock than has a remnant stand of young subcanopy sporophytes, for non-kelp species dominated on completely cleaned surfaces for at least 12 months, while good kelp-recruitment followed when plants with stipes of 250 mm or less were left intact; 250 mm was chosen because such plants are (a) about to enter their second year, (b) predictably becoming very active, and (c) likely to escape the cutter-bar of a simple harvesting machine because of their natural flexibility. CHEKS-1, therefore, incorporated this size-restriction; it was not, however, species-selective in view of the invasive capability of L. pallida. All unwanted crop was to be returned to the sea, where, presumably, it would perform its natural biological role.

Experimental site and cropping periodicity. Maximum commercial harvest in this project is 3000 tons per year (wet) Ecklonia-stipes, at which rate the 250 ha Ecklonia-field at Kommetjie would be worked once in 5,5 years. This allows ample scope for the planned 3-year cropping rotation. Though this periodicity may result in a 12% annual production-loss, we think this will not seriously affect the environment.

Environmental monitoring. A census of commercial catches of the rock-lobster, Jasus lalandii, and of the population of mussels, presumed to feed on kelp-debris (6), should be taken regularly.

CHEKS-1. This was started in September 1979, having been duly formulated as follows:

1. cropping shall occur at Kommetjie, inshore of a line where <u>Ecklonia</u> ceases to be dominant;
2. all kelps with stipes longer than 250 mm shall be cropped;
3. all unwanted croppings shall be returned to the sea;
4. cropping shall be done, in an ordered sequence, of adjacent 10 m wide strips, the first to be normal to the shore;
5. a diary of the harvest shall be kept;
6. the harvest shall be not more than: 750 tons (wet) in the first year; 1500 tons in the second year; and, thereafter, 3000 tons;
7. return cropping shall occur after, (a) three years, (b) four years;
8. the environment shall be monitored.

Results

The harvest-data can be summarised as follows:

1. total harvest: 186 tons;
2. duration of operation: 10,5 months;
3. mean harvest/effort: c.1 ton/diver-hour;
4. total number of operational days: 90;
5. mean biomass of harvest: 7,9 kg/m^2.

Floating weed evaded capture when swept out to sea by conditions that came to be recognized and that arose ever more often, it was claimed, as the cropped area expanded. Attempts to net the drift failed and, in view of this difficulty, the company asked that we consider a different culling-procedure, especially as there was a report that rock-lobster catches had decreased in the cleared area. This last-mentioned claim is to be investigated as were also the unjustified complaints by local residents that beached seaweed litter had increased as a result of kelp-cropping. Without in any way implying that the CHEKS-1 method is biologically unsound or that the company's problems are insurmountable, we will substitute CHEKS-2 for the present programme at the end of August 1980. Monitoring will, however, continue.

The harvest, which consists of Ecklonia stipes only, predicts a total stipe-mass of 19750 tons in the 250 ha field (Formula IV above). Since frond-mass and stipe-mass approximate each other (our data), total standing crop is 39500 tons (cf. 33000 tons in Formula I above).

References

1. Field, J.G., Griffiths, C.L., Griffiths, R.J., Jarman, N., Zoutendyk, P., Velimirov, B. and Bowes, A.: Trans. R. Soc. S. Afr. 44(2),145-203 (1980).
2. Howard, J.A.: Aerial photo-ecology. Faber & Faber, London. 325 + xvii pages. (1970).
3. Mann, K.H., Jarman, N., and Dieckmann, G. Trans. R. Soc. S. Afr. 44(1), 27-41 (1979).
4. Dieckmann, G.S.: Aspects of growth and production of Laminaria pallida (Grev.) J. Ag. off the Cape Peninsula. M.Sc. thesis. Department of Botany, University of Cape Town, Cape Town (1978).
5. Kain, J.M.: J. mar. biol. Ass. U.K. 43, 129-151 (1963).
6. Newell, R.C., Lucas, M.I., Velimirov, B. and Seiderer, L.J.: Mar. Ecol. Prog. Ser. 2, 45-59 (1980).

SEASONAL VARIATION OF AGAR-AGAR PRODUCED IN TAIWAN AREA

Shang-Shyng Yang, Chen-Yen Wang and Hsi-Hua Wang
Department of Agricultural Chemistry, National Taiwan University, Taipei, Taiwan, R.O.C.

Abstract

The quality of agar-agar produced in Taiwan was affected by the cultivation location, weather conditions and manufacturing process. In general, jelly point, jelly strength and 3,6-anhydrogalactose content of the agar-agar produced in the southern part of Taiwan, Chiatung and Tungkong areas, was higher than those produced in the central part of Taiwan, Chiayi area, while the sulfate content of agar-agar was reverse. Jelly strength and calcium content of agar-agar had statistically positive correlation. Especially the correlation between jelly strength and jelly point was significant. On the other hand, jelly strength and agar yield, total chlorophyll content, chlorophyll a content of *Gracilaria*, pyruvate content and sulfate content of agar-agar had negative correlation.

Introduction

Gracilaria is a seaweed has been cultivated in southern part of Taiwan for 22 years and it is as sourced of agar-agar (4). The manufacture of agar-agar in Taiwan began in 1944 during the period of World War II, when the supply of agar-agar from Japan was limited (3,4). At present, six plants and several family-size producers are producing commercial agar-agar in Taiwan (12) and the total daily output capacity was above 700kg(4). There was a number of papers describing the manufacturing process of agar-agar (3,6), no systemic information regarding the seasonal variation of the quality of agar-agar produced in Taiwan. Therefore, studies on quality changes with seasonal variation of agar-agar produced were conducted. In

Xth International Seaweed Symposium
Printed in Germany

addition, the correlations among environmental conditions, properties of *Gracilaria*, the physical and chemical properties of agar-agar were also discussed.

Materials and Methods

1. Seaweed: *Gracilaria verrucosa* (*G. confervoides*) were collected from the Chiayi, Tainan, Tungkong and Chiatung areas, and extraction of agar-agar, jelly strength, agar yield, and 3,6-anhydrogalactose content determination were described in previous (13).

2. Specific viscosity: An Ostwald viscometer was used to measure the viscosity of agar-agar at $65.0^{\circ}C \pm 0.1^{\circ}C$(15).

3. Jelly point (jelly temperature): Jelly point of agar-agar was measured by Kim's (10) and Young's (15) methods.

4. Sulfate content: Sulfate content of seaweed samples was measured after alkaline titration and precipitation by 4-chloro-4'-aminodiphenyl (1,9).

5. Pyruvic acid content: Pyruvate content of agar-agar samples was quantitatively determined by the lactic dehydrogenase method (7).

Results and Discussion

1. Jelly point: Jelly points of agar-agar from all areas were gradually decreased from August 1977 to July 1978, and the jelly point of agar-agar produced in Chiayi area was somewhat lower than those produced in other areas at the same period (Fig.1a). Guiseley (8) showed that jelly point and jelly strength were correlated. The higher the content of the methoxy group in agarose, the higher its jelly point. On the other hand, Yaphe (14) reported that 6-O-methoxyl content of agar-agar was high but the jelly point was low. Whether the agar-agar produced in Chiayi area had a lower jelly point due to a difference in methoxy content remains to be determined.

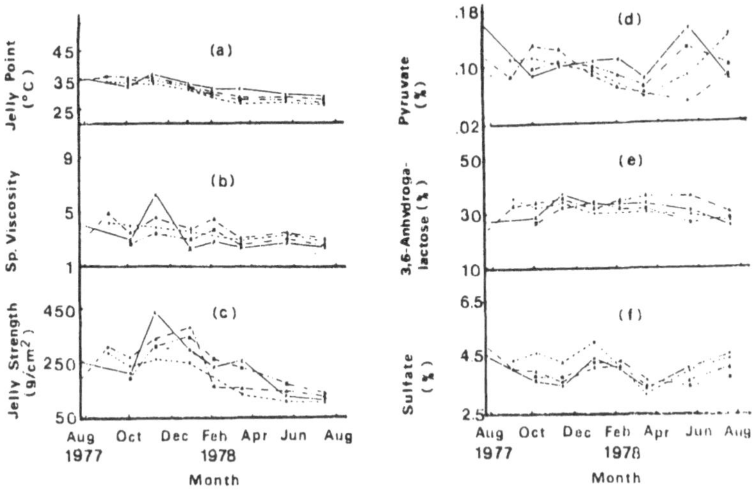

Fig. 1. Seasonal variation of properties of agar-agar produced in Taiwan area:

(a) Jelly point; (b) Specific viscosity;
(c) Jelly strength; (d) Pyruvate content;
(e) 3,6-Anhydrogalactose content; (f) Sulfate content.

•---------•: Chiayi; ▲-·-·-·-▲: Tainan;
x————x: Chiatung; ■-··-··-■: Tungkong.

2. Specific viscosity:Most of the specific viscosity of agar-agar obtained from Chiayi area, Tainan area, Chiatung area and Tungkong area was between 3.0 and 4.0,2.8 and 4.5,2.4 and 4.4,and 2.0 and 2.8, respectively (Fig.1b). Seasonal variation of specific viscosity of agar-agar produced in Chiatung was more significant than those produced in other areas and it had a maximal value 6.1 in November 1977. Specific viscosity of agar-agar produced in Chiayi area, decreased from 4.5 in September 1977 to 3.1 in March 1978. However, in the cases of Tainan and Tungkong areas, the fluctuation was very narrow. The type of seaweed (11) and its growth season (15) has been shown to affect the specific viscosity of agar-agar. In Taiwan, specific viscosity and temperature had a significant negative correlation. It is possible that the physiological properties and quality of agar-agar produced are dependent on how rapidly the *Gracilaria* are growing.

3. Jelly strength: Most of the jelly strength of agar-agar produced in

Chiayi area, Tainan area, Chiatung area and Tungkong area was 200-340 g/cm^2, 200-400g/cm^2, 200-350g/cm^2, 150-250g/cm^2, respectively(Fig.1c). Chiatung area had the maximal jelly strength in November 1977, while the minimal jelly strengths were in October 1977 and July 1978. While Tungkong area had the maximal value in January 1978 and the minimal values were in October 1977 and July 1978. Tainan area also had the maximal value in January 1978 but had the minimal value in February 1978. It is likely that jelly strength of the agar-agar produced in Tainan, Tungkong and Chiatung areas was higher than those produced in Chiayi area. This might be due to the low temperature and short sunshine period in Chiayi. Causey *et al.* (2) and Kim (10) had shown that kinds of light, light intensity, light period affected the growth, photosynthetic and metabolic reactions in algae.

4. Pyruvic acid content: During the period of July 1977 and July 1978, it was found that pyruvic acid content in agar-agar decreased from July 1977 to March 1978 then increased again (Fig.1d). The tendency of pyruvic acid content in agar-agar was reverse the jelly strength. They had maximal pyruvic acid content during June to August. Duckworth reported that the jelly strength of agar-agar decreased as the 3-galactosyl group was substituted by pyruvic acid (7). In Taiwan, pyruvic acid content and jelly strength exhibited a negative correlation.

5. 3-6-Anydrogalactose content: Maximal 3,6-anhydrogalactose content (37%) was found in the agar-agar produced in November 1977, while the minimal value (28%) was in July 1977 and July 1978 (Fig.1e). The content of 3,6-anhydrogalactose increased gradually from July 1977 to December 1977 then the fluculation was very narrow until it decreased in June 1978. Rees (11) showed that the higher the content of 3,6-anhydrogalactose in cell wall polysaccharide of *Gracilaria*, the higher the jelly strength of agar. Our results show that 3,6-anhydrogalactose content and agar yield are negatively correlated with temperature.

6. Sulfate content: The maximal sulfate content (4.6-5.1%) was found in July 1977 and the minimal value (2.8-3.7%) was found in March 1978 (Fig. 1f). Agar-agar produced in Chiayi area had higher sulfate content

than that in other areas. In addition, seasonal variation of sulfate content in Tainan, Tungkong and Chiatung areas was very similar. Generally speaking, low sulfate content in agar-agar was accompanied with high jelly strength and 3,6-anhydrogalactose content. This pattern was also observed by Rees (11) who observed that the jelly strength of agar-agar decreased when its 3,6-anhydrogalactose was substituted by L-galactose-6-sulfate.

7. Correlation among agar-agar properties, *Gracilaria* properties and environmental condition: The importnt results of the correlations can be summarized as follows:(1) agar yield had a positive statistical correlation with sunshine period, while jelly point, jelly strength, specific viscosity, pyruvate content, 3,6-anhydrogalactose content and sulfate content of agar exhibited a negative correlation; (2) Jelly strength had a positive correlation with calcium content, jelly point, specific viscosity, and 3,6-anhydrogalactose content, and a negative correlation with sunshine period, air temperature, pyruvate content and sulfate content of agar extracted; (3) 3,6-anhydrogalactose contents had a negative correlation with pyruvate and sulfate contents; and (4) pyruvate and sulfate contents exhibited a positive correlation.

From the results mentioned, we conclude that the quality of agar-agar produced in Taiwan is primarily determined by environmental conditions, such as sunshine period and temperature. The quality of agar-agar was affected directly by the raw material of *Gracilaria* seaweed being processed. It should be possible to select several appropriate strains or species of *Gracilaria*, each adapted to a particular area, able to tolerate some ranges of salinity, light intensity and temperature. Such strains or species will enable us to make high quality agar-agar more efficiently.

Acknowledgment

The authors express their gratitude to Dr. Y.M Chiang for helpful discussion, and National Science Council of ROC for financial support.

References

1. Belcher, R., Nutten, A.J., Stephen, W.I.: J.Chem. Soc., 1953, 1334-1337 (1953).
2. Causey, N.B., Prytherch, J.P., McCaskill, J., Humm, H.J., Walf, F.A.: Bull. Duke Univ. Mar. Stat., 3, 19-24 (1944).
3. Chen, C.H.:China Fisheries Monthly, 260, 7-13 (1974).
4. Chen, G.C.: JCRR Fisheries Series 25B, 56-67 (1977).
5. Chen, M.S., Chen, W.S.: China Fisheries Monthly, 196, 7-11 (1969).
6. Chiang, Y.M., Chen, W.S.: J. Fish. Soc. Taiwan, 1, 62-64 (1972).
7. Duckworth, M., Yaphe, W.: Chem. and Ind., 747-748 (1970).
8. Guiseley, K.B.: Carbohyd. Res., 13 247-256 (1970).
9. Jones, A.S., Latham, D.S.: Chem. and Ind., 662-663 (1954).
10. Kim, D.H.: Botanica Marina, 13, 149-162 (1970).
11. Rees, D.A.: J.Chem. Soc., 1961, 5168-5171. (1961).
12. Taiwan Fisheries Bureau: Survey Report of Marine Food Processing Factories in Taiwan, Taiwan Provincial Fisheries Bureau, Taiwan, Republic of China (1975).
13. Wang, C.Y., Yang, S.S.: Proc. Natl. Sci. Counc. ROC., 4, 78-86(1980).
14. Yaphe, W.: Analyt. Chem., 32, 1327-1330 (1960).
15. Young, K.S.: Fish. Res. Bd. Canad. Tech. Rept. 454 (1974).

THE SEASONAL VARIATION IN THE BIOMASS AND REPRODUCTIVE STATES OF GRACILARIA IN MANILA BAY

Gavino C. Trono, Jr. and Rhodora Azanza-Corrales
Department of Botany, College of Arts and Sciences, University of the Philippines, Diliman, Quezon City, Philippines

Introduction

Harvesting wild stocks of *Gracilaria* in Manila Bay is a major fishery activity during the "gulamang-dagat" season. Addition of artificial substrates to the natural environment have been tried to enhance its production (1). Taxonomic works have so far identified three species from within the area, namely, *G. verrucosa*, *G. salicornia* and *G. coronopifolia* (2). Its seasonal occurrence and production in the bay were studied in relation to ecological factors during the different seasons (3). Periodicity in the occurrence and variations in the populations of vegetative and reproductive thalli have been noted in Chile and Japan (7,14), although these were not related to any environmental factor.

This paper is the result of our research efforts to gather basic information which could be used as basis for the formulation of a resource management scheme to maintain and enhance the production of *Gracilaria* in the area.

Materials and Methods

Monthly standing crop measurements were done in the study area (14°30' to 14°35' N Lat and 120°55' to 120°59' E Long) from January to March 1977 and from August to February 1978.

Xth International Seaweed Symposium
Printed in Germany

Monthly samplings for the seasonal variation of reproductive states of G. verrucosa were done using the modified methods of Dawes et al. (5) and Mshigeni (6). Monthly temperature, pH and salinity were also recorded. Records of rainfall from the Philippine Weather Bureau were also noted.

Results and Discussion

A. Biomass Production of Gracilaria

Peaks in biomass occurred during the months of November to February when NE Monsoon, characterized by calm water, prevailed over the country. Rainfall was minimal with corresponding high salinity (Figs. 1, 2). Low standing crop was recorded during the months of June to September during the SW Monsoon period characterized by strong wave action, high rainfall and low salinity. Direct physical damage and removal of the lanky and delicate thalli of G. verrucosa by strong waves during this period was apparent. An inverse relationship between biomass and rainfall was recorded during this period. Earlier studies (1,3) have shown that G. verrucosa was absent during the SW Monsoon period.

Although an inverse relation between salinity and biomass is apparent, it is the pronounced wave action which appeared to control the seasonality in biomass production. This euryhaline species (4) could thrive in a wide salinity range of 10-50°/oo with 25-35°/oo as the optimum (7,8). Temperature and pH did not seem to affect the biomass. Grazers were not observed in the study area. Harvesting usually starts in the month of December and ends in February (peak months of growth).

Species competition seemed to exert significant influence on Gracilaria biomass. Acanthophora spicifera was abundant from August to October when biomass of the Gracilaria declined. These associated algal species have been reported to cause

Fig. 1 Record of rainfall, ph, salinity and temperature in the study area from January 1977 to January 1978.

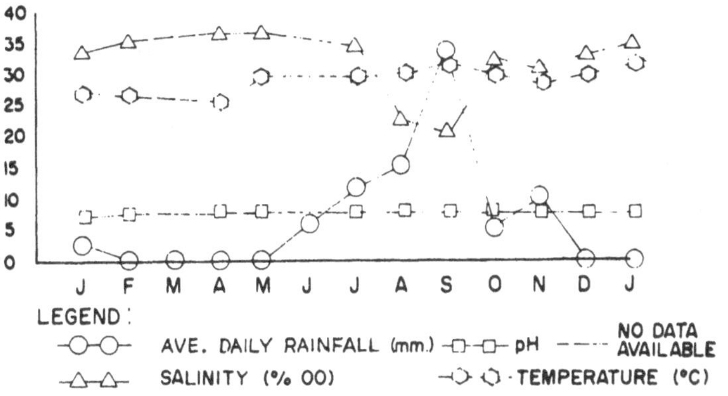

Fig. 2 Record of standing crops and the percentage occurrence of reproductive thalli in G. verrucosa from January 1977 to January 1978.

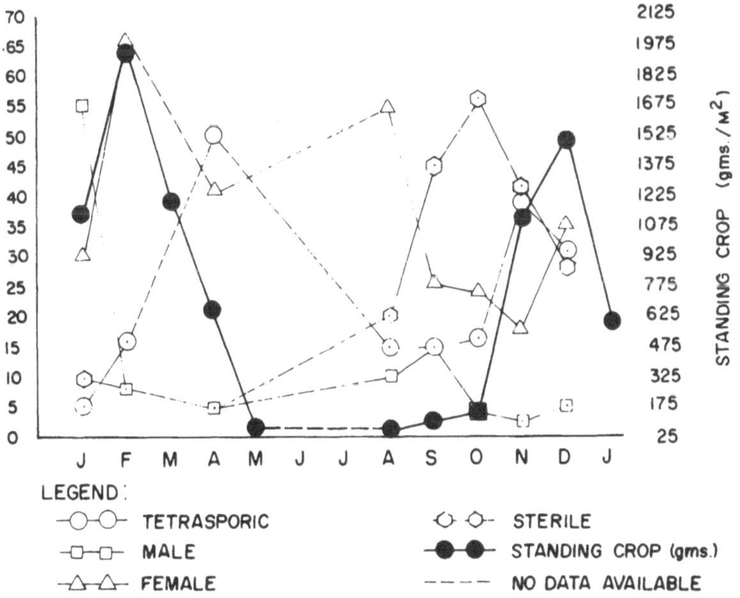

negative effects on the amount of harvest and health of Gracilaria plants (7) and removal of these "weeds" have been shown to encourage luxuriant growth of Gracilaria fragments (9). Variation in the standing crop vary according to species (4). Interactions of any or all of the environmental factors in one area and the effect of these on one species may not be the same on another. A. spicifera seemed to be able to compete with Gracilaria verrucosa particularly during the SW Monsoon period. It appeared to be able to withstand turbulence since its thalli are shorter, tougher, more pliable and thus not easily uprooted or broken thereby giving it an ecological advantage. More work is planned along this line.

This study supports the results of previous works (3,10,11) that algal seasonality and standing crop in the tropics may be attributed principally to seasonal variation of monsoons and the combination of ecological factors prevailing in the area during the different seasons of the year. Addition of suitable substrates was proven to be important in the sustained growth and production of Gracilaria.

B. Reproductive State of G. verrucosa

Changes in the reproductive states of G. verrucosa from January 1977 to January 1978 is also shown in Figure 2. Although fertile thalli were observed almost throughout the year, there appeared to be significant differences in the ratio of the populations of tetrasporophyte, cystocarpic and spermatangial thalli. Monthly samples showed the preponderance of cystocarpic plants and the low occurrence of the male thalli. The low occurrence of male thalli was also reported among the members of Gigartinales (5,6,7,14). In contrast to G. verrucosa, other members of Gigartinales (12,13,14) showed the preponderance of tetrasporophytes over the cystocarpic forms. Mshigeni (6,12) related the above phenomenon to spore size, densities, and other factors being equal, the bigger spore the faster it will sink than the smaller spore. Faster sinking rate offers ecological advantage to the bigger spore. In this study, the tetraspores are bigger (25.0-27.5 μ dia.) than the carpospores (20-24 μ dia.), which could have caused the preponderance of the cystocarpic over the tetrasporic thalli. On the other hand, in Hypnea cervicornis and H. chordacea, the carpospores are bigger than the tetraspores (6), thus the preponderance of the tetrasporophytes. Hypnea was found to repeat the tetrasporic phase by means of detached stichidia, a major means of propagating H. nidifica where sexual thalli have never been observed (12).

The fertile thalli predominated over the sterile thalli during months of high salinity (Figs. 1, 2). Higher salinity (30-35‰) regimes seemed to favor reproduction. The population is mostly vegetative when the salinity is 20-25‰. These results further substantiate previous works (4,8) showing that salinity regimes affects reproductive potential of algae. No fertile materials were observed in G. verrucosa when salinity was below 26‰ (4). Vegetative regeneration seemed to be an important means of sustaining G. verrucosa populations during months of low salinities. This was shown to be true for Eucheuma (5) and Hypnea (6,12). It could therefore be suggested that fragments of the species should be left on the substrates to allow for vegetative regeneration especially during months of low salinity.

Literature Cited

1. Bureau of Fisheries and Aquatic Resources Special Gracilaria Project Progress Report. BFAR, Quezon City, Philippines (1977). (unpublished)
2. Trono, G.C., Jr. and R. Azanza-Corrales: The Taxonomy and Reproductive Morphology of Gracilaria Species in Manila Bay (1979). (unpublished)
3. Vannajan, S. and G.C. Trono, Jr.: Notes on the Seasonal and Spatial Distribution of the Marine Benthic Algae in the Southern Portion of Manila Bay (1975). (unpublished)

4. Hoyle, M.D.: The Literature Pertinent to the Red Algal Genus Gracilaria of Hawaii. Marine Agronomy U.S. Sea Grant Program, Hawaii (1975).

5. Dawes, C.J., A.C. Mathieson and D.P. Cheney: Bull. Mar. Sci. 24, 235 (1974).

6. Mshigeni, K.E.: Bot. Mar. 19, 341 (1976).

7. Kim, D.H.: Bot. Mar. 12, 140 (1970).

8. Conover, J.T.: Pub. Inst. Mar. Sci. Univ. Texas 5, 97 (1958).

9. Raju, P.V. and P.C. Thomas: Bot. Mar. 14, 71 (1971).

10. Doty, M.S.: In Fertility of the Sea (J.D. Costlow, ed.) Gordon and Breach. New York, 25 pp (1973).

11. Saraya, A.: Taxonomy and Ecology of the Marine Benthic Algae of Santiago Is. and Adjacent Areas. Dissertation, University of the Philippines (1976). (unpublished)

12. Mshigeni, K.E.: Bot. Mar. 19, 313 (1976).

13. Isaac, W.E. and F. Hewitt: Univ. J. S. Afr. Bot. 19, 73 (1953).

14. Saito, Y.: Bull. Jap. Soc. Phycol. 6, 87 (1958).

Part III

Special Sessions

METHODS AND TERMINOLOGY IN PHYCOLOGY

A special session on a possible role of the International Seaweed Symposium

Göran Michanek, Dept of Marine Botany, University of Göteborg, Sweden

Introducing new methods and creating new terms in science has a high status. This is a good thing whenever it deepens our knowledge.
Sometimes it does not. Sometimes a scientist in a particular field should profit better from his colleagues' efforts if only these were comparable.

Two examples: I wanted to review available data on the amino acid content in seaweed species. The first nine papers which I consulted were from my own collection of randomly gathered reprints. In each of them the amino acid content was expressed in a different way: 1) g/100 g protein, 2) as mg/g N, 3) g of amino acid nitrogen/100 g of protein nitrogen, 4) percentage of total N in hydrolysate, 5) mg/100 g food, 6) mg/100 g raw material (not indicated whether dry weight or fresh weight), 7) g/100 g dry algae, 8) g N/100 g dry algae, 9) "on the basis of the ratio of amino acid in the sample and fish protein". None of the authors gave any conversion factors.

Biomass data on seaweeds from different parts of the world are hard to compare. There are almost as many assessment methods as there are researchers, and the procedures by which data have been obtained are seldom satisfactorily described if one wants to compare the results. In my review "Seaweed Resources of the Ocean" (Michanek 1975) I quoted two seemingly contradicting sets of data from South Africa. Shuttleworth (1951) estimated the drift weeds of Ecklonia and Laminaria at "many hundreds of thousands, if not millions of tons". Isaac and Molteno (1953) gave an estimate that from five enumerated localities only some 6000 t dry weight of Ecklonia can be collected annually. These localities represent but a small part of the coastline along which kelp grow. Much of the west coast, however, is very inaccessible for commercial collection.

Xth International Seaweed Symposium
Printed in Germany

Dr. R.H. Simons of the Sea Fisheries Institute, Cape Town, has kindly clarified the situation. About 100 000 t wet mass of kelps occurs at Kommetje, a small but densely colonised area on the Cape Peninsula coast. Reckoning this at one-fiftieth part of the total, there are present about 5 million tons of kelp in the cold-temperate waters of the Republic of South Africa; the annual turn-over-rate has been found to be at least 100% so that the potential cast is 5 million tons per year; assuming that only 10% of this potential arrives on the shore the annual yield would be in the vicinity of 500 000 tons; assuming further that only 10% of the coast is commercially accessible a final figure of 50 000 tons materializes; in dried form this amounts to about 5 000 tons (i.e. the normal commercial return); it follows that in effect both Shuttleworth's figures and those of Isaac and Molteno are correct - it depends on the point of view.

In other fields there is no confusion yet, but the high degree of specialization forces us into a demand for mutual efforts. Kinds and characteristics of colloids are just as important for the possible utilization of a resource as its sustainable yield. The chemical analysis, however, calls for a different type of training and other laboratory facilities than most field phycologists possess. Especially there have been recent developments of ^{13}C NMR for characterization of agars and carrageenans. It is now possible to obtain definitive evidence for the type of polysaccharide present in different marine algae. This is a rather sophisticated approach and requires specialization in chemical analysis. Prof. W. Yaphe has envisaged the establishment of international reference centres for the analysis of phycocolloids. The chemistry departments of all developed countries have the instrumentation, it is a question of setting up procedures so that the botanists who collect and identify the raw material can also prepare the phycocolloid for analysis. The International Symposia offer a unique opportunity to initiate cooperative projects and to standardise procedures.

There is no need for further examples of the advantage of standard methods and joint efforts, their number is unlimited. A remainder is rather needed of the importance of investigations outside the standard methods.

The standard procedure for investigating antibacterial activity of seaweed extracts is to check a large number of algae on a small set of test bacteria. Pratt *et al.* (1953) went outside this standard set and tested the Californian *Rhodomela larix* on, among others, *Mycobacterium tuberculosis*, which actually was inhibited. From Puerto Rico to India a number of common marine algae have shown to be strongly active against the tuberculosis bacterium, and pharmaceutical utilization is now suggested. Spore formation conditions were first searched in gradients of temperature and light intensity until the importance of daylength was realized. And so on. Any practise, recommended or not, includes a risk that other possibilities may be overlooked.

When it was decided to raise the question of a possible demand for recommendations on methods and terminology in phycology at a special session during the Xth International Seaweed Symposium, the idea was to be utterly restrictive. Hence recommendations should be proposed only in cases where we are obviously suffering from a lack of coordination of such efforts. As an example it would be an advantage if field phycologists and university teachers knew by which apparatus potential consumers prefer to have gel strength and viscosity described. If one is East, and one is West, it's better the twain shall meet.

A first step could reasonably be to collect, review and distribute already existing recommendations and to make a bibliography of existing handbooks e.g. the FAO and Unesco manuals of methods, those of IBP, Baltic Biologists, Phycological Society of America etc.

During the session, however, it became evident that there is a widespread demand for manuals on entire fields, in particular on that of resource assessment. It was also seen as an advantage when manuals are published by FAO or Unesco, as they are then more easily accessible in developing countries.

A very good procedure for the creation of any manual would be to present a draft version to a small working group, which could discuss and finalize it. Such workshops could with advantage be organized immediately before or after the International Seaweed Symposia, and run parallell to the common semi-touristic phycological excursions.

Even without such distinct aims many participants would welcome small workshops on subjects which could be of a particular interest to a limited number of participants, e.g. the cultivation of Gracilaria, practical demonstration of the phycosociological working method etc. The coming together of phycologists from all over the world could also be used for the completion of bibliographies on determination keys, monographies of genera, local lists of algal floras, references to papers containing phycosociological data, line transects, or data on biomass per unit area.

The suggestion of very small workshops is already being positively considered by the hosts of the 1983 symposium, and anybody who wants to propose a certain theme for a possible workshop should address himself to professor Tseng.

I was asked to serve as a convenor for an open working group on methods and terminology in phycology, and professor Yaphe was asked to act as group leader for those concerned with questions related to chemical methods. No other participants were elected, everybody is invited to contribute as far as his interests may push him.

References

1. Isaac, W.E., Molteno, C.J.: J.S.Afr.Bot. 19(3), 85-92 (1953).
2. Michanek, G.: FAO Fish. Techn. Pap. (138), 1-127 (1975).
3. Pratt, R., Mantner, H.G., Gardner, G.M.: J.Am.pharm.Ass., 42 (1953).
4. Shuttleworth, R.G.: S.Afr.Industr.Chem., 115-117 (1951).

Part IV

PARTICIPANTS

Ahrens, B., c/o C.E. Roeper, Klosterallée 74, D-2000 Hamburg 13, West Germany.

Allen, J.R., Rua de Pedroucos 66-1, P-1900 Lisboa, Portugal.

Ameln, H., Protan & Fagertun, Box 420, N-3001 Drammen, Norway.

Andon, S., 144 E. 44th Street, New York, N.Y. 10017, USA.

Andrade, F.A. de Leite, R. Perpendicular ã R. Jorge Abreau, Lote 2, S Pedro do Estoril, 2765 Estoril Codex, Portugal.

Andrén, M., Uppsala Universitet, Fysiologisk Botanik, Box 540, S-751 21 Uppsala, Sweden.

Anfield, J., University of Stirling, Stirling FK9 4LA, Scotland, UK.

Arasaki, S., 3-33-5, Tokumaru, Tabashiku, Tokyo 175, Japan.

Arasaki, T., 3-33-5, Tokumaru, Tabashiku, Tokyo 175, Japan.

Ardré, F., Muséum, Laboratoire de Cryptogamie, 12 Rue Buffon, F-75005 Paris, France.

Armisén, R., Hispanagar S.A., Pedro de Valdivia 36, Madrid 6, Spain.

Athanasiadis, A., Marine Botanical Institute, Carl Skottsbergs Gata 22, S-413 19 Göteborg, Sweden.

Austin, A., Dept of Biology, University of Victoria, Victoria BC V8W 2Y2, Canada.

Baardseth, E., Institutt for Marinbiologi, N-5065 Blomsterdalen, Norway.

Barilotti, D.C., 2145 East Belt Street, San Diego, Ca 92113, USA.

Barker, D., Pedigree Petfoods, Melton Mowbray, Leicestershire LE13 1BB, UK.

Bekheet, I., Botany dept, Faculty of Science, Alexandria University, Mohanem Bey, Alexandria, Egypt.

Bellion, C., Société CECA S.A., Usine de Baupte, F-50500 Carentan, France.

Bjerre-Petersen, E., Litex Industri, DK-2600 Glostrup, Denmark.

Björn-Rasmussen, S., Marine Botanical Institute, Carl Skottsbergs Gata 22, S-413 19 Göteborg, Sweden.

Black, H.J., Institut für Meereskunde, Düsternbrooker Weg 20, D-2300 Kiel 1, West Germany.

Black, W.A.P., Croftfoot House, Polmont, Falkirk FK2 0PS, Scotland, UK.

Blancho Sanchez, G., Apartado Postal No. 453, Unidad de Ciencias Marinas, Ensenada, Baja California, Mexico.

Blond, J. Le, 112 Rue de Lilly, F-92100 Boulogne sur Seine, France.

Blunden, G., School of Pharmacy, Portsmouth Polytechnic, King Henri 1st Street, Portsmouth, Hants. PO1 2DZ, UK.

Blunn, G., Dept of Plant Sciences, Baines Wing, Leeds University, Leeds, UK.

Boalch, G.T., The Laboratory, Citadel Hill, Plymouth PL1 2PB, UK.

Bonin, D., Station Marine d'Endoume et Centre d'Oceanographie, Rue de la Batterie-des-Lions, F-13007 Marseille, France.

Bonotto, S., Radiobiology Dept, C.E.N./S.C.K., B-2400 Mol, Belgium.

Booth, E., Selborn Villa, Beach Grove, Ashton, Preston PR2 1DX, Lancs., UK.

Boudouresque, C.F., Lab. of Marine Phytobenthos and Pollution, University of Luminy, F-13288 Marseille, Cedex 2, France.

Boulos, H.F., 47 El Zaher Street, Mohanem Bey, Alexandria, Egypt.

Braud, J.-P., Société CECA S.A., Usine de Baupte, F-50500 Carentan, France.

Braun, R. von, Aprilvägen 26, S-183 40 Täby, Sweden.

Bruml, P., T.M. Duché & Sons (U.K.) Ltd, Ford Lane, Pendleton, Salford M6 6PB, UK.

Bryce, A.J., No. 10 Yorktown, Irvine, Ca 92714, USA.

Bula Meyer, G., Apartado Aéreo 890, Santa Marta, Magdalena, Colombia.

Cajipe, G.J.B., P.O. Box 1, Marine Sciences Center, U.P., Quezon City, Philippines.

Callow, J.A., Dept of Plant Sciences, University of Leeds, Leeds LS2 9JT, UK.

Callow, M.E., Dept of Plant Sciences, University of Leeds, Leeds LS2 9JT, UK.

Campo, D.J. del, Hispanagar SA, Pedro de Valdivia 36, Madrid 6, Spain.

Cancino, J., Lab. Zoologia, Univ. Catolica de Chile, Casilla 114-D, Santiago, Chile.

Caram, B., Lab. de Protistologie Marine, Station Biologique, F-06230 Villefranche/Mer, France.

Carlsson, L., Marinbotaniska Lab., Kemicentrum, Box 740, S-220 07 Lund, Sweden.

Chait, A., Moneda 812 of. 1010, Santiago, Chile.

Chapman, A.R.O., Dept of Biology, Dalhousie University, Halifax, N.S. B3H 4J1, Canada.

Chaudhry, S., International Trade Centre UNCTAD/GATT, 4 Route des Morillons, CH-1211 Geneva 22, Switzerland.

Chen, L.C.M., 1411 Oxford Street, Halifax, N.S. B3H 3Z1, Canada.

Cheney, P., Pedigree Petfoods, Melton Mowbray, Leicestershire LE13 1BB, UK.

Chiang, Y.-M., Institute of Oceanography, National Taiwan University, Taipei, Taiwan.

Christensen, J., Langs Hegnet 59 C, DK-2800 Lyngby, Denmark.

Christensen, T., Institut for Sporeplanter, Øster Farimagsgade 2 D, DK-1353 København K, Denmark.

Cinelli, F., Laboratorio di Ecologia Marina, I-80077 Ischia Porto (Napoli), Italy.

Clare, K., Alginate Industries Ltd, 22 Henrietta Street, London WC2E 8NB, UK.

Clayton, M.N., Botany Dept, Monash University, Clayton, Victoria 3168, Australia.

Colina, A., Institut für Meereskunde, Düsternbrooker Weg 20, D-2300 Kiel 1, West Germany.

Cordeiro-Marino, M., Seção de Ficologia, Instituto de Botânica, Caixa Postal 4005, 01000-São Paulo, Brazil.

Cousens, R.D., Dept of Biology, Dalhousie University, Halifax, N.S. B3H 4JI, Canada.

Craigie, J.S., Atlantic Regional laboratory, NRC of Canada, 1411 Oxford Street, Halifax N.S. B3M 1R1, Canada.

Cullinane, J.P., Dept of Botany, University College, Cork, Ireland.

Cuomo, V., CIBA-GEIGY S.P.A., Via Prov. Schiti 131, I-80058 Torre-Annunziata, Italy.

Dakay, B.U., P.O. Box 402, Cebu City, Philippines.

Daniel, G., Biology Dept, King Henry I Street, Portsmouth, Hants., UK.

Delépine, R., 2 Villa Guibert, F-75016 Paris, France.

Dempsey, M., Biology Dept, King Henry I Street, Portsmouth, Hants., UK.

Denizot, M., Institute de Botanique, Rue Broussonnet, Montpellier, France.

Deveau, L.E., Marine Colloids Ltd, P.O. Box 2610, Dartmouth East, Dartmouth, N.S. B2W 4B7, Canada.

Dion, P., Station Biologique, F-29211 Roscoff, France.

Dondain, G., IRANEX S.A., 4 Rue Frederic Passy, F-92200 Neuilly sur Seine, France.

Dörnbrack, G., Mövenstrasse 7, D-2000 Hamburg 60, West Germany.

Doty, M.S., Botany Dept, University of Hawaii, 3190 Maile Way, Honolulu, HI 96822, USA.

Druehl, L., Bamfield Marine Station, Bamfield B.C. VOR 1B0, Canada.

Dugoujon, J., CECA S.A., 11 Av. Morane-Saulnier, F-78140 Velizy-Villacoublay, France.

Durairatnam, M., Apt. 402, Edificio Riomar, Avenida Deodoro-240, Natal RN-59000, Brazil.

Echegary-Taborga, J.E., Calle Pavía 93-95, 6° 1^{a}, Barcelona 28, Spain.

Elman, S., Phillip Rockley Ltd, 1 Dag Hammarskjold Plaza, New York, N.Y. 10017, USA.

Elsworth, J.F., Dept Organic Chemistry, University of Cape Town, Rondebosch 7700, South Africa.

Enright, C.T., 1411 Oxford Street, Halifax, N.S. B3H 3Z1, Canada.

Errico, A.J., P.O. Box 741, Far Hills, N.J. 07931, USA.

Etcheverry D., H., Casilla 3499-Correo 3, Valparaiso, Chile.

Evans, L.V., Dept of Plant Sciences, University of Leeds, Leeds LS2 9JT, UK.

Farnham, W.F., Marine Lab., Ferry Rd, Hayling Island, Hants. PU11 UDG, UK.

Fjeld, A., Biologisk Stasjon, N-1440 Dröbak, Norway.

Fletcher, R.L., Biology Dept, King Henry I Street, Portsmouth, Hants., UK.

Folcher, M., Société CECA S.A., Usine de Baupte, F-50500 Carentan, France.

Fralick, R.A., Plymouth State College, University System of New Hampshire, Plymouth, N.H. 03264, USA.

Frank, J., Gas Research Inst., IOW 35th Street, Chicago, Ill., USA.

Fries, L., Inst. Fysiologisk botanik, Box 540, S-751 21 Uppsala, Sweden.

Gagné, J., Biology Dept, Dalhousie University, Halifax, N.S. B3H 4J1, Canada.

Ganesan, E.K., Institute of Oceanography, Universidad de Oriente, Apartado Postal 195, Cumana 6101 A, Venezuela.

Gerard, V.A., Kerckhoff Marine Lab., 101 Dahlia Street, Corona del Mar, Ca 92625, USA.

Gibor, A., Biology Dept, UCSB, Santa Barbara, Ca 93106, USA.

Glombitza, K.-W., Institut f. Pharmaz. Biologie, Nussallee 6, D-5300 Bonn 1, West Germany.

Granéli, E., Marinbotaniska Lab., Kemicentrum, Box 740, S-220 07 Lund, Sweden.

Guiry, M.D., Dept of Botany, University College, Galway, Ireland.

Gustavsson, U., Marine Botanical Institute, Carl Skottsbergs Gata 22, S-413 19 Göteborg, Sweden.

Guterstam, B., V. Ågatan 47, S-150 13 Trosa, Sweden.

Güven, K.C., Faculty of Pharmacy, University, Istanbul, Turkey.

Hagen, N.T., Fjellveien 4, N-9020 Tromsdalen, Norway.

Hägerdal, B., Applied Biochemistry, Chemical Center, P.O. Box 740, S-220 07 Lund, Sweden

Hägerhäll, B., Ministry of Agriculture, Fack, S-103 20 Stockholm, Sweden.

Haldorsen, H., Institutt for marinbiologi, N-5065 Blomsterdalen, Norway.

Hall, A., 1 Church Terrace, Baslow, Bakewell, Derbyshire DE4 1RY, UK.

Hallsson, S.V., P.O. Box 7063, 127 Reykjavik, Island.

Hanisak, M.D., Harbor Branch Foundation, RRI, Box 196, Fort Piere, Fla 33450, USA.

Hansson, G., Teknisk Mikrobiologi, Kemicentrum, Box 740, S-220 07 Lund, Sweden.

Härdstedt-Roméo, M., Lab. de Physique et Chimie Marine, B.P. 8, La Darse, F-06230 Villefranche-sur-mer, France.

Hardy, F.G., Dept of Plant Biology, The University, Newcastle upon Tyne NE1 7RU, UK.

Harkin, E., Marine Biological Station, Port Erin, Isle of Man, UK.

Heimdal, B.R., Institutt for marinbiologi, N-5065 Blomsterdalen, Norway.

Hemmingsen, P., Litex Industri, Postbox 7, DK-2600 Glostrup, Denmark.

Heslop-Christensen, J., Vandkvalitetsinstituttet, Saralyst Allé 52, DK-Højbjerg, Denmark.

Hofsten, A. von, Fys. bot., Box 540, S-751 21 Uppsala, Sweden.

Holmsgaard, J.E., Litex Industri, P.O. Box 7, DK-2600 Glostrup, Denmark.

Hoppe, H.A., Finkenstrasse 5, D-2150 Buxtehude, West Germany.

Hughes, B., Dept of Science, Bristol Polytechnic, Coldharbour Lane, Bristol, UK.

Hultmark, E., Del. för samordning av havsresursverks., Box 295, S-401 24 Göteborg, Sweden.

Hunter, M.A., T.M. Duché & Sons (U.K.) Ltd, Ford Lane, Pendleton, Salford M6 6PB, UK.

Ikeda, H., Kita Hiyo shityo/magumano, Higashiyamaka, Kyoto, Japan.

Indergaard, M., Institutt for marin biokjemi, N-7034 Trondheim-NTH, Norway.

Jaasund, E., Marine Botanical Institute, Carl Skottsbergs Gata 22, S-413 19 Göteborg, Sweden.

Jacobs, W., 3 Frome Rd, Nunney (near Frome), Somerset, UK.

Jenneborg, L.-H., Marine Botanical Institute, Carl Skottsbergs Gata 22, S-413 19 Göteborg, Sweden.

Jensen, A., Institutt for marin biokjemi, N-7034 Trondheim-NTH, Norway.

Jensen, E.B., Litex Industri, Box 7, DK-2600 Glostrup, Denmark.

Ji, M., 7 Nanhai Road, Institute of Oceanology, Academia Sinica, Qingdao, Shandong, China.

Jones (Kain), J.M., Dept of Marine Biology, Port Erin, Isle of Man, UK.

Joubert, C., IRANEX S.A., 4 Rue F. Passy, F-92200 Neuilly sur Seine, France.

Juliussen, R., Duevegen 10, N-9014 Håpet, Norway.

Kadoya, K., Kadoya & Co, P.O. Box 2154, Boeki Bldg 123-1, Kobe port, Japan.

Kajimura, M., Marine Biological Station, Shimane University, Kamo, Saigo, Oki-gun, 685 Japan.

Kettani, R., P.O. Box 741, Far Hills, N.J., USA.

King, R.J., Botany School, University of N.S.W., P.O. Box 1, Kensington 2033, Australia.

Knoepffler-Peguy, M., Laboratoire Arago, F-66650 Banyuls/Mer, France.

Kornfeldt, R.-A., Marinbotaniska Lab., Kemicentrum, Box 740, S-220 07 Lund, Sweden.

Kremer, B., Seminar f. Biologie, PH Rheinland, Gronewaldstr. 2, D-5000 Köln 41, West Germany.

Kristensen, K., Protan & Fagertun A.S., P.O. Box 420, N-3001 Drammen, Norway.

Kristiansen, A., Institut for Sporeplanter, København Universitet, Ø. Farimagsgade 2 D, DK-1353 København K, Denmark.

Kuylenstierna, M., Marine Botanical Institute, Carl Skottsbergs Gata 22, S-413 19 Göteborg, Sweden.

Laite, P.S., Marine Colloids Div., FMC Corp., P.O. Box 308, Rockland, Maine 04841, USA.

Larsen, B., Inst. of Marine Biochemistry, N-7034 Trondheim-NTH, Norway.

Laserna, E.C., P.O. Box 1, Marine Sciences Center, University of the Philippines, Quezon City, Philippines.

Lawson, G.W., Dept of Biological Sciences, University of Lagos, Lagos, Nigeria.

Lehnberg, W., Institut für Meereskunde, Düsternbrooker Weg 20, D-2300 Kiel, West Germany.

Lestang-Bremond, G. von, 18, Rue de la Préfecture, F-49000 Angers, France.

Levring, T., Marine Botanical Institute, Carl Skottsbergs Gata 22, S-413 19 Göteborg, Sweden.

Lim, J.R., Genu Products Philippines Inc., P.O. Box 568, Cebu City, 6401, Philippines.

Lloyd, N., Genu Products Canada Ltd, Suite 606, Cogswell Tower, Halifax, Nova Scotia, Canada.

Lua, C.Y., P.O. Box 365, Cebu City, Philippines.

Lundberg, B., Dept of Botany, Hebrew University, Jerusalem, Israel.

Lüning, K., Biologische Anstalt Helgoland, D-2192 Helgoland, West Germany.

Luxton, D., Davis Gelatine (N.Z.) Ltd, Connal Street, P.O. Box 19-542, Woolston, Christchurch, New Zealand.

Magne, F., Lab. de Biologie végétale marine, 7 Quai St. Bernard, F-75230 Paris, Cedex 05, France.

Mandura, A., Botany Dept, University of Glasgow, Glasgow, G12 8QQ, Scotland, UK.

Manley, S.L., Kerckhoff Marine Laboratory, 101 Dahlia Street, Corona del Mar, Ca, USA 92625.

Mann, K.H., Dept of Biology, Dalhousie University, Halifax, N.S., Canada B3H 4J1.

Marcelli, E., Ciba-Geigy S.P.A., Via Prov. Schiti, 131, Torre-Annunziata, Italy.

Markham, J.W., Biologische Anstalt Helgoland, D-2192 Helgoland, West Germany.

Masaki, T., Fac. of fisheries, Hokkaido University, 3-1-1 Minato-cho, Hakodate, 041 Japan.

Mathiesen, H., Aarhus universitet, Botanisk Institutt, Nordlandsvej 68, DK-8240 Risskov, Denmark.

Mathiesen, L., Aarhus universitet, Botanisk Institutt, Nordlandsvej 68, DK-8240 Risskov, Denmark.

Mathieson, A.C., Jackson Estuarine Laboratory, University of New Hampshire, Durham, N.H. 03824, USA.

Matulewics, M.C., Dept of Chemistry, Royal Holloway College, University of London-Egham-Surrey TW20 0EX, UK.

Mayer, A.M.S., Patagonia Comercial S.R.L., Junin 558 3° Oficina 302, 1026 Buenos Aires, Argentine.

McCandless, E.L., Dept Biology, McMaster University, Hamilton, Ont. L8S 4K1, Canada.

McDowell, R.H., 20, Oak Tree Close, Virginia Water, Surrey, UK.

McLachlan, J., Atlantic Regional Laboratory, 1411 Oxford Street, Halifax, Canada B3H 3Z1.

McLaughlin, O., Lis-na-carric, Brighton, Road, Foxrock, Dublin 18, Ireland.

McLean, M., Crichie Farm Cottages, Inverurie, Aberdeen AB5 9NP, Scotland, UK.

McNeely, W.H., Kelco, Division Merck & Co., 8355 Aero Drive, San Diego, Ca 92123, USA.

Meer, J. van der, Atlantic Regional Laboratory, National Research Council Canada, 1411 Oxford Street, Halifax, N.S. B3H 3Z1, Canada.

Michanek, G., Marine Botanical Institute, Carl Skottsbergs Gata 22, S-413 19 Göteborg, Sweden.

Moeller, H., 45 St. Andrews Rd, Southampton, New York, USA.

Møller-Pedersen, P., Institut for Sporeplanter, Ø. Farimagsgade 2 D, DK-1353 København K, Denmark.

Mori, H., Tokyo Kasei-Gakuin College, F 102 Sanban-cho 22, Chiyoda-ku, Tokyo, Japan.

Morrice, L.M., Dept of Biochemistry, Univ. of Aberdeen, Marischal Sollege, Aberdeen AB9 1AS, Scotland, UK.

Moss, B., Dept of Plant Biol., The University, Newcastle-upon-Tyne, UK.

Moss, J.R., 6405 Chartres Drive, Rancho Palos Verdes, Ca 90274, USA.

Mshigeni, K.E., P.O. Box 35060, Dar es Salaam, Tanzania.

Mühlberg, L., Multikem Corp., 230 Route 46, Little Ferry, N.J. 07643, USA.

Mukai, L., Dept of Oceanography, Dalhousie University, Halifax N.S., Canada.

Müller, D.G., Fakultät für Biologie der Universität Konstanz, Postfach 7733, D-7750 Konstanz, West Germany.

Munda, I.M., Biological Institute SAZU, Novi trg 3, Ljubljana, Yugoslavia.

Navarette, V., Productos Quimicos Algina Ltd, Edificio Tajamar, Torre A, Depto 2601, Santiago, Chile.

Neushul, M., Marine Science Institute, UCSB, Santa Barbara, Ca 93106, USA.

Niell, F.X., Instituto de Invest. Pesqueras, Muelle de Bouzas, Vigo, Spain.

Nielsen, G.Æ., Miljøstyr. Havforureningslab., Kavalergården 6, DK-2920 Charlottenlund, Denmark.

Nielsen, K., Københavns Pektinfabrik, DK-4623 Lille Skensved, Denmark.

Nielsen, R., Institut for Sporeplanter, Havbiologisk afd., Tordenskjoldsg. 13, DK-9900 Frederikshavn, Denmark.

Niemeck, R.A., Dept of Biology, Sioux Falls College, Sioux Falls, South Dakota, USA 57101.

Nishide, E., 2-32-27-406, Minami-Karasuyama, Setagaya, Tokyo 157, Japan.

Nisizawa, K., 176,10-ban-4, Koyama-3, Nerima-ku, Tokyo, Japan.

North, W.J., California Institute of Technology, Pasadena, Ca 91125, USA.

Norve, H., Genu Products Canada Ltd, Suite 606, Cogswell Tower, Halifax N.S., Canada B3J 3K1.

Nowak-Krawietz, H., Botanischer Garten und Museum Berlin, 1000 Berlin 33, Königin-Luise-Str. 6-8, West Germany.

Nultsch, W., Lehrstuhl für Botanik, Fachbereich Biologie, Lahnberg, D-3550 Marburg, West Germany.

Nygren, S., Marine Botanical Institute, Carl Skottsbergs Gata 22, S-413 19 Göteborg, Sweden.

Nygård, B., Litex Industri, Box 7, DK-2600 Glostrup, Denmark.

Ofstad, R., Institutt for marin biokjemi, N-7034 Trondheim-NTH, Norway.

Øhrstrøm, P.B., Maxicrop USA Inc., P.O. Box 964, Arlington Heights, Illinois 60006, USA.

Okuda, T., Dept of Fisheries, Kyushu University 46-04, Fukuoka 812, Japan.

Oliveira F., E.C. de, Dep. Botânica, Caixa Postal 11461, São Paulo, Brazil.

Olsson, P., Marinbotaniska Lab., Kemicentrum, Box 740, S-220 07 Lund, Sweden.

Opstad, I., H 273 Stud. byen, N-5036 Fantoft, Bergen, Norway.

Ortega, M.M., Instituto de Biologia, Departamento de Botánica, Apda. Postal 70-233, Mexico 20, D.F., Mexico.

Pace, D.R., Occam Marine Technologies Ltd, P.O. Box 5083, Armdale, Halifax Nova Scotia B3L 4M6, Canada.

Peats, S., Marine Colloids, P.O. Box 308, Rockland, Maine 04841, USA.

Pedersen, A., Inst. for Marinbiologi, N-5065 Blomsterdalen, Norway.

Pedersén, M., Inst. för Fysiologisk botanik, Uppsala Universitet, Box 540, S-751 21 Uppsala 1, Sweden.

Pedersen, S.M., Marine Pollution Laboratory, Kavalergården 6, DK-2920 Charlottenlund, Denmark.

Pellegrini, L., Institut de Cytology, Faculté de Luminy, 70, Route L. Lachamp, F-13288 Marseille - Cedex 2, France.

Pellegrini, M., Institut de Cytology, Faculté de Luminy, 70, Route L. Lachamp, F-13288 Marseille - Cedex 2, France.

Penot, M., Physiologie Végétale, Faculté des Sciences, F-29283 Brest Cedex, France.

Percival, E., Bourne Laboratory, Royal Holloway College, Egham Hill, Egham, Surrey GU25 4JF, UK.

Perez, R., Inst. des Peches, B.P. 1049, Rue de l'ile d'Yeu, F-44037 Nantes, France.

Pinheiro-Joventino, F., Universidade Federal do Ceara, Laboratório de Ciências do Mar, Av. da Abolição, 3207, Caixa Postal 1072, 60000 Fortaleza-CE, Brazil.

Pino, H., Chile Exportaciones LTDA, Moneda 812 - 10th and 11th fl., Santiago, Chile.

Pluzek, K.-J., Protein Laboratory, University of Copenhagen, Sigurdsgade 34, DK-2200 København N, Denmark.

Pohl, P., Inst. f. Pharmazeutische Biologie, D-2300 Kiel, Grasweg 9, West Germany.

Polne. M., Bio. Sci. UCSB, Santa Barbara, California 93106, USA.

Porse, H., The Copenhagen Pectin Factory Ltd, 813/814 Supreme House, Penang Road, Singapore 0923, Malaysia.

Price, J.H., Dept of Botany, British Museum (Nat. hist.), Cromwell Road, London SW7 5BD, UK.

Pringle, J.D., Resource Branch, Fisheries & Oceans, P.O. Box 550, Halifax N.S., Canada B3S 2S7.

Probyn, T.A., Dept of Biology, Dalhousie University, Halifax N.S. B3H 4J1, Canada.

Quillet, M., U.C.O., B.P. 808, F-49005 Angers, France.

Rasmussen, T., Strandlovsvej 38, DK-2300 København S, Denmark.

Ravanko, O., Dept of Botany, University of Turku, SF-20500 Turku 50, Finland.

Reed, R.H., Dept of Biological Sciences, University of Dundee, Dundee DD1 4HN, Scotland, UK.

Rex, B., Tjärnö Marinbiologiska Lab., Pl. 2781, S-452 00 Strömstad, Sweden.

Ricohermoso, M.A., Marine Colloids (Phil.) Inc., Looc, Mandaue, Cebu City, Philippines.

Rivera-Carro, H., Unidad de Ciencias Marinas, Apartado Postal 453, Ensenada, Baja California Norte, Mexico.

Rosell, K.G., Dept of Biological Sciences, Simon Fraser University, Burnaby B.C., Canada V5A 1S6.

Rotmann, K., P.O. Box 68327, Bryanston 2021, Randburg, South Africa.

Rueness, J., Inst. for Marine Biology, P.O. Box 1069, Blindern, Oslo 3, Norway.

Rueness, M., Inst. for Marine Biology, P.O. Box 1069, Blindern, Oslo 3, Norway.

Sakagami, Y., Tokyo Institute of Technology, O-Okayama Meguroku Tokyo 152, Japan.

Sander, W., 8225 Aero Drive (KELCO), San Diego, California, USA.

Sanderson, J.L., Alginate Industries Ltd, Grangestone, Girvan, Ayrshire, Scotland, UK.

Santelices, B., Inst. Biologia, Univ. Católica de Chile, Casilla 114-D, Santiago, Chile.

Saunders, R.G., c/o Bamfield Marine Station, P.O. Box 97, Bamfield B.C., Canada V0R 1B0.

Schnetter, R., Botanisches Institut, Justus-Liebig-Universität, Senckenbergstr. 17-21, D-6300 Giessen, West Germany.

Schöldström, K., Björnbärsvägen 2, S-430 90 Öckerö, Sweden.

Schramm, W., Institut für Meereskunde, D-2300 Kiel , Düsternbrooker Weg 20, West Germany.

Seip, W.F., BBL Microbiology System, P.O. Box 243, Cockeysville, MD 21030, USA.

Seoane-Camba, J.A., Departamento de Botánica, Facultad de Farmacia, Av. Gmo. Franco s/n, Universidad de Barcelona, Barcelona, Spain.

Seto, R., 1-31 Taisha-chyo, Nishinomiya-shi Hyogo, Japan 662.

Silverberg, B., Kyrkogatan 31, S-432 00 Varberg, Sweden.

Simons, R.H., Bot. Dept, University of Cape Town, Rondebosch 7700, South Africa.

Simonsson, G.H., Brattebergsskolan, S-430 90 Öckerö, Sweden.

Sjöstedt, B., Marinbotaniska Lab., Kemicentrum, Box 740, S-220 07 Lund, Sweden.

Smith, B.D., Dept of Fisheries, Box 2223, Halifax Nova Scotia, Canada B3J 3C4.

Söderström, J., Rubingatan 45, S-421 62 Västra Frölunda, Sweden.

Soriano, L., Soriano S.A., El Salvador 5161, 1414 Buenos Aires, Argentine.

Sparrow, B.W., Inst. Paint Marine Coatings, Yealm Road, Newton Ferrers, Plymouth PL8 1BL, UK.

Sreenivasa Rao, P., Central Salt & Marine Chemicals Research Institute, Bhavnagar 364002, India.

Srivastava, L.M., Dept of Biol. Sciences, Simon Fraser University, Burnaby B.C., Canada V5A 1S6.

Stancioff, D., 2 Spring Street, Camden, Maine 04843, USA.

Staunton, M., Streamstown, Westport, Ireland.

Stehouwer, C., P.O. Box 67525, Bryanston 2021, Sandton, South Africa.

Steinnes, A., Protan, Box 420, N-3001 Drammen, Norway.

Stokes, R.L., 609 King Street, Aberdeen AB2 1SN, Scotland, UK.

Taasen, J.P., University of Tromsö, Inst. of Biology & Geology, P.O. Box 790, N-9001 Tromsö, Norway.

Taylor, A.R.A., Dept of Biology, University of New Brunswick, Bag Service # 45111, Fredericton N.B., Canada E3B 6E1.

Terry, L.A., Dept of Plant Biology, The University, Newcastle upon Tyne NE1 7RU, UK.

Torgersen, H., Algea Produkter A/S, Postboks 161, N-6501 Kristiansund N, Norway.

Trono Jr, G.C., Dept of Botany, College of Arts and Sciences, University of the Philippines, Diliman, Quezon City 3004, Philippines.

Tseng (Zeng), C.K., Academia Sinica, Institute of Oceanology, 7 Nan-Hai Road, Qingdao, Shandong, China.

Tuno, S., Kyoto Women's University, 17 Imakumano Kita, Hiyoshi-cho, Higashiyama, Kyoto 605, Japan.

Tuntomo, c/o Fa. Lumbung & Co, JL.R.E. Martadinata 130, Ujung Pandang, Indonesia.

Turvey, J.R., School of Physical and Molecular Sciences, University College of North Wales, Bangor, Gwynedd LL57 2UW, UK.

Tveter-Gallagher, E., Jackson Estuarine Laboratory, Durham, N.H. 03824, USA.

Umamaheswara Rao, M., Andhra University, Waltair - 530 003, India.

Urquidi, R., V.P. for Research, Tic Gums, 144 E 44th St., New York, New York 10017, USA.

Uyenco, F.R., Marine Sciences Center, University of the Philippines, Diliman, Quezon City, Philippines.

Vega, J.L., Juan Canalejo, 83-85, La Coruna, Spain.

Velasquez, G.T., Dept of Botany, University of the Philippines, Quezon City, Philippines.

Vreeland, V., Dept of Botany, University of California, Berkeley, Ca 94720, USA.

Wachenfeldt, T. von, Marinbotaniska Lab., Kemicentrum, Box 740, S-220 07 Lund, Sweden.

Wærn, M., Inst. of Plant Ecology, Box 559, S-751 22 Uppsala 1, Sweden.

Wærn, M., Marine Botanical Institute, Carl Skottsbergs Gata 22, S-413 19 Göteborg, Sweden.

Waldemarsson, S., Marinbotaniska Lab., Kemicentrum, Box 740, S-220 07 Lund, Sweden.

Wallentinus, I., Askö Laboratory, University of Stockholm, Box 58, S-150 13 Trosa, Sweden.

Weber, A., Institut für Allgemeine Botanik, Jungiusstr. 6, D-2000 Hamburg 36, West Germany.

Webster, A.G., Dept of Political Science, University of P.E.I., Charlottetown P.E.I., Canada C1A 4P3.

Wee, A., P.O. Box 218, Zamboanga City 7801, Philippines.

Wee, L.H., P.O. Box 241, Manila, Philippines.

Weidner, M., Gyrhofstr. 15, D-5000 Köln, West Germany.

Wennberg, T., Marine Botanical Institute, Carl Skottsbergs Gata 22, S-413 19 Göteborg, Sweden.

Wernwall, S., Alfa-Laval AB, Fack, S-147 00 Tumba, Sweden.

Westermeier, R., Institut für Pflanzenökologie, Heinrich Buff Ring 38, Justus Liebig Universität, D-6300 Giessen, West Germany.

Wettern, M., Botanisches Institut der Technischen Universität, Humboldtstr. 1, D-3300 Braunschweig, West Germany.

Wheeler, W., Marine Science Institute, University of California, Santa Barbara, Ca 93106, USA.

Whelan, P.M., Dept of Botany, University College, Cork, Ireland.

Whitney, L., Kelco, Div. of Merck & Co. Inc., 8355 Aero Drive, San Diego, Ca 92123, USA.

Whyte, J.N.C., Fisheries and Oceans Canada, 6640 N.W. Marine Drive, Vancouver B.C., Canada V6T 1X2.

Wu, C., 7 Nanhai Road, Institute of Oceanology, Academia Sinica, Qingdao, Shandong, China.

Wynne, W.H., American Agar & Chemical Company, 1751 Hancock Street, San Diego, Ca 92110, USA.

Yang, S.-S., Dept of Agricultural Chemistry, National Taiwan University, Taipei, Taiwan.

Yaphe, W., Dept of Microbiology, 3775 University Street, Montreal, P.Q. H3A 2B4, Canada.

Yasuhara, T., 12-23, Nakatamachi, Mishima-shi, Suizuoka, Mishima 441, Japan.

Yoneshigue, Y., Inst. Pesquisas da Marinha, Estação de Biologia Marinha, 28.910 Arraial do Cabo (RJ), Brazil.

Zarmouth, M., P.O. Box 14007, UCSB, Santa Barbara, Ca 93107, USA.

Zavodnik, N., Center for Marine Research, Institute "Ruder Boskovič", 52210 Rovinj, Yugoslavia.

Zoriki, T., Tachibana cho, 1-12-13, Toyonaka City, Japan 560.

SUBJECT INDEX

TAXONOMIC INDEX

Kolibri 360 GmbH
Ferdinand-Jühlke-Straße 7
99095 Erfurt, Deutschland
produktsicherheit@kolibri360.de